D1606942

Function, Phylogeny, and Fossils

Miocene Hominoid Evolution and Adaptations

ADVANCES IN PRIMATOLOGY

Current Volumes in the Series:

ANTHROPOID ORIGINS
Edited by John G. Fleagle and Richard F. Kay

COMPARATIVE BIOLOGY AND EVOLUTIONARY RELATIONSHIPS OF TREE SHREWS
Edited by W. Patrick Luckett

EVOLUTIONARY BIOLOGY OF THE NEW WORLD MONKEYS AND CONTINENTAL DRIFT
Edited by Russell L. Ciochon and A. Brunetto Chiarelli

FUNCTION, PHYLOGENY, AND FOSSILS: Miocene Hominoid Evolution and Adaptations
Edited by David R. Begun, Carol V. Ward, and Michael D. Rose

NEW INTERPRETATIONS OF APE AND HUMAN ANCESTRY
Edited by Russell L. Ciochon and Robert S. Corroccini

NURSERY CARE OF NONHUMAN PRIMATES
Edited by Gerald C. Ruppenthal

PRIMATES AND THEIR RELATIVES IN PHYLOGENETIC PERSPECTIVE
Edited by Ross D. E. MacPhee

SIZE AND SCALING IN PRIMATE BIOLOGY
Edited by William L. Jungers

SPECIES, SPECIES CONCEPTS, AND PRIMATE EVOLUTION
Edited by William H. Kimbel and Lawrence B. Martin

A Continuation Order Plan is available for this series. A continuation order will bring delivery of each new volume immediately upon publication. Volumes are billed only upon actual shipment. For further information please contact the publisher.

Function, Phylogeny, and Fossils

Miocene Hominoid Evolution and Adaptations

Edited by

DAVID R. BEGUN
University of Toronto
Toronto, Canada

CAROL V. WARD
University of Missouri
Columbia, Missouri

and

MICHAEL D. ROSE
University of Medicine and Dentistry of New Jersey
New Jersey Medical School
Newark, New Jersey

Plenum Press • New York and London

Library of Congress Cataloging-in-Publication Data

Function, phylogeny, and fossils : miocene hominoid evolution and adaptations / edited by David R. Begun, Carol V. Ward, and Michael D. Rose.
p. cm. -- (Advances in primatology)
Includes bibliographical references and index.
ISBN 0-306-45457-2
1. Fossil hominids. 2. Human evolution. 3. Paleontology--Miocene. I. Begun, David R. II. Ward, Carol V. III. Rose, Michael D. IV. Series: Advances in primatology (Plenum Press)
GN282F85 1997
599.93'8--dc21

96-40485
CIP

ISBN 0-306-45457-2

A Division of Plenum Publishing Corporation
233 Spring Street, New York, N. Y. 10013

http://www.plenum.com

10 9 8 7 6 5 4 3 2 1

Printed in the United States of America

Contributors

Peter Andrews
Natural History Museum
London SW7 5BD, England

David R. Begun
Department of Anthropology
University of Toronto
Toronto, Ontario M5S 3G3, Canada

Brenda R. Benefit
Department of Anthropology
Southern Illinois University
Carbondale, Illinois 62901

Louis de Bonis
Laboratoire de Gébiologie, Biochronologie, Paléontologie Humaine
86022 Poitiers Cedex, France

Barbara Brown
Department of Anatomy
Northeastern Ohio University College of Medicine
Rootstown, Ohio 44272-0095

Terry Harrison
Department of Anthropology
New York University
New York, New York 10003

Richard F. Kay
Department of Biological Anthropology and Anatomy
Duke University Medical Center
Durham, North Carolina 27710

Jay Kelley
Department of Oral Biology
College of Dentistry
University of Illinois at Chicago
Chicago, Illinois 60612

László Kordos
The Hungarian Geological Museum
H-1143 Budapest, Hungary

George Koufos
Laboratory of Geology
University of Thessaloniki
540 06 Thessaloniki, Greece

Meave Leakey
Division of Palaeontology
National Museums of Kenya
Nairobi, Kenya

Monte L. McCrossin
Department of Anthropology
Southern Illinois University
Carbondale, Illinois 62901

David Pilbeam
Peabody Museum
Harvard University
Cambridge, Massachusetts 02138

Todd C. Rae
Department of Mammalogy
American Museum of Natural History
New York, New York 10024-5192
Present address:
Department of Anthropology
University of Durham
Durham DH1 3HN, England

Lorenzo Rook
Dipartimento di Scienze della Terra
Università di Firenze
Florence, Italy

Michael D. Rose
Department of Anatomy, Cell Biology, and Injury Science
University of Medicine and Dentistry of New Jersey
New Jersey Medical School
Newark, New Jersey 07103

Jeffrey H. Schwartz
Department of Anthropology
University of Pittsburgh
Pittsburgh, Pennsylvania 15260

Peter S. Ungar
Department of Anthropology
University of Arkansas
Fayetteville, Arkansas 72701

Alan Walker
Departments of Anthropology and Biology
The Pennsylvania State University
University Park, Pennsylvania 16802

Carol V. Ward
Anthropology and Pathology & Anatomical Sciences
University of Missouri
Columbia, Missouri 65211

Steve Ward
Department of Anatomy
Northeastern Ohio Universities College of Medicine
Rootstown, Ohio 44272

Myriam Zylstra
Department of Anthropology
University of Toronto
Toronto, Ontario M5S 3G3, Canada

Preface

In 1993, two of the editors of this volume (Ward and Begun) organized a symposium for the Toronto meeting of the American Association of Physical Anthropology. The idea behind the symposium was to gather researchers with long-term expertise in particular aspects of Miocene hominoid paleobiology for a comparison of different perspectives. We were specifically interested in the issue of the interdigitation of functional and phylogenetic analysis in paleobiology. We were also interested in the differing interpretations that result from a focus on a certain taxon versus a particular region of the body. Our own experience with Miocene hominoids and with those who study them suggested this subject area as a useful case study.

The main result of the symposium was that there is much more to discuss. The consensus of the participants was that there is a great deal of value in attempting to compare and integrate differing approaches to Miocene hominoid paleobiology. It was also generally agreed that there was a need to review the Miocene again, given the tremendous number of discoveries and new interpretations that have come to light since the publication in 1983 of the highly influential book in this same series, *New Interpretations of Ape and Human Ancestry,* edited by R. L. Ciochon and R. S. Corruccini. We decided to combine forces with Mike Rose, an alumnus of *New Intrepetations,* to edit a volume that would serve two main purposes. The first was to expand on the theme of the symposium, to explore different perspectives on Miocene hominoids in the hope of reaching a consensus, not so much on what happened, but on how to proceed with the analysis of this complex period in hominoid evolutionary history. The second was to update, in a substantially different format, the Miocene hominoid portions of the venerable Ciochon and Corruccini volume. Whereas they chose to cover a wide array of topics in ape and human evolutionary history and systematics, we decided to focus on a few well-known genera of Miocene hominoids. We felt that these genera were the best candidates for a case study approach, being represented by more fossils from a greater diversity of anatomical regions than most, and thus providing researchers with the best opportunity to cover a range of issues related to both function and phylogeny.

The chapters in this volume were reviewed by at least three referees, in most cases experts who were not themselves contributors. We are very grateful to these referees, who will remain anonymous, for helping us to produce a volume of high quality. We are also grateful to the *Advances in Primatology* Series Editors, John Fleagle and Ross MacPhee, for their support and encouragement, and to Mary Phillips Born, Senior Editor at Plenum, for her guidance and patience.

In the past 15 years there has been a revolution in discovery, analytical techniques, and interpretation in Miocene hominoid paleobiology. A perusal of the literature in this field may lead one to conclude that there is little agreement among researchers and that things are getting worse with each new find. The editors of this book were enthusiastic about bringing together Miocene hominoid scholars because we were confident that there is more agreement than is at first apparent. We do in fact know much more today about the Miocene than we did 15 years ago. Consensus is emerging on the relationships between early and late Miocene hominoids; on the first appearance of great apes in the fossil record; on the relationship between Asian and African great apes and humans; and on the spectacular functional and taxonomic diversity of the Miocene hominoid fossil record as a whole. While there is much more work to be done, we think that this volume contributes to a greater understanding of this diversity. We hope that the chapters it contains will stimulate further discussions and interactions, not only among morphologists, but also between us and behavioral and molecular biologists, ecologists, geologists, and everyone else interested in the origins of adaptations in the Hominoidea.

David Begun
Carol Ward
Mike Rose

Toronto, Canada
Columbus, Missouri
Newark, New Jersey

Contents

Function and Phylogeny in Miocene Hominoids

1

CAROL V. WARD, DAVID R. BEGUN, and MICHAEL D. ROSE

Introduction

The Miocene was the most diverse era in hominoid evolutionary history. Miocene hominoids exhibit a level of taxic, morphological, and biogeographic diversity that far exceeded that of living apes. Over 30 genera of Miocene apes are currently recognized (Begun, 1995a); probably only a small percentage of those that existed. Out of this radiation arose the ancestor of modern apes and humans. The abundance and importance of Miocene hominoid fossils have made them the focus of extensive analysis and reinterpretation over the years.

Despite this intensive study, researchers still present contradictory interpretations of hominoid evolutionary history (e.g., Andrews, 1992; Begun, 1992a,b, 1994, 1995b; Conroy *et al.*, 1992; Dean and Delson, 1992; Moyà-Solà and Köhler, 1993, 1995). Uncertainties about hominoid evolution are partly related to a diverse but fragmentary fossil record (Fleagle, 1983, 1986). Although many taxa are recognized, few are represented by many body parts.

CAROL V. WARD • Anthropology and Pathology and Anatomical Sciences, University of Missouri, Columbia, Missouri 65211. DAVID R. BEGUN • Department of Anthropology, University of Toronto, Toronto, Ontario M5S 3G3, Canada. MICHAEL D. ROSE • Department of Anatomy, Cell Biology, and Injury Science, University of Medicine and Dentistry of New Jersey, New Jersey Medical School, Newark, New Jersey 07103.

Function, Phylogeny, and Fossils: Miocene Hominoid Evolution and Adaptations, edited by Begun *et al.* Plenum Press, New York, 1997.

Even the best-known taxa, such as *Proconsul* and *Oreopithecus*, are not known from all skeletal elements. In addition, many taxa probably remain completely unknown. It is very unlikely than even 50% of the actual diversity of Miocene hominoids have been recovered, and perhaps less than 20%. Taxa critical to understanding hominoid relations are not yet known. For example, there are no known convincing fossil relatives of African apes. Large temporal, morphological, and biogeographic gaps exist between extant and recognized Miocene taxa, obscuring their relations. Recovery of more fossils will fill in some of these gaps, and provide new characters for analysis.

Not all the ambiguity regarding hominoid evolutionary interpretation can be blamed on the fossil record, however. Disagreements about patterns of hominoid evolution result from difficulties inherent in interpreting the systematics of fossil taxa. Understanding the evolutionary relations of taxa is necessary for accurate interpretation of the biology and behavior of fossil species, and for tracing patterns of change throughout the lineages leading to modern apes and humans. Interpreting hominoid evolutionary history relies on two separate yet related considerations: those regarding functional anatomy and those regarding phylogenetic interpretation (Fleagle, 1983, 1986). These two types of consideration are integrally related; neither can be pursued without reference to the other. This inextricability is reflected in most studies of Miocene hominoids, although the degree of emphasis on function or on phylogeny has varied widely.

Studies emphasizing functional interpretation have tended to set aside systematic considerations in favor of a focus on reconstructing behavior (e.g., Rose, 1983, 1988, 1991; Ward, 1993; Begun *et al.*, 1994). Similarly, studies that have concentrated on systematics have tended to employ functional hypotheses or assumptions in a post hoc fashion to justify or illuminate the significance of particular character suites (Walker and Teaford, 1989; Andrews, 1992; Begun, 1992a; Dean and Delson, 1992). Many of these studies have recognized the presence of shared derived morphological, functional, or behavioral features as the criterion for grouping taxa. However, they have varied widely in the degree to which formal techniques of character analysis have been used.

Despite difficulties inherent in integrating function and phylogeny (see below), there have been a number of productive investigations of the phylogenetic implications of functional features of extant and fossil catarrhines. Some studies have concentrated on the phylogenetic implications of features of a particular anatomical region (e.g., Fleagle, 1983; Sarmiento, 1983; Gebo, 1989; Begun, 1993; Sanders and Bodenbender, 1994). Others have concentrated on assessing the phylogenetic position of a single taxon or closely related group of taxa using all available functional information (e.g., Ferembach, 1958; Andrews and Groves, 1976; Begun, 1992b; McCrossin, 1994). A striking recent example of the latter approach is the strong argument, based largely on functional features of the postcranial skeleton, for the position of

Oreopithecus as a hominoid of modern aspect (Harrison, 1986; Sarmiento, 1987; Harrison and Rook, this volume).

Function and Phylogeny

Integrating functional and phylogenetic approaches is complicated but necessary. We must understand the evolutionary relations of a group of taxa in order to reveal the directionality of morphological change within those lineages, which in turn reveals the history of natural selection that shaped each taxon. At the same time, an understanding of function permits evaluation of character used in a phylogenetic analysis. Though knowing the function of a trait is not a prerequisite to its inclusion in (or exclusion from) a phylogenetic analysis, this knowledge does allow researchers to make more informed decisions about the nature of characters, character interdependence, and even about the probability of homoplasy (see below). Although the hypothesized functional significance of a set of characters is often cited as evidence of their evolutionary significance, it is an error to judge the usefulness of a character in a phylogenetic analysis (a methodological consideration) on the basis of its hypothesized function (a theoretical issue).

A phylogenetic analysis reveals character polarities, and thus the vector of selection in a lineage. Where directionality can be determined, the recent history of natural selection is clear and easy to interpret. For example, the presence of numerous apomorphies related to habitual upright posture and locomotion in *Australopithecus afarensis* is easily interpretable as evidence of recent natural selection for habitual bipedality.

The functional significance of primitive characters is less obvious. For example, the presence of curved hand and foot phalanges in *A. afarensis* has been used to suggest (1) that arboreality was an important component of the positional behavior of this taxon and (2) that arboreality was selected for in *A. afarensis*, at least as part of a locomotor repertoire (e.g., Stern and Susman, 1984). However, phylogenetic analysis reveals that all hominoids except *Homo* have curved phalanges, so *A. afarensis* probably evolved from an ancestor with curved phalanges. Because curved phalanges are probably primitive for *A. afarensis*, they did not arise by recent selection in the *A. afarensis* lineage.

There are multiple possible explanations for the presence of primitive traits in any taxon. Primitive characters may be retained as a result of stabilizing selection for maintenance of original function, e.g., that phalangeal curvature implies stabilizing selection on arboreal behaviors and related anatomy (e.g., Susman and Stern, 1991). Alternatively, primitive characters may simply be retained in the absence of negative selection (e.g., Latimer, 1991). Finally, they could be linked genetically or developmentally with other characters that are maintained by stabilizing selection, although they themselves are not.

Testing hypotheses regarding the functional significance of primitive characters is complicated, and not reliably done in many cases. Without identifying the directionality of selection, we cannot make accurate inferences about patterns of natural selection in a lineage. Thus, the hypothesis of evolutionary significance of phalangeal curvature in *A. afarensis,* selection for arboreality, depends not on the functional importance of this set of characters, but on their polarity as revealed by the evolutionary history of early hominids. Because of the dependence on a phylogenetic hypothesis for accurate functional analysis, an important first step is to make an initial attempt to infer relations in a set of taxa. Once a phylogenetic hypothesis has been generated, it can then be used to interpret evolution and the functional anatomy of the taxa.

Another reason to perform phylogenetic analyses first is that functional hypotheses are more complex and difficult to test than cladistic hypotheses. Functional hypotheses rely on analogies to living forms or to functioning anatomical systems. They are reconstructionist in their structure; they depend on successive interpretations of a hierarchy of interrelated and interdependent structures, from parts to wholes, to produce a coherent picture of function. Also, functional hypotheses are process-driven. They depend on assumptions about evolutionary or biological process (usually natural selection) for justification.

Cladistic hypotheses, on the other hand, are simpler. They are typically reductionistic and depend on the definition of character states and the assumption of homology. They do not evoke particular evolutionary or biological processes other than the basic assumption that evolution occurs. Cladistic systematics is pattern recognition while functional anatomy is process identification based on pattern interpretation.

Many published phylogenetic analyses depend, in part, on functional conclusions. Choosing characters and weighting and ordering character states frequently relies on assumptions about functional or ontogenetic relations among morphological structures, whether these assumptions are implicit or explicit. Functional criteria are often used to polarize character states by suggesting predictable outcomes of morphological transformation. It is common in paleontology to assume that multistate characters have predictable trajectories based on presumed functional constraints. For example, it has been assumed that hominoid premaxillae progress from short to longer to longest (Ward and Kimbel, 1983; Ward and Pilbeam, 1983), as if it were somehow functionally or ontogenetically impossible for another sequence of change to have taken place. Similarly, molars are assumed to have gone from small to bigger to biggest, as if constrained by some functional parameter to this particular pattern of change (e.g., Robinson, 1954). Finally, the transformation sequence of enamel thickness is assumed to be constrained functionally in the sense that some changes are considered developmentally more difficult to account for than others, and so are considered less likely (Martin, 1985). Characters are assumed to be primitive if they correspond to one end of the presumed trajectory, derived if they correspond to the other, and homoplasious if their multiple occurrence is considered likely on functional grounds. Begun and Kordos (this volume) suggest an alternative to this procedure that avoids some of the pitfalls of *a priori* character analysis while still integrating functional morphology and systematics.

Parallelism and Convergence

Most ambiguities in the interpretation of systematic relations among hominoids resulting from complexities of the evolutionary process can be partly addressed by available techniques of morphological analysis. Because taxa generally can be defined by morphological criteria, artificial classificatory schemes can be superimposed on actual evolutionary histories to approximate their real patterns. However, there almost certainly were periods in the evolution of most lineages during which complex, rapid, or chaotic sequences of genetic events and/or populational dynamics occurred. Such events may be particularly common during the early phases of the evolution of new forms, but may also occur sporadically during the entire evolutionary history of a lineage. In these instances, it is difficult to delineate taxa in the manner required of modern systematics.

While some aspects of the evolutionary process cannot be uncovered by the methodology of systematics, other aspects are actually revealed post hoc this way. For example, convergences and parallelisms, jointly referred to as homoplasies, are identified using cladistic analysis. Homoplasies are defined as characters that appear more than once in a cladogram, and so must have evolved independently in different lineages.

The existence of widespread homoplasies among the Hominoidea has been recognized for some time, even before the application of cladistic methods (e.g., Tuttle, 1975; Fleagle, 1983, 1986). However, interpretations differ about which character states are homologous and which are homoplasious. Unlike a cladistic identification of homoplasies via parsimony (Stewart, 1993), many studies have used different criteria to identify homoplasies.

The widespread occurrence of homoplasy in hominoid evolution makes it necessary to consider the broadest possible set of anatomical regions in studying Miocene hominoid phylogeny. A focus on just one anatomical region lends itself to the real danger of producing a character tree that may or may not have anything to do with the phylogeny of the included taxa (Begun, 1994; C. V. Ward, this volume). Homoplasy also makes it necessary to consider both the functional and phylogenetic signals of each character included in an analysis. The issue of homoplasy brings together concepts from both systematics and function. A homoplasy is a character in a systematic analysis, but it is also a process in the functional sense.

In the analysis of Miocene hominoids, ad hoc hypotheses are frequently invoked to assess the relative likelihood of convergence based on a series of functional arguments. For example, it is sometimes suggested that postcrania are more likely to be subject to homoplasy than are crania, because they appear to be plastic in their response to environmental stresses, and so are not useful indicators of phylogeny (e.g., Andrews, 1992). Teeth are often thought to be better indicators of phylogeny, because they are more closely constrained by genes, and less affected by nongenetic influences on morphology, apart from wear, infectious disease, and growth abnormalities (Andrews and Martin, 1987).

Similarly, complex characters are thought to be less likely to have evolved more than once than are simple characters (Skelton and McHenry, 1992; Harrison and Rook, this volume). Thus, complex characters are often weighted more heavily in phylogenetic analyses. Some studies advise treating multiple characters that are part of strongly integrated functional complexes as single characters. Examples include features related to a particular dietary strategy or positional behavior (Skelton and McHenry, 1992). Grouping characters this way has the effect of reducing the weight of each component character in a cladistic analysis.

Clearly, researchers must remain aware of the functional assumptions they make when using morphological data to assess the systematic relations of a set of taxa. Similarly, functional studies must consider the phylogenetic history of the taxon in question when making hypotheses about the evolutionary significance of a suite of morphological features.

Content of This Volume

Integrating function and phylogeny continues to be undertaken in numerous ways by students of hominoid evolution, as evidenced by the many approaches taken by authors of chapters in this volume. This volume explores current approaches to integrating functional and phylogenetic interpretations of Miocene hominoids. The Miocene hominoid record provides an ideal forum for exploration of these issues, because of its taxonomic richness, ambiguity of extant and fossil ape relations, and in the diversity of adaptations reflected by its taxa.

The chapters in this volume were written by researchers with extensive experience in the subject they cover. Each author was asked to integrate data and interpretations related to functional anatomy and phylogenetic reconstruction in her or his own way. Each chapter confronts this challenge differently, and as a whole the volume reveals the degree to which there is diversity in approaches to understanding hominoid evolutionary history. Given this diversity, it is perhaps remarkable that there is as much consensus as currently exists in interpretations of hominoid evolutionary history (Begun *et al.*, this volume).

The first chapter, by David Pilbeam, provides historical context on the past 30 years of hominoid evolutionary studies. The shift in ideas that came about from the introduction of cladistic methodology is emphasized, along with a view of the last common ancestor of great apes and humans, the paleobiology of which is largely a logical consequence of the phylogenetic analysis.

Peter Andrews and colleagues cover another overview topic within the context of this volume, the interrelationship among functional anatomy, phylogeny, and the independent evidence of paleoenvironment, the latter serving perhaps not to test, but at least to reinforce interpretations deriving from the former two. Though both functional anatomy and paleoenvironment are subject to substantial ranges of interpretation, it is interesting that conclusions about the functional anatomy of Miocene hominoids, based on morphological analyses independent of paleoenvironmental considerations, are congruent in all cases with the data from paleoenvironmental analysis. When combined with a phylogenetic hypothesis, this work provides a basis for interpreting ecological trends in hominoid evolution and an understanding of the ecological significance of the events of hominoid evolutionary history.

Subsequent chapters focus on more integrated data sets. One group of authors was asked to provide an analysis of a restricted anatomical region from the points of view of function and phylogeny, and the other to analyze a group of related fossils with a similar goal. The results are as varied as the topics.

Todd Rae covers the evolution of the hominoid face, with an emphasis on the earlier phases of hominoid evolution. Rae's approach is primarily phylogenetic, and he illustrates the importance of using fossils to flesh out the evolutionary history of a lineage. This chapter also explores the functional significance of the major changes in the early evolution of the hominoid face, with functional interpretation constrained by the results of the phylogenetic hypothesis presented.

Mike Rose focuses on the evolution of the hominoid forelimb. His survey of the forelimbs spans the Oligocene to the Miocene, and includes a large number of taxa. Rose integrates function and phylogeny in his conclusion about the origins of the Hominoidea and of modern great apes and humans. The characters used to support his phylogenetic hypothesis for the earliest hominoids are functionally related to arboreality in large hominoids with short or missing tails. The characters used to place the late Miocene hominoids in turn are related to a functional shift toward more orthograde postures.

Carol Ward explores issues of function and phylogeny in the evolution of the hominoid trunk and hindlimb. She isolates characters from the hominoid trunk and hindlimb, and produces a character tree from just this set of data. Her tree differs from traditional phylogenetic interpretations, reflecting the different patterns and rates of evolution that characterize hominoid postcrania as compared with skulls, jaws, and teeth. Her study illustrates the type of biases that can arise by using a single anatomical region to infer phylogenetic history of a set of taxa.

Rich Kay and Peter Ungar focus on Miocene hominoid dental evolution. They present intriguing conclusions on the relationship between phylogeny and function in their analysis of occlusal patterns and its functional implications. These conclusions are a cautionary tale for those who would interpret the behavior of fossil taxa from direct comparisons to living forms, without taking into account the important evolutionary differences that can separate living and extinct groups.

Bobbie Brown focuses on the mandible in Miocene hominoid evolution. Her chapter illustrates that although abundant in the fossil record, mandibles have revealed little conclusive information about hominoid evolutionary relations. She notes that it is clear that functional constraints on the hominoid mandible are numerous, going beyond the usual consideration of bite forces and muscles of mastication. She suggests that until the functional constraints governing mandibular form are worked out, the significance of the diversity of mandibular morphology in the Miocene remains difficult to interpret.

Jay Kelley's chapter focuses not on an anatomical region *per se,* but on life history, a specific biological entity nonetheless. The discussion in this chapter is mostly devoted to *Sivapithecus,* but could be applied as well to other large samples of Miocene hominoids with well-preserved juvenile dentitions (e.g., *Dryopithecus* or *Proconsul*). Kelley brings up a systematics issue—how is the Hominoidea to be defined?; and a functional issue—once defined, how can the Hominoidea be functionally described in a meaningful and inclusive manner? The answer, he suggests, is to be found in the analysis of life history rather than the more traditional lists of characters.

The remaining chapters focus on specific genera of Miocene hominoids. They are arranged in rough chronological order of the taxa covered. Alan Walker discusses phylogenetic and function issues in *Proconsul,* the best-known genus of Miocene hominoid. Walker uses a range of data on the anatomy of *Proconsul* to describe the paleobiology of this taxon, and to place this taxon in a phylogenetic context. Walker concludes that *Proconsul* is indeed a hominoid, but a primitive one that, while quite different from Oligocene forms, is linked to later Miocene and living hominoids by a small number of characters and behaviors.

Meave Leakey and Alan Walker discuss *Afropithecus,* and provide a useful summary of material attributed to this taxon. Leakey and Walker describe an interesting pattern in the face and anterior dentition of *Afropithecus* that they relate to a sclerocarp feeding strategy in this fossil genus, most similar to that seen in living pitheciines. This derived anatomy and specialized behavior is superimposed on an otherwise primitive hominoid morphology, making the precise placement of *Afropithecus* ambiguous.

Monte McCrossin and Brenda Benefit provide new data on *Kenyapithecus,* based on their recent work at Maboko Island, Kenya. Based on this work they are able to describe a new set of adaptations in the long-known taxon *Ken-*

yapithecus. They also propose a unique view of the phylogeny of Miocene hominoids. They suggest, among other things, that *Kenyapithecus*, too, was a pitheciine-like sclerocarp feeder [but see Leakey and Walker (this volume) for a different view].

Steve Ward surveys the history and current understanding of the functional and phylogenetic interpretations of *Sivapithecus*, illustrating how analysis of this central taxon in the history of Miocene hominoid studies has come full circle. Ward, in discussing evidence from both the skull and postcranium, comes to the conclusion that relations of this genus to *Pongo* are less clear-cut than previously suspected. Like *Oreopithecus* (Harrison and Rook, this volume), the skull and the postcranium give different phylogenetic signals, and Ward shows just how complicated this situation has become in *Sivapithecus*.

David Begun and László Kordos attempt to integrate phylogenetic and functional approaches to understand the evolutionary relations of *Dryopithecus*. These authors first conduct a phylogenetic analysis and then apply functional criteria to evaluate potentially competing phylogenies, to assess the behavioral history of anatomical change, and to "explain" apparent parallelisms. They propose a method for integrating functional and phylogenetic data, and reveal their own revision of relations among late Miocene and living hominids (great apes and humans).

Louis de Bonis and George Koufos provide an update on *Ouranopithecus*. Their integration of functional and phylogenetic approaches stresses the functional (and ultimately behavioral) and ecological reasons for the similarities they find between *Ouranopithecus* and *Australopithecus*.

Terry Harrison and Lorenzo Rook discuss the function and phylogeny of *Oreopithecus*, perhaps the best Miocene hominoid case study illustrating the complex interrelationship of functional and phylogenetic approaches. In contrast to some of the studies cited above, Harrison and Rook are critical of preconceptions about the relative reliability of cranial versus postcranial characters.

Finally, Jeff Schwartz provides a detailed analysis of a suite of characters relevant to interpretations of the phyletic position of *Lufengpithecus*. He integrates functional and phylogenetic approaches through a detailed assessment of developmental factors related to comparative hominoid facial morphology.

To conclude the volume, Begun, Ward, and Rose assemble a list of 240 characters and their diverse states among hominoids. Characters, character states, and taxa were chosen based on their inclusion in the chapters of this volume, though the character assignments are based on these authors' observations, and not on data presented in this volume. This was done in order to maintain internal consistency, because a character analysis like any analysis is more reliable when the data it contains are actually observed as opposed to being plucked from the literature. However, this does result in an analysis that reflects the interpretations of these authors, and not a more neutral list of traits directly from the chapters of this volume. Begun *et al.* run a series of cladistic analyses on these data in the final chapter. Their resulting phylogenetic hypothesis differs from many interpretations presented in the other chapters in interesting ways, partly because this analysis includes a larger data base than has ever been assembled to assess hominoid evolution.

Goals of This Volume

As is apparent from the brief survey of Miocene hominoid evolution presented in this volume, there are as many approaches to integrating data from the functional and phylogenetic perspectives as there are researchers in hominoid paleobiology. Though approaches and conclusions vary, each author provides important data on comparative hominoid morphology and functional anatomy.

This volume is intended to serve two main purposes. The first is to emphasize both the complex nature of the Miocene hominoid fossil record today and the diversity of approaches currently being applied. Sources of disagreement come from both the intrinsic complexity of the issues and differences in approaches to understanding this complexity.

The second purpose of this volume is to assemble a comprehensive data base of the most completely known large-bodied Miocene hominoids. These data can be used by other researchers to test various hypotheses outlined in this volume, and as a framework for the analysis of new data from new fossils and new methods of data collection, such as image analysis, CT scan, and histomorphology.

Finding more fossils, and extracting more information from the fossils that we have, are some important avenues for progress in understanding of living and fossil hominoids. Researchers must also carefully examine their own assumptions and perspectives on integrating functional studies with phylogenetic interpretation to fully understand the complexities of hominoid evolution. We hope that this volume has brought these issues to the attention of primate paleontologists and paleoanthropologists, and in doing so has furthered our understanding of the patterns of evolution that produced modern apes and humans.

References

Andrews, P. 1992. Evolution and environment in the Hominoidea. *Nature* **360:**641–646.

Andrews, P., and Groves, C. P. 1976. Gibbons and brachiation. *Gibbon and Siamang* **4:**167–218.

Andrews, P., and Martin, L. 1987. Cladistic relationships of extant and fossil hominoids. *J. Hum. Evol.* **16:**101–118.

Begun, D. R. 1992a. Miocene fossil hominids and the chimp–human clade. *Science* **247:**1929–1933.

Begun, D. R. 1992b. Phyletic diversity and locomotion in primitive European hominids. *Am. J. Phys. Anthropol.* **87:**311–340.

Begun, D. R. 1993. New catarrhine phalanges in Rudabánya (northeastern Hungary) and the problem of parallelism and convergence in hominoid postcranial morphology. *J. Hum. Evol.* **24:**373–402.

Begun, D. R. 1994. Relations among the great apes and humans: New interpretations based on the fossil great ape Dryopithecus. *Yearb. Phys. Anthropol.* **37:**11–63.

Begun, D. R. 1995a. Miocene apes. In: C. R. Ember and M. Ember (eds.), *Research Frontiers in Anthropology—Advances in Archeology and Physical Anthropology,* pp. 3–30. Prentice–Hall, Englewood Cliffs, NJ.

Begun, D. R. 1995b. Late Miocene European orang-utans, gorillas, humans, or none of the above? *J. Hum. Evol.* **29:**169–180.

Begun, D. R., Teaford, M. F., and Walker, A. 1994. Comparative and functional anatomy of *Proconsul* phalanges from the Kaswanga primate site, Rusinga Island, Kenya. *J. Hum. Evol.* **26:**89–165.

Conroy, G. C., Pickford, M., Senut, B., and Van Couvering, J. 1992. *Otavipithecus namibiensis,* first Miocene hominid from southern Africa. *Nature* **356:**144–148.

Dean, D., and Delson, E. 1992. Second gorilla or third chimp? *Nature* **359:**676–677.

Ferembach, D. 1958. Les limnopithèques du Kenya. *Ann. Paleontol.* **44:**149–249.

Fleagle, J. G. 1983. Locomotor adaptations of Oligocene and Miocene hominoids and their phyletic implications. In: R. L. Ciochon and R. S. Corruccini (eds.), *New Interpretations of Ape and Human Ancestry,* pp. 301–324. Plenum Press, New York.

Fleagle, J. G. 1986. The fossil record of early catarrhine evolution. In: B. A. Wood, L. Martin, and P. Andrews (eds.), *Major Topics in Primate and Human Evolution,* pp. 130–149. Cambridge University Press, London.

Gebo, D. L. 1989. Locomotor and phylogenetic considerations in anthropoid evolution. *J. Hum. Evol.* **18:**201–233.

Harrison, T. 1986. A reassessment of the phylogenetic relations of *Oreopithecus bambolii* Gervais. *J. Hum. Evol.* **15:**541–583.

Latimer, B. 1991. Locomotor behavior in *Australopithecus afarensis:* The issue of arboreality. In: Y. Coppens and B. Senut (eds.), *Origine(s) de la Bipédie chez les Hominidés. Cahiers de Paléoanthropologie,* pp. 169–176. Editions du CNRS, Paris.

McCrossin, M. L. 1994. *The Phylogenetic Relationships, Adaptations, and Ecology of Kenyapithecus.* Ph.D. dissertation, University of California, Berkeley.

Martin, L. 1985. Significance of enamel thickness in hominoid evolution. *Nature* **314:**260–263.

Moyà-Solà, S., and Köhler, M. 1993. Recent discoveries of *Dryopithecus* shed new light on evolution of great apes. *Nature* **365:**543–545.

Moyà-Solà, S., and Köhler, M. 1995. New partial cranium of *Dryopithecus* Lartet, 1863 Hominoidea, Primates, from the upper Miocene of Can Llobateres, Barcelona, Spain. *J. Hum. Evol.* **29:**101–139.

Robinson, J. T. 1954. Prehominid dentition and hominid evolution. *Evolution* **8:**324–334.

Rose, M. D. 1983. Miocene hominoid postcranial morphology: Monkey-like, ape-like, neither, or both? In: R. L. Ciochon and R. S. Corruccini (eds.), *New Interpretations of Ape and Human Ancestry,* pp. 405–417. Plenum Press, New York.

Rose, M. D. 1988. Another look at the anthropoid elbow. *J. Hum. Evol.* **17:**193–224.

Rose, M. D. 1991. The process of bipedalization in hominids. In: Y. Coppens and B. Senut (eds.), *Origine(s) de la Biédie chez les Hominidés. Cahiers de Paléoanthropologie,* pp. 37–48. Editions du CNRS, Paris.

Sanders, W. J., and Bodenbender, B. E. 1994. Morphometric analysis of lumbar vertebra UMP 67-28: Implications for spinal function and phylogeny of the Miocene Moroto hominoid. *J. Hum. Evol.* **26:**203–237.

Sarmiento, E. E. 1983. The significance of the heel process in anthropoids. *Int. J. Primatol.* **4:**127–152.

Sarmiento, E. E. 1987. The phylogenetic position of *Oreopithecus* and its significance in the origin of the Hominoidea. *Am. Mus. Novit.* **2881:**1–44.

Skelton, R. R., and McHenry, H. M. 1992. Evolutionary relationships among early hominids. *J. Hum. Evol.* **23:**309–350.

Stern, J. T., and Susman, R. L. 1984. The locomotor anatomy of *Australopithecus afarensis. Am. J. Phys. Anthropol.* **60:**279–317.

Stewart, C. B. 1993. The powers and pitfalls of parsimony. *Nature* **361:**603–607.

Susman, R. L., and Stern, J. T. 1991. Locomotor behavior of early hominids: Epistemology and fossil evidence. In: Y. Coppens and B. Senut (eds.), *Origine(s) de la Bipédie chez les Hominidés. Cahiers de Paléoanthropologie,* pp. 121–132. Editions du CNRS, Paris.

Tuttle, R. H. 1975. Parallelism, brachiation, and hominoid phylogeny. In: W. P. Luckett and F. S. Szalay (eds.), *Phylogeny of the Primates,* pp. 447–480. Plenum Press, New York.

Walker, A., and Teaford, M. F. 1989. The hunt for *Proconsul. Sci. Am.* **260:**76–82.

Ward, C. V. 1993. Torso morphology and locomotion in *Proconsul nyanzae. Am. J. Phys. Anthropol.* **92:**291–328.

Ward, S. C., and Kimbel, W. H. 1983. Subnasal alveolar morphology and the systemic position of *Sivapithecus. Am. J. Phys. Anthropol.* **61:**157–171.

Ward, S. C., and Pilbeam, D. R. 1983. Maxillofacial morphology of Miocene hominoids from Africa and Indo-Pakistan. In: R. L. Ciochon and R. S. Corruccini (eds.), *New Interpretations of Ape and Human Ancestry,* pp. 211–238. Plenum Press, New York.

Research on Miocene Hominoids and Hominid Origins

2

The Last Three Decades

DAVID PILBEAM

Introduction

This brief commentary begins with my graduate work with Elwyn Simons which culminated in the 1965 "Preliminary Revision of the Dryopithecinae," then about the most comprehensive review of the paleontological record of the great apes (Simons and Pilbeam, 1965). It reflected a time when that fossil record was dominated by teeth and fragmentary jaws, one decent skull, and a handful of postcranial remains. Much has changed since then, both in the record itself and in how it is interpreted, and what follows are reflections on where we are, how we got here, and where we might go next.

My comments are focused on the last 15 years, although much of what has happened reflects the taxonomically rather austere backdrop of that 1965 "Preliminary Revision," an analysis that reduced the number of Neogene large hominoid genera to only three: *Dryopithecus, Ramapithecus,* and *Gigantopithecus,* with *Ramapithecus* being the earliest well-documented hominid. It was easy three decades ago to infer firm phylogenetic links between Miocene

DAVID PILBEAM • Peabody Museum, Harvard University, Cambridge, Massachusetts 02138.

Function, Phylogeny, and Fossils: Miocene Hominoid Evolution and Adaptations, edited by Begun *et al.* Plenum Press, New York, 1997.

and living hominoids. When the number of taxa was limited, the number of morphological characters small, phylogenetic analysis not rigorous, and when molecular clocks could safely be ignored, there were few impediments to tracing extant lineages well back into the Neogene, or even earlier. Novacek (1993) observed that "G. G. Simpson confessed that things were much less muddled when he and a few others first pioneered studies of Mesozoic mammals in the 1930s. Taxa were few and the number of alternatives limited. With increased activity over more recent decades, the situation is evermore complex." So it was and is for Miocene hominoids.

Because the fossil record was dominated by jaws and teeth until a decade or so ago, discussions of Miocene hominoid evolutionary relationships and paleobiology were inevitably focused on analyses of those parts. I will say little about tooth function except that extensive research over the past several decades has established that most Miocene hominoids were probably frugivores of some kind (Kay, 1977; Teaford and Walker, 1984). Unfortunately, the low diversity of living hominoids and the fact that tooth design differs both among extant forms and between them and almost all Miocene hominoids (Corruccini, 1975) make the interpretation of the relatively subtle inter-Miocene hominoid tooth differences quite difficult. That they were mainly "frugivores" is perhaps not surprising, though it is not so clear how useful that gross characterization is to understanding behavioral differences among Miocene hominoids, or between them and living apes.

The low diversity of extant hominoids, along with the fact that almost all Miocene apes differ quite markedly from the living in many cranial, dental, and postcranial characters, also presents problems in reconstructing phylogenetic relationships. If we had available only the amount and kind of mostly dental information available for the Miocene hominoid record, it would be difficult to work out plausible phylogenetic relationships in almost any mammal group, including hominoids (e.g., Hartman, 1988, 1989). We never thought to address the "drunk's dilemma." Should we search for our lost keys in the dark where we dropped them or in the light where we can see? Could the data in hand address the questions we were asking or not? For the kinds of phylogenetic questions about hominoids we want to address, the answer was, probably no.

Two Decades of Progress

Between the 1960s, and the late 1970s and early 1980s, the consensus changed. *Ramapithecus* was essentially defrocked and there were few prepared to defend its hominid status (Lewin, 1987). This defrocking was, to a very considerable extent, related to the "molecular" studies of Sarich and Wilson (see Sarich, 1983, and Lewin, 1987), although this was for the most part barely or grudgingly acknowledged. At this time, the important cladistic revolution was only just starting to have a major impact on paleoprimatology. I pick three phylogenetic schemes to represent somewhat different approaches in the late

1970s. Greenfield (1979, 1980) had a taxonomically simple proposal, in which *Dryopithecus,* including *Proconsul,* was ancestral to *Sivapithecus,* which included essentially all known middle and late Miocene forms and which was, in turn, the common ancestor of all extant large hominoids. At almost the same time, I (1979) proposed a structurally similar taxonomic scheme, although Greenfield's *Sivapithecus* group was represented by a number of very similar genera grouped in the Ramapithecidae. I presented three alternative phylogenetic schemes, in descending order of probability. The first derived just hominids from ramapithecids; the second, like Greenfield's, derived all large living hominoids from ramapithecids; the third saw the extant great apes and hominids as evolving not from ramapithecids, but from an unknown later Miocene ape. A little later Kay and Simons (Kay, 1981; Kay and Simons, 1983), like Greenfield, included all Miocene large hominoids except *Proconsul* and *Dryopithecus* in a single genus, *Sivapithecus,* and saw this as a hominid exclusively ancestral to australopithecines and *Homo.*

What these schemes shared in common was the lumping of taxa (individuals, species, or genera) mainly on the basis of a few dental traits, and in particular thick occlusal enamel (Simons and Pilbeam, 1972; Kay, 1981) (although some of the included taxa do not in fact have thick enamel). Again, this is understandable given the nature of the fossil record. But, again, we never asked ourselves whether, for any reasonably diverse group of living primates, we could have reconstructed their relationships if all we had as morphological characters were those represented by the pre-1980 Miocene hominoid record.

The fossil record began to improve in the late 1970s with the recovery of the first Miocene facial and cranial remains since Mary Leakey's discovery in 1948 of the Rusinga *Proconsul* skull (Clark and Leakey, 1951). The period between 1976 and 1985 saw the discovery, publication, and initial interpretation of significant craniofacial specimens of *Sivapithecus meteai* (specimen originally discovered in 1967 though not published until 1980) (Andrews and Tekkaya, 1980), *Sivapithecus sivalensis* (Pilbeam, 1982), *Ouranopithecus* (Bonis and Melentis, 1977; also Bonis *et al.,* 1990), *Dryopithecus hungaricus* (from Rudabánya) (Kordos, 1987), *Lufengpithecus* (Lu *et al.,* 1981; Wu *et al.,* 1981), *Kenyapithecus* (Pickford, 1985), and the regrettably still unnamed new genus from the Samburu Hills (Ishida *et al.,* 1984). It was also during this time that Ward's research on patterns of morphological variation in the hominoid premaxillary–palatal region was published (Ward and Pilbeam, 1983; Ward and Kimbel, 1983).

These critical new specimens had two major impacts. First, they demonstrated that middle and especially late Miocene hominoids were truly more diverse than most of us had believed, with at least five late Miocene genera in Europe and Asia and two or three in Africa—the latter surely and the former less surely an underestimate. For the middle Miocene hominoids sampling remains poor and any diversity estimates at either genus or species level, based as they are almost entirely on jaw fragments and teeth, are likely to be artificially low. Nevertheless, diversity may approach that of the early Miocene (Harrison, 1992). For the early Miocene we have seen a growth in the number of species similar to the Euro-Asian late Miocene, though in a geographically much smaller area of east Africa: between the 1940s and 1993 new craniofacial discoveries have quadrupled the number of genera and close to tripled the number of species. So, the Miocene hominoid world was much more diverse than we had imagined even 15 years ago.

Cranio-Dental Remains

Steven Ward's work, expanding that of Peter Andrews (Andrews and Cronin, 1982), on subnasal anatomy of living hominoids, played a major role in the development of a new generation of phylogenetic hypotheses, especially built around detailed facial similarities between *Sivapithecus* (from Indo-Pakistan) and *Pongo* (Ward and Brown, 1986). The two genera differed from all other extant and fossil hominoids in certain facial features which could be plausibly interpreted as synapomorphies. *Sivapithecus* came to be seen both as clearly different from other Miocene hominoid genera and as firmly linked to a living ape. This link came to serve as a key anchor point in morphological/genetic syntheses (see Pilbeam, 1986, 1989). Other Miocene-to-Recent links gained somewhat less widespread acceptance, though sister-group relationships were proposed between *Ouranopithecus* and hominids (Bonis and Melentis, 1985; Bonis *et al.,* 1990), and the Samburu maxilla and *Gorilla* (Pilbeam, 1986; Pickford, 1986; Andrews, 1986).

Lawrence Martin's work (1985) on enamel thickness and histology, developing earlier suggestions in more sophisticated and systematic ways, was widely influential throughout this period. It reinforced the consensus that thick enamel was primitive for extant large hominoids and contributed what became a key morphological similarity linking *Pan* and *Gorilla* as sister taxa: relatively thin enamel that had apparently been laid down in two depositional phases, faster and then slower.

Remember, we never asked whether these characters could be used to infer plausible phylogenies in living groups. It is worth noting that in very extensive listings of characters for a phylogenetic analysis of living hominoids (Groves and Paterson, 1991; Shoshani *et al.,* 1996), there are few that are preserved in the Neogene record of heads and teeth. And it is further notable that one that is, enamel histology and thickness, is now controversial in the sense that competent workers can disagree about both character definition and developmental processes (Martin, 1985; Beynon *et al.,* 1991).

An important focus for future research should be to develop well-supported and corroborated hypotheses concerning relationships in a range of extant speciose higher taxa, particularly primates. I will reveal my prejudice by arguing that this would best be achieved by obtaining nucleotide sequence data on a sufficient number of nuclear and mitochondrial regions. This should then help to clarify the utility of morphological data where the definition and polarity of "characters" is often unclear (Pilbeam, 1996). Cladograms can best be used both to examine the evolution of physiological or morphological characters (Block *et al.,* 1993), particularly of the kind preserved in the relevant fossils, when these are mapped into the cladogram, and to attempt to gain some agreement on what are indeed "good" characters. Let us hope that such an exercise yields some set of "general" characters: for example, that a particular feature of the face is "always" or more likely "frequently" a "good" character. However, if Sarich (1993) is right, and the unbounded nature of morphological data means that a character useful in one situation is not necessarily useful in another, we are in trouble. More likely, we will be left struggling somewhere between our earlier kinds of discourse dominated by "experts" and a more desirable and objective one in which we can all agree on morphological data and analytical procedures (in the way that, providing nucleotide sequences are accurately determined, everyone using a particular analytic procedure will achieve the same result).

Postcranial Remains

While a great deal of taxonomic and phylogenetic activity, along with some functional interpretation, was going on with heads and teeth, postcranial analyses were focusing predominately on issues concerning function. We have recently seen important new discoveries and fresh interpretations, including additional data on the 1948 partial skeleton of *Proconsul heseloni* (previously *africanus*) from Rusinga (Walker *et al.*, 1993; Walker and Pickford, 1983), an important partial skeleton of *P. nyanzae* from Mfwanganu (Ward *et al.*, 1993), material from the other early Miocene genera *Afropithecus* and *Turkanapithecus* (Leakey and Leakey, 1986a,b), and further work on the intriguing lumbar vertebra from Moroto (Walker and Rose, 1968; Ward, 1993). With the exception of Moroto, the other genera resemble each other closely in preserved parts, and although they exhibit a few characters that are possible synapomorphies with living hominoids, they are for the most part rather unlike extant apes with their broad and shallow chests, short lumbar regions, and forelimbs adapted to suspensory behaviors. *P. nyanzae,* in contrast, has a deep and narrow thorax, a long lumbar region (six vertebrae), and forelimbs, hindlimbs, and extremities adapted mainly to palmigrade, noncursorial, above-branch quadrupedal climbing and running (Ward, 1993; Walker *et al.*, 1993). The other Miocene taxa, though less well sampled, were probably similar. Although no doubt broadly ancestral to living hominoids, the two groups differ very markedly cranially and postcranially, and the early Miocene apes almost certainly predate and are collectively distant from the extant radiation.

The Moroto vertebra is, as noted, an interesting exception. Its anatomy can be linked functionally to a trunk with a short lumbar region, a broad, shallow thorax, and with orthograde positional behaviors (Filler, 1986; Kelley, 1986; Ward, 1993). It is unusual among Miocene hominoids in resembling the living forms in a significant way. It is unlikely that Moroto represents *Afropithecus* (Ward, 1993), given postcranial similarities of *Proconsul* and *Afropithecus,* differences between the *Proconsul* and the Moroto lumbar vertebrae, and the differences between *Afropithecus* and the Moroto face.

For the middle Miocene, new and largely unpublished *Kenyapithecus* post-cranial material apparently differs little from *Proconsul* (Rose, personal communication; McCrossin and Benefit, this volume). Of note is the interesting association between a distal humeral articular surface that resembles generally those of living apes and a proximal shaft that is strongly retroflexed in lateral view and laterally convex in anterior view. Such shafts occur mainly in extant forms that are active quadrupedal leapers and runners (Begun, 1992).

For the later half of the Miocene, only *Dryopithecus* and *Sivapithecus* are known from more than one or two postcranial elements, except for the well-documented *Oreopithecus*. New material of *Dryopithecus* (Moyà-Solà and Köhler, 1996) has significantly expanded the sample. For *Sivapithecus*, sampling is still inadequate in that only bits of limbs and extremities are known, but these resemble other Miocene taxa as much as or more than living hominoids. Again, distal humeri resemble those of extant apes, while shafts and proximal joints differ markedly (Begun, 1992, 1993; Pilbeam *et al.*, 1980, 1990; Rose, 1983, 1988, 1989; Spoor *et al.*, 1991).

Along with the Moroto vertebra and possibly *Dryopithecus, Oreopithecus* stands out as exceptional among Miocene hominoids. Known since 1872, a partial skeleton was discovered in 1958 and initially described in the early 1960s (see Straus, 1963). The skeleton was reanalyzed by Sarmiento (1987), Harrison (1986), and Jungers (1987) in the mid-1980s, analyses that made clear the many close postcranial resemblances of *Oreopithecus* and living apes. Its hominoid status is still by no means widely accepted, although I agree with Sarmiento, Harrison, and Jungers that *Oreopithecus* is a hominoid of (post-cranially) modern aspect. Indeed, its postcranial similarities are the key to this identification. To refute this relationship would involve very substantial non-size-related homoplasy throughout the postcranium. Why did it take so long to recognize the hominoid status of *Oreopithecus*? Perhaps because its teeth and skull look so unlike hominoids (Harrison, 1986; Delson, 1986; Andrews, 1992), and we lived, indeed still do live, in a "head-dominated" world.

In addition to a much better fossil record and greatly improved functional interpretations, we have seen many other important changes over the past 15 years. Our field has been absorbed into modern evolutionary biology, and there is a strong case to be made that the integration of neontological and paleontological (morphological) studies—taxonomic, functional, and phylogenetic—with behavioral ecology and molecular genetics has gone as far in evolutionary anthropology as it has anywhere (Pilbeam, 1989).

It is noteworthy that only relatively recently have computer-assisted phylogenetic analyses been possible. Earlier cladistic studies were often listings of supposed synapomorphies supporting a preferred cladogram, without an exhaustive comparative review of all possible alternative cladograms (see listings in Groves and Paterson, 1991). Now that PAUP, PHYLIP, MacClade, and other personal computer programs are widely available, it becomes clearer from more comprehensive analyses—that is, those including the preferred characters of others as well as one's own, and those examining all possible cladograms—that morphological character incongruence is at least as abundant as at the nucleotide level and generally more so. This is due to homoplasy or poor character definition or both (Pilbeam, 1996). Again, not to sound like a broken record, we need to look carefully first for morphological characters that work well in resolving relationships in living groups, which requires that we have a well-supported phylogeny, praying that they are generally recognizable, and then—if they are—applying them to the fossil record (see Chamberlain and Gooder, 1993). We also need to be careful not to fall back too quickly on homoplasy to explain away inconvenient similarities, not to use it too readily as a *deus ex machina* to save a favored hypothesis.

Fossil Hominoid Relationships: Current Views

This cannot be a comprehensive history of Neogene hominoid evolutionary studies, so I begin this section by focusing on just one important recent review, that of Peter Andrews (1992), and start by noting an interesting contradiction in that paper. At the beginning (p. 641, Box 1) Andrews notes the large number of features (synapomorphies) uniting extant hominoids, especially postcranial ones. Indeed, hominoids, and particularly large hominoids, are characterized by an extensive suite of adaptations to orthograde positional behavior, however overprinted these may be by specific modifications (Washburn, 1963). Later in the paper when discussing *Oreopithecus,* Andrews (1992, p. 645) notes the (many) postcranial similarities of *Oreopithecus* to "the ancestral hominoid morphotype" but states that the highly specialized *Oreopithecus* dental/cranial morphology "must cast some doubt" on its attribution as a hominoid. The many postcranial features linking *Oreopithecus* to living hominoids are, of course, those synapomorphies shared among living hominoids. If these similarities are not in fact phylogenetically informative, they are homoplasies (or symplesiomorphies), and postcranial support for a monophyletic Hominoidea evaporates. I shall return later to this point, that, when in conflict, postcranial similarities seem more often to be explained away by our *deus* than are cranial/dental features.

Until recently (and Box 1 in Andrews indeed reflects this) most of us agreed that many of the postcranial similarities of living hominoids are indeed synapomorphies reflecting common ancestry. Despite some differences, extant apes are basically similar postcranially. The African apes are to a considerable extent allometric expressions of "the same" animal (see among many Shea, 1984; Jungers and Susman, 1984) and, although orangutans differ somewhat from *Pan* and *Gorilla*, especially in hands and feet (Tuttle, 1974), it seems unlikely that there are many homoplasies among the similarities between African and Asian apes. Sarmiento's (1985) comparative study of the morphology of captive and wild orangutans reveals that the more quadrupedal (though non-knuckle-walking) captive orangutans showed a number of convergent similarities (for example, in the wrist) to African apes, underlining the morphogenetic similarities among large hominoids.

The extant great apes also show many synapomorphies in cranial, facial, mandibular, and dental features (see Corruccini and McHenry, 1990; Andrews, 1992). To be sure, there are differences, yet relative to most Miocene hominoids the living species share many features. Of particular interest is a detailed and neglected morphometric study by Corruccini and McHenry (1980) demonstrating that the great apes share many features that are probably homologous; a sample of Miocene hominoids (*Proconsul, Dryopithecus,* and *Sivapithecus* including *Ramapithecus*) are also similar to each other, differ collectively from the extant apes, and are inferred to diverge "earlier from the hominid–pongid common ancestral lineage than living chimpanzees, gorillas and orangutans, which all share a number of derived traits not detected in any Miocene ape" (Corruccini and McHenry, 1980, p. 217).

Phylogenetic analyses of Miocene hominoids have been almost invariably "head-oriented" (for understandable reasons), but what if instead we began with bodies? Except for *Oreopithecus,* possibly *Dryopithecus,* and the Moroto vertebra, Miocene hominoids that are reasonably well characterized postcranially are more notable for their differences from extant apes than for their similarities. The postcrania of *Proconsul heseloni* and *P. nyanzae* are important because of their relative completeness: they show how good material can clarify and transform (Walker *et al.,* 1993; Ward, 1993). Indeed, we perhaps need a moratorium on describing more teeth until the other Miocene apes "catch up" and are as well characterized as *Proconsul!* Although *Proconsul* species resemble hominoids in a few features (Ward *et al.,* 1991; Rose, 1992; Beard *et al.,* 1986), morphologically and adaptively they are clearly different kinds of animals. They are "hominoids of archaic aspect," cranially, dentally, and postcranially, as opposed to the living "hominoids of modern aspect." I predict that when more complete remains of *Kenyapithecus, Sivapithecus, Ouranopithecus, Lufengpithecus,* and *Otavipithecus* are recovered, they will resemble more closely *Proconsul,* especially axially, than living apes.

The limbs and extremities of those Miocene forms that are known reasonably well, except for *Oreopithecus,* possibly *Dryopithecus,* and Moroto, differ in most characters from great apes, the exception being the distal humerous, which resembles extant hominoids (known in *Dryopithecus* from Rudabánya, *Sivapithecus,* and *Kenyapithecus*). However, in both *Kenyapithecus* and *Sivapithecus* the shaft is known and is distinctly *un*like extant hominoids (Begun, 1992; Pilbeam *et al.,* 1990). I believe that the jury is still out on *Dryopithecus,* in part because not enough is known postcranially about the samples from Klein Hadersdorf, St. Gaudens, and Rudabánya to know whether one, two, or three taxa are represented. However, the humeral shaft from Klein Hadersdorf is of "primitive aspect," and the St. Gaudens diaphysis is rather crushed anteriorly, which may contribute to its straightness in lateral view (Begun, 1992). Rose (this volume) has reconstructed a non-apelike posteriorly directed head, while in anterior view the shaft has a clear lateral convexity (Begun, 1992), both features suggesting a thorax and probably vertebral column different from those of apes. Hands and fingers, where known, also differ from extant apes (Begun, 1993). New material from Spain (Moyà-Solà and Köhler, 1993, 1996) is still not sufficiently complete to determine how like or unlike *Oreopithecus* it is postcranially.

Cranially and dentally (with three possible exceptions, *Ouranopithecus,*

Sivapithecus, Dryopithecus) the Miocene apes differ quite clearly from the living. In the case of *Ouranopithecus* Louis de Bonis and his associates have proposed that dental resemblances to hominids in occlusal morphology and canine size support an evolutionary relationship (Bonis and Melentis, 1977; Bonis *et al.*, 1990). In a number of ways, *Ouranopithecus* teeth differ from homologues of, for example, *Sivapithecus* or *Dryopithecus*, and are phenetically closer to early hominids. However, the possibility that this is related to the hyperthick enamel of *Ouranopithecus* needs to be carefully examined (L. Martin, personal communication). Cranially, *Ouranopithecus* does not particularly resemble any extant hominoid, sharing this pattern of general nonresemblance with *Dryopithecus* (Moyà-Solà and Köhler, 1993) and *Lufengpithecus* (Schwartz, 1990). Their relationships to each other, to other Neogene hominoids, and to extant apes are all unclear for several reasons, primarily the lack of a consensus on phylogenetically useful cranial and dental characters in hominoids.

Sivapithecus and Oreopithecus

In *Sivapithecus* we have a Neogene ape that raises a different set of very interesting issues. Here is an Asian Miocene hominoid in the "right" temporal and geographic position to be ancestral to a living ape, the orangutan (Kappelman *et al.*, 1991), and it also shows close and detailed similarities to the orangutan in several facial and palatal features in which both differ from all other known primates (Shea, 1985; Ward and Brown, 1986). However, as we have slowly discovered more about the postcranial anatomy of *Sivapithecus* (and the point was made quite forcefully to me with the realization that the very apelike distal humeri of *Sivapithecus* were attached to very nonmodern shafts), it became clear that, with the exception of the distal humerus, *Sivapithecus* not only lacks many postcranial features that might be plausible synapomorphies with *Pongo*, but also fails to reveal many traits that could be confidently identified as resembling the ancestral large hominoid morphotype. Whether *Sivapithecus* is or is not linked with *Pongo*, some very interesting convergences are clearly involved (Pilbeam *et al.*, 1990).

Oreopithecus is an interesting case in reverse. Strong similarities to living apes postcranially stand in marked contrast to cranial and dental differences. Of particular note is the possibly very small brain volume (Szalay and Berzi, 1973; Harrison, 1989) and a most unusual occlusal morphology (Harrison, 1986). As noted already, *Dryopithecus* from Spain (Moyà-Solà and Köhler, 1993, 1996), although better known than previously, is in my opinion not sufficiently complete to determine its relationships.

As the number of Miocene taxa has increased and as they become better characterized nondentally, phylogenetic analysis grows more rather than less complicated. The extent of homoplasy (because of our inability to define adequate characters in many cases) is considerable. And as "everyone's preferred characters" are added for objective analysis of Miocene relationships by PAUP, MacClade, and so forth, we can expect to see some very interesting analytical challenges. Whichever phylogenetic hypothesis is preferred, homoplasy is likely to be considerable. Again, will we seem happier invoking postcranial rather than cranial or dental homoplasy? That we have done so in the past has at least as much to do with the history of discovery in our field (lots of teeth came before good postcranial material) as with what is biologically plausible. Of course, cases of cranial, dental, and postcranial homoplasy can all be cited. I hope we can avoid getting into "cranial versus postcranial" controversies; any plausible phylogenetic scheme is likely to have "good" (and not so good) characters distributed throughout the skeleton.

Alternative Interpretations

We can now frame two starkly opposed phylogenetic hypotheses (although the truth may well lie in between). In one, particular hominoids of archaic aspect (e.g., *Sivapithecus, Kenyapithecus, Dryopithecus, Ouranopithecus*) are seen as sister taxa of particular lineages of living hominoids. In the other, archaic hominoids form a monophyletic sister clade to hominoids of modern aspect. The former hypothesis has been strongly favored over the past three decades, whatever the variation in individual details [note the continuity from Simons and Pilbeam (1965) to Andrews (1992)]. However, my hunch is that as further Miocene postcranial material is recovered, we may well see this as being an oversimplification, and shift toward the second hypothesis (McCrossin and Benefit, 1994; Pilbeam, 1996).

Historically we have viewed Miocene apes as "on their way to being living apes," and our presentist perspective (Stocking, 1968) has tended to downplay the differences between the two groups. I am arguing here for a more historicist approach that acknowledges the differences, regardless of their phylogenetic significance, and that sees the archaic hominoids as uniquely interesting animals, regardless of their relationship(s) to the living. Whether some of these archaic hominoids are related to specific living apes, or whether they are monophyletic and not directly related at all, they are of considerable interest in their own right and deserve study as uniquely distinct catarrhines, not just for what they are not. They are postcranially as well as dentally unique. They are found with faunas that often differ from those associated with living apes in ways that are difficult to interpret. And they disappear, along with many other taxa, during the great late Miocene faunal shake-up.

With the exception of *Proconsul,* Miocene large hominoids are so poorly sampled postcranially that it is not surprising that attempts to reconstruct their positional repertoires have been relatively limited. I have suggested here that when more is known of their axial skeletons, most will be seen to resemble *Proconsul.* That is, they were predominantly arboreal quadrupeds with mainly above-branch patterns of movement: forelimb morphology, especially the distal humerus, indicates behaviors in which stability in flexion–extension and pronation–supination was essential, though probably not the significantly suspensory patterns that would be associated with straight humeral shafts.

Perhaps this implies slow and deliberate climbing involving "arm-pulling" along with episodes of more active quadrupedal moving and leaping, rather than bouts of suspension. Inferring positional behavior of archaic hominoids more fully will be important to understanding the evolution of the modern hominoid postcranial pattern, because the latter is surely derived from the former, even if only once.

In many features of tooth design, the Miocene hominoids resemble each other and differ from living apes (although *Ouranopithecus* is a possible exception in need of further study). Two features show variation among Miocene hominoids: cingular expression and enamel thickness. The functional significance of neither is well understood, although it is widely assumed that enamel thickness is related in some way to physical properties of food. Whether or not these properties refer to rare but important or to common food items is unclear. Phylogenetically these features are also not well understood. For the brief period in which it has been competently studied, enamel thickness has come to be seen as an important phylogenetic character. However, the recent recognition that the otherwise rather similar early Miocene east African hominoids *Proconsul* and *Afropithecus* differ markedly in enamel thickness, as do the late Miocene European genera *Dryopithecus* and *Ouranopithecus* (L. Martin, personal communication), suggests that thickness may have changed frequently and should therefore be used with care as a character. Benyon and colleagues' (1991) work on enamel histology has also presented a different developmental model to Martin's (1985), one in which the thin enamel of living African apes can be interpreted as primitive for the African hominoid clade rather than as derived in *Pan* or *Gorilla* as in Martin's model. Regardless of who is right, or whether both are wrong, this is a more labile character than had been believed earlier. Disagreement among very competent workers demonstrates the problem of defining morphological characters and assigning polarity.

Hominid Origins

What can a study of the Miocene apes tell us about hominid origins? Historically, our tendency to see archaic forms as phases in the evolution of particular extant lineages, the consensus that thick enamel was both derived and phylogenetically informative as a character, the belief that aspects of the great ape postcranium were too derived to represent the ancestral hominid condition and that certain features of the hominid postcranium were primitive, has produced a consensus view that the common ancestral morphotype for hominids was unlike any living ape, with thick-enameled cheek teeth, "generalized" extremities, and an axial skeleton—particularly lumbar region—different from those of great apes (Pilbeam, 1996). An alternative has coexisted with this majority view, owing its clearest articulation over the past 30 years to Sherwood Washburn (1963, 1968, 1982; also Sarich, 1971). Based on comparative anatomy (drawing attention to the many similarities among living hominoids in the forelimb and trunk) and comparative genetics (which made the *Pan–Gorilla–Homo* relationship a trichotomy), this alternative saw the common ancestor as being basically chimplike, with a short lumbar region, plantigrade feet, long and perhaps knuckle-walking hands, and thin-enameled teeth. This has been very much a minority position. However, several recent developments should, I think, cause us to revisit this model.

The genetically inferred relationship among chimpanzees, gorillas, and humans is indeed a very close one, and the precise branching sequence makes little practical difference to the reconstruction of the ancestral hominid morphotype (Pilbeam *et al.*, 1990). Nonetheless, there have been and continue to be strongly expressed disagreements over the interpretation of genetic, karyotypic, and morphological data. The disagreement over whether *Pan–Gorilla* or *Pan–Homo* are sister taxa is not, repeat not, a case of "genetics versus morphology." Within each category of data (DNA sequences, chromosomes, anatomy), there are advocates of both perspectives (see Groves and Paterson, 1991; Stanyon, 1992; Marks, 1992; Bailey, 1993; Rogers, 1993; Ruvolo, 1994; 1995; Shoshani *et al.*, 1996). However, morphologists and paleontologists have historically almost invariably argued that because humans are very distinct anatomically they are also anciently derived. Many human morphological features have also been seen as "primitive," while those of the great apes were "specialized." So apes were often said to be derived from an ancestor that was more "humanlike" (or more recently more "australopithecine").

It is harder to sustain this position now. Several studies of the arm and hand show resemblances between African apes and hominids, including *Australopithecus,* which imply a period in hominid ancestry when the forelimb was used as a support during quadrupedal locomotion (Washburn, 1968; Corruccini, 1978; Lewis, 1989). Whether this ancestor was or was not a chimp-type knuckle-walker is irrelevant. Indeed, many forelimb features of knuckle-walking African apes may not be strictly knuckle-walking adaptations at all, but adaptations for stability when the limb is used as a strut. Some characters such as those of the distal humerus even resemble those in decidedly non-knuckle-walkers like *Sivapithecus* (Pilbeam *et al.*, 1980; McHenry and Corruccini, 1983).

The axial skeleton provides little support for the notion that hominids could *not* have passed through an apelike (short-backed) phase. Although humans differ from apes in the *average* number of lumbar vertebrae (humans have 5.0, *Pongo* 4.0, *Pan* and *Gorilla* 3.6, smaller gibbons 5.1, the larger siamang 4.4), the number of vertebral patterns within hominoid species is highly variable (Schultz, 1961). Evolving a "longer" from a "shorter" lumbar region need not be seen as a morphogenetic "problem." As far as tooth morphology is concerned, we have been in a period in which thick enamel has been seen as primitive for large hominoids, with hominids having thick enamel and African apes thin. However, there is now as good a case to be made that thickness is labile, and that the hominid condition is derived from a *Pan*-like ancestor.

The morphological evidence can obviously be read in several ways. The reading that firmly eliminates a chimplike phase from hominid ancestry has some underpinnings from an earlier perspective that sees apes as "specialized" and hominids as "primitive," and it is clearly not the only plausible reading. It is perhaps worth repeating what I wrote in 1989 about the *Pan–*

Homo versus *Pan–Gorilla* controversy: "Either way, the ancestor of hominids and apes is likely to have been a large, stiff-backed, long-armed, short-legged, arm-swinging arboreal form that was also a quadrupedal climber and walker in the trees and on the ground. At least some knuckle-walking or fist-walking is likely" (Pilbeam, 1989, p. 128). It probably lived in Africa, close to or in the rain forest. If this is indeed the case, the known Miocene hominoid record has little or perhaps nothing to tell us about hominid origins. However, hominids wouldn't have evolved had they not once been rather like chimps, and understanding better the extant great ape (including chimp) adaptations depends on finding their elusive ancestors. And it also involves learning more about their still enigmatic archaic cousins.

References

Andrews, P. 1986. Fossil evidence on human origins and dispersal. *Cold Spring Harbor Symp. Quant. Biol.* **51:**419–426.

Andrews, P. 1992. Evolution and environment in the Hominoidea. *Nature* **360:**641–646.

Andrews, P. J., and Cronin, J. E. 1982. The relationship of *Sivapithecus* and *Ramapithecus* and the evolution of the orangutan. *Nature* **297:**541–546.

Andrews, P., and Tekkaya, I. 1980. A revision of the Turkish Miocene hominoid *Sivapithecus meteai. Palaeontology* **23:**85–95.

Bailey, W. J. 1993. Hominoid trichotomy: A molecular overview. *Evol. Anthropol.* **2:**101–108.

Beard, K. C., Teaford, M. F., and Walker, A. 1986. New wrist bones of *Proconsul africanus* and *P. nyanzae* from Rusinga Island, Kenya. *Folia Primatol.* **47:**97–118.

Begun, D. R. 1992. Phyletic diversity and locomotion in primitive European hominids. *Am. J. Phys. Anthropol.* **87:**311–340.

Begun, D. R. 1993. New catarrhine phalanges from Rudabánya (northeastern Hungary) and the problem of parallelism and convergence in hominoid postcranial morphology. *J. Hum. Evol.* **224:**373–402.

Beynon, A. D., Dean, M. C., and Reid, D. J. 1991. On thick and thin enamel in hominoids. *Am. J. Phys. Anthropol.* **86:**295–309.

Block, B. A., Finnerty, J. R., Stewart, A. F. R., and Kidd, J. 1993. Evolution of endothermy in fish: Mapping physiological traits on a molecular phylogeny. *Science* **260:**210–214.

Bonis, L. de, and Melentis, J. 1977. Un nouveau genre de primate hominoide dans le Vallesien (Miocene superieur de Macedoine). *C. R. Acad. Sci.* **284:**1393–1396.

Bonis, L. de, and Melentis, J. 1985. La place du genre *Ouranopithecus* dans l'evolution des Hominides. *C. R. Acad. Sci.* **300:**429–432.

Bonis, L. de, Bouvrain, G., Geraads, D., and Koufos, G. 1990. New hominid skull material from the late Miocene of Macedonia in northern Greece. *Nature* **345:**712–714.

Chamberlain, A. T., and Gooder, S. J. 1993. Phylogenetic analysis of morphometric data: Applications to *Cercopithecus* and *Australopithecus* species groups. *Am. J. Phys. Anthropol. Suppl.* **16:**69.

Clark, W. E. L., and Leakey, L. S. B. 1951. The Miocene Hominoidea of East Africa. *Br. Mus. Nat. Hist. Fossil Mamm. Afr.* **1:**1–117.

Corruccini, R. 1975. *A Metrical Study of Crown Component Variation in the Hominoid Dentition.* Ph.D. dissertation, University of California, Berkeley.

Corruccini, R. S. 1978. Comparative osteometrics of the hominoid wrist joint, with special reference to knucklewalking. *J. Hum. Evol.* **7:**307–321.

Corruccini, R. S., and McHenry, H. J. 1980. Cladometric analysis of Pliocene hominids. *J. Hum. Evol.* **9:**209–221.

Delson, E. 1986. An anthropoid enigma: Historical introduction to the study of *Oreopithecus bambolii. J. Hum. Evol.* **15:**523–531.

Filler, A. G. 1986. *Axial Character Seriation in Mammals: An Historical and Morphological Exploration of the Origin, Development, Use and Current Collapse of the Homology Paradigm*. Ph.D. dissertation, Harvard University.

Greenfield, L. D. 1979. On the adaptive pattern of "*Ramapithecus*." *Am. J. Phys. Anthropol.* **50:**527–548.

Greenfield, L. D. 1980. A late divergence hypothesis. *Am. J. Phys. Anthropol.* **58:**351–365.

Groves, C. P., and Paterson, J. D. 1991. Testing hominoid phylogeny with the PHYLIP programs. *J. Hum. Evol.* **20:**167–183.

Harrison, T. 1986. A reassessment of the phylogenetic relationships of *Oreopithecus bambolii Gervais*. *J. Hum. Evol.* **15:**541–583.

Harrison, T. 1989. New estimates of cranial capacity, body size and encephalization in *Oreopithecus bambolii*. *Am. J. Phys. Anthropol.* **78:**237.

Harrison, T. 1992. A reassessment of the taxonomic and phylogenetic affinities of the fossil catarrhines from Fort Ternan, Kenya. *Primates* **33:**501–522.

Hartman, S. E. 1988. A cladistic analysis of hominoid molars. *J. Hum. Evol.* **17:**489–501.

Hartman, S. E. 1989. Stereophotogrammetric analysis of occlusal morphology of extant hominoid molars: Phenetics and function. *Am. J. Phys. Anthropol.* **80:**145–166.

Ishida, H., Pickford, M., Nakano, H., and Nakano, Y. 1984. Fossil anthropoids from Nachola and Samburu Hills, Samburu District, northern Kenya. In: H. Ishida, S. Ishida, and M. Pickford (eds.), *Study of the Tertiary Hominoids and Their Palaeoenvironments in East Africa: 2*, pp. 87–132. African Study Monographs, Kyoto University.

Jungers, W. L. 1987. Size and morphometric affinities of the appendicular skeleton in *Oreopithecus bambolii* (IGF 11778). *J. Hum. Evol.* **16:**445–456.

Jungers, W. L., and Susman, R. L. 1984. Body size and allometry in African apes. In: R. L. Susman (ed.), *The Pygmy Chimpanzee*, pp. 131–178. Plenum Press, New York.

Kappelman, J., Kelley, J., Pilbeam, D., Sheikh, K. A., Ward, S., Anwar, M., Barry, J. C., Brown, B., Hake, P., Johnson, N. M., Raza, S. M., and Shah, S. M. I. 1991. The earliest occurrence of *Sivapithecus* from the middle Miocene Chinji Formation of Pakistan. *J. Hum Evol.* **21:**61–73.

Kay, R. F. 1977. Diets of early Miocene African hominoids. *Nature* **268:**628–630.

Kay, R. F. 1981. *Sivapithecus simonsi*, a new species of Miocene hominoid, with comments on the phylogenetic status of the Ramapithecinae. *Int. J. Primatol.* **3:**113–173.

Kay, R. F., and Simons, E. L. 1983. A reassessment of the relationship between later Miocene and subsequent Hominoidea. In: R. L. Ciochon and R. S. Corruccini (eds.), *New Interpretations of Ape and Human Ancestry*, pp. 577–623. Plenum Press, New York.

Kelley, J. 1986. *Paleobiology of Miocene Hominoids*. Ph.D. dissertation, Yale University.

Kordos, L. 1987. Description and reconstruction of the skull of *Rudapithecus hungaricus* Kretzoi (Mammalia). *Ann. Hist. Nat. Mus. Natl. Hung.* **79:**77–88.

Leakey, R. E., and Leakey, M. G. 1986a. A new Miocene hominoid from Kenya. *Nature* **324:**143–145.

Leakey, R. E., and Leakey, M. G. 1986b. A second new Miocene hominoid from Kenya. *Nature* **324:**146–148.

Lewin, R. 1987. *Bones of Contention*. Simon & Schuster, New York.

Lewis, O. J. 1989. *Functional Morphology of the Evolving Hand and Foot*. Clarendon Press, Oxford.

Lu, Q., Xu, Q., and Zheng, L. 1981. Preliminary research on the cranium of *Sivapithecus yunnanensis*. *Vertebr. Palasiat.* **19:**101–107.

McCrossin, M., and Benefit, B. 1994. Maboko Island and the evolutionary history of Old World monkeys and apes. In: R. S. Corruccini and R. L. Ciochon (eds.), *Integrative Paths to the Past*, pp. 95–122. Prentice–Hall, Englewood Cliffs, NJ.

McHenry, H. M., and Corruccini, R. S. 1983. The wrist of *Proconsul africanus* and the origin of hominoid postcranial adaptations. In: R. L. Ciochon and R. S. Corruccini (eds.), *New Interpretations of Ape and Human Ancestry*, pp. 353–367. Plenum Press, New York.

Marks, J. 1992. The promises and problems of molecular anthropology in hominid origins. In: T. Nishida, W. C. McCrew, P. Marler, M. Pickford, and F. B. M. de Waal (eds.), *Topics in Primatology*, pp. 441–454. University of Tokyo Press, Tokyo.

Martin, L. 1985. Significance of enamel thickness in hominoid evolution. *Nature* **314:**260–263.

Moyà-Solà, S., and Köhler, M. 1993. Recent discoveries of *Dryopithecus* shed new light on evolution of great apes. *Nature* **365:**543–545.
Moyà-Solà, S., and Köhler, M. 1996. The first *Dryopithecus* skeleton: Origins of great ape locomotion. *Nature* **379:**156–159.
Novacek, M. J. 1993. Reflections on higher mammalian phylogenetics. *J. Mamm. Evol.* **1:**3–30.
Pickford, M. 1985. *Kenyapithecus:* A review of its status based on newly discovered fossils from Kenya. In: P. V. Tobias (ed.), *Hominid Evolution,* pp. 107–112. Liss, New York.
Pickford, M. 1986. Hominoids from the Miocene of east Africa and the phyletic position of *Kenyapithecus. Z. Morphol. Anthropol.* **76:**117–130.
Pilbeam, D. 1979. Recent finds and interpretations of Miocene hominoids. *Annu. Rev. Anthropol.* **8:**333–352.
Pilbeam, D. 1982. New hominoid skull material from the Miocene of Pakistan. *Nature* **295:**232–234.
Pilbeam, D. 1986. Distinguished lecture: Hominoid evolution and hominoid origins. *Am. Anthropol.* **88:**299–312.
Pilbeam, D. 1989. Human fossil history and evolutionary paradigms. In: M. K. Hecht (ed.), *Evolutionary Biology at the Crossroads,* pp. 177–138. Queens College Press, New York.
Pilbeam, D. 1996. Genetic and morphological records of the Hominoidea and hominid origins: A synthesis. *Mol. Phylogenet. Evol.* **5:**155–168.
Pilbeam, D., Rose, M. D., Badgley, C., and Lipschutz, B. 1980. Miocene hominoids from Pakistan. *Postilla* **181:**1–94.
Pilbeam, D., Rose, M. D., Barry, J. C., and Shah, S. M. I. 1990. New *Sivapithecus* humeri from Pakistan and the relationship of *Sivapithecus* and *Pongo. Nature* **348:**237–239.
Rogers, J. 1993. The phylogenetic relationships among *Homo, Pan and Gorilla:* A population genetics perspective. *J. Hum. Evol.* **22:**201–215.
Rose, M. D. 1983. Miocene hominoid postcranial morphology: Monkey-like, ape-like, neither, or both? In: R. L. Ciochon and R. S. Corruccini (eds.), *New Interpretations of Ape and Human Ancestry,* pp. 405–417. Plenum Press, New York.
Rose, M. D. 1986. Further hominoid postcranial specimens from the late Miocene Nagri Formation of Pakistan. *J. Hum. Evol.* **15:**333–367.
Rose, M. D. 1988. Another look at the anthropoid elbow. *J. Hum. Evol.* **17:**193–224.
Rose, M. D. 1989. New postcranial specimens of catarrhines from the Middle Miocene Chinji Formation, Pakistan: descriptions and a discussion of proximal humeral functional morphology in anthropoids. *J. Hum. Evol.* **18:**131–162.
Rose, M. D. 1992. Kinematics of the trapezium–first metacarpal joint in extant anthropoids and Miocene hominoids. *J. Hum. Evol.* **22:**255–266.
Ruvolo, M. 1994. Molecular evolutionary processes and conflicting gene trees: The hominoid case. *Am. J. Phys. Anthropol.* **94:**89–114.
Ruvolo, M. 1995. Seeing the forest and the trees. *Am. J. Phys. Anthropol.* **98:**218–232.
Sarich, V. M. 1971. A molecular approach to the question of human origins. In: P. Dolhinow and V. M. Sarich (eds.), *Background for Man,* pp. 60–81. Little, Brown, Boston.
Sarich, V. M. 1983. Retrospective on hominoid macromolecular systematics. In: R. L. Ciochon and R. S. Corruccini (eds.), *New Interpretations of Ape and Human Ancestry,* pp. 137–150. Plenum Press, New York.
Sarich, V. M. 1993. Mammalian systematics: Twenty-five years among their albumins and transferrins. In: F. S. Szalay, M. J. Novavek, and M. C. McKenna (eds.), *Mammal Phylogeny,* pp. 103–114. Springer-Verlag, Berlin.
Sarmiento, E. 1985. *Functional Differences in the Skeleton of Wild and Captive Orangutans and Their Adaptive Significance.* Ph.D. dissertation, New York University.
Sarmiento, E. 1987. The phylogenetic position of *Oreopithecus* and its significance in the origin of the Hominoidea. *Am. Mus. Novit.* **2881:**1–44.
Schultz, A. H. 1961. Vertebral column and thorax. *Primatologia* **4(5):**1–66.
Schwartz, J. H. 1990. *Lufengpithecus* and its potential relationship to an orang-utan clade. *J. Hum. Evol.* **19:**591–605.
Shea, B. T. 1984. An allometric perspective on the morphological and evolutionary relationships

between pygmy (*Pan paniscus*) and common (*Pan troglodytes*) chimpanzees. In: R. L. Susman (ed.), *The Pygmy Chimpanzee,* pp. 89–130. Plenum Press, New York.

Shea, B. T. 1985. On aspects of skull form in African apes and orangutans, with implications for hominoid evolution. *Am. J. Phys. Anthropol.* **68:**329–342.

Shoshani, J., Groves, C. P., Simons, E. L., and Gunnell, G. F. 1996. Primate phylogeny: Morphological vs. molecular results. *Mol. Phylogenet. Evol.* **5:**101–153.

Simons, E. L., and Pilbeam, D. 1965. Preliminary revision of the Dryopithecinae (Pongidae, Anthropoidea). *Folia Primatol.* **3:**81–152.

Simons, E. L., and Pilbeam, D. R. 1972. Hominoid paleoprimatology. In: R. Tuttle (ed.), *The Functional and Evolutionary Biology of Primates,* pp. 36–61. Aldine, Chicago.

Spoor, C. F., Sondaar, P. Y., and Hussain, S. T. 1991. A hominoid hamate and first metacarpal from the Late Miocene Nagri Formation of Pakistan. *J. Hum. Evol.* **21:**413–424.

Stanyon, R. 1992. How polymorphisms and homoplasy can be informative about the evolution and phylogeny of humans and apes. In: T. Nishida, W. C. McGrew, P. Marler, M. Pickford, and F. B. M. de Waal (eds.), *Topics in Primatology,* Vol. 1, pp. 423–440. University of Tokyo Press, Tokyo.

Stocking, G. W. 1968. *Race, Culture, and Evolution.* Free Press, New York.

Straus, W. L. 1963. The classification of *Oreopithecus.* In: S. L. Washburn (ed.), *Classification and Human Evolution,* pp. 146–177. Viking Press, New York.

Szalay, F. S., and Berzi, A. 1973. Cranial anatomy of *Oreopithecus. Science* **180:**183–185.

Teaford, M. F., and Walker, A. 1984. Quantitative differences in dental microwear between primate species with different diets and a comment on the presumed diet of *Sivapithecus. Am. J. Phys. Anthropol.* **64:**191–200.

Tuttle, R. 1974. Darwin's apes, dental apes, and the descent of man: Normal science in evolutionary anthropology. *Curr. Anthropol.* **15:**367–398.

Walker, A. C., and Rose, M. D. 1968. Fossil hominoid vertebra from the Miocene of Uganda. *Nature* **217:**980–981.

Walker, A. C., and Pickford, M. 1983. New postcranial fossils of *Proconsul africanus* and *Proconsul nyanzae.* In: R. L. Ciochon and R. S. Corruccini (eds.), *New Interpretations of Ape and Human Ancestry,* pp. 325–351. Plenum Press, New York.

Walker, A., Teaford, M. F., Martin, L., and Andrews, P. 1993. A new species of *Proconsul* from the early Miocene of Rusinga/Mfangano Islands, Kenya. *J. Hum. Evol.* **25:**43–56.

Ward, C. V. 1993. Torso morphology and locomotion in *Proconsul nyanzae. Am. J. Phys. Anthropol.* **92:**291–328.

Ward, C. V., Walker, A., and Teaford, M. F. 1991. *Proconsul* did not have a tail. *J. Hum. Evol.* **21:**215–220.

Ward, C. V., Walker, A., and Teaford, M. F., and Odhiambo, I. 1993. Partial skeleton of *Proconsul nyanzae* from Mfangano Island, Kenya. *Am. J. Phys. Anthropol.* **90:**77–111.

Ward, S. C., and Brown, B. 1986. The facial skeleton of *Sivapithecus indicus.* In: D. R. Swindler and J. Erwin (eds.), *Systematics, Evolution, and Anatomy,* pp. 413–451. Liss, New York.

Ward, S. C., and Kimbel, W. H. 1983. Subnasal alveolar morphology and the systematic position of *Sivapithecus. Am. J. Phys. Anthropol.* **61:**157–171.

Ward, S. C., and Pilbeam, D. R. 1983. Maxillofacial morphology of Miocene hominoids from Africa and Indo-Pakistan. In: R. L. Ciochon and R. S. Corruccini (eds.), *New Interpretations of Ape and Human Ancestry,* pp. 211–238. Plenum Press, New York.

Washburn, S. L. 1963. Behavior and human evolution. In: S. L. Washburn (ed.), *Classification and Human Evolution,* pp. 190–203. Viking Press, New York.

Washburn, S. L. 1968. *The Study of Human Evolution.* Condon Lectures, Oregon State System of Higher Education, Eugene.

Washburn, S. L. 1982. Human evolution. *Perspect. Biol. Med.* **25:**583–602.

Wu, R., Han, D., Xu, Q., Lu, Q., Pan, Y., Zhang, X., Zheng, L., and Xiao, M. 1981. *Ramapithecus* skulls found first time in the world. *Vertebr. Palasiat.* **26:**1018–1021.

Interrelationships between Functional Morphology and Paleoenvironments in Miocene Hominoids

3

PETER ANDREWS, DAVID R. BEGUN, and MYRIAM ZYLSTRA

Introduction

Previous chapters in this book have focused on the interaction between functional and phylogenetic analysis in interpreting fossil hominoid taxa or anatomical regions. The point has been made repeatedly that functional and phylogenetic analysis inform one another. Both functional analysis and phylogenetic analysis in turn are informed by paleoenvironmental data. Reconstructions of positional behavior and diet in fossil hominoids based on their functional anatomy are tested or at least reinforced by reconstructions of paleoenvironments. Paleoenvironmental reconstructions in turn are based in

PETER ANDREWS • Natural History Museum, London SW7 5BD, England. DAVID R. BEGUN and MYRIAM ZYLSTRA • Department of Anthropology, University of Toronto, Toronto, Ontario M5S 3G3, Canada.

Function, Phylogeny, and Fossils: Miocene Hominoid Evolution and Adaptations, edited by Begun *et al.* Plenum Press, New York, 1997.

part on the behavioral implications (ecomorphology) of fossil taxa, including hominoids. Paleoenvironmental patterns during hominoid evolution also represent responses to environmental conditions, and as such can be treated as characters relevant to phylogenetic analysis (e.g., Andrews, 1982). In this chapter we summarize the paleoenvironmental data from most of the localities that have produced the fossils examined in this book. The functional anatomy and behavioral implications of the hominoid fossils are also summarized in the context of now-current paleoenvironmental reconstructions.

Methods vary for reconstructing paleoenvironments, with different approaches appropriate for different cases. Most methods rely on comparisons of past floras and faunas with those living today. Taxonomic comparisons are the most common in paleontology, with inferences on paleoenvironment being made on the basis of relationships of fossil with living taxa. There is growing emphasis also on morphological comparisons, whereby functional morphologies of fossil animals can be interpreted by reference to those of living animals, with the ecological consequences of the morphologies inferred from these. Similar analyses of fossil plants make use of such features as leaf size or morphology of leaf margins.

Total species diversity of fossil biotas can also provide limited ecological information, particularly when plant diversity is taken into account. Taphonomic processes, however, may bias diversity patterns to such an extent that such information may be misleading. Diversity may also be analyzed by single ecological parameters such as size distributions of animal species or of tree leaves. Ecological diversities of animal faunas may be analyzed by univariate statistics or combined in multivariate functions to provide more complete information on the structure of whole communities, and these analyses may also be manipulated by rarefaction to simulate particular taphonomic biases in fossil faunas or to attempt to reconstruct past communities that have no living counterpart today.

To apply any of these methods to fossil biota is always difficult and often impossible because of the fossilization process itself. In this chapter, therefore, only a limited number of sites that have good stratigraphic and taphonomic control will be considered, and the number will be further restricted to include only the more significant fossil hominoid localities. A list of the sites and their respective ages and fossil hominoids is given in Table I.

Songhor and Koru, Kenya

The Songhor and Koru deposits have been dated at between 19 and 20 Ma (million years ago) (Pickford, 1983). The Songhor sediments correlate with the youngest of the Koru formations (Pickford, 1981), the Chamtwara Formation, with the Legetet Formation slightly older and the Koru Formation older still (Pickford and Andrews, 1981). The paleoenvironments of Songhor

Table I. Hominoid Localities with Number of Species (Where Known), Age, and the Hominoid Species Present[a]

Locality	Age	Fossil hominoids
Songhor (N=57)	Early Miocene	*Proconsul africanus, P. major, Rangwapithecus gordoni, Nyanzapithecus vancouveringi*
Koru (N=47)	Early Miocene	*Proconsul africanus, P. major*
Rusinga Island (N=27)	Early Miocene	*Proconsul heseloni, P. nyanzae, Nyanzapithecus vancouveringi*
Kalodirr	Early Miocene	*Afropithecus turkanensis, Turkanapithecus kalakolensis*
Moroto	?Middle Miocene	*?Afropithecus turkanensis*
Maboko Island (N=29)	Middle Miocene	"*Kenyapithecus*" *africanus, ?Proconsul* sp., *Nyanzapithecus pickfordi*
Fort Ternan (N=45)	Early Miocene	*Kenyapithecus wickeri, ?Proconsul* sp.
Paşalar (N=58)	Middle Miocene	*Griphopithecus alpani*
Rudabánya (N=59)	Late Miocene	*Dryopithecus brancoi, Anapithecus hernyaki*
Can Llobateres (N=67)	Late Miocene	*Dryopithecus laietanus*
Can Ponsic (N=49)	Late Miocene	*Dryopithecus crusafonti*
Siwaliks zone 6 (N=44)	Late Miocene	*Sivapithecus indicus*
Ravin de la Pluie (N=19)	Late Miocene	*Graeopithecus freybergi (Ouranopithecus macedoniensis)*
Lufeng, China (N=67)	Late Miocene	*Lufengpithecus lufengensis*

[a]The "small apes" (Harrison, 1987, 1989) are excluded from this list.

and Koru have been described in detail in Andrews *et al.* (1979) and Evans *et al.* (1981). For Songhor, the fauna from a single sedimentary unit has been analyzed in addition to the whole fauna, and for Koru, the faunas from the three formations have been described separately. Two types of weighted averages have been used on these faunas (Evans *et al.*, 1981). Habitat spectra for the faunas from Songhor and the Legetet and Chamtwara formations from Koru are very similar to each other and indicate closest similarity with tropical forest faunas today. The taxonomic habitat index shows the same thing for these faunas. The Koru Formation, the earliest in the Koru sequence, differs in both analyses in having a more open woodland component, implying a greater degree of seasonality in the climate.

The community structures of the Songhor and Legetet/Chamtwara faunas are again similar and strongly indicate tropical forest conditions. The faunas are dominated by small mammals, with frugivorous and insectivorous species common, and many arboreal and semiarboreal species. The fauna from the Koru Formation differs again in having a greater component of large mammals, terrestrial in spatial niche and browsing in dietary niche. Proportions of grazing and carnivorous species are particularly low in all faunas, in common with present-day tropical rain forest faunas (Evans *et al.*, 1981). This evidence is consistent with that from the weighted averages analyses in showing the presence of wet tropical forest at Songhor and the Legetet and Chamtwara Formations of Koru. The presence of *Proconsul africanus, P. major,* and *Rangwapithecus gordoni* in some abundance at these sites suggest that these hominoid species were living in this kind of habitat. The first two species are also present in the Koru Formation, but much less abundant.

The Songhor and Koru faunas have been analyzed by multivariate analysis of the ecological diversity data (Andrews, 1996). This employs the same kind of ecological data (size, diet, and locomotion) as have been used in the diversity histograms previously published (Andrews *et al.*, 1979; Evans *et al.*, 1981; Andrews, 1990a), but instead of analyzing the data separately, they are combined in multivariate functions. A variety of methods have been used, including principal coordinates, correspondence analysis, cluster analysis, and several distance statistics, but the results of only the first of these are shown in Fig. 1. Both the Songhor and Koru faunas are seen to group intermediate between the present-day African deciduous forest faunas and those from wet evergreen forests from Africa and Asia.

Euclidean distances calculated for the Songhor and Koru faunas are again consistent with interpretations of forest habitats at these sites. In this case it has been possible to modify the analyses of recent faunas by rarefaction, by which the recent faunas are reduced in size to the sizes of the fossil ones (Dreyer, 1984). The reductions have not been done randomly, as would be more usual in the rarefaction procedure, but have been done systematically and directed at specific parts of the faunas as a modified form of stratified random sampling (Andrews, 1992b, 1996). Figure 2 shows an analysis in which rarefaction has been applied to a recent fauna from an African tropical evergreen forest (Irangi). Figure 2A shows the distances of 23 recent faunas from a variety of recent African habitats from the Irangi fauna at the origin of the figure. Panels B and C show the same fauna from which have been removed 32% of the smallest species (B) and 32% of the largest species (C). The analysis of the Irangi fauna reduced by loss of small species shows the coalescence of all of the habitats with which it is compared, although the lowland tropical forest faunas are still the closest to the origin (B), i.e., the distances from the Irangi fauna at the origin are still smallest for other tropical forest faunas. Loss of large species has produced a much smaller difference (C), maintaining the same pattern as the unaltered fauna. The Songhor fauna (D) is intermediate between the two, showing some similarity to both, but on the whole it is closer to the rarefaction analysis with reduction of large species. It would appear to be a forest fauna that has lost many large species and a few small species during the time the fauna was being accumulated and fossilized.

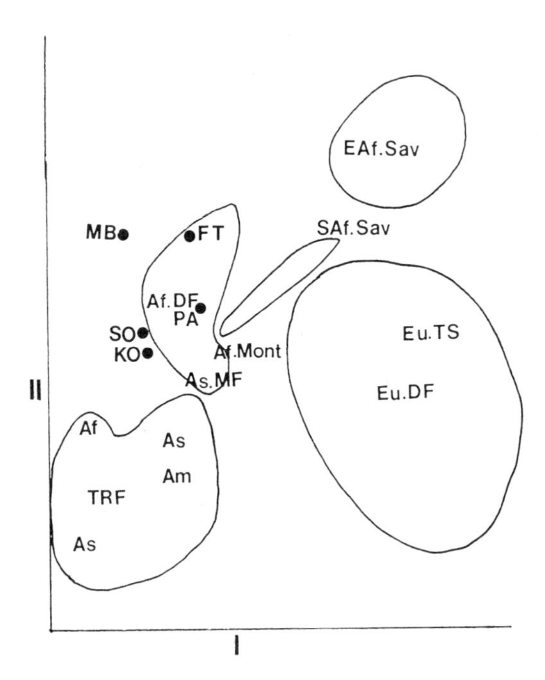

Fig. 1. Principal coordinates analysis of 83 modern and 5 fossil mammalian faunas. The first principal coordinate is shown on the horizontal axis, and the second on the vertical axis. The fossil faunas are shown individually as follows: SO, Songhor; KO, Koru; FT, Fort Ternan; MB, Maboko Island; PA, Paşalar. The living faunas are shown as distributions of collecting localities by geographic area and habitat type: EAf.Sav, savanna woodlands from East Africa; SAF.Sav, savanna woodlands from South Africa; Eu.TS, Eurasian temperate steppes; Eu.DF, deciduous forests from temperate Eurasia; TRF, tropical rain forests (wet, evergreen) from As—Asia, Af—Africa, Am—Central America; As.MF, monsoon forests from subtropical Asia; Af.DF, deciduous tropical forests from Africa; Af.Mont, montane forests from tropical East Africa.

This bias was predicted by independent taphonomic analysis (Pickford and Andrews, 1981).

The same analysis has been done for the fauna from the Koru Chamtwara Formation (panel E). The fauna is smaller than the Songhor one (Table I), but the patterns are similar and similar also to the Legetet Formation fauna not shown here. They are most similar to lowland forest faunas but are evidentially more modified as shown by the greater collapse downward and to the right of the distributions. In this case there has probably been greater loss of small species during fossilization. In general it may be concluded that tropical forest conditions pertained over most of the time and area of Songhor and Koru during the early Miocene.

The sample of hominoid limb bones from Songhor includes a number of forelimb and hindlimb specimens, but relatively few of these are published in any detail (e.g., Clark and Leakey, 1951; Pilbeam, 1969b; Langdon, 1986; Lewis, 1989). Most of these represent taxa of diverse body sizes, from *Micro-*

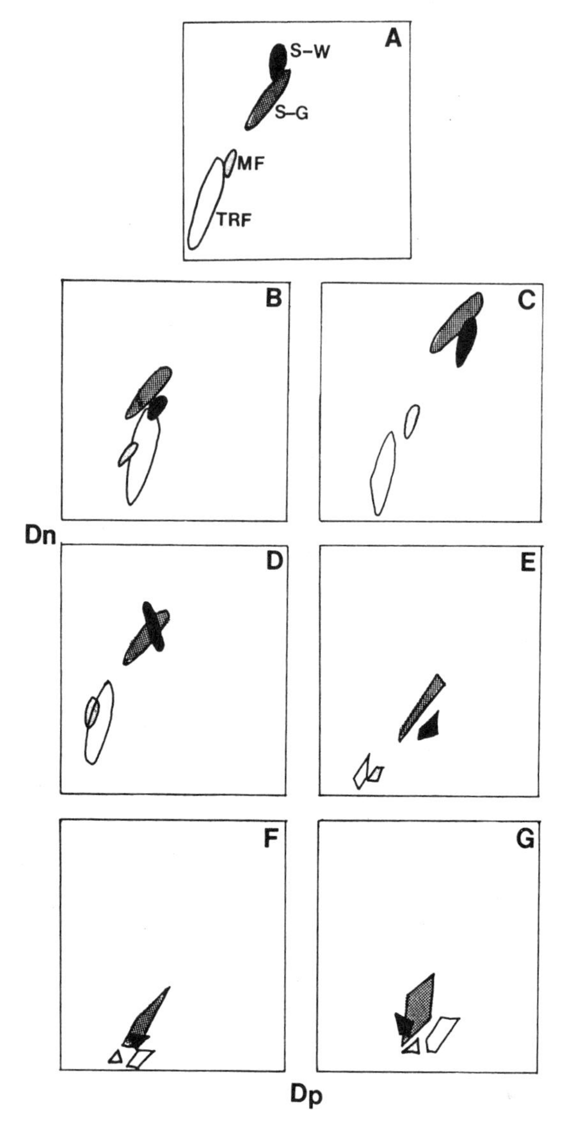

Fig. 2. Euclidean distances of fossil and recent faunas. Distances are measured from the origin at the bottom left-hand corner of every panel using two methods of calculating distances, based on proportions on the horizontal axes (dp) and on absolute numbers on the vertical axes (dn). In each panel, the reference assemblage at the origin is shown compared with the distributions of 23 modern African faunas grouped as follows: TRF, tropical rain forest; MF, montane wet forest; S-W, savanna woodland; S-G, savanna grassland. Following are the seven reference faunas: A, lowland wet evergreen forest faunal assemblage from Irangi, Zaire, $N = 76$; B, the same forest fauna reduced by 32% by removal of the smallest species; C, the same forest fauna reduced by 32% by removal of the largest species; D, the total Songhor fauna; E, the Koru fauna from the Chamtwara Formation; F, the Rusinga Island fauna from Kaswanga, levels KG and KB; G, the total Maboko Island fauna. (Data from Andrews and Van Couvering, 1975, and Andrews *et al.*, 1979.)

pithecus, the size of small cercopithecids (4 kg), to *Proconsul major,* the size of a female gorilla or male orang, possibly up to 80 kg (C. Ruff, personal communication). Other postcranial elements from a similar diversity of taxa are also known from nearby localities of the same age (Legetet, Chamtwara, and Mteitei Valley) and one radial fragment is known from Koru. These mostly represent smaller forms, the largest among them being *Rangwapithecus* and *Proconsul africanus* at roughly 15 to 18 kg (Fleagle, 1988), though a few of the phalanges from Legetet, Chamtwara, and Songhor are the size of small chimp to small gorilla phalanges. Apart from some of the hand and foot bones, most of these specimens are quite fragmentary and none are associated with other postcrania or with cranial remains. Recently recovered specimens (Nengo and Rae, 1992) are said to be more similar to living hominoids than is the later-occurring Rusinga Island *Proconsul.*

There are few clear indications from much of this sample of positional behavior different from that suggested for *Proconsul* from Rusinga Island. The Rusinga Island interpretations are based on much larger samples of more complete and associated postcranial remains (see below). The hand and foot bones share with Rusinga specimens a general absence of specialized features of cercopithecids or living hominoids, for the most part indicative of reduced ranges of joint mobility in the former and increased ranges in the latter (Napier and Davis, 1959; Schön and Ziemer, 1973; Morbeck, 1975, 1983; Lewis, 1980, 1985, 1989; McHenry and Corruccini, 1983; Rose, 1983, 1988, 1992; Walker and Pickford, 1983; Langdon, 1986; Beard *et al.,* 1986; Ward *et al.,* 1993). The long bones of the upper limb lack most of the characteristics in modern hominoids and middle to late Miocene hominoids associated in living forms with higher frequencies of suspensory positional behaviors. The morphology of these postcrania is consistent with an interpretation of generalized arboreality, and are therefore congruent with paleoenvironmental interpretations of tropical forest being widespread at these sites. In addition, there are differences in characters of the tarsus and forelimb that are suggestive of somewhat increased tarsal and carpal mobility, and the more powerful antebrachial and pedal digital flexor muscles are also suggestive of more powerful grasping capabilities than are present in either Old World or New World monkeys (Walker and Pickford, 1983; Beard *et al.,* 1986; Langdon, 1986; Lewis, 1989; Begun *et al.,* 1993; Ward *et al.,* 1993). The latter is also indicated by the morphology of the hand and foot phalanges.

The phalanges from Songhor and nearby Tinderet localities, though similar to better-known material from Rusinga (Napier and Davis, 1959; Begun *et al.,* 1993), differ in a number of subtle traits that might be related to minor ecological differences between the older and more recent localities. Regardless of size or taxonomic affiliation, the Songhor and Koru phalanges are generally smaller in transverse dimensions, more curved, and with more strongly developed fibrous flexor sheath ridges, ventral midline structures, and deeper, narrower trochlear. The distal articular ends also tend to be more circular in lateral view, as opposed to the more teardropped shape of the

Rusinga specimens, and they have deeper collateral ligament pits. In modern anthropoids these characteristics are usually associated with forms that move more rapidly in an arboreal milieu, with more leaping and running along branches, as opposed to slower moving, more deliberate climbing. Interestingly, Lewis (1985) noted a difference between Songhor and Rusinga *Proconsul* calcanei, in which the calcaneo-cuboidal articulation was more orang-like at Songhor and more African ape-like at Rusinga. These differences, especially in above-branch arboreal quadrupeds like most Miocene hominoids, may correspond at least in part to the structural differences between wetter forests at Songhor and drier or more open forests at Rusinga (see below). Songhor and Koru are richer in hominoid taxa, with seven in the former and six to eight in the latter (Harrison, 1989), compared to five for Rusinga. Songhor and Koru are also more diverse in hominoid body size distribution than Rusinga. These differences might also be causally related to the differences in paleoenvironment of the two localities (see below).

Rusinga Island, Kenya

The paleoenvironments of Rusinga Island have yet to be documented in any detail. The deposits are 1–2 million years younger than those from Songhor and Koru (Drake *et al.*, 1988), but the depositional environment is very different, representing flood plain to riverine conditions. There is great variation in both depositional conditions and in the faunas that are preserved, with major distinctions in faunas both laterally at the same stratigraphic level and from one level to another (Andrews and Van Couvering, 1975), and until these variations are recorded in any detail it is not possible to write any comprehensive account of Rusinga paleoenvironments.

One site from Rusinga Island contains a very rich seed and fruit flora, which provides good evidence for high-diversity seasonal woodland. This probably indicates a seasonal climate similar to that in western Kenya today, where the natural vegetation, before its modification by agriculturally developed humans, would have been closed canopy deciduous woodland but not tropical rain forest (Collinson, in preparation). On the other hand, a small fauna described in Andrews and Van Couvering (1975) shows some similarity to forest faunas in its taxonomic habitat index (Evans *et al.*, 1981), but the community structure differs in having more large mammals, browsing herbivores, and terrestrial species than are typically present in African forest faunas. Greater seasonality and drier conditions than were present in the Songhor and Koru forests are indicated by these differences. There are many species known from the Rusinga fauna that are related to living species that today occupy wet tropical forest habitats (Andrews and Van Couvering, 1975), but there is no unequivocal evidence for either the presence or absence of such habitats at Rusinga (Andrews *et al.*, 1979).

Euclidean distances for one of the Rusinga Island faunas are shown in Fig. 2F. This fauna is very small (Table I), and the distribution pattern of distances is compressed toward the *x* axis. This indicates extensive loss of species, but the pattern may well have been derived from one where the closest similarities were with present-day evergreen forest faunas, either lowland or montane, since these are closest to the origin. This must remain a very tentative conclusion in view of the strong bias evident in the distribution pattern. The most that can be said for the Rusinga paleoenvironments, and for the hominoids associated with the floras and faunas, is that there was some form of wooded habitat present over most of the area for most of the time. At certain times the climate may have been wetter, supporting forest, and at other times it may have been drier and more seasonal, supporting drier types of deciduous woodland and forest.

The Rusinga Island sample of *Proconsul* is the most extensive postcranial collection for any Miocene hominoid. The most abundant species are *Proconsul heseloni* (Walker *et al.*, 1993) and *P. nyanzae,* with the hominoidlike *Dendropithecus macinnesi* also abundant (*D. macinnesi* is also known from Tinderet localities, but from fewer, more fragmentary postcranial remains). The latter is similar in size to *Nyanzapithecus vancouveringi,* for which postcrania are unknown. The evidence for most elements of the limbs and axial skeleton strongly suggest some form of arboreal quadrupedalism with indications of enhanced joint excursion ranges but for the most part with little indication of a suspensory component, except perhaps in *Dendropithecus* (see references above and Rose, this volume; C. Ward, this volume). Rusinga *Proconsul* species may have differed from above-branch arboreal quadrupedalism of living primates in enhanced grasping capabilities, especially of the hallux, while *Dendropithecus* had elongated forelimbs suggestive of at least some suspensory behavior (Rose, this volume). All of this is quite consistent with the reconstruction of the paleoenvironment at Rusinga (and Tinderet) as a forested one. The functional anatomy of some of the Rusinga sample differs slightly from homologues in the samples from Songhor, Chamtwara, and Legetet, possibly in connection with the paleoenvironmental differences suggested above.

Kalodirr, Kenya, and Moroto, Uganda

The Kalodirr deposits are dated at between 16 and 18 Ma based on faunal correlations with the Buluk and Rusinga faunas (Leakey and Leakey, 1986). The fauna has not yet been described in any detail, but it is said to be similar in all respects to the Hiwegi Formation faunas from Rusinga Island (Leakey and Leakey, 1986). No conclusions can be drawn on the paleoenvironment based on this, although the presence of three hominoid species in the fauna is suggestive of wooded conditions having been present. One of these species is *Afropithecus turkanensis* (Leakey and Leakey, 1986), and although the postcranial remains of *Afropithecus* from Kalodirr have yet to be analyzed in detail, certain conclusions on patterns of positional behavior are possible. Hand, foot, and forelimb bones strongly resemble those of *Proconsul nyanzae* from Rusinga, suggesting similar patterns of positional behavior (Leakey *et al.*, 1988; Ward, 1993). The phalanges of the hand and foot from Kalodirr attributed to *Afropithecus* are very similar to those of Rusinga, and more distinct from those of Tinderet sequence localities (see above).

The hominoid from Moroto, originally assigned to *Proconsul major* (Pilbeam, 1969a, Andrews, 1978), has for some time been informally removed from *Proconsul* (Martin, 1981; Kelley and Pilbeam, 1986), and is now either grouped with *Afropithecus* (Leakey *et al.*, 1988; Andrews, 1992a) or considered unique (Ward, 1993; Sanders and Bodenbender, 1994; Begun, 1994). One postcranial element from Moroto, a lumbar vertebra, has recently been described as strongly indicative of more modern hominoidlike forelimb-dominated positional behavior than other African Miocene hominoid specimens (Ward, 1993, this volume; Sanders and Bodenbender, 1994).

Maboko Island and Fort Ternan, Kenya

The Fort Ternan and Maboko Island deposits are dated at between 14 and 15 Ma, some 2 million years later than the Rusinga Island deposits (Pickford, 1985). The fossil fauna from Fort Ternan comes from several paleosols (Retallack, 1991), and because the specimens are relatively intact, both in terms of individual specimens and as partial skeletons, the fauna does not appear to have been transported any distance (Shipman *et al.*, 1981). The fauna is mixed, however, and a small part of it was transported from farther away (Shipman, 1982). Maboko Island has a series of riverine flood plain deposits varying from fine clays to sands, and the fossils are much more broken. Some levels consisted of lakeside trampled silts (Andrews *et al.*, 1981), while the sands contained a transported assemblage consisting largely of isolated teeth. The level in which occurred hominoid limb bones is probably a calcrete soil developed on overbank silts, and it is likely that the fossils were associated with the soil formation rather than transported with the silts (Benefit, 1994; Benefit & McCrossin, 1995).

The paleoenvironment of both Fort Ternan and Maboko Island has been discussed in a number of papers, with conflicting results. Early results using weighted averages and community ecology (Andrews *et al.*, 1979; Andrews and Evans, 1979; Evans *et al.*, 1981) concluded that the habitat of both sites was thick closed woodland with some forest elements. Large terrestrial mammals were found to dominate the fauna, as in all nonforest faunas, and the highest dietary category was herbivorous browsing, but the proportions of arboreal and frugivorous species were too high for any of the more open-country faunas. Principal coordinates analyses of these ecological data (Fig. 1) locate the Fort Ternan fauna at the dry end of the forest distribution. Correspondence analysis using body weight data (Bonis *et al.*, 1992) also groups the Fort Ternan fauna with tropical forest faunas from Africa, although size is the least diagnostic of the ecological data available (Andrews *et al.*, 1979). Some form of seasonal woodland or single-canopy forest is indicated by this evidence.

Analysis of the Fort Ternan gastropods was said to indicate forest conditions at Fort Ternan (Pickford, 1985, 1987), but this work was criticized by Shipman (1986), who proposed ecotonal conditions at Fort Ternan between closed woodland and more open environments. Pickford's earlier (1983) conclusion of woodland to dry forest at Fort Ternan is more consistent with Shipman's work. Shipman's conclusions were based on the relative abundances of the faunal elements, for the most abundant species are those that have the terrestrial browsing adaptations mentioned above. In the weighted averages analyses, the uncommon species carry as much weight as the most common species, which could bias the results, and it is more probable that Shipman's conclusions are correct since they incorporate relative abundances. On the other hand, taphonomic bias is likely to be much greater when numbers of individuals are used as opposed to numbers of species (Andrews, 1990a), and Shipman has also shown (1982, Shipman *et al.*, 1981) that the Fort Ternan fauna was derived from mixed sources. She has identified the main source as the ecotonal woodland conditions, with some of the hominoids coming from a more distant forest source.

Functional interpretations of the common fossil antelopes have been shown to indicate the presence of woodland adaptations in these species (Kappelman, 1991). This is consistent with Shipman's results and further supports closed woodland and/or woodland ecotone environments at Fort Ternan. The size distribution of the fauna analyzed by cenogram also shows closed woodland affinities (Cerling *et al.*, 1992). Retallack (1992) has claimed on the basis of the paleosols present at Fort Ternan that the environment was open grassland, and this conclusion is supported by the presence of fossil grasses at certain levels in the site. Stable isotope analysis, however, has revealed that the grasses were C_3 grasses which grow under tree cover, not the open-country grasses of the African savannas of today (Cerling *et al.*, 1992).

The conclusion for Fort Ternan paleoenvironments must be that the area was dominated by closed-canopy woodlands, with open-country ecotone and forests in the vicinity. The *Kenyapithecus* fossils are taphonomically consistent with the majority of the fauna (Shipman, 1986), and as such may be considered an inhabitant of the main habitat, but the few fragmentary *Proconsul* fossils are differently preserved and may have been derived from one of the outlier habitats, probably forest.

There has been just as much controversy over the Maboko Island fauna as for the Fort Ternan fauna. Principal coordinates analysis of the ecological community data (based on size, locomotor, and dietary distributions of the fauna) separates the Maboko fauna from all of the recent and fossil faunas beyond the limits of the driest type of African deciduous forest (Fig. 1). The woodland affinities based on weighted averages have already been mentioned. The gastropods indicate drier conditions than were present at Fort Ternan (Pickford, 1983), and the correspondence analysis of Bonis *et al.* (1992) separates the Maboko fauna from rain forest faunas, so that it is intermediate between these and savanna woodland faunas. Dry woodland is suggested as most likely, but Euclidean distances calculated for the Maboko fauna (Fig. 2G) are ambiguous and either indicate a mixed assemblage or one that is so reduced as to give no clear indication of ecological affinities.

Kenyapithecus wickeri from Fort Ternan and "*Kenyapithecus*" *africanus* from Maboko Island are represented by a number of postcranial specimens, overall quite similar to those of *Proconsul* from Rusinga (Clark and Leakey, 1951; Morbeck, 1983; Begun, 1992a). A recently recovered proximal humeral fragment and other newly discovered material are said to suggest possible terrestrial adaptations (McCrossin and Benefit, 1993; McCrossin and Benefit, this volume). However, overall, there are few indications in the postcrania of any differences in habitat preferences from *Proconsul* or *Afropithecus*. "*Kenyapithecus*" femora from Maboko differ from *Proconsul* in their relative shaft length and in certain details of proximal morphology, both consistent with the view that it was basically an arboreal form, but possibly with enhanced hip joint mobility (Aiello, 1981; Ruff *et al.*, 1989). Hominoid phalanges from Maboko are generally more like Rusinga phalanges than are the Songhor and Koru specimens (see above), despite the fact that Rusinga and Maboko are more separated in time and share many fewer taxa than Rusinga and Songhor. This suggests similar activities for the Rusinga and Maboko hominoids, and therefore similar habitats, supporting the conclusions that these localities may have been somewhat drier overall than the Songhor and Koru localities (see above). Dental differences, in particular those related to enamel thickness, may be indicative of food choice and therefore possible habitat differences. However, thickly enameled hominoid teeth such as those of *Kenyapithecus* (and *Afropithecus*) are known from forested to more open environments (Andrews and Martin, 1991). Enamel thickness is therefore probably not a very reliable indicator of paleoenvironment.

Ecological and functional data are therefore consistent in finding little change from the late part of the early Miocene of Rusinga Island to the early part of the middle Miocene of Maboko Island and Fort Ternan, for there is little change in either environment or climate and little change also in the postcranial adaptations of the hominoids associated with them. There is, however, a phylogenetic difference currently recognized, that between the stem hominoid lineage represented by *Proconsul* and the stem hominid lineage represented by *Kenyapithecus*.

Paşalar, Turkey

The Paşalar deposits are the same age as the Maboko Island sediments, about 15 Ma (Bernor and Tobien, 1990). They are the earliest non-African deposits to contain hominoid primates and are presently situated in the warm temperate zone of the eastern Mediterranean. Depositional conditions of the main fossiliferous beds consisted of rapid flood accumulation of locally derived material (Andrews and Alpagut, 1980; Bestland, 1990). This was followed by the formation of paleosols, which indicate tropical to subtropical conditions during the Miocene with a strongly seasonal (wet–dry) climate (Bestland, 1990). Stable isotope analyses of both soils and tooth enamel from nine of the mammalian species show the presence of C_3 vegetation (Quade *et al.*, 1994), indicating closed cover conditions. Analyses of size-based cenograms and community structure of the Paşalar fauna (Andrews, 1990b) also indicate woodland conditions with closest similarities to subtropical deciduous forest faunas of north central India, and the same is indicated by rarefaction analysis of Euclidean distances.

More detailed analyses of the ecological composition of the Paşalar fauna point to similar conclusions. The analysis of carnivore guild structure (Viranta and Andrews, 1994) shows the presence of two insectivorous carnivores which distinguishes the Paşalar fauna from temperate faunas, but the presence of only two omnivorous carnivores distinguishes it from tropical forest faunas. The proportions of grazing versus carnivorous mammals shown here in Fig. 3 confirm this reconstruction, placing the Paşalar fauna with the Asian monsoon forest faunas. The proportions of both carnivores and grazers are lower than is usual in nonforest faunas. Similarly, the principal coordinates analysis in Fig. 1 groups the Paşalar fauna with the seasonal African deciduous forest faunas and close also to the Asian monsoon forests. Rarefaction analyses show the same thing (Andrews, 1992b, 1996), indicating moreover that the Paşalar fauna has been biased taphonomically by loss of very small species.

The Paşalar fauna was also examined by Bonis *et al.* (1992) who found that it was placed intermediate between recent rain forest faunas and flood plain faunas. All of this work agrees in rejecting tropical wet evergreen forest as being represented at Paşalar, suggesting instead seasonally dry closed woodland conditions.

There is good taphonomic evidence that the greater part of the Paşalar fauna was derived from a limited geographic region and accumulated over a short period of time (Andrews and Ersoy, 1990). As one of the most abundant elements, *Griphopithecus alpani* formed part of the mammalian fauna as a whole, and it is concluded that it was associated with the environmental reconstruction proposed here, that is, subtropical seasonal forests. Work in preparation by Ayhan Ersoy on the postcranial remains from Paşalar suggests that *Griphopithecus* was an arboreal species with above-branch specialization like arboreal cercopithecines today. Postcranial remains from Klein Hadersdorf, thought to be at least congeneric with the Paşalar forms (Begun, 1992a), suggest a similar pattern of positional behavior. To this generalized arboreal quadrupedalism might have been superimposed some reduction in limb length and some enhancement of grasping power, given the large size of these forms relative to most primate above-branch arboreal quadrupeds (Begun, 1992a).

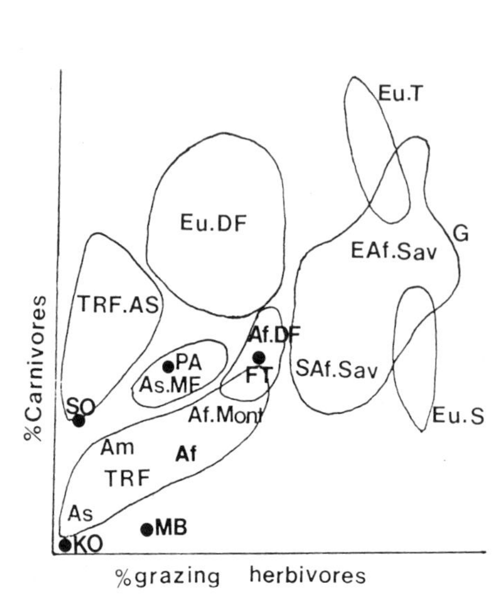

Fig. 3. The proportions of mammalian species with carnivorous adaptations are shown here plotted against the proportions of species with grazing adaptations for a series of recent and fossil faunas. There are 83 modern and 5 fossil mammalian faunas. The fossil faunas are shown separately, as follows: SO, Songhor; KO, Koru; FT, Fort Ternan; MB, Maboko Island; PA, Paşalar. The living faunas are shown as distributions of collecting localities by geographic area and habitat type: EAf.Sav, savanna woodlands from East Africa; SAf.Sav, savanna woodlands from South Africa; Eu.T, Eurasian cold tundra habitats; Eu.S, Eurasian cold temperate steppes; Eu.DF, deciduous forests from temperate Eurasia; TRF, tropical rain forests (wet, evergreen) from As–Asia, Af–Africa, Am–Central America; As.MF, monsoon forests from subtropical Asia; AS, seasonal forest from eastern Africa (China and Burma); Af.DF, deciduous tropical forests from Africa; Af.Mont, montane forests from tropical East Africa.

Rudabánya, Hungary

The Rudabánya deposits are dated to roughly 10 Ma (Kordos, 1991). The sedimentary environment is very different from any of the preceding sites in that it consists of lignites, black muds and marls deposited on the edge of a fluctuating lake. The most fossiliferous sediments are the black muds formed beneath or within the lignites, and there are plant remains in growth position within these deposits. The plant remains, fauna, and sediments indicate some form of swamp vegetation (Kordos, 1982). Fossil plants from several Rudabánya localities have been described (Kretzoi *et al.*, 1976), providing evidence of subtropical forest conditions, with the species composition of the flora similar to those of forests in southern China today.

The community structure of the fauna is similar to that of Paşalar and Sansan (Fig. 4), and it is similar also to the modern tropical deciduous forest faunas and subtropical seasonal forests to which Paşalar has already been

Fig. 4. Ecological diversity histograms for four European Miocene faunas described in the text (from Artemiou, 1984). Percentage numbers of species are shown on the vertical axis for three ecological categories: size, locomotion, and diet. The size classes are A, 0–100 g; B, 100–1000 g; C, 1–10 kg; D, 10–45 kg; E, 45–90 kg; F, 90–180 kg; G, 180–360 kg; H, > 360 kg. The locomotor classes are L, terrestrial; S, semiarboreal/terrestrial; Sc, scansorial; A, arboreal; Q, aquatic; E, aerial; and F, fossorial. The dietary classes are I, insectivorous; F, frugivorous; B, herbivorous browsing; G, herbivorous grazing; C, carnivorous; and O, omnivorous.

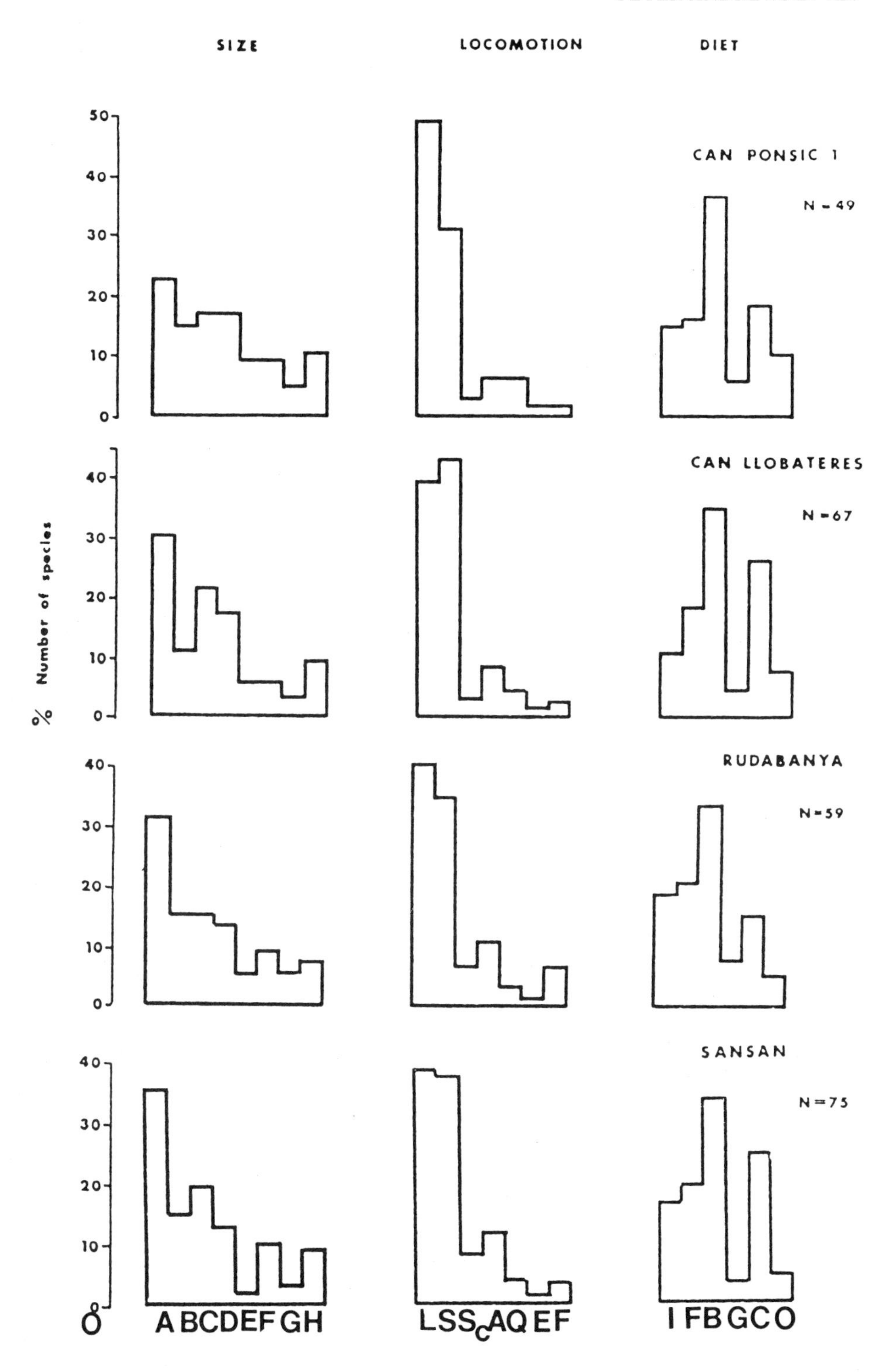
SIZE
LOCOMOTION
DIET
CAN PONSIC 1
N = 49
CAN LLOBATERES
N = 67
RUDABANYA
N = 59
SANSAN
N = 75
% Number of species
O
A B C D E F G H
L S S$_C$ A Q E F
I F B G C O

compared. Large terrestrial mammals are the most abundant group, although semiarboreal species are also common. Similarly, browsing herbivores are the most abundant dietary group, while frugivores and insectivores are also common. This pattern is distinct from the tropical forest patterns (Andrews *et al.*, 1979), but it is also distinct from open-country patterns from either tropical or temperate zones. It can be said, therefore, that the mammalian fauna community structure supports the conclusions from the flora in suggesting subtropical seasonal forest as the habitat of Rudabánya.

One hominoid (*Dryopithecus*) and a nonhominoid pliopithecid (*Anapithecus*) are known from Rudabánya. Both are present together in the upper black mud, from which the greater part of the large and small mammalian fauna is derived (Kretzoi *et al.*, 1976; Kordos, 1982). Since there is little evidence of transport in the accumulation of these fossils, it is concluded that both primates are associated with this environment.

Both primate taxa from Rudabánya are represented by postcranial remains with unambiguous indications of arboreality, most likely including some degree of suspensory positional behavior, but not rapid suspensory locomotion (Morbeck, 1983; Rose, 1983, this volume; Begun, 1988, 1992a, 1993; Begun and Kordos, this volume). Some form of suspensory positional behavior is evident from details of the anatomy of the elbow and phalanges, while arboreality more generally is suggested by the anatomy of the tarsal bones. The digits generally indicate powerful grasping and some degree of below-branch arboreality. The elbow also provides evidence of joint stability throughout a wide range of flexion–extension and pronation–supination, associated in living hominoids with suspensory positional behaviors. *Dryopithecus* and *Anapithecus* seem to converge in details of phalangeal morphology related to suspensory positional behavior (Begun, 1993). This would suggest fairly strong ecological constraints on positional behavior, imposed by an arboreal milieu in which roughly 8- to 35-kg primates did not frequently descend to the ground.

Can Llobateres and Can Ponsic, Spain

These two Spanish sites are close to the same age as Rudabánya, but possibly slightly younger than 10 Ma (Begun *et al.*, 1990; Begun, 1994). Little has been written about their paleoenvironments except for some anecdotal data and the analyses of community structure done by Artemiou (1984). The latter are reproduced here in Fig. 4 compared with Rudabánya and Sansan. The fossils preserved at both sites are fragmentary and crushed, but the presence of ribs and vertebrae together with skulls and mandibles suggests absence of transport in the accumulation of the fossils.

The ecological diversity spectra of the mammalian faunas show a high degree of consistency with other middle Miocene faunas (Fig. 4). The size spectra are undiagnostic, as in all Neogene faunas. In terms of spatial niche, terrestrial and semiarboreal species dominate, with the latter more abundant at Can Llobateres and the former more abundant at Can Ponsic. Arboreal and scansorial species are rare. In the dietary analyses, herbivorous browsers are dominant in both faunas, but frugivores and insectivores are also common. Carnivores are more abundant in the Can Llobateres fauna, similar to the case of Sansan. These distribution patterns exclude both tropical forest and temperate woodlands as possible reconstructions of the habitat for the Spanish sites. As for Rudabánya and Paşalar, the ecological structure of the two sites is closest to that of the subtropical seasonal forests of India or the tropical seasonal woodlands of China, and this is accepted provisionally as the most likely interpretation for the environment of the species of *Dryopithecus* contained in the faunas. More work obviously needs to be done to substantiate this conclusion.

Postcrania from these localities are less well studied than at Rudabánya, but there are no substantial differences between the three samples, all attributed to *Dryopithecus* (Begun *et al.*, 1990; Begun, 1992b; Begun and Kordos, 1993, this volume). Phalangeal and carpal specimens from both Spanish sites show indications of suspensory postures and substantial midcarpal mobility, without showing more specialized great ape characters related to specific modes of positional behavior in the modern forms (Begun, 1994). A distal tibia from Can Llobateres is broadly similar to *Proconsul*, as is the talus and to a lesser extent the entocuneiform from Rudabánya, all suggestive of arboreality, without the increased mobility of the hominoid tarsus. Based on the current evidence, there are no indications of significant differences in positional behavior between the Spanish and Hungarian sites, suggesting broadly similar paleoecologies as well.

Siwaliks, Pakistan

Hominoids are known from several parts of the long Siwalik section between 7 and 12 Ma (Kappelman *et al.*, 1991). The biostratigraphic and taphonomic work on these deposits by the Harvard group is without parallel in the study of hominoid sites (e.g., Barry *et al.*, 1985), but unfortunately there has been little published on ecological reconstructions of the fossil ape localities. Depositional environments have been described in great detail (Behrensmeyer, 1988), and they indicate a wide range of fluvial to flood plain environments, particularly the latter. Many of the fossil assemblages appear to be diachronous, i.e., from more than one time and source (Badgley *et al.*, 1986). Hominoids were a minor constituent of faunas that were dominated by bovids and equids, most of which were browsing forms, not grazing. The predominance of browsing herbivores and their wide range of body size suggests leafy forage was present through a considerable vertical range (Badgley,

1989). From this is inferred that the vegetation contained areas of shrub and woodland or forest, but no attempt has been made to go further than this because most fossils preserved in the fossil sites represent nonarboreal mammals.

Barry *et al.* (1990) studied a series of faunas from the Pakistan Siwalik sequence ranging from 16 to 7 Ma. They showed that there were periods of great stability in the Siwalik faunas, e.g., the period before 12 Ma, from 12 to 10 Ma and between 9 and 8 Ma. Before this time, species diversity was higher, and after 8 Ma there was a considerable turnover in species, but within this time span the strong faunal similarities appear to indicate the presence of a distinctive faunal community. Divergences from this community type can all be explained taphonomically (Barry *et al.*, 1990; Badgley, 1989).

The community structure of a composite fauna was shown to differ from present-day rain forest faunas by the dominance of terrestrial browsing herbivores (Andrews, 1983). On the other hand, it was seen to have close similarities to other Miocene faunas and to seasonal woodland to subtropical deciduous forests, but the value of analysis of such a mixed and time-averaged fauna must be questioned.

Stable isotope analyses for the Siwalik succession show that all of the early sites had C_3 vegetation, and there was a transition to C_4 at about 7 Ma (Quade *et al.*, 1989). This was the time of a major faunal turnover, including the disappearance of *Sivapithecus indicus* from the fossil record. It can be concluded that the C_3 vegetation present before 7 Ma was probably some form of closed canopy woodland or forest, with tropical to subtropical climate, and this would probably have been seasonal in nature.

Siwalik hominoid postcrania are relatively numerous but somewhat contradictory in their indications of positional behavior. Hand, foot, and forelimb bones are generally similar to other Miocene forms, but often with additional characters found only in living hominoids and related to increases in ranges of joint excursions (Pilbeam *et al.*, 1980; Morbeck, 1983; Rose, 1983, 1984, 1986, 1988, 1989; Raza *et al.*, 1983; Spoor *et al.*, 1991). In this regard the hand and foot bones are most like the phalanges, carpals, and tarsals of *Dryopithecus*. As in *Dryopithecus*, the elbow also provides evidence of joint dynamics associated in living hominoids with suspensory positional behaviors. The upper part of the forelimb, however, is more similar to large cercopithecids in shaft cross-sectional morphology, curvature, and deltopectoral attachment (Pilbeam *et al.*, 1990; Rose, this volume). If the proximal half of the humerus of *Sivapithecus* suggests that this taxon was not or was only infrequently suspensory, then it may have been occasionally terrestrial (Andrews, 1983) or more generally quadrupedal (Rose, this volume). Given their relatively large body size compared with broadly similar *Dryopithecus*, and given the difficulties involved in large, clawless, tailless mammals moving above branches (Grand, 1978; Cartmill, 1985), *Sivapithecus* may have been forced to the ground, perhaps more often than are living, suspensory great apes (Andrews, 1983). This may be related in part to the prevalence of terrestrial browsers in the Siwaliks fauna. On the other hand, the phalanges and hallucal skeleton of *Sivapithecus* do suggest powerful grasping capabilities, and the carpal and tarsal bones more mobility than typical of terrestrial quadrupeds, overall more consistent with arboreality. As suggested for *Griphopithecus* from Klein Hadersdorf (Begun, 1992a), this combination of characters may be indicative of a form of arboreal quadrupedalism that placed a premium on broadening of the base of support, decreasing the distance of the center of mass from the support, and generating high friction coefficients, all in response to the potentially very high torques generated by a large body above a narrow support.

Ravin de la Pluie and Xirochori, Greece

The Ravin de la Pluie (RP1) site consists of fluviatile gravels and red clays of late Vallesian age, 9.5 to 10 Ma (Bonis *et al.*, 1986; Bonis and Koufos, this volume). There is an impoverished fauna dominated by one or two species, although it is not evident whether this is a taphonomic artifact or truly representative of the environment (Bonis *et al.*, 1992). Bovids make up nearly three-quarters of the fauna (MNI = 130). Using faunal resemblance indices, Bonis *et al.* (1992) found that the RP1 site is taxonomically distinct from the other Macedonian Miocene faunas, but this is partly because it is earlier in time as well as perhaps having an ecological component. Size rank abundance curves are said to show similarities with curves established for savanna faunas (Bonis *et al.*, 1992, Fig. 9) resulting from the steepness of the line and the presence of a break for medium-sized animals. However, it must be said that the steepness of the line is nothing more than a reflection of species paucity (only 12 species are known), and the break in the line is a feature of seasonal habitats ranging from seasonal tropical forests to savanna. Moreover, in the absence of any theoretical explanation of the breaks, other than the empirical observation that they simply occur, the interpretation of savanna similarities in the cenogram may be questioned.

Also using body weight data, the RP1 fauna was compared with recent and other fossil faunas by multivariate analysis (Bonis *et al.*, 1992). Principal components analysis separated the RP1 fauna from the other Miocene sites (although no comparison was made with living faunas), but correspondence analysis grouped it with the other Greek Miocene faunas distinct from any of the wide range of recent comparative faunas. There is evidently taxonomic overprinting on the ecological data in the latter analysis, something that arises commonly with multivariate analyses (Andrews, 1996). In this case there is minimal ecological information in the analysis, so that neither multivariate analysis supports (or contradicts) the earlier work. All that can be said of this is that some form of seasonal climate is indicated, and this could range from seasonal forests to more open environments.

A single hominoid species has been described from these sites. This is referred to *Ouranopithecus macedoniensis* by Bonis (Bonis and Melentis, 1977), but others have referred it to the prior-named *Graecopithecus freybergi* (von Koenigswald, 1972; Martin and Andrews, 1984). No postcrania have been described for this hominoid, but it may be ecologically significant that molar enamel thickness has been found to be very great, thicker than that found on any living species of primate and comparable in thickness to that of robust australopithecines (Andrews and Martin, 1991). This may be related to increasing degrees of abrasion in the diet related to the use of foods from terrestrial sources, and/or to the incorporation of particularly tough food such as may be present in more strongly seasonal and harsher environments.

Baccinello, Italy

The fossil fauna from Baccinello, Italy, estimated to be about 8 Ma (Harrison and Harrison, 1989), has a high level of endemism and low diversity, both of which make it difficult to reconstruct the paleoenvironment. The fauna is associated with pollen, however, and this indicates an environment with lowland mesophytic forest consisting of mixed broad-leaved and coniferous species with a rich understory of bushes, small trees, and ferns (Harrison and Harrison, 1989). The abundant ferns and aquatic seed plants indicate the presence of water and moist substrates, and the development of the lignites in which the fossils are preserved indicates that these were developed in swamp conditions. Grass pollen is rare, so that open-country environments were rare to absent. The closed forest developed in swamp conditions is similar to that inferred for Rudabánya, and the closest modern analogue is given by Harrison and Harrison (1989) as the mixed mesophytic forests of the Yangtze river valley of east-central China. Warm temperate to subtropical climates with summer rainfall are inferred for both Baccinello and Rudabánya.

Among the fauna from Baccinello is the highly specialized hominoid *Oreopithecus* (Hürzeler, 1958). The dental anatomy of *Oreopithecus* has been described as that of a highly specialized folivore, while its postcranial anatomy is clearly that of a large-bodied powerful climber and specialized below-branch arboreal quadruped (for recent comprehensive literature reviews see Harrison, 1986; Harrison and Rook, this volume; Kay and Ungar, this volume; Sarmiento, 1987). This functional complex is completely consistent with the reconstruction of the ecology of Baccinello. Debate continues concerning the phyletic affinities of *Oreopithecus* (Harrison and Rook, this volume), but whether monkey, ape, or neither, there is much homoplasy in the anatomy of *Oreopithecus*, probably as a consequence at least in part of the limited range of options available to large-bodied highly arboreal primates.

Lufeng, China

The fossil fauna from Lufeng has been correlated with the finely documented Siwalik sequence, and shown by this means to be about 8 Ma (Flynn and Guo-qin, 1982). The fossils were accumulated in swamp deposits in an upland valley in a part of China that now has a subtropical to tropical climate. There is a good flora from the site, which is dominated by arboreal pollen, and this indicates a forested swamp with thickly vegetated margins (Badgley *et al.*, 1988). Plant material was probably autochthonous in the lignite swamps, as it was at Rudabánya, and like that site there was alternation between carbonaceous muds and lignites, suggesting sediment deposition in standing or slow-moving water of varying depth. At Lufeng there are five complete cycles of upward organic enrichment (Badgley *et al.*, 1988), whereas at the Rudabánya main locality there are only two. Living representatives of the flora suggest moist but not wet tropical forest with open glades (Badgley *et al.*, 1988). Tree ferns, understory ferns, and epiphytes are all present, indicating moist to moderately dry habitats.

On the basis of present-day comparisons, it is claimed that several of the mammals present at Lufeng were forest dwellers (Badgley *et al.*, 1988). Both small mammals like flying squirrels and larger ones like tragulids and deer, are taken to indicate forest conditions. Aquatic mammals are also present, and there is great abundance of small mammals, many with affinities with European taxa rather than with those from the Siwalik succession. This is of paleogeographic interest, for the European floras of the middle Miocene have been shown to have closest taxonomic affinities with the present-day Chinese one (Kovar-Eder, 1996). The great majority of herbivorous forms present in the fauna are browsers, with a great range of body sizes (Guo-qin, 1993).

The fossil hominoid *Lufengpithecus lufengensis* is known from beds 2–6 in section D at Lufeng (Guo-qin, 1993). Out of a total of 3200 mammalian specimens collected, 1185 represent this hominoid, approximately one-third of the entire collection. It is thought that there was a climatic shift up through the section, passing from warm and humid to drier, and then to cool and humid, but *Lufengpithecus* was present throughout these changes (Guo-qin, 1993).

Primate postcranial remains from Lufeng are not plentiful but do seem indicative of habitat preferences predictable on the basis of the paleoenvironmental reconstruction of the locality. The phalanges of the adapid and the primitive catarrhine from Lufeng (*Sinoadapis* and *Laccopithecus*) show indications of strong arboreality (Meldrum and Pan, 1988), and are consistent with a wet forested habitat. A few more postcranial specimens are available for the hominoid *Lufengpithecus*. Scapular, clavicular, radial, and manual phalangeal fragments are described as bearing some similarity to extant great apes, and are certainly consistent with arboreality. Powerful grasping digits and more apelike than monkeylike shoulder mobility seem to be indicated, possibly related to specialized arboreality, with some suspensory capabilities (Xiao, 1981; Wu and Xu, 1986; Lin *et al.*, 1987).

Discussion

The principal hominoid localities of the Old World have been briefly analyzed here to determine their paleoenvironments. These have been related to the functional attributes of the hominoid species included in the faunas in order to investigate the interrelationships between function and environment. As a final stage, we now superimpose this ecomorphologic pattern onto a system of phylogenetic relationships, not to contribute to phylogeny as such but rather to reach a better understanding of the evolutionary changes in the Hominoidea. It may eventually be possible to incorporate conclusions concerning paleoenvironment and ecomorphology into the development of a historical ecology of the Hominoidea (*sensu* Brooks and McLennan, 1991). However, at this point, with so many questions left unanswered, we chose to employ these ecological data post hoc, to add an ecological covering to the framework based primarily on traditional systematics.

The relationships of the fossil hominoids discussed in this chapter are not entirely clear, and in fact there is disagreement among the authors of this chapter on some of the details. One possible phylogeny is shown in Fig. 5 based on Andrews (1992a). The living apes and humans are shown along the top of the figure, with solid bars indicating the known time ranges of fossil hominoids along the vertical time scale. The earliest fossil hominoids are known only from tropical Africa, and of these the Proconsulidae are shown as a stem hominoid group with a time range from 25 to 17 Ma. There is general agreement about this placement (although see Harrison, 1987; Walker and Teaford, 1989). After the divergence of the gibbons comes a trichotomy leading to three stem hominid groups, the Afropithecini, still African, the Dryopithecini of Europe, and the Kenyapithecini known from both Europe and Africa. These three taxa are recognized at tribe level and form a paraphyletic group in the Hominidae. This is controversial and is a matter of temporary convenience since the interrelationships between these taxa are unclear, as is their possible relationship with any living taxa. It is probable that *Dryopithecus* is more closely related to the extant hominids (the great ape and human clade) than are the afropithecins or kenyapithecins (Begun, 1992a,c; Martin and Andrews, 1992), but there is some disagreement over the precise placement of *Dryopithecus* among the great apes and humans (Begun, 1994, 1995; Begun and Kordos, this volume; Dean and Delson, 1992; Moyà-Solà and Köhler, 1993). There is also disagreement over the placement of the *Graecopithecus* (or *Ouranopithecus*) (Begun, 1992c, 1994, 1995; Dean and Delson, 1992; Bonis and Koufos, 1993), and *Oreopithecus* (Harrison and Rook, this volume), which are therefore excluded from Fig. 5. The evidence for a link between

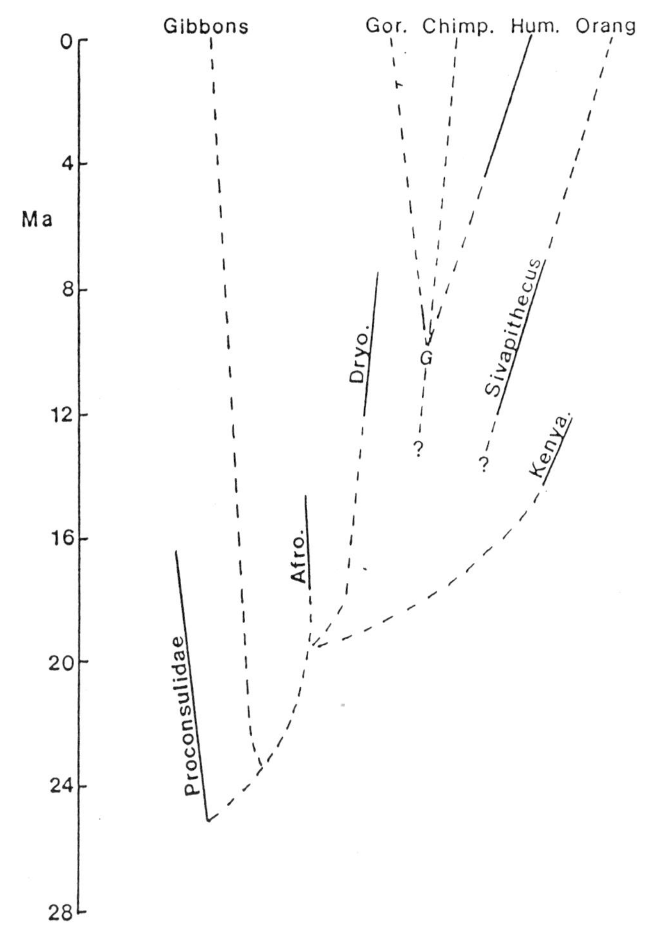

Fig. 5. Relationships of the fossil and living Hominoidea. The vertical scale shows time in millions of years. Solid lines show the known time ranges of the fossil taxa, and dashed lines show their possible evolutionary relationships. Afro., Afropithecini; Dryo., Dryopithecini; Kenya., Kenyapithecini; G, *Graecopithecus*.

Sivapithecus and the orangutan is stronger (Andrews, 1992a; although see Pilbeam *et al.*, 1990), and included here for our purposes is the Chinese hominoid *Lufengpithecus* (Schwartz, 1990, this volume).

The earliest records of the Proconsulidae associate the fossil apes with African tropical rain forest, e.g., at Songhor and Koru. The later species of this group, from Rusinga Island, are associated with more open forests, almost certainly more deciduous in a more seasonal climate. Both groups appear to have been generalized arboreal quadrupeds, probably with greater carpal and tarsal mobility than catarrhine monkeys but not as great as in living apes. There are differences, however, between the Songhor species of *Procon-*

sul and the Rusinga species in a number of modifications of the postcrania that are associated in present-day catarrhine primates with species that move more rapidly and are more agile in trees, with greater degrees of leaping and running. The Songhor/Koru species associated with the richer tropical forests are also better adapted to running and leaping in trees, while the Rusinga species associated with more open forest are less well-adapted arborealists and are more generalized above-branch quadrupeds.

The slightly later species of *Afropithecus* from Kenya and Uganda have postcrania similar to the Rusinga species of *Proconsul,* and it is particularly unfortunate that so little is known of the paleoenvironments of the sites from which they come. On the other hand, the paleoenvironments of the middle Miocene kenyapithecin sites are well known and are reconstructed as dry seasonal forest or woodland, with ecotonal variation to more open conditions. The postcrania of "*Kenyapithecus*" from Maboko Island and *Kenyapithecus wickeri* from Fort Ternan show many similarities with the Rusinga species morphology, which is consistent with the paleoenvironmental interpretation, and in addition there is some suggestion of terrestriality in the Maboko humerus, both the proximal articulation and the curvature of the shaft. It is inferred that these middle Miocene forms retained similar postcranial morphologies in similar habitats to the Rusinga forms, with some evidence for greater terrestriality on the one hand and drier, more open, and more seasonal conditions on the other. These similarities are in contrast to the change in molar enamel thickness, which in proconsulids is thin and in afropithecins and kenyapithecins is thick.

Emigration from Africa led hominoids to new paleoenvironmental contexts, and contributed to some significant morphological changes. The paleoenvironment of the earliest non-African hominoid, from Paşalar in Turkey, has been reconstructed as subtropical seasonal forest similar to the monsoon forests of India. The physiognomy of this forest has similarities with tropical seasonal forest-woodlands, but its nontropical nature is evident in greater seasonal variations with more open canopy structure of the forest. Primates living today in such environments are necessarily partly terrestrial because of this, but evidence available at this stage on the Paşalar hominoid indicates that it retained above-branch arboreal specializations similar to those from Rusinga Island, with no direct evidence of terrestriality.

Dryopithecus from central Europe is associated with subtropical seasonal forest of a wetter and less seasonal nature than was present at Paşalar. For the first time there is evidence of suspensory behavior in the postcrania of *Dryopithecus,* although this behavior was evidently less specialized than in present-day great apes. It is identified here as below-branch posture without rapid below-branch locomotion. The even more highly specialized below-branch arboreal quadruped *Oreopithecus* is also associated with a wet subtropical seasonal forest but with a more strongly aquatic setting than at Rudabánya. The highly specialized and homoplastic anatomy of *Oreopithecus* may represent one of the better examples in the Miocene of a dramatic impact of specific environmental conditions on hominoid functional anatomy. The very wet, insular setting at Baccinello (Harrison and Harrison, 1989) may have isolated *Oreopithecus* and imposed unusual but not unique environmental conditions, contributing to the development of specialized hominoidlike positional behavior, folivorous primatelike feeding strategies, and possibly large body size, the last of which is also observed among some of the Baccinello micromammals (Harrison and Harrison, 1989).

A combination of suspensory adaptations with similarities to terrestrial cercopithecids is also seen in *Sivapithecus,* and this is associated again with seasonal subtropical forest probably similar to that present at the *Dryopithecus* localities.

The evidence from *Dryopithecus, Oreopithecus,* and *Sivapithecus* shows that from the latter part of the middle Miocene, great-ape-like suspensory behavior had evolved in at least three groups of hominoid in seasonal forest environments in subtropical climatic regimes. What may have been happening in tropical regions of Africa is still unknown, but we venture to predict that the increasing degree of terrestriality of African apes is more compatible with seasonal open forest environments than with closed forest environments of the early Miocene.

References

Aiello, L. C. 1981. Locomotion in the Miocene Hominoidea. In: C. B. Stringer (ed.), *Aspects of Human Evolution,* pp. 63–98. Taylor & Francis, London.

Andrews, P. 1978. A revision of the Miocene Hominoidea of East Africa. *Bull. Br. Mus. Nat. Hist. Geol.* **30:**85–224.

Andrews, P. 1982. Ecological polarity in primate evolution. *Zool. J. Linn. Soc.* **74:**233–244.

Andrews, P. 1983. The natural history of *Sivapithecus.* In: R. L. Ciochon and R. Corruccini (eds.), *New Interpretations of Ape and Human Ancestry,* pp. 441–463. Plenum Press, New York.

Andrews, P. 1990a. *Owls, Caves and Fossils.* Natural History Museum, London.

Andrews, P. 1990b. Paleoenvironment of the Miocene fauna from Paşalar, Turkey. *J. Hum. Evol.* **19:**569–582.

Andrews, P. 1992a. Evolution and environment in the Hominoidea. *Nature* **360:**641–646.

Andrews, P. 1992b. Community evolution in forest habitats. *J. Hum. Evol.* **22:**423–438.

Andrews, P. 1996. Palaeoecology and hominoid palaeoenvironments. *Biol. Rev.* **71:**257–300.

Andrews, P., and Alpagut, B. 1990. Description of the fossiliferous units at Paşalar, Turkey. *J. Hum. Evol.* **19:**343–361.

Andrews, P., and Ersoy, A. 1990. Taphonomy of the Miocene bone accumulations at Paşalar, Turkey. *J. Hum. Evol.* **19:**379–396.

Andrews, P., and Evans, E. M. N. 1979. The environment of *Ramapithecus* in Africa. *Paleobiology* **5:**22–30.

Andrews, P., and Martin, L. 1991. Hominoid dietary evolution. *Philos. Trans. R. Soc. London Ser. B* **334:**199–209.

Andrews, P., and Van Couvering, J. H. 1975. Paleoenvironments in the East African Miocene. In: F. S. Szalay (ed.), *Approaches to Primate Paleobiology,* pp. 62–103. Karger, Basel.

Andrews, P., Lord J. and Evans, E. M. N. 1979. Patterns of ecological diversity in fossil and modern mammalian faunas. *Biol. J. Linn. Soc.* **11:**177–205.

Andrews, P., Meyer, G. E., Pilbeam, D. R., Van Couvering, J. A., and Van Couvering, J. A. H. 1981. The Miocene fossil beds of Maboko Island, Kenya: Geology, age, taphonomy and palaeontology. *J. Hum. Evol.* **10:**35–48.

Artemiou, C. 1984. Mammalian community paleoenvironment. *Paleontol. Contrib. Montpellier* **2:**91–109.

Badgley, C. 1989. Community analysis of Siwalik mammals from Pakistan. *J. Vert. Paleontol.* **9:**11A.

Badgley, C., Tauxe, L., and Bookstein, F. L. 1986. Estimating the error of age interpolation in sedimentary rocks. *Nature* **319:**139–141.

Badgley, C., Guo-qin, Q., Wanyong, C., and Defen, H. 1988. Paleoenvironment of a Miocene Tropical Upland fauna: Lufeng, China. *Nat. Geogr. Res.* **4:**178–195.

Barry, J. C., Johnson, N. M., Raza, S. M., and Jacobs, L. L. 1985. Neogene mammalian faunal change in southern Asia. *Geology* **13:**617–640.

Barry, J. C., Flynn, L. J., and Pilbeam, D. R. 1990. Faunal diversity and turnover in a Miocene terrestrial sequence. In: R. M. Ross and W. D. Allan (eds.), *Causes of Evolution,* pp. 381–421. University of Chicago Press, Chicago.

Beard, K. C., Teaford, M. F., and Walker, A. 1986. New wrist bones of *Proconsul africanus* and *P. nyanzae* from Rusinga Island, Kenya. *Folia Primatol.* **47:**97–118.

Begun, D. R. 1988. Catarrhine phalanges from the Late Miocene (Vallesian) of Rudabánya, Hungary. *J. Hum. Evol.* **17:**431–438.

Begun, D. R. 1992a. Phyletic diversity and locomotion in primitive European hominids. *Am. J. Phys. Anthropol.* **87:**311–340.

Begun, D. R. 1992b. *Dryopithecus crusafonti* sp.nov. a new Miocene hominoid species from Can Ponsic (northeastern Spain). *Am. J. Phys. Anthropol.* **87:**291–309.

Begun, D. R. 1992c. Miocene fossil hominids and the chimp–human clade. *Science* **257:**1929–1933.

Begun, D. R. 1993. New catarrhine phalanges from Rudabánya (northeastern Hungary) and the problem of parallelism and convergence in hominoid postcranial morphology. *J. Hum. Evol.* **24:**373–402.

Begun, D. R. 1994. Relations among the great apes and humans: New interpretations based on the fossil great ape *Dryopithecus. Yearb. Phys. Anthropol.* **37:**11–63.

Begun, D. R. 1995. Late Miocene European orang-utans, gorillas, humans, or none of the above? *J. Hum. Evol.* **29:**169–180.

Begun, D. R., and Kordos, L. 1993. Revision of *Dryopithecus brancoi* Schlosser, 1910, based on the fossil hominid material from Rudabánya. *J. Hum. Evol.* **25:**271–286.

Begun, D. R., Moyà-Solà, and Köhler, M. 1990. New Miocene hominoid specimens from Can Llobateres (Valles Penedes, Spain) and their geological and paleoecological context. *J. Hum. Evol.* **9:**255–268.

Begun, D. R., Teaford, M. F., and Walker, A. 1993. Comparative and functional anatomy of *Proconsul* phalanges from the Kasawanga Primate Site, Rusinga Island, Kenya. *J. Hum. Evol.* **25:**89–165.

Behrensmeyer, A. K. 1988. Vertebrate preservation in fluvial channels. *Palaeogeogr. Palaeoclimatol. Palaeoecol.* **63:**183–199.

Benefit, B. R. 1994. Phylogenetic, paleodemographic, and taphonomic implications of *Victoriapithecus* deciduous teeth. *Am. J. Phys. Anthrop.* **95:**277–331.

Benefit, B. R., and McCrossin, M. L. 1995. Miocene hominoids and hominid origins. *Ann. Rev. Anthrop.* **24:**237–256.

Bernor, R. L., and Tobien, H. 1990. The mammalian geochronology and biogeography of Paşalar (middle Miocene, Turkey). *J. Hum. Evol.* **19:**551–568.

Bestland, E. 1990. Sedimentology and paleopedology of Miocene alluvial deposits at the Paşalar hominoid site, western Turkey. *J. Hum. Evol.* **19:**363–378.

Bonis, L. de, and Koufos, G. 1993. The face and mandible of *Ouranopithecus macedoniensis:* Description of new specimens and comparisons. *J. Hum. Evol.* **24:**469–491.

Bonis, L. de, and Melentis, J. 1977. Un nouveau genre de primate hominoide dans le Vallesien (Miocene superieur) de Macedoine. *C. R. Acad. Sci.* **284:**1393–1396.

Bonis, L. de, Bouvrain, G., Koufos, G., and Melentis, J. 1986. Succession and dating of the late

Miocene primates of Macedonia. In: J. G. Else and P. C. Lee (eds.), *Primate Evolution*, Vol. 1, pp. 107–114. Cambridge University Press, London.

Bonis, L. de, Bouvrain, G., Geraads, D., and Koufos, G. 1992. Diversity and paleoecology of Greek late Miocene mammalian faunas. *Palaeogeogr. Palaeoclimatol. Palaeoecol.* **91:**99–121.

Brooks, D. R., and McLennan, D. A. 1991. *Phylogeny, Ecology and Behavior.* University of Chicago Press, Chicago.

Cartmill, M. 1985. Climbing. In: M. Hildebrand, D. M. Bramble, K. F. Liem, and D. B. Wake (eds.), *Functional Vertebrate Morphology,* pp. 73–88. Belknap Press, Cambridge.

Cerling, T. E., Kappelman, J., Quade, J., Ambrose, S. H., Sikes, N. E., and Andrews, P. 1992. Reply to comments on the paleoenvironment of *Kenyapithecus* at Fort Ternan. *J. Hum. Evol.* **23:**371–377.

Clark, W. E. L., and Leakey, L. S. B. 1951. The Miocene Hominoidea of East Africa. *Br. Mus. Nat. Hist. Fossil Mamm. Af.* **1:**1–117.

Dean, D., and Delson, E. 1992. Second gorilla or third chimp? *Nature* **359:**676–677.

Drake, R. L., Van Couvering, J. A., Pickford, M., Curtis, G. H., and Harris, J. A. 1988. New chronology for the early Miocene mammalian faunas of Kisingiri, western Kenya. *J. Geol. Soc. London* **145:**479–491.

Dreyer, S. D. 1984. *The Theory and Use of Methods for the Study of Mammalian Paleoenvironment.* Ph.D. thesis, University of London.

Evans, E. M. N., Van Couvering, J. H., and Andrews, P. 1981. Paleoenvironment of Miocene sites in western Kenya. *J. Hum. Evol.* **10:**35–48.

Fleagle, J. G. 1988. *Primate Adaptation and Evolution.* Academic Press, New York.

Flynn, L. J., and Guo-qin, Q. 1982. Age of the Lufeng, China, hominoid locality. *Nature* **298:**746–747.

Grand, T. I. 1978. Adaptations of tissue and limb segments to facilitate moving and feeding in arboreal folivores. In: G. G. Montgomery (ed.), *The Ecology of Arboreal Folivores,* pp. 231–241. Smithsonian Institution Press, Washington, DC.

Guo-qin, Q. 1993. The environmental ecology of the Lufeng hominoids. *J. Hum. Evol.* **24:**3–11.

Harrison, T. 1986. A reassessment of the phyletic relationships of *Oreopithecus bambolii* Gervais. *J. Hum. Evol.* **15:**541–583.

Harrison, T. 1987. The phylogenetic relationships of the early catarrhine primates: A review of the current evidence. *J. Hum. Evol.* **16:**41–80.

Harrison, T. 1989. A new species of *Micropithecus* from the middle Miocene of Kenya. *J. Hum. Evol.* **18:**537–557.

Harrison, T. S., and Harrison, T. 1989. Palynology of the late Miocene *Oreopithecus*-bearing lignite from Baccinello, Italy. *Palaeogeogr. Palaeoclimatol. Palaeoecol.* **76:**45–65.

Hürzeler, J. 1958. *Oreopithecus bambolii* Gervais: A preliminary report. *Verh. Naturforsch. Ges. Basel* **69:**1–48.

Kappelman, J. 1991. The paleoenvironment of *Kenyapithecus* at Fort Ternan. *J. Hum. Evol.* **20:**95–129.

Kappelman, J., Kelley, J., Pilbeam, D., Sheikh, K. A., Ward, S., Anwar, M., Barry, J. C., Brown, B., Hake, P., Johnson, N. M., Raza, S. M., and Shah, S. M. I. 1991. The earliest occurrence of *Sivapithecus* from the middle Miocene Chinji Formation of Pakistan. *J. Hum. Evol.* **21:**61–73.

Kelley, J., and Pilbeam, D. 1986. The dryopithecines: Taxonomy, comparative anatomy, and phylogeny of Miocene large hominoids. In: D. R. Swindler and J. Erwin (eds.), *Comparative Primate Biology, Vol. 1: Systematics, Evolution, and Anatomy,* pp. 361–411. Liss, New York.

Kordos, L. 1982. The prehominid locality of Rudabánya (NE Hungary) and its neighbourhood: A palaeogeographic reconstruction. *Mag. All. Foldt. Intez. Evi Jel.* **1980:**395–406.

Kordos, L. 1991. Le *Rudapithecus hungaricus* de Rudabánya (Hongrie). *L'Anthropologie* **95:**343–362.

Kovar-Eder, J. 1996. In: R. Bernor, V. Fahlbusch, and S. Rietschel (eds.), *Evolution of Neogene Continental Biotopes in Central Europe and the Eastern Mediterranean.* Columbia University Press, New York.

Kretzoi, M., Kroloppe, E., Lörincz, H., and Pálfalvy, I. 1976. Floren- and faunnenfunde der altpannonischen Prähominiden-Fauna von Rudabánya and ihre stratigraphische Bedeutung. *Mag. All. Foldt. Intez. Evi Jel.* **1974:**365–394.

Langdon, J. H. 1986. Functional morphology of the Miocene hominid foot. *Contrib. Primatol.* **22:**1–255.
Leakey, R. E., and Leakey, M. G. 1986. A new Miocene hominoid from Kenya. *Nature* **324:**143–145.
Leakey, R. E., Leakey, M. G., and Walker, A. C. 1988 Morphology of *Afropithecus turkanensis* from Kenya. *Am. J. Phys. Anthropol.* **76:**289–307.
Lewis, O. J. 1980. The joints of the evolving foot. Part III. The fossil evidence. *J. Anat.* **131:**272–298.
Lewis, O. J. 1985. Derived morphologies of the wrist articulations and theories of hominoid evolution. Part II. The midcarpal joints of higher primates. *J. Anat.* **142:**151–172.
Lewis, O. J. 1989. *Functional Morphology of the Evolving Hand and Foot.* Clarendon Press, Oxford.
Lin, Y., Wang, S., Gao, Z., and Zhang, L. 1987. The first discovery of the radius of *Sivapithecus lufengensis* in China. *Geol. Rev.* **33:**1–4.
McCrossin, M. L., and Benefit, B. R. 1993. New *Kenyapithecus* postcrania and other primate fossils from Maboko Island, Kenya. *Am. J. Phys. Anthropol. Suppl.* **16:**55–56.
McHenry, H. M., and Corruccini, R. S. 1983. The wrist of *Proconsul africanus* and the origin of hominoid postcranial adaptations. In R. L. Ciochon and R. S. Corruccini (eds.), *New Interpretations of Ape and Human Ancestry,* pp. 353–367. Plenum Press, New York.
Martin, L. 1981. New specimens of *Proconsul* from Koru, Kenya. *J. Hum. Evol.* **10:**139–150.
Martin, L., and Andrews, P. 1984. The phyletic position of *Graecopithecus freybergi* Koenigswald. *Courier Forschungs. Senckenberg.* **69:**25–40.
Martin, L., and Andrews, P. 1992. Renaissance of Europe's ape. *Nature* **365:**494.
Meldrum, D. J., and Pan, Y. 1988. Manual proximal phalanx of *Laccopithecus robustus* from the latest Miocene site of Lufeng. *J. Hum. Evol.* **17:**719–731.
Morbeck, M. E. 1975. *Dryopithecus africanus* forelimb. *J. Hum. Evol.* **4:**39–46.
Morbeck, M. E. 1983. Miocene hominoid discoveries from Rudabánya: Implications from the postcranial skeleton. In: R. L. Ciochon and R. S. Corruccini (eds.), *New Interpretations of Ape and Human Ancestry,* pp. 369–404. Plenum Press, New York.
Moyà-Solà, S., and Köhler, M. 1993. Recent discoveries of *Dryopithecus* shed new light on evolution of great apes. *Nature* **365:**543–545.
Napier, J. R., and Davis, P. R. 1959. The forelimb skeleton and associated remains of *Proconsul africanus. Br. Mus. Nat. Hist. Fossil Mamm. Af.* **16:**1–69.
Nengo, I., and Rae, T. C. 1992. New hominoid fossils from the early Miocene site of Songhor, Kenya. *J. Hum. Evol.* **23:**423–429.
Pickford, M. 1981. Preliminary miocene mammalian biostratigraphy for western Kenya. *J. Hum. Evol.* **10:**73–97.
Pickford, M. 1983. Sequence and environments of the lower and middle Miocene hominoids of western Kenya. In R. L. Ciochon and R. S. Corruccini (eds.), *New Interpretations of Ape and Human Ancestry,* pp. 421–439. Plenum Press, New York.
Pickford, M. 1985. A new look at *Kenyapithecus* based on recent discoveries in western Kenya. *J. Hum. Evol.* **14:**113–143.
Pickford, M. 1987. Fort Ternan (Kenya) paleoenvironment. *J. Hum. Evol.* **16:**305–309.
Pickford, M., and Andrews, P. 1981. The Tinderet Miocene sequence in Kenya. *J. Hum. Evol.* **10:**11–33.
Pilbeam, D. R. 1969a. Tertiary Pongidae of East Africa: Evolutionary relationships and taxonomy. *Bull. Peabody Mus. Nat. Hist.* **31:**1–185.
Pilbeam, D. R. 1969b. Possible identity of Miocene fossil tali from Kenya. *Nature* **223:**648.
Pilbeam, D. R., Rose, M. D., Badgley, C., and Lipschutz, B. 1980. Miocene hominoids from Pakistan. *Postilla* **181:**1–94.
Pilbeam, D. R., Rose, M. D., Barry, J. C., and Shah, S. M. I. 1990. New *Silvapithecus* humeri from Pakistan and the relationship of *Sivapithecus* and *Pongo. Nature* **348:**237–239.
Quade, J., Cerling, T. E., and Bowman, J. R. 1989. Development of Asian monsoon revealed by marked ecological shifts during the latest Miocene in northern Pakistan. *Nature* **342:**163–166.
Quade, J., Cerling, T. E., Andrews, P., and Alpagut, B. 1994. Palaeodietary reconstruction of

Miocene fauna from Paşalar, Turkey, using stable carbon and oxygen isotopes of fossil tooth enamel. *J. Hum. Evol.* **28:**377–384.
Raza, S. M., Barry, J. C., Pilbeam, D. R., Rose, M. D., Shah, S. M. I., and Ward, S. 1983. New hominoid primates from the middle Miocene Chinji Formation, Potwar Plateau, Pakistan. *Nature* **306:**52–54.
Retallack, G. 1991. *Miocene Paleosols and Ape Habitats of Pakistan and Kenya.* Oxford University Press, London.
Retallack, G. 1992. Comment on the paleoenvironment of *Kenyapithecus* at Fort Ternan. *J. Hum. Evol.* **23:**365–371.
Rose, M. D. 1983. Miocene hominoid postcranial morphology: Monkey-like, ape-like, neither, or both? In: R. L. Ciochon and R. S. Corruccini (eds.), *New Interpretations of Ape and Human Ancestry,* pp. 405–417. Plenum Press, New York.
Rose, M. D. 1984. Hominoid postcranial specimens from the Middle Miocene Chinji Formation, Pakistan. *J. Hum. Evol.* **13:**503–516.
Rose, M. D. 1986. Further hominoid postcranial specimens from the late Miocene Nagri Formation of Pakistan. *J. Hum. Evol.* **15:**333–367.
Rose, M. D. 1988. Another look at the anthropoid elbow. *J. Hum. Evol.* **17:**193–224.
Rose, M. D. 1989. New postcranial specimens of catarrhines from the Middle Miocene Chinji Formation, Pakistan: Descriptions and a discussion of proximal humeral functional morphology in anthropoids. *J. Hum. Evol.* **18:**131–162.
Rose, M. D. 1992. Kinematics of the tapezium–1st metacarpal joint in extant anthropoids and Miocene hominoids. *J. Hum. Evol.* **22:**255–266.
Ruff, C. B., Walker, A., and Teaford, M. F. 1989. Body mass, sexual dimorphism and femoral proportions of *Proconsul* from Rusinga and Mfangano Islands, Kenya. *J. Hum. Evol.* **18:**515–536.
Sanders, W. J., and Bodenbender, B. E. 1994. Morphometric analysis of lumbar vertebra UMP 67-28: Implications for spinal function and phylogeny of the Miocene Moroto hominoids. *J. Hum. Evol.* **26:**203–237.
Sarmiento, E. E. 1987. The phylogenetic position of *Oreopithecus* and its significance in the origin of the Hominoidea. *Am. Mus. Novit.* **2881:**1–44.
Schön, M. A., and Ziemer, L. K. 1973. Wrist mechanism and locomotor behavior of *Dryopithecus (Proconsul) africanus. Folia Primatol.* **20:**1–11.
Schwartz, J. 1990. *Lufengpithecus* and its potential relationship to an orang-utan clade. *J. Hum. Evol.* **19:**591–605.
Shipman, P. 1982. Reconstructing the paleoecology and taphonomic history of *Ramapithecus wickeri* at Fort Ternan, Kenya. *Museum Briefs* 26, Museum of Anthropology, Columbia.
Shipman, P. 1986. Paleoecology of Fort Ternan reconsidered. *J. Hum. Evol.* **15:**193–204.
Shipman, P., Walker, A., Van Couvering, J. A., Hooker, P. J., and Miller, J. A. 1981. The fort Ternan hominoid site, Kenya: Geology, age, taphonomy and paleoecology. *J. Hum. Evol.* **10:**49–72.
Spoor, C. F., Sondaar, P. Y., and Hussain, S. T. 1991. A hominoid hamate and first metacarpal from the later Miocene Nagri Formation of Pakistan. *J. Hum. Evol.* **21:**413–424.
Viranta, S., and Andrews, P. 1994. Carnivore guild structure in the Paşalar Miocene fauna. *J. Hum. Evol.* **28:**359–372.
von Koenigswald, G. H. R. 1972. Ein Unterkiefer eines fossilen Hominoiden aus dem Unterpliozän Griechenlands. *K. Ned. Akad. Wet.* **75:**385–394.
Walker, A., and Pickford, M. 1983. New postcranial fossils of *Proconsul africanus* and *Proconsul nyanzae.* In R. L. Ciochon and R. S. Corruccini (eds.), *New Interpretations of Ape and Human Ancestry,* pp. 325–351. Plenum Press, New York.
Walker, A., and Teaford, M. 1989. The hunt for *Proconsul. Sci. Am.* **260:**76–82.
Walker. A., Teaford, M., Martin, L., and Andrews, P. 1993. A new species of *Proconsul* from the early Miocene of Rusinga/Mfangano Islands, Kenya. *J. Hum. Evol.* **25:**43–56.
Ward, C. V. 1993. Torso morphology and locomotion in *Proconsul nyanzae. Am. J. Phys. Anthropol.* **92:**291–328.

Ward, C. V., Walker, A., Teaford, M. F., and Odhiambo, I. 1993. Partial skeleton of *Proconsul nyanzae* from Mfangano Island, Kenya. *Am. J. Phys. Anthropol.* **90**:77–111.

Wu, R., and Xu, Q. 1986. Relationship between Lufeng *Sivapithecus* and *Ramapithecus* and their phylogenetic position. *Acta Anthropol.* **5**:26–30.

Xiao, M. 1981. Discovery of a fossil hominoid scapula at Lufeng. In: *Collected Papers of the 30th Anniversary of the Yunnan Provincial Museum,* pp. 41–44. Yunnan Provincial Museum.

The Early Evolution of the Hominoid Face

4

TODD C. RAE

Introduction

In phylogenetic analyses of hominoid primates, the face as an anatomical unit has often been considered to be of secondary importance in comparison with dental and postcranial evidence (e.g., Andrews and Martin, 1987), related both to the abundance of teeth in the fossil record and to the predominance of postcranial characteristics in diagnoses of the superfamily (e.g., Harrison, 1987). Recent fossil discoveries and the adoption of more rigorous phylogenetic methods, however, provide an opportunity to document more fully the evolution of the hominoid facial skeleton and to decipher the phylogenetic signal contained within this important anatomical region.*

Many well-preserved fossil faces of Miocene primates are known from throughout the Old World (MacInnes, 1943; Clark and Leakey, 1951; Fleagle, 1975; Andrews and Tekkaya, 1980; Pilbeam, 1982; Wu and Pan, 1985; Leakey *et al.*, 1988a,b; Teaford *et al.*, 1988; Bonis *et al.*, 1990; Moyà-Solà and Köhler, 1993). Details of the facial skeleton have proved significant in studies of some of the middle/late Miocene hominoids (Preuss, 1982; Ward and Pil-

*Rak (1983) has noted that the face as an anatomical region is in fact poorly defined. Here, Rak's informal designation is implied by the use of the word *face*.

TODD C. RAE • Department of Mammalogy, American Museum of Natural History, New York, New York 10024-5192. *Present address:* Department of Anthropology, University of Durham, Durham DH1 3HN, England.

Function, Phylogeny, and Fossils: Miocene Hominoid Evolution and Adaptations, edited by Begun *et al.* Plenum Press, New York, 1997.

beam, 1983; Begun, 1992). These taxa, however, are generally considered to be great apes and, as such, do not help to illuminate the emergence and early evolution of hominoids.

Anthropoids from the early Miocene (23–17 Ma), on the other hand, may be critical for understanding the emergence of hominoid primates, as they have been positioned by many workers at or near the base of the hominoid radiation (Andrews, 1985; Fleagle, 1986, 1988). The craniodental and postcranial characters, however, have been cited to support widely divergent phyletic placements of the early Miocene taxa (cf. Szalay and Delson, 1979; Harrison, 1987). Although some of fossil faces from the early Miocene have been analyzed phenetically (Bilsborough, 1971; Corruccini and Henderson, 1978), this material had never been examined in a phylogenetic context. The present chapter is a discussion of a series of cladistic analyses performed on facial data from extant and extinct catarrhine taxa to test between the alternative phylogenetic hypotheses proposed for the early Miocene forms.

As the main focus of the study was to test the phylogenetic position of the early Miocene catarrhines, the emphasis here is on the earlier evolution of the hominoid face. The results, however, suggest some interesting patterns in catarrhine facial evolution and are examined here in the context of hominoid evolution throughout the Miocene; hominoids from the middle/late Miocene and some possible functional correlates of the pattern of change discerned for some characters are discussed separately.

For the sake of simplicity, the non-bilophodont catarrhine primates from the early Miocene of east Africa and Saudi Arabia will be referred to throughout this chapter as "proconsulids," a purely informal term that is *not* meant to imply the existence of the family Proconsulidae as defined by Leakey (1963).

The fossil record of the early Miocene in Afro-Arabia documents an explosive radiation of catarrhine diversity. As many as 11 genera and 16 species of non-bilophodont catarrhines are known from this time period (see Table I). The "proconsulids" represent a diversity of adaptations, ranging from the *Cebus*-sized *Micropithecus clarki* to *Proconsul major*, which had an estimated body weight about that of a male chimpanzee (Fleagle, 1988). These fossil taxa also demonstrate a diversity of locomotor and positional behaviors. Although most "proconsulid" species are interpreted as arboreal quadrupeds (Rose, 1983), some species possess characteristics indicating more suspensory (*Dendropithecus macinnesi*) or possibly terrestrial (*P. nyanzae*) adaptations (Fleagle, 1988). The abundance and diversity of "proconsulids" and the rarity of cercopithecoids during this time stand in striking contrast to the paucity of hominoid genera and profusion of Old World monkeys in the Holocene (Andrews, 1986).

Four distinct groups of hypotheses of possible phylogenetic placement of the "proconsulids" have been proposed previously. Until relatively recently, workers postulated that "proconsulid" taxa were direct ancestors of living hominoid genera (Fig. 1a; Hopwood, 1933; Clark and Leakey, 1951; Simons and Pilbeam, 1965). With more fossil material, particularly postcranial re-

Table I. Anthropoid Taxa Examined for the Study[a]

Extant taxa	"Proconsulids"
Cebus capucinus	*Proconsul africanus* (includes *P. heseloni*) (Walker *et al.*, 1993)
Samiri sciureus	*Proconsul nyanzae*
Lagothrix lagotricha	*Proconsul major*
Callicebus moloch	*Rangwapithecus gordoni*
Colobus polykomos	*Nyanzapithecus vancouveringorum*
Presbytis obscura	*Limnopithecus legetet*
Nasalis larvatus	*Limnopithecus evansi*
Cercopithecus neglectus	*Kalepithecus songhorensis*
Macaca mulatta	*Dendropithecus macinnesi*
Miopithecus talapoin	*Micropithecus clarki*
Hylobates hoolock	*Afropithecus turkanensis*
Pongo pygmaeus	*Turkanapithecus kalakolensis*
Pan troglodytes	*Simiolus enjiessi*
	Stem catarrhine species
	Pliopithecus vindobonensis[b]
	Aegyptopithecus zeuxis

[a]For the living species, ten males and ten females were examined. For the extinct taxa, all published specimens were examined, except where indicated. Individual fossil specimens are listed in Rae (1993).
[b]The most complete cranial specimen of *P. vindobonensis*, Individual II, was unavailable for the present study.

mains, it became increasingly obvious that taxa such as *Proconsul* and *Dendropithecus* were unlike modern hominoids (and more primitive) in many aspects of their anatomy (Clark and Thomas, 1951; Napier and Davis, 1959). As a result, many workers (Andrews, 1978; Szalay and Delson, 1979; Walker and Teaford, 1989) considered the "proconsulids" (although often exclusive of the smaller forms) to be hominoids that diverged from the lineage leading to modern great apes sometime after the hylobatid/great ape split (Fig. 1b). Andrews later placed the proconsulid group below the great ape/gibbon divergence, considering the early Miocene forms to be stem hominoids (for a discussion of the stem lineage concept, see Ax, 1985) that share some but not all of the derived features of living apes (Andrews, 1985; Fig. 1c), a position supported by Fleagle (1986). A fourth alternative, with "proconsulids" removed even further from a close relationship with living hominoids, was proposed by Harrison (1982, 1987, 1988). He concluded that none of the "proconsulids" share the diagnostic derived postcranial characters of the living Hominoidea and that these taxa are best considered stem catarrhines, having shared a common ancestor with living catarrhines before the divergence of cercopithecoids and hominoids (Fig. 1d).

The picture that has emerged is that "proconsulids" resemble hominoids primarily in ancestral characteristics and that they lack some, if not all, of the diagnostic postcranial features of living hominoids. Recently, however, some of these same postcranial synapomorphies of living apes also have been found

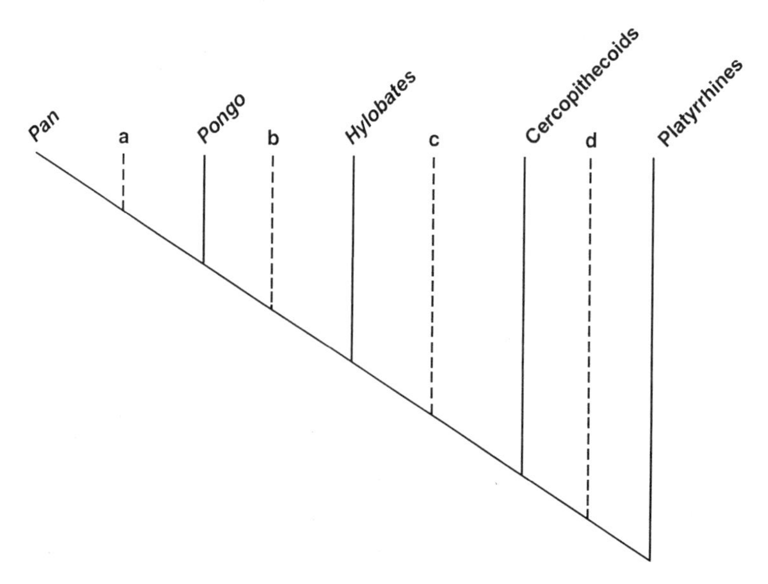

Fig. 1. Proposed phylogenetic positions of "proconsulids": (a) related exclusively to living genera, (b) stem great apes, (c) stem hominoids, and (d) stem catarrhines.

to be lacking in undoubted fossil hominoids such as *Sivapithecus* (Pilbeam *et al.*, 1990) and *Kenyapithecus* (Benefit and McCrossin, personal communication). These revelations have led some workers to doubt the homology of the forelimb morphology of living apes, despite the functional similarity (Larson, 1992). This, in turn, underscores the importance of examining other anatomical systems of "proconsulids" for possible synapomorphic resemblances with extant catarrhines.

To address this controversy, a series of facial characteristics was examined in "proconsulids," a sample of extant catarrhine taxa, and an outgroup consisting of extant platyrrhines and fossil stem catarrhine species. These characters were subjected to parsimony analysis and the results compared with published hypotheses of proconsulid phylogenetic position. In this way, the phylogenetic pattern indicated by an independent character set from a different anatomical system was used as a critical test of the alternative hypotheses outlined above (Fig. 1). The resulting pattern of hominoid facial evolution also presents some interesting implications for the functional interpretation of evolving character states and the analysis of "functional complexes."

Materials

The "proconsulid" taxa analyzed (see Table I) are from sites in east Africa. These sites are primarily associated with extinct volcanic centers (Bishop, 1963; Pickford, 1986) and have been dated to between 20 and 17 Ma (Bishop *et al.*, 1969; McDougall and Watkins, 1985; Drake *et al.*, 1988; Boschetto *et al.*, 1992). Unless otherwise indicated, character states were determined by examination of original specimens. Miocene taxa from the Middle East and Asia that have been considered closely related to the "proconsulids," including *Afropithecus* ("*Heliopithecus*") (Andrews and Martin, 1987), *Dionysopithecus* (Li, 1978), and *Platydontopithecus* (Gu and Lin, 1983), are too incomplete for the present analysis, as is the possibly distinct African genus "*Xenopithecus*" (Hopwood, 1933; Madden, 1980).

In addition to "proconsulids," two other fossil catarrhine species were examined: *Pliopithecus vindobonensis* from the middle Miocene of Europe and *Aegyptopithecus zeuxis* from the Oligocene of the Fayum, Egypt. These taxa were used as successive outgroups in the analysis of the fossil material (see Methods below). Several extant anthropoid taxa were also analyzed (Table I) to provide an evolutionary framework within which to place the "proconsulids," as an understanding of the pattern of character evolution in the facial skeleton among extant catarrhines is essential for addressing phylogenetic and functional hypotheses in the evolution of the "proconsulids."

Characters

Twelve characters of the facial skeleton were chosen for the analysis. Previous analyses (Rae and Simons, 1992; Rae, 1993) indicate that these characters (listed below) diagnose most extant catarrhine clades with a high degree of confidence. Measurements for metric traits were taken to the nearest 0.1 mm with Mitutoyo Digimatic digital calipers. Contours (medial alveolus, nasal projection, zygomatic orientation) were recorded as tracings from an industrial contour guide (Schmitt, 1991) and were coded meristically.

All of the raw measurements for the metric characteristics were converted into ratios to focus comparisons on the shape of the areas in question, rather than on their absolute size (see below for character-by-character discussion). The denominator for each ratio was taken from a topographically related structure, rather than using a generic body-size surrogate (e.g., cheek tooth area, M_1 length). For the extant taxa, none of the ratios are significantly correlated ($p < 0.05$) with mean species (mixed sex) body weights taken from Fleagle (1988) using Spearman's rank order correlation method (Sokal and Rohlf, 1981). Many of the metric samples did not meet the assumptions of analysis of variance (Sokal and Rohlf, 1981). Following Conover and Iman (1981), the data were converted to ranks and compared with the multiple comparisons test (MCPAIR) contained in the computer package BIOM (Rohlf, 1982). The GT2 test was used to determine significance, due to the presence of heterogeneous variances in the samples (Sokal and Rohlf, 1981).

Homogeneous subset coding (Simon, 1983) was employed to convert the metric (continuous) data into discrete codes for phylogenetic analysis. Coding proceeds by comparing all taxon means to one another; those that do not differ significantly are grouped in a homogeneous subset. Those taxa that belong to the same subsets are coded as identical. Where overlap occurs, such as when a taxon belongs to two adjacent homogeneous subsets, the taxon is coded as intermediate. Fossil taxa, for which sample sizes are small, were assigned the same code as the extant taxon whose mean the fossils most closely approximate.

1. *Nasoalveolar height*—Measured from prosthion to the anterior nasal spine. Maximum width of the premaxilla is used as the denominator here to evaluate nasoalveolar height relative to a topographically related measurement of the width of the face. Outgroup comparisons indicate that the primitive eucatarrhine condition is intermediate, coded as 2, with extremes found in some Old World monkeys (0) and in great apes (4).
2. *Piriform aperture width*—The maximum width of the nasal aperture, divided by the maximum width of the premaxilla (again used as a measure of facial width). The outgroup condition is 4 (or wide), most taxa are intermediate (2), and some cercopithecines are narrow (0).
3. *Interorbital width*—Measured as the bidacryonic width between the orbits, and again scaled relative to the maximum width of the premaxilla; other facial width measurements (e.g., bizygomatic width) were unavailable for nearly all of the fossil taxa. Outgroup analysis indicates that the primitive condition is relatively wide (5), with *Colobus* possessing a wider (6) interorbital region, and most other taxa are narrower (1–2) to much narrower (*Cercopithecus* = 0).
4. *Anterior palate width*—The width of the palate at the canine divided by the width at M^2. M^2 width is used as the denominator for this character to distinguish between more V-shaped palates (narrow at C, wide at M^2) from U-shaped dental arcades. The outgroup node is coded 4 (intermediate), while hominoids are generally wider (5–6) in this region, and cercopithecoids narrower (3–0).
5. *Incisive foramen/canal*—Either the premaxilla ends before the maxilla and thus produces an incisive foramen (code = 0), or the posterior margin of the premaxilla overlaps the anterior aspect of the maxilla, forming an incisive canal (1). The incisive foramen is the condition at the outgroup node.
6. *Premaxillary superior extension*—The suture between the premaxilla and the maxilla meets the nasal bone (a) high on the nasal bone, sometimes reaching superiorly to the frontal (code = 3), (b) near the midpoint of nasal length (2), (c) low on the nasal, near the piriform aperture (1), or (d) the suture is fused in the adult (0). The code 2 corresponds to the primitive condition.
7. *Medial alveolar orientation*—The palate shows a medial alveolar margin that is primitively parabolic, or sloping, in sagittal section (0), or it is steep-sided (1).

8. *Maxillary sinus size*—The paranasal spaces can be laterally extensive (0), as in the outgroup and in living hominoids, or are laterally restricted (1), a derived condition seen in Old World monkeys.
9. *Zygomatic root position*—The position of the root of the zygomatic was noted, relative to tooth position, as an expression of facial prognathism. Outgroup analysis suggests that the original condition in catarrhines is to have the zygomatic arch arise more posteriorly (at M^2 or beyond; code = 1), while more derived taxa have the arch placed more anteriorly (0).
10. *Piriform aperture shape*—This character concerns the vertical position of the widest part of the piriform aperture. If the widest point is near the middle of the aperture, the aperture is termed oval (0); this is the outgroup condition. When the widest point is low, the aperture is triangular (1).
11. *Nasal projection*—The nasal bones either project anteriorly at the midline, forming a smooth bridge of the nose that is parabolic in transverse section (0), or the interorbital region lacks such a parabolic projection and lies flat across the bridge of the nose (1). Projecting nasals are present in the outgroup.
12. *Zygomatic orientation*—The body of the zygomatic is either set perpendicular to the occlusal plane, such that it rises vertically from the alveolar margin (0), or the zygomatic is oblique to the occlusal plane (1) and the orbital margin is anterior of the root of the zygomatic arch. These morphologies are referred to as "vertical" and "sloping" throughout. This character is ambiguous at the outgroup mode, using the Maddison *et al.* (1984) method of polarity determination.

Methods

All multistate characters are treated as ordered (Pogue and Mickevich, 1990; Wilkinson, 1992; Slowinski, 1993). Ordering character states is an implicit weighting function; parsimony algorithms treat ordered (or additive) characters such that a change between adjacent character states is considered more likely than changes between the extremes of the range. For metric characters, where there is a demonstrable order to states, and for morphological characteristics with two extremes and obvious intermediates, ordering is highly recommended (Slowinski, 1993).

Multiple outgroups were used because of the wide diversity of platyrrhine facial morphology and because the use of multiple outgroups may also increase the probability that the resulting cladograms are globally parsimonious (Maddison *et al.*, 1984; although see Nixon and Carpenter, 1993). The three successive outgroups for this analysis were *Pliopithecus, Aegypto-*

pithecus, and the platyrrhines. The relationships of these taxa to the ingroup are well established (Andrews, 1985; Harrison, 1987).

The outgroup node was reconstructed using the methods outlined in Maddison *et al.* (1984) and the platyrrhine phylogeny of Ford (1986). Only one character (zygomatic orientation) was ambiguous by the multiple outgroup method and was coded as unknown for the reconstructed outgroup.

The data matrix (Table II) was analyzed using the maximum parsimony algorithm contained in the computer program Hennig86 (Farris, 1988). This algorithm is a fast and effective means for identifying minimum length topologies (Platnick, 1989). Resultant most parsimonious cladograms were evaluated with reference to the retention index (RI) (Farris, 1989a). The traditionally reported consistency index (CI) of Kluge and Farris (1969) is sensitive to numbers of taxa and characters in an analysis (Archie, 1989) and is subsequently difficult to compare across different studies. The retention index, on the other hand, is a measure of the amount of homoplasy contained within a particular topology that is independent of tree length (Farris, 1989b) and therefore RI can be used as a more accurate relative measure of the amount of homoplasy present.

The fossil taxa were not simply interpolated into a topology based entirely on extant taxa, as it has been shown that fossil taxa can have a profound effect on the topology of "extant only" trees (Gauthier *et al.*, 1988). The effect

Table II. Data Matrix for Extant/Extinct Analysis

	Character[a]											
Taxon	1	2	3	4	5	6	7	8	9	10	11	12
Outgroup	2	4	5	4	0	2	0	0	1	0	0	?
Cercopithecus	2	0	0	4	0	3	1	1	1	0	0	0
Colobus	3	0	6	2	0	3	0	1	0	0	0	1
Hylobates	1	4	5	5	0	0	1	0	1	0	1	1
Macaca	1	3	0	1	0	1	1	1	1	0	0	0
Miopithecus	0	0	0	4	0	2	1	1	0	0	0	1
Nasalis	0	1	3	2	0	2	1	1	1	0	0	0
Pan	4	3	4	6	1	0	1	0	1	1	1	0
Pongo	4	2	0	6	1	0	1	0	1	1	1	0
Presbytis	0	2	3	2	0	3	0	1	0	0	0	1
Afropithecus	3	2	5	6	0	1	1	0	?	0	1	0
Proconsul nyanzae	1	2	?	6	0	1	1	0	1	0	1	0
Turkanapithecus	1	?	5	6	?	1	1	?	0	0	1	0
Dendropithecus	1	?	?	6	0	1	1	0	1	0	?	?
Micropithecus	?	?	?	?	0	1	0	0	1	?	?	1
Rangwapithecus	2	2	?	6	0	1	1	0	?	?	?	?
Kalepithecus	3	?	?	?	?	?	?	0	?	0	?	?
Simiolus	3	?	?	?	0	1	?	?	?	?	?	?

[a]Characters as numbered in text.

is related primarily to the fact that fossil taxa display unique combinations of character states that are not duplicated in living organisms. These unique character combinations, in addition to potentially altering the pattern of cladogenesis determined from only living taxa, can also affect the distribution of character change on topologies that remain the same after the addition of fossil taxa. The addition of fossils to an "extant only" tree is useful, therefore, as a test of the presumed homologies expressed by the codes assigned to taxa. Characters that had been treated as synapomorphies based on analyses of only living taxa may be revealed to have been derived independently.

Several taxa were excluded from the complete analysis. After comparison with the outgroup node, *Limnopithecus legetet, L. evansi, Proconsul africanus* (including *P. heseloni*), *P. major,* and *Nyanzapithecus vancouveringorum* were excluded because none of these taxa possess derived features of the face that would unequivocally place them with any of the catarrhine clades without severely diminishing the resolution of the consensus trees.

Results

The maximum parsimony analysis resulted in 180 equally parsimonious minimum length trees. Each has a length of 59 steps and an RI of 67. A strict consensus cladogram for this analysis is shown in Fig. 2. The large number of equally parsimonious solutions is related almost entirely to the preponderance of missing data for the fossils that form the two major polytomies.

One major recognized catarrhine group, the Colobinae, was not found to be monophyletic for the facial characters analyzed. There is a basal cercopithecoid polytomy, with *Colobus, Presbytis,* and a (*Nasalis* + cercopithecines) group equidistant from one another. This arrangement reflects the distribu-

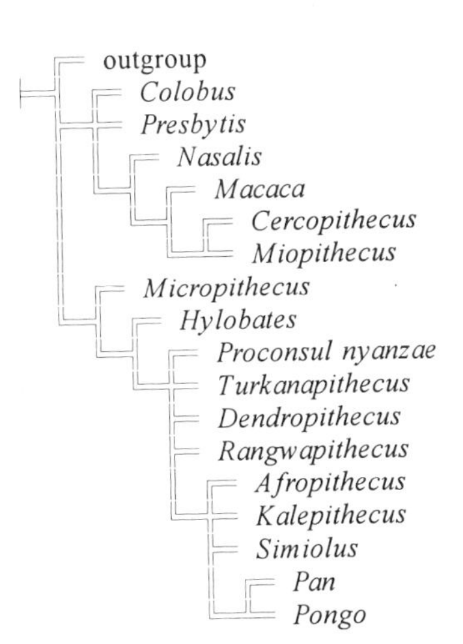

Fig. 2. Most parsimonious consensus cladogram of the living and extinct taxa. The cladogram shown is the strict consensus of 180 equally parsimonious solutions. Length = 59 steps, RI = 67.

tion of medial alveolus orientation (Character 7) and zygomatic orientation (Character 12). *Colobus* and *Presbytis* retain the inferred primitive catarrhine condition for each of these characters, while *Nasalis* shares the derived conditions of a vertical zygoma and a steep-sided palate with the cercopithecines and, convergently, with the great apes.

Although the main focus of the study was the pattern of cladogenesis in hominoid primates, that the characters examined here do not support a monophyletic Colobinae deserves further study. There are sufficient synapomorphies of the bony skeleton, teeth, and soft tissues of colobines (for a review see Strasser and Delson, 1987) to suggest that the nonmonophyletic Colobinae result seen here may be anomalous. Interesting in this regard is the fact that this portion of the analysis (i.e., Cercopithecoidea) consists entirely of extant taxa. The addition of fossil Old World monkeys may well serve to eliminate the inconsistencies between the facial data and those of the rest of the body. All other extant catarrhine clades (including Cercopithecinae) were found to be monophyletic and diagnosed by several derived character states.

The distribution of derived character states for the hominoids is shown in Fig. 3. Several characters are uninformative phylogenetically, in terms of larger groups of catarrhines. Interorbital width, for example, is ambiguous at the base of the Hominoidea and Cercopithecoidea. Thus, the width of the

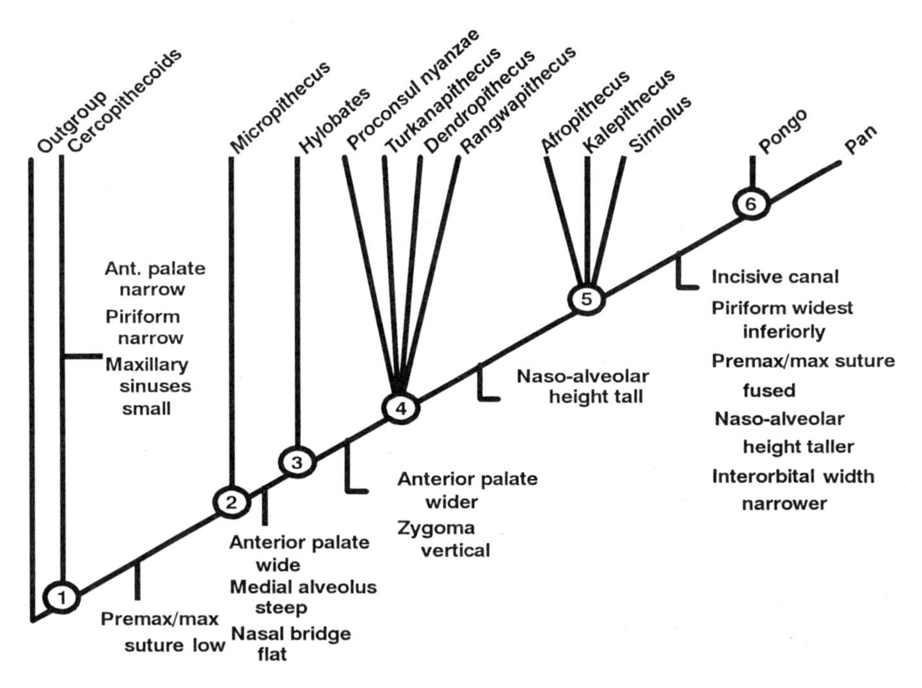

Fig. 3. Most parsimonious consensus cladogram with the unambiguous synapomorphies shown at the internodes at which they evolved. Details of the cercopithecoid clade are not shown.

interorbital septum cannot be interpreted as having a synapomorphous condition anywhere within Catarrhini, except at the extant great ape node, where a synapomorphic reduction is inferred.

Hominoids as a whole are characterized by wider anterior palates, while Old World monkeys share the alternate derived condition of narrower anterior palates. This conclusion emphasizes the importance of outgroup analysis; without comparison with outgroup taxa, the ancestral condition could have been interpreted as either wide or narrow. The more probable explanation is that both wide and narrow anterior palates are derived, albeit in opposite directions, from a primitively intermediate condition. Previous analyses (e.g., Harrison, 1987) had listed a wide anterior palate as a synapomorphy of apes. The analysis reported here suggests that a moderately wide anterior palate evolved at the basal hominoid node and that this condition increased again at the base of the great ape clade. This arrangement of nested synapomorphies allows better differentiation of hominoid groups.

Another result seen clearly in the consensus tree is that the "proconsulids" do not form a homogeneous group. The characters analyzed position *Micropithecus* as a stem hominoid, while the remaining taxa are located after the hylobatid/great ape divergence. The remaining "proconsulids" are also paraphyletic, with three taxa (*A. turkanensis, K. songhorensis,* and *S. enjiessi*) more closely related to the living great apes than the other four.

The first polytomy is found at the base of the great ape crown group (Fig. 3, node 4). Four fossil taxa (*P. nyanzae, T. kalakolensis, R. gordoni,* and *D. macinnesi*) are clustered there, all sharing the great ape synapomorphies of a wider anterior palate and vertical zygomatic, but lacking the tall nasoalveolar clivus that characterizes the taxa at the next node. This next node (Fig. 3, node 5) is a polytomy as well, with *A. turkanensis, K. songhorensis, S. enjiessi,* and the living great ape clade (*Pan* and *Pongo*) all equidistant from one another, and all sharing an increase in the height of the nasoalveolar clivus. The *Pan*/*Pongo* clade (Fig. 3, node 6) is diagnosed by the derived conditions of an incisive canal, a triangular piriform aperture, and a fused maxillopremaxillary suture that is shared convergently with living hylobatids. Living great apes, in this analysis, are also characterized by a narrower interorbital region, relative to the preceding node.

The distribution of several of the characters on this topology is different from that seen in analyses of only extant taxa (see Rae, 1993). Fusion of the maxillopremaxillary suture, for example, is interpreted as a synapomorphy of living Hominoidea when only living taxa are analyzed (Rae, 1993), but is diagnosed here as convergent between the *Pan*/*Pongo* clade and gibbons when fossil taxa are considered.

Similarly, four characters of the anterior palate are entirely congruent in extant great apes: tall nasoalveolar clivus, incisive canal, inferiorly wide piriform aperture, and a very wide anterior palate (Rae, 1993). These characters were treated as independent, even though it is conceivable that a single change in a hypothetical "functional complex" of the anterior region of the face might have been responsible. The combined fossil/extant analysis reported here demonstrates that these characters did not evolve simultaneously. The anterior palate widened first, followed by an evolutionary splitting event, then nasoalveolar height increased before another speciation, after which the incisive foramen became a canal, the maximum piriform width migrated inferiorly, and nasoalveolar height increased again. The nonsynchronous, mosaic evolution of the face in great apes outlined here highlights the fundamental importance of including fossils in phylogenetic analyses. Not only do the unique character combinations found in extinct taxa force us to reanalyze the polarity of characters, but the pattern and order in which derived states appear can also lead to better hypotheses of adaptation, as large sets of topologically congruent characters in extant clades can be seen as a series of steplike transitions with the inclusion of fossils. The explicit mapping of character change across topologies can only lead to better and more precise explanations of evolutionary change in "functional complexes."

Other Taxa

Evolution of the hominoid face did not stop at the end of the early Miocene. In fact, the results suggest that several features did not appear until the last common ancestor of living great apes. For the specific predictions of character state change in the phylogenetic arrangement presented above to be corroborated, fossil hominoids more closely related to extant great apes should show the development of some or all of the derived characters diagnosed as synapomorphies for the living great ape (*Pan*/*Pongo*) node. On the basis of published data, it can be argued that some taxa from the middle/late Miocene do share some of the synapomorphies of the living great apes suggested by this analysis.

An abundance of new facial material of the middle/late Miocene hominoid *Dryopithecus* from Europe (Begun, 1987) has been uncovered recently (Begun, 1992; Moyà-Solà and Köhler, 1993). Of the characters that are interpreted as synapomorphies of the *Pan*/*Pongo* group (node 6), *Dryopithecus* shares the presence of an incisive canal, a taller nasoalveolar clivus, and a piriform aperture that is wide inferiorly (Begun, 1992). In addition, *Dryopithecus* may share the condition of an ethmoidal frontal sinus (Begun, 1992) with *Turkanapithecus* (Leakey *et al.*, 1988a), *Afropithecus* (Pilbeam, 1969; Leakey *et al.*, 1988b), *Proconsul africanus* (including *P. heseloni*) (Walker and Teaford, 1989), and Recent African apes and humans. Unfortunately, the homology of the frontal sinus (i.e., its derivation from the ethmoidal sinus) is difficult to demonstrate in fossil taxa. The same characters can be cited to support the great ape status of *Ouranopithecus*, based on the facial evidence published to date (Bonis *et al.*, 1990). The characters analyzed here, however, cannot distinguish between alternative arrangements of these fossil taxa relative to the living apes (Andrews and Martin, 1987; Bonis *et al.*, 1990; Begun, 1992; Moyà-Solà and Köhler, 1993).

Many of the same derived characters shared by *Dryopithecus* and the *Pan/Pongo* group are also seen in *Sivapithecus* (Ward and Kimbel, 1983; Ward and Pilbeam, 1983) from the late Miocene of Asia. In particular, the interorbital area of the specimen GSP 15000 (seen clearly in casts and photos) is nonprojecting, a synapomorphy of hominoids, and the presence of a triangular piriform aperture, an incisive canal, and the tall nasoalveolar clivus argue for the inclusion of *Sivapithecus* in the clade with extant great apes. The "Asian" pattern of the subnasal region in this taxon (Ward and Kimbel, 1983; Ward and Pilbeam, 1983) might then be interpreted, in this topology, as a derived feature shared with the living orangutan *Pongo,* although the polarity of the "Asian" and "African" patterns is difficult to determine. Other Asian fossil faces, such as *Lufengpithecus,* are damaged severely and the figures and descriptions cannot satisfactorily resolve questions concerning the character states present. Casts of *Laccopithecus robustus,* however, show that this taxon possesses nasal bones that are nonprojecting, as in living hominoids, "proconsulids," and other fossil apes.

Functional Implications

In addition to contributing to our knowledge of the evolutionary relationships of the "proconsulids," the pattern of evolution reported here suggests hypotheses on functional aspects of hominoid facial evolution as well. Perhaps even more importantly, character-by-character analysis of change in an anatomical region across a cladogram, especially when fossils are included in the data matrix, allows a more detailed explanation of adaptation to emerge, one that might be overlooked in discussions of "functional complexes" examined in living taxa only.

As discussed above, the widening of the anterior palate seen in hominoids corresponds exactly with other changes in the front of the face when only living apes are examined (Rae, 1993). The more detailed analysis discussed in this chapter, including data from fossil taxa, demonstrates that the anterior palate changed independently of other facial characteristics and in a mosaic fashion during the evolution of hominoids. This result suggests that the characters of the anterior portion of the face did not change simultaneously, as one might expect given an interpretation of the anterior face as a "functional complex."

From the analysis of only living taxa it also might be tempting to view the wide anterior palate as an epiphenomenon of the increased size of the anterior dentition seen in Recent hominoids (e.g., Harrison, 1982), since an increase in the mesiodistal size of the incisors would necessitate a larger premaxilla to hold them. The inclusion of the fossil taxa demonstrates, however, that the anterior palate became wider *before* the characteristic hypertrophied incisors of living apes appeared. In other words, the individual characters of the "functional complex" of the anterior face did not evolve simultaneously. This mosaic pattern of change must be taken into account in discussions of the functional aspects of facial evolution in catarrhines.

The process of mastication is the most obvious and perhaps most important functional influence on facial anatomy. Many aspects of the viscerocranium are undoubtedly linked to the function of chewing, although attempts to summarize the inherent complexity of the facial masticatory apparatus into a comprehensible system amenable to biomechanical analysis have met with varied success (for a review see Rak, 1983). Many of the characters discussed here, however, have little if any direct functional relationship with the masticatory mechanism. For example, the interorbital region may contribute to the buttressing of the face during mastication, but bone strain studies (Hylander and Johnson, 1992) have shown that the forces distributed in the superior interorbital region are small when compared with the strains seen in the maxillae and zygoma. These results indicate a much-reduced stress-bearing role of the interorbital region during mastication.

The zygomatic, conversely, probably plays a major role in mastication, as one of the major jaw-closing muscles (the masseter) attaches to this bone. Indeed, the strains on the zygomatic are characteristically higher than those found in other regions of the face (Hylander and Johnson, 1992). Changes in the position of the zygomatic arch root (Character 9) and the orientation of the zygomatic bone (Character 12) might have a profound effect on the action of the masseter muscle and thus on mandibular movements and masticatory forces. These kinds of changes could seriously affect power stroke dynamics of the chewing cycle. The position of the zygomatic origin, however, is invariable or ambiguous at all of the catarrhine nodes; thus, no higher-level taxa are characterized by a change in the position of the zygomatic arch. Zygomatic orientation, on the other hand, shows a different pattern.

Two groups, the (cercopithecines + *Nasalis*) group and the (great apes + most "proconsulids") group, share the presence of a vertical zygoma convergently. Linking this character with specific behaviors or dietary patterns shared by the groups in question, even if only gross differences are considered, is difficult. While most of the taxa included in these groups are either known (or inferred, for extinct taxa) frugivores, there are significant exceptions (*Nasalis* and *Rangwapithecus* are folivorous), and the inferred ancestral condition (by outgroup analysis) would have been frugivory, as well. It is impossible to link zygomatic orientation unambiguously with gross dietary categories.

An alternative hypothesis is that the change to a vertical zygoma is a structural change effected to preserve biomechanical equivalence. The taxa positioned below the "vertical zygoma" nodes (*Colobus, Presbytis, Micropithecus,* and *Hylobates*) are relatively orthognathic and their zygomata slope inferoposteriorly from the coronal plane. If the length of the face increased in the (cercopithecines + *Nasalis*) group and the (great apes + most "proconsulids") group, the inferior border of the zygoma may have tracked the change in the jaws to maintain a similar average functional vector for the masseter. In this case, the change in shape may have been effected to preserve a preexisting topographical relationship between the orientation of the zygomatic (and, thus, the average vector of the masseter) and the mandible, preserving functional "equivalence" in taxa with different degrees of prognathism.

The evolutionary functional hypothesis outlined above is only possible given explicit knowledge about the order of appearance of derived character states. In this manner, a more detailed understanding of the process of adaptation and the interaction between function and phylogeny can be gleaned.

Conclusions

The results of the analysis reported here support the hypothesis that most "proconsulids" belong to the great ape clade (Andrews, 1978; Szalay and Delson, 1979; Walker and Teaford, 1989), corresponding to position (b) shown in Fig. 1. The "proconsulids" do not form a homogeneous group; *Micropithecus* is stem hominoid, the remaining "proconsulids" are a paraphyletic assemblage of stem great apes. These phylogenetic hypotheses for the early Miocene taxa are supported by several uniquely derived characteristics of the face and suggest that change in hominoid facial characteristics was asynchronous and mosaic in pattern. These results also imply that many of the postcranial similarities between living hylobatids and great apes may have arisen in parallel.

The observations and results described above allow a detailed reconstruction of the series of steps involved in the evolution of the viscerocranium in hominoids. The last common ancestor of living catarrhines possessed a face characterized by a medium-sized anterior palate, a medium-width and oval piriform aperture, a short nasoalveolar clivus, large maxillary sinuses, a patent maxillopremaxillary suture that contacts the nasals about halfway along their length, and projecting nasal bones. The subsequent change in the morphology of the face in early hominoids was characterized by at least five splitting events, corresponding to the nodes described in Fig. 3:

Node 2: low position of the maxillopremaxillary suture/nasal contact
Node 3: wide anterior palate, flattened nasals, and a steep medial alveolus
Node 4: wider anterior palate and vertical zygoma
Node 5: tall nasoalveolar clivus
Node 6: taller nasoalveolar clivus, incisive canal, triangular piriform aperture, and a fused maxillopremaxillary suture

Judging from the paleontological evidence, the first five splitting events in hominoid evolution occurred in the late Oligocene or early Miocene, that is, before about 20 Ma. The last splitting event noted here (node 6), and the emergence of great apes that closely resemble living forms in facial characteristics, probably occurred by the middle Miocene. None of the early Miocene forms, however, had yet achieved the diagnostic postcranial adaptations of the Recent great apes. These postcranial changes occurred later and probably in parallel with those in the hylobatid lineage. Some of the characters examined can be linked to the function of the face as a masticatory element, and the evolutionary changes seen in these features on the most parsimonious tree may be interpreted as preserving functional equivalence across different facial topologies. These results highlight the importance of facial morphology in the systematic study of catarrhine evolution and the importance of fossils for the explication of phylogenetic pattern.

ACKNOWLEDGMENTS

My sincere thanks go to Drs. D. Begun, C. Ward, and M. Rose for inviting me to present these findings at the "Function, Phylogeny, and Fossils" symposium. The finished product benefited greatly from discussions with them and other symposium participants.

My thanks are also extended to those individuals who allowed me access to collections of primates in their care and aided me in their study: G. Musser, W. Fuchs, I. Tattersall, and J. Brauer (A.M.N.H.); B. Patterson and W. Stanley (F.M.N.H.), R. Thorington and L. Gordon (N.M.N.H.); P. Andrews (N.H.M., London); B. Engesser and F. Wiedenmayer (N.H.M., Basel); M. Leakey and E. Mbutu (K.N.M., Nairobi); J. Sebedduka and J. Sikkintu (Uganda Museum, Kampala); and E. Simons and P. Chatrath (D.P.C.). I would also like to thank J. W. Wanjohi, of the Office of the President, Kenya, for permission to study the fossils in Kenya.

I am grateful for many illuminating discussions with M. E. Lewis, and Drs. L. B. Martin, W. Jungers, D. Krause, S. Ward, J. Farris, and especially J. Fleagle, although they are in no way responsible for the views expressed here. Dr. E. Delson and an anonymous reviewer provided many important comments and clarifications, for which I am grateful.

The research reported here was supported by grants from NSF (BNS 91119226 to T. C. R., BNS 9012154 to J. Fleagle), the LSB Leakey Foundation, and the Department of Anthropology, SUNY at Stony Brook.

References

Andrews, P. 1978. A revision of the Miocene Hominoidea of East Africa. *Bull. Br. Mus. Nat. Hist. Geol.* **30**:85–224.

Andrews, P. 1985. Family group systematics and evolution among catarrhine primates. In: E. Delson (ed.), *Ancestors: The Hard Evidence*, pp. 14–22. Liss, New York.

Andrews, P. 1986. Fossil evidence on human origins and dispersal. *Cold Spring Harbor Symp. Quant. Biol.* **51:**419–426.
Andrews, P., and Martin, L. 1987. The phyletic position of the Ad Dabtiyah hominoid. *Bull. Br. Mus. Nat. Hist. Geol.* **41:**383–393.
Andrews, P., and Tekkaya, I. 1980. A revision of the Turkish Miocene hominoid *Sivapithecus meteai. Paleontology* **23:**85–95.
Archie, J. 1989. Homoplasy excess ratios: New indices for measuring levels of homoplasy in phylogenetic systematics and a critique of the consistency index. *Syst. Zool.* **38:**253–269.
Ax, P. 1985. Stem species and the stem lineage concept. *Cladistics* **1:**279–287.
Begun, D. 1987. *A Review of the Genus Dryopithecus.* Ph.D. dissertation, University of Pennsylvania.
Begun, D. 1992. Miocene fossil hominids and the chimp–human clade. *Science* **257:**1929–1933.
Bilsborough, A. 1971. Evolutionary change in the hominoid maxilla. *Man* **6:**473–485.
Bishop, W. 1963. The later Tertiary and Pleistocene in Eastern Equatorial Africa. In: F. Howell and F. Bourliere (eds.), *African Ecology and Human Evolution,* pp. 246–275. Aldine, Chicago.
Bishop, W., Miller, J., and Fitch, F. 1969. New potassium-argon age determinations relevant to the Miocene fossil mammal sequence in east Africa. *Am. J. Sci.* **267:**669–699.
Bonis, L. de, Bouvrain, G., Geraads, D., and Koufos, G. 1990. New hominid skull material from the late Miocene of Macedonia in northern Greece. *Nature* **345:**712–714.
Boschetto, H., Brown, F., and McDougall, I. 1992. Stratigraphy of the Lothidok Range, northern Kenya, and K/Ar ages of its Miocene primates. *J. Hum. Evol.* **22:**47–71.
Clark, W. E. L., and Leakey, L. 1951. The Miocene Hominoidea of East Africa. *Br. Mus. Nat. Hist. Fossil Mamm. Afr.* **1:**1–117.
Clark, W. E. L., and Thomas, D. 1951. Associated jaws and limb bones of *Limnopithecus macinnesi. Br. Mus. Nat. Hist. Fossil Mamm. Afr.* **3:**1–27.
Conover, W. J., and Iman, R. L. 1981. Rank transformations as a bridge between parametric and nonparametric statistics. *Am. Stat.* **35:**124–129.
Corruccini, R., and Henderson, A. 1978. Palato-facial comparison of *Dryopithecus* (*Proconsul*) with extant catarrhines. *Primates* **19:**35–44.
Drake, R., Van Couvering, J., Pickford, M., Curtis, G., and Harris, J. 1988. New chronology for the Early Miocene mammalian faunas of Kisingiri, western Kenya. *J. Geol. Soc. London* **145:**479–491.
Farris, J. 1988. *Hennig86,* computer program and manual. Port Jefferson Station (distributed by author).
Farris, J. 1989a. The retention index and homoplasy excess. *Syst. Zool.* **38:**406–407.
Farris, J. 1989b. The retention index and the rescaled consistency index. *Cladistics* **5:**417–419.
Fleagle, J. 1975. A small gibbon-like hominoid from the Miocene of Uganda. *Folia Primatol.* **24:**1–15.
Fleagle, J. 1986. The fossil record of early catarrhine evolution. In: B. Wood, L. Martin, and P. Andrews (eds.), *Major Topics in Primate and Human Evolution,* pp. 130–149. Cambridge University Press, London.
Fleagle, J. 1988. *Primate Adaptation and Evolution.* Academic Press, New York.
Ford, S. 1986. Systematics of the New World monkeys. In: D. Swindler and J. Erwin (eds.), *Comparative Primate Biology, Vol. 1: Systematics, Evolution, and Anatomy,* pp. 73–135. Liss, New York.
Gauthier, J., Kluge, A., and Rowe, T. 1988. Amniote phylogeny and the importance of fossils. *Cladistics* **4:**105–209.
Gu, Y., and Lin, Y. 1983. First discovery of *Dryopithecus* in east China. *Acta Anthropol. Sin.* **2:**305–314.
Harrison, T. 1982. *Small-Bodied Apes from the Miocene of East Africa.* Ph.D. dissertation, University of London.
Harrison, T. 1987. The phylogenetic relationships of the early catarrhine primates: A review of the current evidence. *J. Hum. Evol.* **16:**41–80.
Harrison, T. 1988. A taxonomic revision of the small catarrhine primates from the early Miocene of East Africa. *Folia Primatol.* **50:**59–108.
Hopwood, A. 1933. Miocene primates from British East Africa. *Ann. Mag. Nat. Hist.* **11:**96–98.

Hylander, W., and Johnson, K. 1992. Strain gradients in the craniofacial region of primates. In: Z. Davidovitch (ed.), *The Biological Mechanisms of Tooth Movement and Craniofacial Adaptation*, pp. 559–569. Ohio State University College of Dentistry, Columbus.

Kluge, A., and Farris, J. 1969. Quantitative phyletics and the evolution of anurans. *Syst. Zool.* **18:**1–32.

Larson, S. 1992. Are the similarities in forelimb morphology among the hominoid primates due to parallel evolution? *Am. J. Phys. Anthropol. Suppl.* **14:**106.

Leakey, L. 1963. East African fossil Hominoidea and the classification within this superfamily. In: S. Washburn (ed.), *Classification and Human Evolution*, pp. 32–49. Aldine, Chicago.

Leakey, R., Leakey, M., and Walker, A. 1988a. Morphology of *Afropithecus turkanensis* from Kenya. *Am. J. Phys. Anthropol.* **76:**289–307.

Leakey, R., Leakey, M., and Walker, A. 1988b. Morphology of *Turkanapithecus kalakolensis* from Kenya. *Am. J. Phys. Anthropol.* **76:**277–288.

Li, C. 1978. A Miocene gibbon-like primate from Shihhung, Kiangsu Province. *Vertebr. Palasiat.* **16:**187–192.

McDougall, I., and Watkins, R. 1985. Age of hominoid-bearing sequence at Buluk, northern Kenya. *Nature* **318:**175–178.

MacInnes, D. 1943. Notes on the east African Miocene primates. *J. East. Afr. Uganda Nat. Hist. Soc.* **18:**141–181.

Madden, C. 1980. New *Proconsul* (*Xenopithecus*) from the Miocene of Kenya. *Primates* **21:**241–252.

Maddison, W., Donoghue, M., and Maddison, D. 1984. Outgroup analysis and parsimony. *Syst. Zool.* **33:**83–103.

Moyà-Solà, S., and Köhler, M. 1993. Recent discoveries of *Dryopithecus* shed new light on evolution of great apes. *Nature* **365:**543–545.

Napier, J., and Davis, P. 1959. The fore-limb skeleton and associated remains of *Proconsul africanus*. *Br. Mus. Nat. Hist. Fossil Mamm. Afr.* **16:**1–69.

Nengo, I. O., and Rae, T. 1992. New hominoid primates from the early Miocene site of Songhor, Kenya. *J. Hum. Evol.* **23:**423–429.

Nixon, K., and Carpenter, J. 1993. On outgroups. *Cladistics* **9:**413–426.

Pickford, M. 1986. Cainozoic paleontological sites of western Kenya. *Muencher Geowiss. Abh. Ser. A* **8:**1–151.

Pilbeam, D. 1969. Tertiary Pongidae of east Africa: Evolutionary relationships and taxonomy. *Bull. Peabody Mus. Nat. Hist.* **31:**1–185.

Pilbeam, D. 1982. New hominoid skull from the Miocene of Pakistan. *Nature* **295:**232–234.

Pilbeam, D., Rose, M., Barry, J., and Shah, S. M. I. 1990. New *Sivapithecus* humeri from Pakistan and the relationship of *Sivapithecus* and *Pongo*. *Nature* **348:**237–239.

Platnick, N. 1989. An empirical comparison of microcomputer parsimony programs, II. *Cladistics* **5:**145–161.

Pogue, M., and Mickevich, M. 1990. Character definitions and character state delineation: The bête noire of phylogenetic inference. *Cladistics* **6:**319–361.

Preuss, T. 1982. The face of *Sivapithecus indicus:* Descriptions of a new, relatively complete specimen from the Siwaliks of Pakistan. *Folia Primatol.* **38:**141–157.

Rae, T. 1993. *Phylogenetic Analysis of Proconsulid Facial Morphology*. Ph.D. dissertation, State University of New York at Stony Brook.

Rae, T., and Simons, E. 1992. The significance of the facial morphology of *Aegyptopithecus*. *Am. J. Phys. Anthropol. Suppl.* **14:**134.

Rak, Y. 1983. *The Australopithecine Face*. Academic Press, New York.

Rohlf, F. 1982. *BIOM: A package of statistical programs to accompany the text Biometry*. Stony Brook (distributed by author).

Rose, M. 1983. Miocene hominoid postcranial morphology: Monkey-like, ape-like, neither, or both? In: R. Ciochon and R. Corruccini (eds.), *New Interpretations of Ape and Human Ancestry*, pp. 405–417. Plenum Press, New York.

Schmitt, D. 1991. Strepsirhine humeral head morphology and its implications for notharctine locomotion. *Am. J. Phys. Anthropol. Suppl.* **12:**159.

Simon, C. 1983. A new coding procedure for morphometric data with an example from periodical cicada wing veins. In: J. Felsenstein (ed.), *Numerical Taxonomy,* pp. 378–382. Springer-Verlag, Berlin.

Simons, E., and Pilbeam, D. 1965. Preliminary revision of the Dryopithecinae (Pongidae; Anthropoidea). *Folia Primatol.* **3:**81–152.

Slowinski, J. 1993. "Unordered" versus "ordered" characters. *Syst. Biol.* **42:**155–165.

Sokal, R., and Rohlf, F. 1981. *Biometry,* 2nd ed. Freeman, San Francisco.

Strasser, E., and Delson, E. 1987. Cladistic analysis of cercopithecid relationships. *J. Hum. Evol.* **16:**81–99.

Szalay, F., and Delson, E. 1979. *The Evolutionary History of the Primates.* Academic Press, New York.

Teaford, M., Beard, K., Leakey, R., and Walker, A. 1988. New hominoid facial skeleton from the early Miocene of Rusinga Island, Kenya, and its bearing on the relationship between *Proconsul nyanzae* and *Proconsul africanus. J. Hum. Evol.* **17:**461–477.

Walker, A., and Teaford, M. 1989. The hunt for *Proconsul. Sci. Am.* **260:**76–82.

Walker, A., Teaford, M., Martin, L., and Andrews, P. 1993. A new species of *Proconsul* from the early Miocene of Rusinga/Mfangano Islands, Kenya. *J. Hum. Evol.* **25:**43–56.

Ward, S., and Kimbel, W. 1983. Subnasal alveolar morphology and the systematic position of *Sivapithecus. Am. J. Phys. Anthropol.* **61:**157–171.

Ward, S., and Pilbeam, D. 1983. Maxillofacial morphology of Miocene hominoids from Africa and Indo-Pakistan. In: R. Ciochon and R. Corruccini (eds.), *New Interpretations of Ape and Human Ancestry,* pp. 211–238. Plenum Press, New York.

Wilkinson, M. 1992. Ordered versus unordered characters. *Cladistics* **8:**375–385.

Wu, R., and Pan, Y. 1985. Preliminary observation on the cranium of *Laccopithecus robustus* from Lufeng, Yunnan with reference to its phylogenetic relationship. *Acta Anthropol. Sin.* **4:**7–12.

Functional and Phylogenetic Features of the Forelimb in Miocene Hominoids

5

MICHAEL D. ROSE

Introduction

The forelimb is the most versatile part of the locomotor system in primates. As such, its functional morphology varies widely among taxa with different locomotor specializations. It is therefore an attractive region for investigations of phylogeny, function, and their interrelationships. In addition, forelimb elements are relatively abundant in the catarrhine fossil record. This is especially true for the humerus, which will form the focus of what follows. In this discussion, only major references are cited: more complete bibliographies are included in Rose (1993a, 1994). Characters identified in the text by parenthesized letters, e.g., (a) or (a′), are listed in Table I, and their states in the taxa discussed are listed in Table II. Characters identified with a prime indicate the derived state. Many of the humeral characters are illustrated in Fig. 1.

MICHAEL D. ROSE • Department of Anatomy, Cell Biology, and Injury Science, University of Medicine and Dentistry of New Jersey, New Jersey Medical School, Newark, New Jersey 07103.

Function, Phylogeny, and Fossils: Miocene Hominoid Evolution and Adaptations, edited by Begun *et al.* Plenum Press, New York, 1997.

Primitive Catarrhines

A number of Oligocene and Miocene catarrhines share no forelimb features uniquely with extant hominoids. They are considered, without further justification here, to be primitive, noncercopithecoid, nonhominoid catarrhines (see, e.g., Rose *et al.*, 1992). They are included for two reasons. The first is that many of their features are retained in Miocene taxa that also exhibit additional features that are shared with extant hominoids. The second is that some primitive catarrhines exhibit specialized features that are most likely to be convergent on features found in some extant hominoids.

The *Aegyptopithecus* humerus is short and robust for an animal of its size (Fleagle and Simons, 1982). There is no torsion (p) of the relatively small, ovoid humeral head (q). The tuberosities are set at a small angle to each other (o), and are separated by a broad, shallow bicipital groove (r). These features imply that movement at the shoulder joint is fairly limited except for flexion–extension movements. The proximal shaft, although bearing a convex deltoid plane, is rather angular in cross section (s). It inclines posteriorly (m) and medially (n), as in numerous quadrupedal mammals. Laterally, the distal half of the shaft bears a pronounced brachialis flange, while a medial flange is

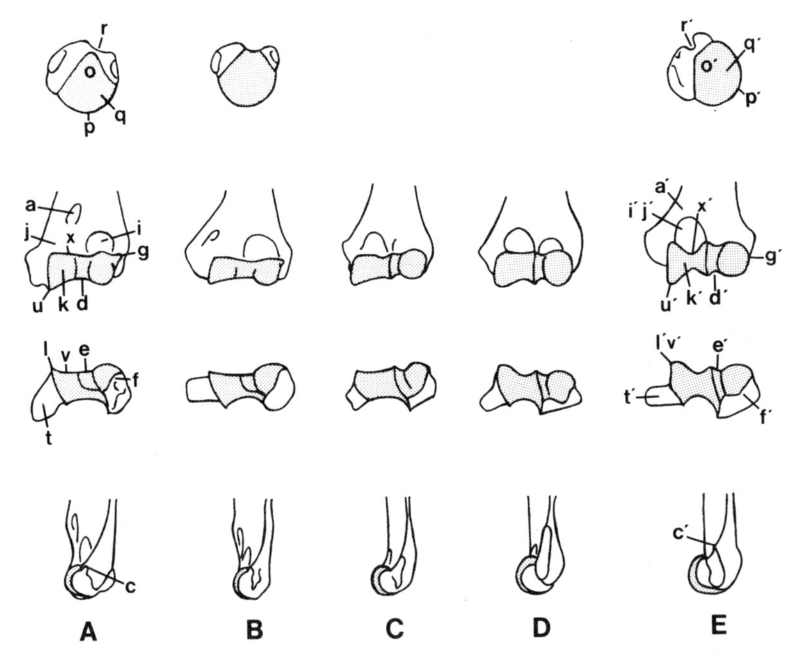

Fig. 1. Humeri of (A) *Cebus,* (B) *Pliopithecus,* (C) *Proconsul,* (D) *Sivapithecus,* and (E) *Pan.* From top to bottom, proximal view of the head, and anterior, distal, and lateral views of the distal end. The labeled characters are defined in Table I. However, not all of the features listed in Table I are illustrated in this figure.

Table I. Character States Used in This Analysis[a]

	Humeral characters
a	Entepicondylar foramen present
a′	Entepicondylar foramen absent
b	Shallow olecranon fossa
b′	Deep olecranon fossa
c	Lateral epicondyle does not extend proximal to capitulum
c′	Lateral epicondyle extends proximal to capitulum
d	Broad zona conoidea
d′	Narrow zona conoidea
e	Shallow zona conoidea
e′	Deep zona conoidea
f	Capitulum does not extend onto distolateral surface of humerus
f′	Capitulum extends onto distolateral surface of humerus
g	Capitular tail present
g′	Capitular tail absent
h	Articular surface does not substantially extend into lateral wall of olecranon fossa
h′	Articular surface substantially extends into lateral wall of olecranon fossa
i	Radial fossa larger than coronoid fossa
i′	Coronoid fossa larger than radial fossa
j	Shallow coronoid fossa
j′	Deep coronoid fossa
k	Cylindrical or conical trochlea
k′	Trochleiform trochlea
l	No strong keel on medial trochlea, or strong keel anteriorly and distally
l′	Strong keel all around medial trochlea
m	Proximal shaft bowed, concave posteriorly
m′	Shaft straight
n	Proximal shaft bowed, concave medially
n′	Shaft straight in anterior view
o	Intertuberosity angle less than 90°
o′	Intertuberosity angle greater than 90°
p	Head torsion less than 15°
p′	Head torsion greater than 15°
q	Head ovoid
q′	Head spherical
r	Bicipital groove broad and shallow
r′	Bicipital groove narrow and deep
s	Proximal shaft angular anteriorly, or trapezoidal in cross section
s′	Proximal shaft rounded in cross section
t	Strongly retroflexed medial epicondyle
t′	Weakly or not retroflexed medial epicondyle
u	Trochlea not trochleiform or, if trochleiform, with sub-equally developed medial and lateral trochlear keels
u′	Trochleiform trochlea with medial lip emphasized in anterior view

(*continued*)

Table I. (*Continued*)

v	Trochlea not trochleiform or, if trochleiform, with sub-equally developed medial and lateral trochlear keels
v′	Trochleiform trochlea with medial lip emphasized in distal view
w	Medial epicondyle does not merge with posterior part of medial side of medial trochlear keel
w′	Medial epicondyle merges with posterior part of medial side of medial trochlear keel
x	Superior border of trochlea straight
x′	Superior border of trochlea notched to accommodate coronoid process
	Ulnar characters
y	Olecranon process not abbreviated
y′	Olecranon process abbreviated
z	Radial notch faces anterolaterally
z′	Radial notch faces laterally
A	Brachialis insertion area unbuttressed medially
A′	Brachialis insertion area buttressed medially
B	Weakly developed supinator ridge
B′	Strongly developed supinator ridge
C	Proximal shaft anteroposteriorly deep
C′	Proximal shaft anteroposteriorly shallow
D	Styloid articulates with pisiform and/or triquetral
D′	Styloid does not articulate with triquetral and pisiform
E	Trochlear notch faces anteriorly
E′	Trochlear notch faces anteroproximally (markedly projecting coronoid process)

[a]Primes indicate derived states.

pierced by an entepicondylar foramen (a). The medial epicondyle is moderately retroflexed (t) but does not merge with the medial wall of the trochlea posteriorly (w). The lateral epicondyle is at the same transverse level as the proximal border of the capitulum (c). On the anterior surface, the radial fossa is larger (i) and deeper (j) than the coronoid fossa. The distal articular surface is anteroposteriorly narrow. The trochlea is relatively wide and approximately cone-shaped (k, u), without a well-developed medial keel (l), and with a straight border forming the inferior margin of the coronoid fossa (x). Posterolaterally, the trochlear articular surface is developed into a modest lip that only minimally extends into the lateral wall (h) of the shallow olecranon fossa (b). The *zona conoidea* is broad (d), shallow (e), and distally facing. The capitulum is only modestly inflated, does not extend onto the distolateral surface of the humerus (f), and extends as a capitular tail anteroproximally (g). Most features of the head and shaft are similar to those of arboreal quadrupedal strepsirrhines. However, the morphology of the distal articular surface and adjacent regions are more similar to those of large-bodied quadrupedal platyrrhines, particularly *Alouatta*. The anteriorly facing trochlear notch (E) of the proximal ulna (Conroy, 1976) mirrors the morphology of the trochlea.

The proximally directed olecranon process is relatively long (y), the radial notch faces anterolaterally (z), and the brachialis insertion area is buttressed medially (A′). The supinator ridge is only weakly developed (B) on the lateral side of the anteroposteriorly deep proximal shaft (C). *Aegyptopithecus* was evidently a deliberately moving arboreal quadruped that progressed with relatively flexed limbs, thus keeping its center of mass close to the support, and taking firm manual grips on the branches on which it was moving.

Most forelimb bones are known, at least in part, for *Pliopithecus* (Zapfe, 1960). The clavicle is most similar to those of colobine monkeys, while the preserved parts of the scapula are similar to those of quadrupedal nonhominoids. *Pliopithecus* is intermediate between *Alouatta* and *Ateles* in the degree of elongation of its forelimb, and resembles a number of quadrupedal monkeys of similar body size in having a brachial index of 105. The humeral head is similar to that of *Aegyptopithecus* in having an oval articular surface, a small intertuberosity angle, and minimal torsion. The distal humerus is also like that of *Aegyptopithecus* in having a cone-shaped trochlea and a broad, shallow zona conoidea. However, the humeral shaft is strikingly different. It is relatively long, straight (m′, n′), gracile, and rounded in cross section (s′). Also, the medial epicondyle is medially rather than posteromedially directed (t′). Compared with *Aegyptopithecus,* the forearm bones are long and gracile. The radial head is markedly oval in shape, due to the presence of a lateral lip (reflecting humeral features d and e). This feature relates to habitual use of the forearm in a fully pronated position, where the lip fully engages with the broad, nonrecessed zona conoidea (Rose, 1988, 1993b). The morphology of the radioulnar joints indicates a moderately extensive pronation–supination range, as in quadrupedal platyrrhines (Robertson, 1985; Sarmiento, 1985). The proximal ulna lacks the buttressing of the brachialis insertion area found in *Aegyptopithecus* (A). The distal ulna articulates with both the pisiform and the triquetral in the wrist (D), as in nonhominoids. However, the hamatrotriquetral joint faces as much medially as it does proximally. These features indicate a pattern of loading onto the medial side of the hand similar to that of quadrupedal nonhominoid anthropoids, combined with a slightly greater ulnar deviation capability than is usual in nonhominoid anthropoids. The articulations of the forelimb indicate predominantly quadrupedal locomotor behavior in *Pliopithecus.* However, the elongated, straight long bones and the medially directed medial epicondyle suggest a significant suspensory component to the locomotor repertoire.

Associated forelimb specimens are known for *Dendropithecus macinnesi* and the morphologically similar *Simiolus enjiessi.* Other forelimb specimens from East Africa, either of indeterminate attribution, or probably belonging to *Limnopithecus* and *Kalepithecus* are similar to those of *Dendropithecus,* where comparisons are possible (Clark and Thomas, 1951; Ferembach, 1958; Simons and Fleagle, 1973; McHenry and Corruccini, 1975; Andrews and Simons, 1977; Aiello, 1981; Harrison, 1982; Fleagle, 1983; Gebo *et al.,* 1988; Senut, 1989; Rose *et al.,* 1992).

Dendropithecus is similar to *Pliopithecus* in the degree of elongation of its forelimb, in the gracility of its shaft, and in its brachial index. A proximal humerus, possibly of *Dendropithecus* (but also possibly of *Proconsul heseloni*), is quite similar to that of *Pliopithecus,* except that the head is slightly more inflated. Proximal humeri, possibly of *Dionysopithecus* (Rose, 1989), and *Nyanzapithecus* (McCrossin, 1994) are also generally similar to that of *Pliopithecus.* In *Dendropithecus* the humeral shaft resembles that of *Pliopithecus* except that the deltoid plane is flatter, the distal shaft is more anteroposteriorly compressed, and there is no entepicondylar foramen distomedially (a′). The rest of the distal humerus is generally similar to that of *Pliopithecus.* However, relative to *Pliopithecus* the distal articular surface is narrower compared to the length of the shaft, the trochlea has a more pronounced conical shape, there is a modestly developed lateral trochlear lip, and the zona conoidea is in the form of a shallow trough anteriorly (but not distally). The zona conoidea is even more anteriorly shallow in *Simiolus.* Similarities to *Pliopithecus* extend to the radius and ulna. However, the ulnar trochlear notch is more convex dorsally (corresponding to the humeral trochlear morphology) and the radial head is more tilted and has a more pronounced oval shape. The forearm bones are more compressed mediolaterally, and more bowed than in *Pliopithecus.* The bones of the wrist and hand are similar to those of extant quadrupedal nonhominoid anthropoids, particularly cercopithecids. Although the morphology of the forelimb of *Dendropithecus* and related taxa differs from that of *Pliopithecus* in a number of respects, the similarities are close enough to suggest a similar pattern of positional behavior.

Turkanapithecus

Turkanapithecus forelimb specimens include an ulna, a partial radius, and some bones of the hand (Leakey *et al.,* 1988a; Rose, 1993a, 1994). The ulna is similar to that of *Pliopithecus,* but has a shorter olecranon process and a brachialis insertion area similar to that of *Aegyptopithecus* (A′). The latter feature indicates the capability for powerful flexion at the elbow, as used in climbing and hoisting activities. A well-developed supinator crest extends distally from the radial notch (B′). As in *Pliopithecus,* the ulnocarpal joint is of the nonhominoid type. The radial head has only a modestly developed lateral lip, and is beveled in side view, suggesting that the humeral zona conoidea is narrow and recessed, as in extant hominoids (d′, e′). The surface for articulation with the ulna at the proximal radioulnar joint is relatively deep and extends all the way around the head, suggesting an extensive pronation–supination range. The distal radial articular surface faces distoanteriorly rather than anteriorly, is anteroposteriorly longer, and is more deeply socketed than in *Pliopithecus.* While these features are not developed to the degree seen in extant hominoids, they suggest a greater mobility at the radiocarpal joint than is usual in nonhominoid anthropoids. The first metacarpal has a saddle-shaped proximal articular surface, similar to that found in large extant hominoids, suggesting that the trapezium–first metacarpal joint might be quite mobile. Other metacarpals and phalanges are similar to those of quadrupedal nonhominoid anthropoids. The combination of powerful elbow flexion, enhanced forearm rotation, a stabilized but still mobile wrist (especially in the radial deviation range), and a mobile thumb all suggest a climbing and/or hoisting capability in addition to a basically quadrupedal locomotor pattern.

Proconsul and Afropithecus

Most of the forelimb can be completely reconstructed for *Proconsul heseloni,* formerly *P. africanus* (Napier and Davis, 1959; Walker and Pickford, 1983; Beard *et al.*, 1986; Begun *et al.*, 1993). The known forelimb specimens of *P. nyanzae* and *P. major* (with one possible exception—Nengo and Rae, 1992), and also of *Afropithecus turkanensis* (Leakey *et al.*, 1988b), differ in size from those of *P. heseloni,* but are similar in most of their known morphological features. All of this material will therefore be considered together. The shoulder region is incompletely known. The lateral third of the clavicle is generally similar to that of *Pliopithecus* and of cercopithecids. The scapula is known from an incomplete and distorted specimen. Its preserved parts share resemblances with the scapulae of nonsuspensory platyrrhines. In particular, the angle between the glenoid and the lateral border is large, the acromion does not markedly overhang the glenoid, the neck is relatively long, the groove on the lateral border faces inferiorly, and the great scapular notch includes a groove on the root of the spine. None of these features are found in extant hominoids. The long bones of the forelimb are relatively shorter than in *Pliopithecus* and *Dendropithecus* and the brachial index is 96. *Proconsul* most resembles large-bodied, nonsuspensory New and Old World monkeys in these long-bone features.

In the humerus, proximal shaft features suggest that the head and its degree of torsion are similar to those of *Aegyptopithecus* and *Pliopithecus.* In its curvature (m, n?) and cross-sectional morphology (s), the shaft is more like *Aegyptopithecus* than *Pliopithecus* or *Dendropithecus.* The triangular cross section of the shaft is the result of there being a flat deltoid plane, bounded by well-developed deltopectoral and deltotriceps crests. The medial epicondyle is retroflexed, even more than in *Aegyptopithecus,* and the posterior part of the medial epicondyle merges with the medial (external) face of the trochlea (w′). *Proconsul* resembles cercopithecids in both of these features. The distal humerus also differs from those described so far in that the lateral epicondyle extends proximal to the level of the capitulum, as in extant hominoids (c′, Senut, 1989). While the trochlea is cone shaped, like those described above, its articular surface extends into the lateral wall (h′) of the relatively deep olecranon fossa (b′). In *P. heseloni* a lateral trochlear lip is slightly more inflated than in *Dendropithecus.* The lip is apparently not present in *P. nyanzae.* The zona conoidea is narrow (d′) and deep (e′). The capitulum, which lacks a tail (g′), extends onto the distolateral surface of the humerus (f′).

The distal humeral features are all matched by corresponding features of the ulnar trochlear notch and the radial head. In particular, the ulnar trochlear notch has a well-developed, laterally facing, proximolateral lip that articulates with the articular surface in the lateral wall of the humeral olecranon fossa when the elbow is in an extended position. As in the previously described ulnae, the radial notch faces anterolaterally. The olecranon process is similar to that of *Turkanapithecus,* except that, in *P. nyanzae* at least, it is retroflexed, as in terrestrial cercopithecids. The proximal articular surface of the radial head is beveled in side view, allowing it to engage with the recessed humeral zona conoidea. The central depression on the proximal surface is quite deep, and engages with the inflated capitulum. Both of these features relate to the stabilization of the humeroradial joint in all positions of flexion–extension and pronation–supination. However, as in *Pliopithecus* and *Dendropithecus,* there is a lateral lip on the proximal surface of the radial head in *Proconsul.* This indicates a position of particular stability in pronation, as in most quadrupedal nonhominoid anthropoids. The presence of articular surface, for articulation with the ulna, all around the periphery of the radial head indicates an extensive pronation–supination capability in the forearm. The distal articular surface, for the radiocarpal joint, is relatively flat as in *Pliopithecus* and *Dendropithecus,* although it is relatively deeper anteroposteriorly.

In *P. heseloni* the ulna articulates with both the pisiform and the triquetral. However, the morphology of the articulation is unique. In nonhominoid anthropoids the confluent articular surfaces of the triquetral and pisiform form a distomedially oriented groove that is limited distally by ridges linked by a medial pisotriquetral ligament. The grooved shape of the articulation, and especially the distal osteoligamentous ridge engage the ulnar styloid and stabilize it, while limiting ulnar deviation. In *Proconsul* the bony ridges on the triquestral and pisiform are not present, and the triquestral component of the groove for the ulnar styloid is more elongated. Thus, while a modicum of stability is provided by the grooved shape of the articulation, the ulnar styloid can travel farther distomedially, providing a larger ulnar deviation excursion than in nonhominoid anthropoids. In addition, the hamatotriquetral joint is linear, with the hamate facet facing medioproximally, as in *Pliopithecus* and extant hominoids other than *Gorilla.* Thus, ulnar deviation is amplified by movement of the triquetral on the hamate. The ulnar side of the wrist evidently combines relative stability with an even greater ulnar deviation capability than suggested for *Pliopithecus.* The possibility that *P. major* might have a radically different ulnocarpal region from that of other *Proconsul* species is too large a problem to pursue in detail here (Nengo and Rae, 1992). *Proconsul* resembles extant hominoids in having a strongly trochleiform joint between the trapezium and the first metacarpal, and a dorsal tubercle on the trapezium that is associated with a well-developed ligament passing between the trapezium and the first metacarpal. These features underlie an enhanced mobility of the thumb utilizable in both grasping and manipulatory activities. The second to fifth metacarpals and the phalanges are generally similar to those of quadrupedal, nonhominoid anthropoids. Forelimb morphology of *Proconsul* is that of a predominantly quadrupedal animal. However, *Proconsul* differs from the taxa considered so far in having a humeroradial joint that is stabilized in all positions, moderately extensive pronation–supination of the forearm and ulnar deviation in the wrist, and a relatively mobile thumb. It is possible that these derived features (together with some in other parts of the postcranium) constitute functional complexes that enable a relatively large-bodied but tailless animal to maneuver successfully in an arboreal environment (Ward *et al.,* 1991; Kelley, 1995; Rose, 1996).

Kenyapithecus

A number of humeral, ulnar, wrist, and hand specimens are known from a number of sites, and have been attributed to at least two species, *K. wickeri* (distal humerus only) and *K. africanus* (Clark and Leakey, 1951; McHenry and Corruccini, 1975; Andrews and Walker, 1976; Feldesman, 1982; Senut, 1989). Recently recovered specimens from Maboko Island are discussed by McCrossin (1994) and McCrossin and Benefit (this volume) and are not considered here, although features discussed by them are included in Table I. The shaft and distal end of the humerus are generally similar to the *Proconsul* humerus, except that the shaft is more gracile. The proximal part of the shaft inclines posteriorly and probably medially as well. The distal humerus differs from that of *Proconsul* in a number of features. The lateral epicondyle is anteroposteriorly deeper than in *Proconsul* and the medial epicondyle is more posteriorly directed. Also, the coronoid fossa is larger than the radial fossa (i′), although it is quite shallow (j). The articular surface in the lateral wall of the olecranon fossa (h′) projects posteriorly as the lateral surface of a well-developed flange.

The distal parts of the ulnar trochlear notch is similar to those of the previously described specimens, reflecting the cone shape of the humeral trochlea. The brachialis insertion area is buttressed and there is a well-developed supinator crest. In addition, the radial notch faces directly laterally (z′). A partial lunate, metacarpal heads, and proximal and middle phalanges are, as in *Proconsul,* most similar to those of quadrupedal nonhominoid anthropoids. *Kenyapithecus* is quite similar to *Proconsul* in most features of its forelimb morphology. However, the presence of a laterally facing ulnar surface for the proximal radioulnar joint, a feature shared with extant hominoids, may reflect a greater pronation–supination capability in the forearm.

Austriacopithecus

Austriacopithecus weinfurti is only known from a humeral shaft and a partial ulna (Ehrensberg, 1938; Zapfe, 1960; Begun, 1992). The humeral shaft is similar to that of *Kenyapithecus* in most features of shaft morphology, but differs in having a flat posterior surface. The ulnar shaft is bowed, medial concave, as in extant large hominoids (great apes and humans). However, although damaged, the olecranon process is apparently similar to that of *Proconsul nyanzae* (y). However, its degree of retroflexion cannot be determined. The preserved part of the trochlear notch is similar to those described above, indicating that the humeral trochlea is probably cone shaped. However, the coronoid process is apparently somewhat more protuberant than in the previously described ulnae. Other features are similar to those of *Kenyapithecus*. The available evidence thus suggests positional capabilities similar to those of *Kenyapithecus*.

Sivapithecus

The *Sivapithecus* forelimb is known from specimens of the humerus, radius, and bones of the wrist and hand. This material is from *S. parvada* and *S. indicus* (Pilbeam *et al.*, 1980, 1990; Rose, 1983, 1984, 1986, 1989; Spoor *et al.*, 1992). The humeral proximal shaft inclines both posteriorly and medially, as in *Kenyapithecus*, and has a flat posterior surface, as in *Austriacopithecus*. The bicipital groove is broad and flat, indicating that the head is probably of the nonhominoid type described above. In the distal humerus, the region of the epicondyles and of the olecranon fossa are mostly similar to that described for *Proconsul* and *Kenyapithecus*. However, unlike in *Proconsul*, *Kenyapithecus*, and cercopithecids, the posterior part of the medial epicondyle does not blend with the external wall of the trochlea (w). As in *Kenyapithecus* and extant hominoids, the coronoid fossa is larger than the radial fossa, but is also deep (j′). The capitulum is similar to those of *Proconsul* and *Kenyapithecus*, while the zona conoidea is more incised. The trochlea is unlike those described above. It resembles the trochlea of extant hominoids in being spool shaped (k′), i.e., truly trochleiform, rather than conical. Well-developed medial (l′) and lateral keels flank a central gutter. The medial keel of the spool is more anteriorly projecting than the lateral keel (v′), as in extant large hominoids. However, the trochlea differs from those of large hominoids in that the medial keel does not extend as far distally as it does in extant large hominoids (u), and the superior border of the trochlea is not notched (x). Nevertheless, this conformation of the humeroulnar joint stabilizes it against forces tending to abduct, adduct, or translate the ulna on the humerus during flexion–extension.

A juvenile radial shaft is quite similar to that of *Proconsul*. The capitate resembles those of large extant hominoids in having an irregular surface for articulation with the third metacarpal, but otherwise is most similar to the capitates of nonhominoid quadrupeds. The hamate has a surface for the triquetral that faces more proximally than medially. In this feature it resembles the hamates of nonhominoid anthropoids, and also *Gorilla*. It thus differs from the hamates of *Pliopithecus, Proconsul,* and extant large hominoids other than *Gorilla*. The articular surface is also abbreviated and trough like, suggesting that the hamatotriquetral joint provides more stability than mobility to the ulnar side of the wrist. The *Sivapithecus* hamate also resembles those of nonhominoids (and also of *Pliopithecus* and *Proconsul*) in having a small hamulus. As in *Proconsul, Afropithecus,* and extant large hominoids, the first metacarpal is saddle shaped, indicating that the thumb is quite mobile around all axes of movement. The proximal phalanges of the hand are relatively long and have well-developed secondary features. The proximal articular surface faces proximodorsally and is oval in shape, as in nonhominoid anthropoids and other Miocene apes for which this feature is known. *Sivapithecus* differs from the taxa described so far in having both the humeroulnar and humeroradial joints well stabilized throughout all movements of the elbow and forearm. However, there is no evidence for the extensive use of locomotor activities other than quadrupedalism and climbing. In fact, the emphasized angulations of the humeral shaft, and an ulnar side of the wrist that is probably load-bearing and relatively immobile, suggest a primacy of quadrupedalism in the positional repertoire.

Lufengpithecus

Lufengpithecus lufengensis is known from partial specimens of the scapula, clavicle, radius, and phalanges (Xiao, 1981; Wu *et al.*, 1986; Lin *et al.*, 1987; Meldrum and Pan, 1988). The scapula and clavicle are described as being comparable to those of great apes. The radial head is circular, with a deep surface for the proximal radioulnar joint, as in extant hominoids. The proximal articular surface of the proximal phalanges is similar to those of *Sivapithecus* and nonhominoid anthropoids.

Dryopithecus

Dryopithecus forelimb specimens include a partial humerus, radius, and ulna, together with a lunate, a hamate, and some phalangeal specimens, from the species *D. fontani* (humeral shaft), *D. laietanus* (wrist bones), and *D. brancoi* (Pilbeam and Simons, 1971; Kretzoi, 1975; Morbeck, 1983; Begun, 1987,

1988, 1992, 1993, 1994; Begun and Kordos, this volume). The humeral shaft inclines medially, but is apparently straight in medial or lateral view (m′), as in extant hominoids. The damaged proximal shaft terminates in the region of the base of the tuberosities. However, there are several morphological features that suggest that the head region is of the same general type as described above, and lacks the torsion of the head characteristic of extant hominoids. Thus, the buttress at the base of the lesser tuberosity is proximodistally aligned along the medial border of the bone, indicating that the lesser tuberosity is not reflected anteriorly and the intertuberosity angle is likely to be small. The swelling of the shaft that buttresses the posterior pole of the head is in the midline posteriorly, not displaced medially as it would be if there was torsion of the head. The junction of the deltoid plane and the base of greater tuberosity faces anteriorly, not anterolaterally. Finally, the bicipital groove is shallow, and most probably wide also.

In most of its features the distal humerus resembles that of *Sivapithecus*. However, in *D. brancoi*, the olecranon fossa is relatively shallow (b), there is a minimal presence of articular surface in its lateral wall (h), and the buttress lateral to the fossa is relatively narrow mediolaterally. On the ulna, the trochleiform shape of the humeral trochlea is mirrored in the saddle-shaped morphology of the ulnar trochlear notch. The radial notch, and brachialis and supinator attachment areas are similar to those of *Kenyapithecus*. However, unlike the ulnae discussed so far, the proximal shaft is anteroposteriorly shallow (C′). The radial head is circular and strongly beveled for articulation with the humeral zona conoidea. The phalanges are similar to those of large-bodied arboreal anthropoids. In most functional features, especially of the elbow region, *Dryopithecus* is similar to *Sivapithecus*. The straight humeral shaft may indicate a greater amount of suspensory behavior in the positional repertoire. As in extant hominoids, the lunate has a relatively large facet for articulation with the capitate and hamate, but is otherwise most similar to non-hominoids. The hamate bears a medioproximally facing facet for the triquetral, as in *Pliopithecus, Proconsul,* and most extant hominoids. The head is rounded, as in hominoids, and the hamulus is moderately long and palmarly directed. The robust phalanges are strongly curved, as in extant hominoids. While the evidence from the shoulder region remains equivocal, the elbow, forearm, and hand of *Dryopithecus* are quite similar to those of extant large hominoids. A positional repertoire including quadrupedalism, orthograde activities, and possibly some suspension is indicated.

Oreopithecus

Most of the bones of the forelimb are known for *Oreopithecus bambolii* (Schultz, 1960; Straus, 1963; Harrison, 1986, 1991; Harrison and Rook, this volume; Sarmiento, 1987, 1988). However, almost all of them have been distorted during fossilization. Despite this, it is possible to tell that the *Oreopithecus* forelimb shares many features with those of extant hominoids. Thus, the clavicle is relatively long and bears a pronounced conoid tubercle. On the scapula, the glenoid is in the form of a dished ovoid. The coracoid is long and robust, while the root of the acromion is situated relatively superiorly in lateral view. The angle between the glenoid and the axillary border is low, while the subscapular buttress is situated superior to the axillary border. The humerus has a spherical head (q′), probably exhibiting medial torsion (p′). The bicipital groove is made narrow and deep (r′) by an anteriorly rotated lesser tuberosity (implying that there is a wide intertuberosity angle, o′). The distal humerus is most like that of *Sivapithecus*. However, it differs in having a medially directed medial epicondyle (t′), and a trochlea that has a markedly protuberant medial trochlear keel (u′). In all of these humeral features *Oreopithecus* resembles extant large hominoids, particularly great apes. As in *Dryopithecus*, the saddle-shaped ulnar trochlear notch reflects the morphology of the humeral trochlea, while the circular, beveled radial head reflects the morphology of the humeral capitulum and zona conoidea. The ulnar trochlear notch faces anteroproximally (E′), and the olecranon process is abbreviated (y′). There is no direct contact between the ulna and the carpus (D′), and the hamate facet for the triquetral faces more laterally than medially. In addition, the hamulus of the hamate, and by implication the pisiform, is distally oriented. All of these features are similar to those of extant large hominoids and relate to considerable mobility in the shoulder, elbow (particularly extension), forearm, and wrist (particularly ulnar deviation). The humeroulnar joint is stabilized against movements other than flexion–extension, while the humeroradial joint is stabilized against movements other than the spinning of the radial head that accompanies forearm pronation–supination. These are all features of an animal for which climbing and suspensory behaviors are an important part of the positional repertoire.

Phylogenetic Considerations

Extant hominoids share a large number of functional morphological features of the forelimb that underlie their extensive use of, especially, climbing and suspensory activities (Table II, Fig. 2). These features clearly differentiate them from cercopithecids and from platyrrhines other than large-bodied atelines, which have converged on the extant hominoid condition in features of, especially, the humeral head and shaft (m′–t′).

The gracile, cylindrical shafts and laterally directed medial epicondyles of *Pliopithecus*, *Dendropithecus*, and *Simiolus* (m′, n′, s′, t′) are considered here to be features related to suspensory habits that, as in atelines, are convergent on the hominoid condition. *Proconsul* shares many long bone features in common with nonsuspensory platyrrhines and with *Aegyptopithecus* that are most likely

Table II. Humeral and Ulnar Characters in Extant Anthropoids and Oligocene and Miocene Catarrhines

Cebus	a	b	c	d	e	f	g	h	i	j	k	l	m	n	o	p	q	r	s	t	u	v	w	x	y	z	**A′**	B	C	D	E
Alouatta	**a′**	b	c	d	e	f	g	h	i	j	k	l	m	n	o	p	q	r	s	**t′**	u	v	w	x	y	z	A	B	C	D	E
Ateles	**a′**	b	c	d	e	**f′**	g	h	i	j	k	l	**m′**	**n′**	**o′**	**p′**	**q′**	**r′**	**s′**	**t′**	u	v	w	x	y	z	A	B	C	D	E
Cercopithecus	**a′**	b	c	d	e	f	g	**h′**	i	j	k	l	m	n	o	p	q	r	s	t	u	v	**w′**	x	y	z	A	B	C	D	E
Colobus	**a′**	b	c	d	e	f	g	**h′**	i	j	k	l	m	n	o	p	q	r	s	t	u	v	**w′**	x	y	z	A	B	C	D	E
Hylobates	**a′**	**b′**	**c′**	d	**e′**	**f′**	**g′**	**h′**	**i′**	**j′**	**k′**	**l′**	**m′**	**n′**	**o′**	**p′**	**q′**	**r′**	**s′**	**t′**	u	v	w	x	**y′**	**z′**	A	**B′**	C	**D′**	E
Pongo	**a′**	**b′**	**c′**	**d′**	e	**f′**	**g′**	**h′**	**i′**	**j′**	**k′**	**l′**	**m′**	**n′**	**o′**	**p′**	**q′**	**r′**	**s′**	**t′**	**u′**	**v′**	w	**x′**	**y′**	**z′**	**A′**	**B′**	**C′**	**D′**	**E′**
Pan	**a′**	**b′**	**c′**	**d′**	**e′**	**f′**	**g′**	**h′**	**i′**	**j′**	**k′**	**l′**	**m′**	**n′**	**o′**	**p′**	**q′**	**r′**	**s′**	**t′**	**u′**	**v′**	w	**x′**	**y′**	**z′**	**A′**	**B′**	**C′**	**D′**	**E′**
Gorilla	**a′**	**b′**	**c′**	**d′**	**e′**	**f′**	**g′**	**h′**	**i′**	**j′**	**k′**	**l′**	**m′**	**n′**	**o′**	**p′**	**q′**	**r′**	**s′**	**t′**	**u′**	**v′**	w	**x′**	**y′**	**z′**	**A′**	**B′**	**C′**	**D′**	**E′**
Aegyptopithecus zeuxis	a	b	c	d	e	f	g	h	i	j	k	l	m	n	o	p	q	r	s	t	u	v	w	x	y	z	**A′**	B	C	-[a]	E
Pliopithecus vindobonensis	a	b	c	d	e	f	g	h	i	j	k	l	**m′**	**n′**	o	p	q	r	**s′**	**t′**	u	v	w	x	y	z	A	B	C	D	E?[b]
Dendropithecus macinnesi[c]	**a′**	b	c	d	e	f	g	h	i	j	k	l	**m′**	**n′**	o	p?	q	r	**s′**	**t′**	u	v	w	x	y	z	A	B	C	D	E?
?*Dionysopithecus* sp.	-	-	-	-	-	-	-	-	-	j	-	-	-	-	o	p?	q	r	**s′**	-	-	-	-	-	-	-	-	-	-	-	-
Simiolus enjiessi	**a′**	b	c	d	e	f	g	h	i	j	k	l	**m′**	**n′**	-	-	-	-	**s′**	**t′**	u	v	w	x	-	-	-	-	-	-	-
Nyanzapithecus pickfordi	-	-	-	-	-	-	-	-	-	-	-	-	-	-	o	p?	q	r	s	-	-	-	-	-	-	-	-	-	-	-	-
Turkanapithecus kalakolensis	-	-	-	**d′?**	**e′?**	-	-	**h′?**	-	-	k?	l?	-	-	-	-	-	-	-	-	u?	-	-	-	y	z	**A′**	**B′**	C	D?	E?
Afropithecus turkanensis	**a′**	**b′**	-	**d′**	**e′**	-	-	-	-	-	-	-	-	-	-	-	-	-	-	-	-	-	-	-	-	-	-	-	C	-	-

Proconsul heseloni	**a′**	**b′**	**c′**	**d′**	**e′**	**f′**	**g′**	**h′**	i	j	k	l	m	n?	-	p?	-	r	s	t	u	v	**w′**	x	y	z	**A′**	**B′**	C	D	E
Proconsul nyanzae	**a′**	**b′**	**c′**	**d′**	**e′**	**f′**	**g′**	**h′**	-	j	-	l	-	-	-	-	-	-	-	-	-	-	-	x	y	z	**A′**	**B′**	C	-	-
Proconsul major	**a′**	**b′**	-	-	-	-	-	-	-	j	-	-	-	-	-	-	-	-	-	-	-	-	-	-	-	-	-	-	-	-	-
Kenyapithecus africanus[d]	**a′**	-	-	-	-	-	-	-	-	-	-	-	m	n?	o	p	q	r	s	-	-	-	-	-	y	**z′**	**A′**	**B′**	C	-	E?
Kenyapithecus wickeri	**a′**	**b′**	**c′**	**d′**	**e′**	**f′**	**g′**	**h′**	**i′**	j	k	l	-	-	-	-	-	-	-	t	u	v	**w′**	x	-	-	-	-	-	-	-
Austriacopithecus weinfurti	-	**b′?**	-	-	-	-	-	-	-	-	k?	l?	m	n	-	-	-	-	s	-	u?	-	-	x	y	**z′**	**A′**	**B′**	C	-	-
Sivapithecus parvada	**a′**	**b′**	**c′**	**d′**	**e′**	**f′**	**g′**	**h′**	-	**j′**	-	-	m	n	-	p?	-	r?	s	-	-	-	-	x	-	-	-	-	-	-	E?
Sivapithecus indicus	**a′**	**b′**	**c′**	**d′**	**e′**	**f′**	**g′**	**h′**	**i′**	**j′**	**k′**	**l′**	-	n	-	p?	-	r	s?	t	u	**v′**	w	-	-	-	-	-	-	-	-
Dryopithecus fontani	**a′**	-	-	-	-	-	-	-	-	-	-	-	**m′**	n	o?	p?	-	r	s	-	-	-	-	-	-	-	-	-	-	-	-
Dryopithecus brancoi	**a′**	b	**c′**	**d′**	**e′**	-	**g′**	h	**i′**	**j′**	**k′**	**l′**	-	-	-	-	-	-	-	t	u	**v′**	w	x	-	**z′**	**A′**	**B′**	**C′**	-	-
Oreopithecus bambolii	**a′**	**b′**	-	**d′**	**e′**	**f′?**	**g′**	**h′?**	-	-	**k′**	**l′**	**m′?**	**n′?**	**o′**	**p′?**	**q′**	**r′**	**s′?**	**t′**	**u′**	**v′**	w	-	**y′**	**z′**	**A′**	**B′**	**C′**	**D′**	**E′?**

[a]-, feature not known for available fossils.
[b]?, feature probable on the basis of damaged specimens or of reciprocal features of another bone.
[c]Includes KNM RU 17376.
[d]Includes material from Baragoi.

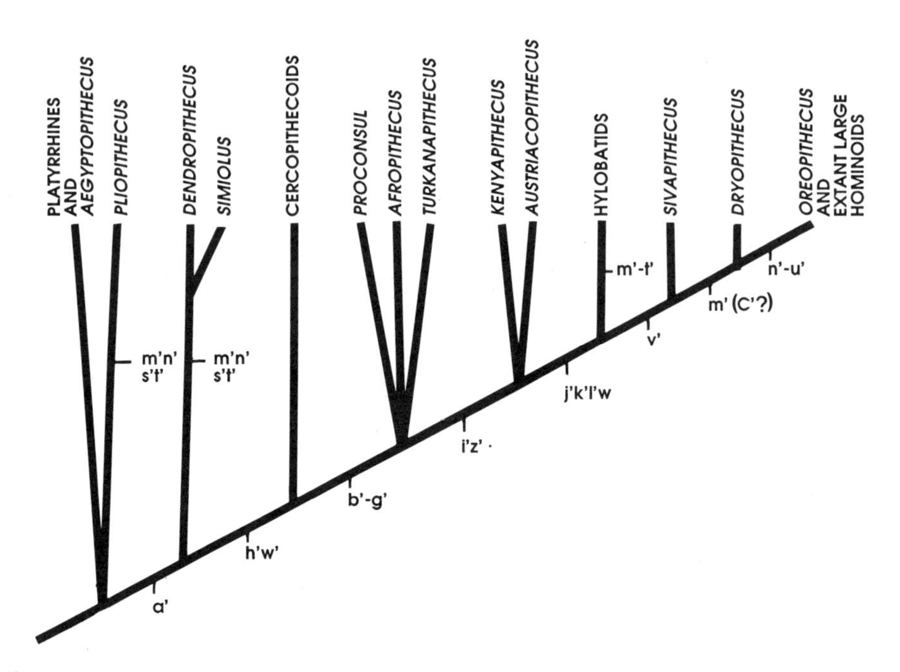

Fig. 2. Cladogram based on humeral and ulnar characters. Distribution of alphabetized characters is given in Table II.

to be retained features that are primitive for catarrhines (i–m, r–v, x–z). Two humeral characters, also present in *Kenyapithecus,* are shared only with cercopithecids (w′), or with cercopithecids and hominoids (h′) and may represent retained characters primitive for the cercopithecid and hominoid clade. In *Proconsul* a number of humeral (and associated radial and ulnar) features are uniquely shared with extant hominoids (b′–g′). Functionally these stabilize the head of the radius as it spins on the capitulum (d′, e′) as part of pronation–supination movements that can take place even in quite extended elbow positions (f′, g′). In these extended positions the ulna is stabilized within the humeral olecranon fossa (b′, h′). The proximal placement of the lateral epicondyle (c′) is less easily explicable. Some or all of these derived features are also present in *Afropithecus* and *Kenyapithecus,* and can be inferred for *Turkanapithecus.* It establishes the hominoid status of these taxa. A strongly buttressed area for the insertion of brachialis on the ulna (A′) is somewhat equivocal in both its functional and phylogenetic implications. It is a feature of extant large hominoids (but also some platyrrhines, e.g., *Cebus*) and many of the fossil taxa (including *Aegyptopithecus*). It is not found in the highly suspensory hylobatids or *Ateles,* or in *Pliopithecus* or *Dendropithecus,* which also may have included suspensory activities in their positional repertoire. The feature presumably is associated with powerful flexion of the forearm, possibly as part of climbing or hoisting activities, but not of the type associated with arm-swinging or brachiation.

In addition to the derived features it shares with *Proconsul*, *Kenyapithecus* also has two additional features suggesting that it may be more derived within hominoids than *Proconsul*. In extant hominoids, a laterally facing radial notch of the ulna (z′) is associated with a large pronation–supination excursion. *Austriacopithecus* resembles *Kenyapithecus* in possessing this feature. The relative smallness of the radial fossa (i′) in *Kenyapithecus* may reflect the narrowness of the zona conoidea. The retention of many primitive features, particularly in the humeroulnar joint, suggests that *Turkanapithecus*, *Afropithecus*, *Proconsul*, *Kenyapithecus*, and *Austriacopithecus* are primitive hominoids, and less derived within hominoids than hylobatids.

In addition to the derived features mentioned above, *Sivapithecus* and *Dryopithecus* also share with extant hominoids the presence of a trochleiform humeral trochlea (k′, l′), associated with a coronoid fossa that is deep (j′) as well as large (i′). They are thus more derived within hominoids than the five fossil taxa discussed above. In addition, the anterior projection of the humeral medial trochlear keel is markedly greater than that of the lateral keel (v′) and, in *Dryopithecus*, the proximal ulnar shaft is anteroposteriorly shallow (C′). These features are shared with large extant hominoids to the exclusion of hylobatids, raising the possibility that *Sivapithecus* and *Dryopithecus* are more derived within hominoids than are hylobatids. However, hylobatids share numerous humeral features with large hominoids to the exclusion of *Sivapithecus* and *Dryopithecus*. These include features of the head (o′–r′), shaft (m′, n′, s′), and medial epicondyle (t′). The fossil taxa retain the primitive hominoid condition for features of the shaft (except, possibly, for m′ in *D. fontani*), the medial epicondyle, and (on the basis of indirect evidence) the head. Thus, unless hylobatids are convergent on large hominoids in these features, the fossil taxa are less derived within hominoids than are hylobatids. There are some indications that hylobatids may indeed be convergent on large hominoids in some of these features. Thus, the morphological features underlying torsion of the humeral head in hylobatids are different from those in large hominoids (Larson, 1988; Rose, 1989). The distal humeral articular surface is also distinctive in that there is a high cubital angle, the surface for articulation with the radius is relatively broad (due mainly to a particularly broad zona conoidea), the capitulum is more cylindrical than globular, the medial trochlear keel is not markedly developed in either anterior or distal view, and the posterior part of the trochlea extends laterally posterior to the capitulum and zona conoidea. The fact that large-bodied atelines have converged on hominoids in many of the same features shows that this constellation of features can be independently acquired in primates that practice specialized forelimb suspensory activities. The independent acquisition of these features in *Pliopithecus*, *Dendropithecus*, and *Simiolus* was suggested above.

While the evidence for the forelimb is equivocal, the majority of the evidence from other regions suggests that *Sivapithecus* and *Dryopithecus* are more derived within hominoids than hylobatids, and they are shown with this status in Fig. 2. The presence of a straight humeral shaft (m′) in *Dryopithecus* suggests that it may be more derived than *Sivapithecus* within large hominoids. Despite the fact that *Dryopithecus* and *Sivapithecus* share numerous similarities, the evidence from the humeral shaft, hamate, and the phalanges suggests a greater contribution of orthograde and suspensory activities in the *Dryopithecus* locomotor repertoire. The evidence of yet-to-be-described specimens may enable these locomotor differences to be more clearly delineated.

For all determinable features of the forelimb long bones, *Oreopithecus* shares the extant large hominoid condition. These include the features n′–u′, not present in *Sivapithecus* or *Dryopithecus*. *Oreopithecus* is thus the most derived of the fossil taxa considered here, and the only one that resembles a large hominoid of modern aspect. As none of the fossil taxa share distinctive forelimb features with particular large hominoids, the differences and possible convergences among large hominoid forelimb bones will not be discussed, although such considerations become necessary when the forelimb evidence is considered in conjunction with the evidence of other parts of the skeleton.

Discussion

While the identification of apomorphies, and their functional and phylogenetic significance form the focus of attention for this chapter and for the book in general, plesiomorphies also deserve some attention, especially when function, and changes in function are concerned. In Table II, most of the characters seen in *Cebus* and *Alouatta* are also found in *Aegyptopithecus, Pliopithecus,* and *Dendropithecus.* Many of them are also found in cercopithecids. In all of these taxa, these characters are most likely to have been retained from the primitive anthropoid condition. If the postcranium as a whole is considered, not just the forelimb features considered here, there are very many such plesiomorphies retained from the anthropoid morphotype, even in such Miocene hominoids as *Afropithecus, Proconsul,* and *Kenyapithecus.* In combination, these features form a functional complex that seems to underlie a locomotor capability most likely expressed predominantly in the form of above-branch arboreal quadrupedalism, for which platyrrhines such as *Cebus* and *Alouatta* are the closest living analogues (Fleagle, 1986; Rose, 1994, 1996). The persistence of so many of these features in a wide variety of Oligocene and Miocene taxa is an indication of the efficacy and adaptability of this functional complex.

As far as the forelimb is concerned, two major trends are evident in the acquisition of features that are derived with respect to the plesiomorph anthropoid pattern. The first involves the elongation of the forelimb, with associated changes, presumable reflecting the addition of some sort of suspensory and/or climbing capability to the positional repertoire. As interpreted here, this probably happened convergently several times during anthropoid evolution: in atelines, *Pliopithecus*, "small apes" such as *Dendropithecus* and *Simiolus*, hylobatids, and at least once in the clade that contains *Dryopithecus*, *Oreopithecus*, and extant large hominoids (Fig. 2).

It would seem that the derived features that define the earliest hominoids are not associated with marked suspensory capabilities. In the forelimb they constitute functional complexes that allow the stabilization of the humeroulnar and humeroradial joints, an increase in the mobility of the forearm and wrist, and an increase in manual dexterity. Some or all of these complexes can be identified in *Afropithecus*, *Kenyapithecus*, *Proconsul*, and *Turkanapithecus*. As suggested above, these functions might be associated with a predominantly quadrupedal negotiation of the arboreal support complex by relatively large, short-tailed or tailless hominoids. In *Dryopithecus*, *Oreopithecus*, and extant hylobatids and great apes, the addition of further derived features to these complexes produces some or all of a further stabilization of the humeroulnar joint, the addition of a hyperextension capability to the elbow joint, and a further increase in the mobility of the forearm and wrist. However, especially in *Oreopithecus* and extant apes, these functional complexes, initially associated with pronograde activities, are combined with other complexes of derived features (such as forelimb elongation, enhanced mobility of the shoulder, and broadening of the thorax) indicating the use of the forelimb in suspensory or other orthograde activities. Nevertheless, it is clear from, for example, the activities of African apes that many of these forelimb functions still enable effective pronograde, quadrupedal activities to be performed. The particular combinations of features found in *Dryopithecus*, *Oreopithecus*, and extant apes clearly differentiate them from the less emphasized suspensory specializations of atelines and some primitive catarrhines.

ACKNOWLEDGMENTS

I thank L. C. Aiello, D. R. Begun, and C. V. Ward for their comments on the submitted version of this chapter. The research reported here was funded by NSF grant SBR 9222526.

References

Aiello, L. C. 1981. Locomotion in the Miocene Hominoidea. In: C. B. Stringer (ed.), *Aspects of Human Evolution*, pp. 63–97. Taylor & Francis, London.

Andrews, P., and Simons, E. L. 1977. A new Miocene gibbon-like genus, *Dendropithecus* (Hominoidea, Primates) with distinctive postcranial adaptations: Its significance to origin of Hylobatidae. *Folia Primatol.* **28:**161–169.

Andrews, P., and Walker, A. 1976. The primate and other fauna from Fort Ternan, Kenya. In: G. L. Isaac and E. McCown (eds.), *Human Origins,* pp. 279–304. Benjamin, Menlo Park, CA.

Beard, K. C., Teaford, M. F., and Walker, A. 1986. New wrist bones of *Proconsul africanus* and *P. nyanzae* from Rusinga Island, Kenya. *Folia Primatol.* **47:**97–118.

Begun, D. R. 1987. *A Review of the Genus Dryopithecus.* Ph.D. dissertation, University of Pennsylvania.

Begun, D. R. 1988. Catarrhine phalanges from the Late Miocene (Vallesian) of Rudabánya, Hungary. *J. Hum. Evol.* **17:**413–438.

Begun, D. R. 1992. Phyletic diversity and locomotion in primitive European hominids. *Am. J. Phys. Anthropol.* **87:**311–340.

Begun, D. R. 1993. New catarrhine phalanges from Rudabánya (northeastern Hungary) and the problem of parallelism and convergence in hominoid postcranial morphology. *J. Hum. Evol.* **24:**373–402.

Begun, D. R. 1994. Relations among the great apes and humans: New interpretations based on the fossil great ape *Dryopithecus. Yearb. Phys. Anthropol.* **37:**11–63.

Begun, D. R., Teaford, M. F., and Walker, A. 1993. Comparative and functional anatomy of *Proconsul* phalanges from the Kasawanga Primate Site, Rusinga Island, Kenya. *J. Hum. Evol.* **25:**89–165.

Clark, W. E. L., and Leakey, L. S. B. 1951. The Miocene Hominoidea of East Africa. *Br. Mus. Nat. Hist. Fossil Mamm. Afr.* **1:**1–117.

Clark, W. E. L., and Thomas, D. P. 1951. Associated jaws and limb bones of *Limnopithecus macinnesi. Br. Mus. Nat. Hist. Fossil Mamm. Afr.* **3:**1–27.

Conroy, G. C. 1976. Primate postcranial remains from the Oligocene of Egypt. *Contrib. Primatol.* **8:**1–134.

Ehrensberg, K. 1938. *Austriacopithecus,* ein neuer menschen-affenartiger Primate aus dem Miozan von Klein-Hadersdorf bei Poysdorf in Niederösterreich (Nider-Donau). *Sitzungsber. Oesterr. Akad. Wiss. Math. Naturwiss. Kl.* **147:**71–110.

Feldsman, M. R. 1982. Morphometric analysis of the distal humerus of the some Cenzoic catarrhines: The late divergence hypothesis revisited. *Am. J. Phys. Anthropol.* **59:**173–195.

Ferembach, D. 1958. Les limnopithèques du Kenya. *Ann. Paleontol.* **44:**149–249.

Fleagle, J. G. 1983. Locomotor adaptations of Oligocene and Miocene hominoids and their phyletic implications. In: R. L. Ciochon and R. S. Corruccini (eds.), *New Interpretations of Ape and Human Ancestry,* pp. 301–324. Plenum Press, New York.

Fleagle, J. G. 1986. The fossil record of early catarrhine evolution. In: B. Wood, L. Martin, and P. Andrews (eds.), *Major Topics in Primate and Human Evolution,* pp. 130–149. Cambridge University Press, London.

Fleagle, J. G., and Simons, E. L. 1982. The humerus of *Aegyptopithecus zeuxis:* A primitive anthropoid. *Am. J. Phys. Anthropol.* **59:**175–193.

Gebo, D. L., Teaford, M. F., Walker, A., Larson, S. G., Jungers, W. L., and Fleagle, J. G. 1988. A hominoid proximal humerus from the Early Miocene of Rusinga Island, Kenya. *J. Hum. Evol.* **17:**393–401.

Harrison, T. 1982. *Small-Bodied Apes from the Miocene of East Africa.* Ph.D. dissertation, University of London.

Harrison, T. 1986. A reassessment of the phylogenetic relationships of *Oreopithecus bambolii* Gervais. *J. Hum. Evol.* **15:**541–583.

Harrison, T. 1991. The implications of *Oreopithecus bambolii* for the origins of bipedalism. In: Y. Coppens and B. Senut (eds.), *Origine(s) de la Bipédie chez les Hominidés. Cahiers de Paléoanthropologie,* pp. 235–244. Editions du CNRS, Paris.

Kelley, J. 1995. A functional interpretive framework for the early hominoid postcranium. *Am. J. Phys. Anthropol.* Suppl. **20:**205.

Kretzoi, M. 1975. New ramapithecines and *Pliopithecus* from the lower Pliocene of Rudabánya in north-eastern Hungary. *Nature* **257:**578–581.
Larson, S. G. 1988. Subscapularis function in gibbons and chimpanzees: Implications for interpretation of humeral head torsion in hominoids. *Am. J. Phys. Anthropol.* **76:**449–462.
Leakey, R. E., Leakey, M. G., and Walker, A. C. 1988a. Morphology of *Turkanapithecus kalakolensis* from Kenya. *Am. J. Phys. Anthropol.* **76:**277–288.
Leakey, R. E., Leakey, M. G., and Walker, A. C. 1988b. Morphology of *Afropithecus turkanensis* from Kenya. *Am. J. Phys. Anthropol.* **76:**289–307.
Lin, Y., Wang, S., Gao, Z., and Zhang, L. 1987. The first discovery of the radius of *Sivapithecus lufengensis* in China. *Geol. Rev.* **33:**1–4.
McCrossin, M. L. 1994. *The Phylogenetic Relationships, Adaptations, and Ecology of Kenyapithecus.* Ph.D. dissertation, University of California, Berkeley.
McHenry, H. M., and Corruccini, R. S. 1975. Distal humerus in hominoid evolution. *Folia Primatol.* **23:**227–244.
Meldrum, D. J., and Pan, Y. 1988. Manual proximal phalanx of *Laccopithecus robustus* from the Latest Miocene site of Lufeng. *J. Hum. Evol.* **18:**719–731.
Morbeck, M. E. 1983. Miocene hominoid discoveries from Rudabánya: Implications from the postcranial skeleton. In: R. L. Ciochon and R. S. Corruccini (eds.), *New Interpretations of Ape and Human Ancestry,* pp. 369–404. Plenum Press, New York.
Napier, J. R., and Davis, P. R. 1959. The forelimb skeleton and associated remains of *Proconsul africanus. Br. Mus. Nat. Hist. Fossil Mamm. Afr.* **16:**1–69.
Nengo, I. O., and Rae, T. C. 1992. New hominoid fossils from the early Miocene site of Songhor, Kenya. *J. Hum. Evol.* **23:**423–429.
Pilbeam, D. R., and Simons, E. L. 1971. Humerus of *Dryopithecus* from Saint Gaudens, France. *Nature* **229:**406–407.
Pilbeam, D. R., Rose, M. D., Badgley, C., and Lipschutz, B. 1980. Miocene hominoids from Pakistan. *Postilla* **181:**1–94.
Pilbeam, D. R., Rose, M. D., Barry, J. C., and Shah, S. M. I. 1990. New *Sivaphithecus* humeri from the Chinji Formation of Pakistan and the relationship of *Sivapithecus* and *Pongo. Nature* **348:**237–239.
Robertson, M. 1985. A comparison of pronation–supination mobility in *Proconsul* and *Pliopithecus vindobonensis. Am. J. Phys. Anthropol.* **66:**219.
Rose, M. D. 1983. Miocene hominoid postcranial morphology: Monkey-like, ape-like, neither, or both? In: R. L. Ciochon and R. S. Corruccini (eds.), *New Interpretations of Ape and Human Ancestry,* pp. 405–417. Plenum Press, New York.
Rose, M. D. 1984. Hominoid specimens from the Middle Miocene Chinji Formation, Pakistan. *J. Hum. Evol.* **13:**503–516.
Rose, M. D. 1986. Further hominoid postcranial specimens from the Late Miocene Nagri Formation of Pakistan. *J. Hum. Evol.* **15:**333–367.
Rose, M. D. 1988. Another look at the anthropoid elbow. *J. Hum. Evol.* **17:**193–224.
Rose, M. D. 1989. New postcranial specimens of catarrhines from the middle Miocene Chinji Formation, Pakistan: Descriptions and a discussion of proximal humeral functional morphology in anthropoids. *J. Hum. Evol.* **18:**131–162.
Rose, M. D. 1993a. Locomotor anatomy of Miocene hominoids. In: D. Gebo (ed.), *Postcranial Adaptation in Nonhuman Primates,* pp. 252–272. Northern Illinois University Press, De Kalb.
Rose, M. D. 1993b. Functional anatomy of the primate elbow and forearm. In: D. Gebo (ed.), *Postcranial Adaptation in Nonhuman Primates,* pp. 70–95. Northern Illinois University Press, De Kalb.
Rose, M. D. 1994. Quadrupedalism in Miocene hominoids. *J. Hum. Evol.* **26:**387–411.
Rose, M. D. 1996. Functional morphological similarities in the locomotor skeleton of Miocene catarrhines and platyrrhine monkeys. *Folia Primatol.* **66:**7–14.
Rose, M. D., Leakey, M. G., Leakey, R. E. F., and Walker, A. C. 1992. Postcranial specimens of

Simiolus enjiessi and other primitive catarrhines from the early Miocene of Lake Turkana, Kenya. *J. Hum. Evol.* **22:**171–237.

Sarmiento, E. E. 1985. *Functional Differences in the Skeleton of Wild and Captive Orang-Utans and their Adaptive Significance.* Ph.D. dissertation, New York University.

Sarmiento, E. E. 1987. The phylogenetic position of *Oreopithecus* and its significance in the origin of the Hominoidea. *Am. Mus. Novit.* **2881:**1–44.

Sarmiento, E. E. 1988. Anatomy of the hominoid wrist joint: Its evolutionary and functional implications. *Int. J. Primatol.* **9:**281–345.

Schultz, A. H. 1960. Einege Beobachtungen und Masse am Skelett von *Oreopithecus* im vergleich mit anderem catarrhinen Primaten. *Z. Morphol. Anthropol.* **50:**136–149.

Senut, B. 1989. *Le Coude Chez les Primates Hominoïdes: Anatomie, Fonction, Taxonomie et Évolution.* Cahiers de Paléoanthropologie, CNRS, Paris.

Simons, E. L., and Fleagle, J. 1973. The history of extinct gibbon-like primates. In: D. M. Rumbaugh (ed.), *Gibbon and Siamang,* Vol. 2, pp. 167–218. Karger, Basel.

Spoor, C. F., Sondaar, P. Y., and Hussain, S. T. 1992. A hominoid hamate and first metacarpal from the Late Miocene Nagri Formation of Pakistan. *J. Hum. Evol.* **21:**413–424.

Straus, W. L., Jr. 1963. The classification of *Oreopithecus.* In: S. L. Washburn (ed.), *Classification and Human Evolution,* pp. 146–177. Viking Press, New York.

Walker, A. C., and Pickford, M. 1983. New postcranial fossils of *P. africanus* and *P. nyanzae.* In: R. L. Ciochon and R. S. Corruccini (eds.), *New Interpretations of Ape and Human Ancestry,* pp. 325–351. Plenum Press, New York.

Ward, C. V., Walker, A., and Teaford, M. F. 1991. *Proconsul* did not have a tail. *J. Hum. Evol.* **21:**215–220.

Wu, R., Xu, Q., and Lu, Q. 1986. Relationship between Lufeng *Sivapithecus* and *Ramapithecus* and their phylogenetic position. *Acta Anthropol. Sin.* **5:**1–30.

Xiao, M. 1981. Discovery of fossil hominoid scapula at Lufeng, Yunnan. In: Yunnan Provincial Museum (ed.), *Collected Papers of the 30th Anniversary of the Yunnan Provincial Museum,* pp. 41–44. Yunnan Provincial Museum, Yunnan.

Zapfe, H. 1960. Die Primatenfunde aus der Miozžanen spaltenfüllung von Neudorf an der March (Devinská Nová Ves) Tschechoslowakei. *Schweiz. Palaeontol. Abh.* **78:**1–293.

Functional Anatomy and Phyletic Implications of the Hominoid Trunk and Hindlimb

6

CAROL V. WARD

Introduction

Traditional hominid phylogenies are based on craniodental and mandibular characteristics primarily because these elements comprise the bulk of the hominoid fossil record. Few postcranial elements are known for most fossil hominoids, complicating intertaxic comparisons. Another reason postcrania tend to be neglected in phylogenetic studies is the assumption that they are more responsive to selective and ontogenetic pressures than skulls and teeth, and thereby are more likely to reflect homoplasy and obscure phylogenetic conclusions. Explanations of homoplasy are generally invoked when postcranial analyses do not support phyletic schemes constructed using craniodental characters, e.g., in the cases of *Sivapithecus* (Pilbeam *et al.*, 1990) and *Oreopithecus* (Harrison, 1986; Sarmiento, 1987).

All of these factors have led some to believe that postcrania are less useful for phylogenetic analyses than are craniodental traits (but see Begun, 1993;

CAROL V. WARD • Anthropology and Pathology & Anatomical Sciences, University of Missouri, Columbia, Missouri 65211.

Function, Phylogeny, and Fossils: Miocene Hominoid Evolution and Adaptations, edited by Begun *et al.* Plenum Press, New York, 1997.

Harrison, 1987; McCrossin and Benefit, 1994; Sarmiento, 1987). However, because many primate lineages are characterized by postcranial apomorphies, postcrania can provide useful and important data for phylogenetic studies.

This chapter considers functional and phyletic implications of trunk and hindlimb morphology of extant and Miocene hominoids. First, I review the functional anatomy of the trunk and hindlimb of large-bodied Miocene hominoids. Second, I create a character tree using only data from the trunk and hindlimb for comparison with traditional hypotheses. Finally, I discuss implications of postcranial data for interpreting evolution of the hominoid skeleton in light of other recently proposed phylogenies based on craniodental and forelimb material.

Miocene Hominoid Anatomy and Its Functional Implications

Proconsul

One of the earliest and best-known Miocene hominoids is *Proconsul* (Table I). Four species are known, and postcranial fossils can be attributed to three (Kelley, 1986; Rafferty *et al.*, 1995; Ward *et al.*, 1993). All *Proconsul* species are similar in preserved parts, differing primarily in body size (Rafferty *et al.*, 1995; Teaford and Walker, 1993), and are treated here as functionally equivalent.

Both *P. heseloni* and *P. nyanzae* possessed a similar, monkeylike torso. The vertebral column was relatively long, as in most monkeys, with elongate vertebral bodies and probably six lumbar segments (Ward, 1991, 1993; Ward *et al.*, 1993). This elongated vertebral column was most likely accompanied by powerful spinal musculature, reflected in the comparatively wide iliac tuberosities and ventrally situated lumbar vertebral transverse processes. Elongate lumbar columns with powerful spinal musculature are associated with short ilia in *Proconsul* and most extant primates, enhancing flexibility of the torso. *Proconsul* differed from most monkeys in that its lumbar accessory processes were relatively small, and more similar to those of hylobatids (Kelley, 1986; Ward *et al.*, 1993). Reduced accessory process length may be a correlate of a lumbar column that had been reduced in number from seven to six segments.

In addition, the *Proconsul* torso was dorsoventrally deep and mediolaterally narrow. *Proconsul* had narrow, laterally facing iliac blades, a short pubis, and lumbar vertebral transverse processes arising from the vertebral body. Qualitative examination reveals rib curvature that best resembles that of cercopithecid monkeys (Walker and Pickford, 1983; personal observation), reflecting a narrow torso (Ward, 1991, 1993; Ward *et al.*, 1993). *Proconsul* also exhibits humeral shaft retroflexion and probably lacked a medially rotated humeral head (Rose, 1983). Among extant primates, these morphologies are associated with ventrally facing scapular glenoid fossae and scapulae that are

Table I. Medium to Large-Bodied Miocene Ape Taxa for Which Trunk and Hindlimb Fossils Have Been Attributed[a]

Taxon	Part of Miocene	Region	Mass (kg)	rib	sc/cl	hum	vert	pelv	fem	pat	tibia	fib	talus	calc	nav	cub	cun	MT	phal
Proconsul	early	E. Africa																	
P. heseloni			10.9[1]	*	*	*	*	*	*	*	*	*	*	*	*	*	*	*	*
P. nyanzae			35.6[1]				*	*	*		*	*	*	*					
P. major			75.1[1]								*		*	*					
Afropithecus	early	E. Africa																	
A. turkanensis			30–35						*			*	*			*		*	*
Rangwapithecus	early	E. Africa																	
R. gordoni			15[2]						*				*	*	*				
Turkanapithecus	early	E. Africa																	
T. kalakolensis			10[2]							*			*			*			
Sp. *incertae sedis*	middle	E. Africa	35–40				*												
Kenyapithecus	mid-late	E. Africa																	
K. africanus			11–22[3]			*	*	*	*	*		*	*	*				*	*
Dryopithecus	mid-late	Europe																	
D. brancoi			27–35[4,5]			*							*						*
D. laietanus			15–25[5]								*								
D. crusafonti			23–32[6]															*	
Pliopithecus	mid-late	Europe																	
P. vindobonensis			7[2]	*	*	*	*	*	*	*	*		*	*			*	*	*
Oreopithecus	late	Europe																	
O. bambolii			30[2]	*	*	*	*	*	*	*	*	*	*	*	*	*	*	*	*
Sivapithecus	late	Europe & Asia																	
S. indicus			50[7]			*			*				*	*			*	*	*
S. parvada			69[7]			*								*		*	*	*	

[a]Asterisk indicates that at least fragments of these elements are preserved for that species. Abbreviations: sc/cl = scapulae and clavicles; hum = humerus; vert = vertebrae; pelv = pelvis; fem = femur; pat = patella; fib = fibula; calc = calcaneus; nav = navicular; cub = cuboid; cun = cuneiforms; MT = metatarsals; phal = phalanges. Body mass estimates from: [1]Rafferty *et al.* (1995); [2]Fleagle (1988); [3]Rose *et al.* (1996); [4]Morbeck (1983); [5]Begun (personal communication); [6]Begun (1992b); [7]Kelley (1988). Others represent rough estimates from qualitative comparison with other taxa by the author.

placed in parasagittal planes on the lateral side of a narrow thoracic cage (Benton, 1969, 1974; Rose, 1993; Ward, 1991, 1993; Ward *et al.*, 1993).

All evidence of *Proconsul* torso morphology reflects a primitive, monkeylike torso with long, flexible vertebral column and narrow thoracic cage. These morphologies are mechanical adaptations to habitual pronograde. There are no obvious anatomical adaptations for habitual forelimb-dominated orthograde arboreal locomotion in the *Proconsul* torso.

Further evidence for habitual pronogrady in *Proconsul* is found in the even width of its acetabular lunate surfaces, apparent in both *P. heseloni* and *P. nyanzae* (Ward, 1991). In more habitually orthograde primates, the cranial aspect of the lunate surface is expanded, presumably reflecting greater loads passing through this portion of the hip joint (Miller and Gunnell, 1992; Ward, 1991, 1992; but see MacClatchy, 1995).

Proconsul lacked an external tail, as do extant hominoids, demonstrated by an ultimate sacral vertebra of *P. heseloni* from Rusinga Island (Ward *et al.*, 1991; contra McCrossin, 1994; McCrossin and Benefit, 1992). *Proconsul* also lacked callosity-bearing ischial tuberosities (Ward *et al.*, 1989, 1993; Ward, 1991, 1993; contra McCrossin, 1994a). Callosities are absent in most anthropoids, but are present in cercopithecids and hylobatids (Rose, 1974).

In the hindlimb, *Proconsul* exhibits joints adapted to a wide range of postures during weight-bearing activities. The femoral neck–shaft angle is high and greater trochanter is relatively low, permitting a wide range of hip joint abduction (Ward, 1992). The fovea capitis is located at the most proximomedial part of the head, and the head articular surface does not extend significantly onto the dorsal aspect of the neck. The femoral head is subspherical unlike that of extant hominoids, nor does its superior margin project as far from the surface of the neck. In femoral head morphology, *Proconsul* most closely resembles arboreal colobines (Rose, 1983; Ward, 1991). There is a prominent tubercle on the dorsal aspect of the neck (Rose, 1983). The function of this tubercle is unclear, but it is found in most platyrrhines, cercopithecids, and hylobatids. The trochanteric fossa opens posteriorly, and does not exhibit the deeply excavated morphology of hominoids and to a lesser extent atelines.

The *Proconsul* knee joint was functionally much like that of hominoids and platyrrhines. Despite the distorted nature of most of the relevant remains, the femoral condyles are roughly circular in lateral view. The condyles are only slightly asymmetrical, resembling those of monkeys and hylobatids. Circular condyles represent an adaptation to knee joints adapted to a variety of flexion–extension postures, and are found in hominoids and atelines. As in many cursorial and leaping mammals, most monkeys have anteroposteriorly deep condyles. This condition maximizes quadriceps moment about the knee in flexion, but this moment rapidly decreases as the knee extends (Lovejoy, 1990). Deep condyles are found in animals that habitually rely on powerful knee extension from a flexed position (Lovejoy, 1990). Hominoids, on the other hand, load their knees in a wider variety of postures, and their circular condyles provide an equivalent quadriceps moment at all positions of flexion–extension. *Proconsul* resembles all other hominoids in having knee joints adapted to load-bearing in a variety of postures.

Proconsul patellae resemble those of all extant hominoids, as well as *Oreopithecus, Kenyapithecus,* and *Pliopithecus* (Ward *et al.*, 1995). Its patellae are roughly equal in mediolateral and superoinferior dimensions. They are thin anteroposteriorly with little nonarticular surface area on the posterior aspect. This morphology confirms conclusions based on condylar geometry, indicating that running and leaping were not emphasized in *Proconsul* locomotion (Ward *et al.*, 1995).

In the ankle and foot, *Proconsul* is also generalized (Langdon, 1986; Clark and Leakey, 1951; Rose, 1983, 1993). Its talocrural joint morphology indicates habitually plantigrade foot postures and quadrupedal progression. The talocrural joint lacks the deeply grooved morphology of cercopithecids, but is not as flat as in extant apes and atelines. *Proconsul* also lacks an anteroposteriorly compressed talar facet on the tibia that characterizes these taxa. A narrow articular area on the great ape tibia may permit a wider range of flexion and extension at the talocrural joint. The *Proconsul* talar trochlea is slightly asymmetrical as in most platyrrhines and parapithecids, but not as extreme as in the condition found in cercopithecids.

The calcaneocuboid joint has a prominent cuboid peg and correspondingly deep pit on the distal calcaneus. This morphology reflects a high degree of midtarsal mobility, and although it is equaled only by *Pongo* among extant primates, is found in many Miocene taxa. The calcaneocuboid joint surface is of uniform width, indicating habitual loading in all degrees of inversion–eversion, rather than being specialized for any particular foot posture (Rose, 1986).

The subtalar joints in *Proconsul* are less like those of extant apes, with a pattern of subtalar mobility intermediate between that of hominoids and cercopithecids (Langdon, 1986; Rose, 1986; Sarmiento, 1983). The posterior talar facet on the calcaneus is large as in extant hominoids, but with a more transversely oriented axis like that of anthropoids outside the African ape–human clade. The anterior talocalcaneal joint is more tightly curved than that of cercopithecids, but less curved than that of extant hominoids. The kinematics of these joints are described in Rose (1986), and like *Sivapithecus, Proconsul* was probably characterized by a pattern of subtalar mobility and function like that of colobines (Rose, 1986; Sarmiento, 1983).

The naviculocuneiform joints of *Proconsul* are also intermediate in shape and orientation between those of extant hominoids and monkeys, but more closely resemble those of hominoids, implying adaptation to grasping foot postures during arboreal climbing and bridging and to plantigrade postures used during quadrupedalism on the ground or large supports. The tarsal bones are intermediate in length and overall morphology between that of propliopithecids and most ceboids on one hand, and that of hominoids and *Ateles* on the other. The cuboid of *Proconsul nyanzae* appears to be slightly longer and/or narrower than that of *P. africanus* (Teaford and Walker, 1993), which is contrary to the predicted relationship based on size differences. However, many catarrhine species exhibit similar amounts of intraspecific variation. Also, the rest of the *P. nyanzae* and *P. africanus* pedal skeletons are so similar the functional explanation for this apparent difference remains unclear. Cuboid mediolateral wedging is moderate in both species of *Proconsul;* in-between that of monkeys and extant hominoids. Pronounced cuboid wedging and short tarsals decrease the load arm of the pedal skeleton (Schultz, 1963; Strasser, 1992), and are associated with positioning the lateral aspect of the foot for hallucal grasping in extant hominoids and atelines. The tarsal skeleton of *Proconsul* suggests that the heel process did not contact the substrate during quadrupedal locomotion as in extant monkeys (Gebo, 1992, 1993; Preuschoft, 1973).

The hallux of *Proconsul,* as with all other Miocene hominoids, was robust and divergent, capable of powerful grasping (Fleagle, 1983; Langdon, 1976; Preuschoft, 1973; Rose, 1983). The hallux was proportionally much larger than in nonhominoid anthropoids. *Proconsul* exhibits a large groove for flexor hallucis longus on the calcaneus, robust fibula with inclined talar articular surface, medially oriented and saddle-shaped hallucal tarsometatarsal joint, and torsion of the hallucal and second metatarsal shafts. A large attachment area for peroneus longus is distinct on the base of the hallucal metatarsal. However, the hallucal metatarsal was not as robust as that of extant apes, *Sivapithecus* or *Oreopithecus.* In addition, the distal articular surface is not as extensive plantarly, and the sesamoid grooves are more pronounced. In these features, the *Proconsul* hallux is intermediate between the extant hominoid and primitive anthropoid conditions.

Afropithecus, Turkanapithecus, and Rangwapithecus

These three monospecific taxa are all known from the early Miocene of east Africa, as is *Proconsul* (Table I). However, none is as well known. They are distinct craniodentally, implying divergent dietary and ecological specializations (see Kay and Ungar, this volume, and Walker, this volume). There is also some variation in the forelimb among these taxa (Rose, this volume), but their preserved hindlimb anatomy is quite similar.

The hindlimb remains of *Afropithecus* are strikingly similar to those of *Proconsul nyanzae* (Leakey *et al.,* 1988; Rose *et al.,* 1996). In all apparent aspects of hindlimb joint morphology and skeletal proportions, *Afropithecus* appears functionally equivalent to *P. nyanzae* (see also Rose, this volume, and Walker, this volume), implying similar locomotor behaviors. The postcranial similarity stands in contrast to pronounced differences in cranial anatomy and inferred function.

Several postcranial specimens are known from a single *Turkanapithecus kalakolensis* individual (Leakey and Leakey, 1986; Leakey *et al.,* 1988). Of the few preserved hindlimb remains, only the femur differs substantially from that of *Proconsul* and *Afropithecus*. The *Turkanapithecus* femur appears less robust and may have been proportionally longer than in *Proconsul* (Rose, 1993). It has less expanded articular surfaces, and the proximal portion of the shaft is mediolaterally broader. Rose (1993, 1994) notes similarities to the femur of *Alouatta*. The talus and cuboid are morphologically equivalent to those of *Proconsul* and *Afropithecus*. The *Turkanapithecus* forelimb suggests a greater reliance on climbing (Rose, 1993, 1994), but the hindlimb was generally like those of other early Miocene apes.

Like *Afropithecus* and *Turkanapithecus*, *Rangwapithecus* closely resembles *Proconsul* in hindlimb morphology. One apparent difference is that *Rangwapithecus* appears to have a flatter anterior talar facet on the calcaneus than *Proconsul*, more like extant hominoids, despite being roughly the size of the smaller *P. nyanzae* specimens. However, *Rangwapithecus* also has a slightly more asymmetrical talar trochlea than *Proconsul*, *Afropithecus*, or *Turkanapithecus*, like some ceboid monkeys and primitive anthropoids. These differences are slight, and may more closely reflect individual variation than taxon-specific functional patterns.

Kenyapithecus

Several new postcranial fossils from Maboko Island, Kenya, attributed to *Kenyapithecus africanus* (McCrossin, 1994a; McCrossin and Benefit, 1994, this volume) and some from Nachola, Kenya, that remain unpublished (Rose *et al.*, 1996), expand our knowledge of the trunk and hindlimb anatomy of *Kenyapithecus* (Table I). No trunk or hindlimb fossils can be attributed to *K. wickeri* (Harrison, 1992). All *Kenyapithecus* fossils are morphologically similar (McCrossin and Benefit, this volume), so they will be discussed at the genus level.

Thoracic, lumbar, and sacral vertebrae of *Kenyapithecus* are all morphologically similar to those of *Proconsul* (Rose *et al.*, 1996). They have long bodies with ventral median keels and transverse processes originating from the centrum; features found only among pronograde extant primates. The sacrum is also much like that of *Proconsul*. The humeral shaft from Maboko appears to be retroflexed, as in *Proconsul*, *Sivapithecus*, and most monkeys (Pilbeam *et al.*, 1990; Rose, this volume), suggesting that *Kenyapithecus* proximal humeral morphology would have been similar to humeri of these other taxa as well, with a posteriorly directed humeral head. Vertebral and humeral morphology of *Kenyapithecus* reflects a mediolaterally narrow thoracic cage and craniocaudally elongate torso, like that found in *Proconsul* and *Sivapithecus*.

Femoral morphology of *Kenyapithecus* is generally similar to that of *Proconsul*. The neck appears to be slightly shorter than that of *Proconsul* and it is more gracile, but other features are functionally equivalent. A slightly shorter neck may indicate a greater emphasis on terrestriality in *Kenyapithecus*, a conclusion supported by analyses of other postcranial elements (McCrossin,

1994a,b; McCrossin and Benefit, 1994, this volume; Rose, personal communication). The patella is short, broad, and thin, suggesting knee morphology and function similar to *Proconsul* and all other extant and fossil nonhuman hominoids (McCrossin, 1994a; Ward *et al.*, 1995).

The talocrural joint was morphologically and functionally like that of *Proconsul nyanzae*. *Kenyapithecus* has a talar articular facet that is as broad anteroposteriorly as mediolaterally (McCrossin and Benefit, this volume; Rose *et al.*, 1996). It also exhibits a distinct median keel for articulation with a deeply grooved talar trochlea, also primitive for hominoids.

The first metatarsal is reported to be more robust than that of *Proconsul* (McCrossin, 1994a,b; McCrossin and Benefit, this volume). The larger hallux of *Kenyapithecus* does not reflect enhanced hallucal grasping, however, because the flat, distally oriented metatarsal facet on the entocuneiform reveals the presence of a habitually adducted hallux like that found in baboons (McCrossin, 1994a; McCrossin and Benefit, this volume). The morphology of the first metatarsal head also differs from that of extant hominoids with well-developed hallucal grasping (Rose *et al.*, 1996). An adducted hallux may be evidence that *Kenyapithecus* was more terrestrial than any other Miocene ape described so far. However, the large size of the hallucal metatarsal, robust fibula with hominoidlike talofibular orientation, strong grooves for hallucal flexor musculature, and presence of a medial heel process on the calcaneus (McCrossin, 1994a; McCrossin and Benefit, this volume; Rose *et al.*, 1996) are found in extant anthropoids with well-developed hallucal grasping, and suggest that *Kenyapithecus* may have retained significant pedal grasping capabilities for arboreal locomotion.

Moroto

A well-known palate from Moroto, Uganda, has been attributed to *Proconsul* (Allbrook and Bishop, 1963; Andrews, 1978; Pilbeam, 1969) and was later suggested informally to be more closely allied with *Afropithecus* (Leakey *et al.*, 1988). However, the Moroto palate differs from both genera in aspects of maxillary and dental anatomy (Kelley and Pilbeam, 1986; Leakey, 1963; Leakey *et al.*, 1988; Martin, 1981), rendering its current taxonomic status indeterminate.

A beautifully preserved lumbar vertebra is thought to be associated with the palate (Rose and Walker, 1968). The Moroto vertebra is much more modern apelike in details of vertebral morphology than those of *Proconsul* (Sanders and Bodenbender, 1994; Ward, 1991, 1993). It has a relatively shorter body, more posteriorly oriented transverse processes, and lacks accessory processes. These morphologies imply a shorter, stiffer lumbar spine and broad torso, features associated with forelimb-dominated climbing and suspension in extant anthropoids (Ward, 1991, 1993). The pattern of vertebral anatomy seen in the Moroto vertebra is correlated with a broad thorax, transversely oriented scapulae with laterally facing glenoid fossae, long clavicles, a high costal angle, broad, posteriorly facing iliac blades, and long ilia in extant hominoids (Benton, 1969, 1974; Ward, 1991, 1993). The Moroto vertebra is more similar to those of large hominoids than hylobatids or atelines, and implies a well-developed emphasis on forelimb-dominated arboreality. The Moroto fossils document the earliest evidence for hominoidlike torso anatomy in the primate fossil record.

Because *Proconsul* and *Afropithecus* are strikingly similar in most aspects of postcranial anatomy, and the Moroto vertebra implies a very different functional pattern than either, it is unlikely that the Moroto vertebra belongs to either genus (Ward, 1993). The Moroto fossils probably belonged to a species more derived postcranially and behaviorally than any early Miocene form. This Moroto taxon may be more closely related to extant great apes than were the earlier Miocene hominoids.

Sivapithecus

Sivapithecus is widely considered to be the sister taxon of *Pongo* based on numerous derived features of the craniofacial skeleton (see S. Ward, this volume). Postcranially, however, *Sivapithecus* and *Pongo* differ in functionally important ways. The proximal humerus of *Sivapithecus* is retroflexed, with a strong deltopectoral crest and, probably, a posteriorly directed humeral head (Pilbeam *et al.*, 1990). This morphology would place the humeral head in a position to articulate with ventrally facing glenoid fossae and narrow scapulae lying in parasagittal planes on a transversely narrow thorax. This humeral morphology implies the presence of a monkeylike thoracic cage and torso, given the strong correlation among these morphologies among extant primates (Rose, 1983; Ward, 1991, 1993). This torso anatomy, in turn, represents an adaptation to habitual pronograde. *Pongo*, on the other hand, resembles other hominoids with a wide thorax and apelike shoulder joints, reflecting its emphasis on forelimb-dominated arboreal locomotion.

Unlike the admittedly sparse indicators of torso anatomy, the *Sivapithecus* hindlimb is more derived than that of *Proconsul* and other early Miocene hominoids. The femur is similar to all large-bodied hominoids, with a hip joint that permitted considerable mobility and a knee adapted to weight-bearing in a wide range of flexion and extension (Rose, 1993, 1994).

The anterior calcaneus, cuboid, and cuneiforms are proximodistally shorter than those of early Miocene apes. The cuboid is as short as that of extant African apes; shorter than that of *Pongo*. The cuboid is not as strongly wedged as any extant hominoids, and the distal calcaneus is slightly longer (Rose, 1986). Short tarsals decrease the load arm of the lateral aspect of the foot for hallucal grasping (Schultz, 1963; Strasser, 1992). Enhanced pedal grasping in *Sivapithecus* is also implied by a robust hallux. The hallucal metatarsal and phalanges are more robust than those of earlier hominoids, with a more plantarly extensive distal articular surface on the hallucal metatarsal and smaller sesamoid grooves (Pilbeam *et al.*, 1980; Rose, 1993, 1994). This anatomy implies an emphasis on hallucal grasping in *Sivapithecus* like that of extant great apes. Enhanced grasping may suggest a greater reliance on climbing and perhaps suspension in *Sivapithecus*, but the long, narrow torso implies that habitually pronograde posture was retained.

Dryopithecus

Dryopithecus torso anatomy is known postcranially from a partial humerus lacking only its head from St. Gaudens, France (Begun, 1992a, 1994; Lartet, 1856). Although the proximal end is missing, Rose (1993, 1994, this volume) notes the medial inclination of the shaft when viewed anteriorly, and the base of the lesser trochanter that is aligned with the shaft. He interprets this to mean that the humeral head was posteriorly directed, as in early Miocene hominoids, *Sivapithecus*, and most nonhominoid anthropoids. However, this specimen has a straight shaft in lateral view, and Begun and Kordos (Chapter 14, this volume) and others (Evans and Krahl, 1945) infer that it had a medially rotated head, as do extant hominoids. A straight humeral shaft does not co-occur with a posteriorly directed head in any known anthropoid. Until more fossils are discovered, the shoulder anatomy of *Dryopithecus* cannot be inferred with confidence.

The talocrural joint of *Dryopithecus* was similar to that of *Proconsul* and early Miocene hominoids, with a keeled trochlea, a distally projecting medial malleolus, and an anteroposteriorly expanded talar facet on the tibia (Begun, 1994). The talus and entocuneiform are also like those of primitive hominoids, suggesting a frequently adducted hallux, but the metatarsals and phalanges are more like those of extant great apes. The metatarsals are robust, and most are reported to have apelike articular morphology. The phalanges are cured, and exhibit well-developed flexor ridges (Begun, 1994). These distal elements may imply a greater emphasis on pedal grasping in *Dryopithecus* when compared with earlier hominoids, but more needs to be known about the *Dryopithecus* skeleton before definite conclusions can be made.

Oreopithecus

Oreopithecus bambolii had the most derived great ape-like postcranium known for any Miocene hominoid (Gervais, 1872; Harrison, 1986; Hürzeler, 1949, 1960, 1968; Sarmiento, 1983, 1987; Schultz, 1960; Straus, 1961, 1963). In body proportions and overall morphology, it resembles extant great apes, although it lacks some extremely specialized morphologies (Harrison, 1986;

Sarmiento, 1987). *Oreopithecus* had broad scapulae, long clavicles, an acute costal angle, and broad iliac blades that flared laterally. This morphology reflects a mediolaterally broad torso, an adaptation to enhanced forelimb abduction–adduction. There were five sacral and lumbar vertebrae with no tail, as in hylobatids and hominids (Schultz, 1961; Straus, 1963). Despite the additional segment, *Oreopithecus* vertebrae lacked accessory processes, and appear to have transverse processes arising from the neural arch rather than the vertebral bodies like extant large hominoids.

The hindlimb reflects many of the generalized hominoid morphologies shared by earlier hominoids. The posterior talar facet is more mediolaterally directed and the talar trochlea was not as shallow as in extant apes. A mediolateral plantar arch, a feature found in all anthropoids except great apes and hylobatids, may not have been present in *Oreopithecus*. However, the hallux is robustly constructed, the talar head medially directed, and cuboid wedged, the tarsal bones short, and the foot axis passes through the second digit. These morphologies are adaptations to strong hallucal grasping as in extant hominoids, stronger than in earlier apes.

Despite the presence of some primitive features, in all important functional aspects, the hindlimb and trunk morphology of *Oreopithecus* was strikingly similar to those of extant great apes. *Oreopithecus* was adapted for climbing and below-branch arboreal locomotion as are extant hominoids.

Methods

In order to assess the phylogenetic implications of the variation in hindlimb and torso anatomy seen in Miocene hominoids, a cladistic analysis was performed using only characters from these anatomical regions. This exercise is not meant to demonstrate actual phyletic relations among extant and fossil hominoid taxa. Instead, it provides a character tree to compare with published phylogenetic hypotheses to explore apparently divergent patterns of cranial and postcranial evolution within the Hominoidea.

The analysis follows methodology described by Begun and Kordon (Chapter 14, this volume). The most parsimonious phylogenies are constructed using all available data from the trunk and hindlimb, treating all characters equally. Functional implications of each likely scheme are considered, and used to evaluate these phylogenies. Based on this integration of phylogenetic and functional anatomical data, the evolution of hominoid trunk and hindlimb anatomy is considered.

Miocene hominoid postcranial elements reflecting trunk and hindlimb anatomy are attributed to *Proconsul, Afropithecus, Turkanapithecus, Rangwapithecus, Kenyapithecus, Pliopithecus, Dryopithecus, Oreopithecus,* and *Sivapithecus* (Table I). The Moroto hominoid is not included in the analysis because of the paucity of characters available for this taxon. Fossil taxa are treated at the generic level because of the scarcity of postcranial elements for many of the taxa, and the fact that little species-level variability has been demonstrated within these genera (but see Teaford and Walker, 1993). All extant hominoid genera are included. Hominids include only *Australopithecus* and *Homo*.

Characters chosen for analysis (Table II) reflect important functional complexes in the trunk and hindlimb, and/or have previously-demonstrated taxonomic relevance. Characters chosen are apparent in at least one fossil taxon. Uninformative characters and those suspected of varying simply because of body size are ignored. Characters are independent of one another and uniform within genera. Multistate characters are treated as unordered to remain conservative.

For all extant and Kenyan fossil taxa, character states were collected from original specimens. In other cases, casts were examined and/or character states reported in the literature were used. Character states identified from casts were checked against available written descriptions of the original material. Character states are listed in Table II for all taxa.

Polarity is decided using outgroup analysis. Homoplasies and homologies are determined by analyzing the preferred hypothesis generated by character analysis. Finding a suitable outgroup among extant anthropoids for this analysis is problematic. Cercopithecids have derived postcranial morphology (Harrison, 1989), which could obscure accurate determination of character polarity. Similarly, ceboids are variable postcranially, and many are considered to be highly derived and frequently homoplastic with extant hominoids (Fleagle, 1983; Ford, 1988; Rose, 1974, 1983). Propliopithecidae from the Oligocene are considered to exhibit postcranial anatomy that is primitive for anthropoids (Conroy, 1976; Fleagle, 1975; Fleagle and Simons, 1982; see review in Gebo, 1993), so they represent an appropriate outgroup for studying hominoids. Postcrania are known for two propliopithecid genera, *Aegyptopithecus* and *Propliopithecus*. These taxa have functionally equivalent postcranial anatomy, and can be lumped to represent all Propliopithecidae (Gebo, 1993; Gebo and Simons, 1987). However, some skeletal elements are not known for propliopithecids, complicating their use as an outgroup. For the present analysis, missing character states are filled in by determining the primitive anthropoid condition from character distribution among all living and fossil primates. This method proves satisfactory because for all missing characters, polarity determination was straightforward.

Relationships were calculated using PAUP version 3.1 (Swofford, 1993), which was instructed to retain only maximally parsimonious trees. Consistency indices were calculated and character evolution analyzed using MacClade version 3.0 (Maddison and Maddison, 1992). The only constraint imposed was monophyly of the chimp–gorilla–human clade. This procedure was necessary because of numerous unique hominid morphologies of hominids related to habitual bipedality.

Results of Phylogenetic Analysis of Trunk and Hindlimb Data

The phylogenetic analysis produced five equally parsimonious trees. A consensus tree was calculated that summarizes relationships consistent across all trees (Fig. 1). The trees have a length of 121+ steps, a consistency index of 68%, and a retention index of 77%. These results presented here are based on a limited number of characters from a subset of the skeleton. Thus, as discussed below, they do not necessarily represent realistic hypotheses of relationships among hominoids. Only a complete analysis with all possible characters could best reflect true evolutionary relationships (see Begun, Ward, and Rose, Chapter 18, this volume). The present analysis is a character tree designed only to examine trunk and hindlimb evolution in hominoids.

Despite the limitations of this exercise, this scheme provides important information. All Miocene taxa under investigation in this study share characters derived with respect to propliopithecids, supporting their position in the stem-based clade Hominoidea (*sensu* Williams and Kay, 1995) (Andrews, 1992; Walker and Teaford, 1988).

The five equally parsimonious trees reflect uncertainty of the relations between *Proconsul, Afropithecus, Kenyapithecus,* and *Turkanapithecus* at node 4 (Fig. 1). The trunk and hindlimb show no derived traits linking any of these early Miocene hominoids specifically with any extant taxa. These taxa are grouped within one clade on the basis of their intermediate level of distal fibular robusticity and flexor ridges on the phalanges, and on the pronounced sesamoid grooves on the hallucal metatarsal heads. Had this analysis been performed using ordered characters, characters setting these four taxa apart from the rest of hominoids would correctly be interpreted as morphologies intermediate between the primitive catarrhine condition and that of extant hominoids. *Rangwapithecus* differs from the other early Miocene apes in only one character, namely, the retention of a more asymmetrical talar trochlea. However, the rest of its morphologies are equivalent to these other early Miocene taxa.

Kenyapithecus appears to exhibit autapomorphic terrestrial adaptations not seen in other early Miocene apes (McCrossin, 1994a,b; McCrossin and Benefit, this volume), as may *Rangwapithecus,* although most of its trunk and hindlimb characters are like the other early Miocene apes.

All other nodes are more robust, requiring at least two additional evolutionary steps to switch the position of any taxa. *Pliopithecus* is placed as the sister taxon of extant hominoids, based on its many similarities with hylobatids (Table II). Many of these similarities are also found in atelines, and may represent homoplasy rather than homology (see discussion below). *Sivapithecus* and *Dryopithecus* are linked at node 8 by characters that reflect their intermediate development of pedal grasping. Neither shares more synapomorphies with extant apes.

Although not included in the numeric analysis, the lack of lumbar ver-

Table II. Characters Used in Phylogenetic Analysis[a]

	Character	Hominid	*Pan*	*Gorilla*	*Pongo*	Hylo
1	humeral head orientation (0 = posterior, 1 = medial)	1	1	1	1	1
2	humeral shaft (0 = flexed, 1 = straight)	1	1	1	1	1
3	costal angle (0 = low, 1 = high)	1	1	1	1	1
4	vertebral body height (0 = tall, 1 = intermediate, 2 = short)	2	2	2	2	1
5	accessory processes (0 = large, 1 = small, 2 = absent)	2	2	2	2	1
6	transverse processes (0 = ventral, 1 = intermediate, 2 = dorsal)	2	2	2	2	1
7	lumbar vertebral count (actual numbers listed)	5	4	4	4	5
8	sacral vertebral count (actual numbers listed)	5	6	6	6	5
9	presence of a tail (0 = present, 1 = absent)	1	1	1	1	1
10	sternebrae (0 = narrow, 1 = intermediate, 2 = broad)	2	2	2	2	1
11	iliac blade breadth (0 = narrow, 1 = intermediate, 2 = wide)	2	2	2	2	1
12	iliac blade angle (0 = low, 1 = intermediate, 2 = high)	2	2	2	2	2
13	lower iliac height (0 = short, 1 = intermediate, 2 = long)	0	2	2	2	2
14	cranial lunate surface (0 = narrow, 1 = intermediate, 2 = wide)	2	2	2	2	1
15	pubic length (0 = short, 1 = medium, 2 = long)	2	2	2	2	1
16	trochanteric fossa (0 = open, 1 = intermediate, 2 = dogs, 3 = unique)	3	2	2	2	1
17	femoral head (0 = cylinder, 1 = intermediate, 2 = sphere)	2	2	2	2	2
18	tubercle on femoral neck (0 = present, 1 = absent)	1	1	1	1	0
19	AP depth of femor condyles (0 = deep, 1 = shallow)	0	1	1	1	1
20	femoral condyles (0 = symmetrical, 1 = intermediate, 2 = asymmetrical)	2	2	2	2	0
21	femoral robusticity (0 = gracile, 1 = intermediate, 2 = robust)	0	1	1	1	1

[a]Abbreviations: Hominid = Hominidae, Hylo = Hylobatidae, Prop = hypothetical ancestor based on Propliopithecidae (see below and text), *Proc* = *Proconsul*, *Afro* = *Afropithecus*, *Rang* = *Rangwapithecus*, *Keny* = *Kenyapithecus*, *Turk* = *Turkanapithecus*, Moro = Moroto hominoid, *Dryo* = *Dryopithecus*, *Plio* = *Pliopithecus*, *Siva* = *Sivapithecus*, *Oreo* = *Oreopithecus*. Hypothetical ancestor (Prop) was reconstructed using *Propliopithecus* and *Aegyptopithecus*, which probably exhibit primitive anthropoid postcranial morphology (Conroy, 1976); Fleagle, 1983; Gebo, 1993). Character states not known for any propliopithecid taxon were filled in by inferring ancestral anthropoid character states from character distribution among extant primates. Character states from all taxa were determined by examination of original specimens when possible. In all other cases, casts were examined and/or character states were taken from the literature.

Table II. (*Continued*)

	Prop	*Proc*	*Afro*	*Rang*	*Keny*	*Turk*	Moro	*Dryo*	*Plio*	*Siva*	*Oreo*
1	0	0	?	?	0	?	?	?	0	0	1
2	0	0	?	?	0	?	?	1	1	0	1
3	0	0	?	?	?	?	?	?	?	?	1
4	0	0	?	?	0	?	2	?	1	?	2
5	0	1	?	?	?	?	2	?	0	?	2
6	0	0	?	?	0	?	2	?	1	?	2
7	0	1	?	?	?	?	?	?	0	?	2
8	0	?	?	?	?	?	?	?	0	?	1
9	0	1	?	?	?	?	?	?	0	?	1
10	0	0	?	?	?	?	?	?	1	?	2
11	0	0	?	?	?	?	?	?	0	?	2
12	0	0	?	?	?	?	?	?	0	?	2
13	0	0	?	?	?	?	?	?	0	?	2
14	0	0	?	?	?	?	?	?	?	?	?
15	0	0	?	?	?	?	?	?	?	?	2
16	0	0	?	0	0	0	?	?	0	1	1
17	0	1	?	1	1	1	?	?	1	1	2
18	0	0	?	0	0	?	?	?	0	?	?
19	0	1	?	?	?	1	?	?	1	?	1
20	0	1	?	?	?	1	?	?	0	?	2
21	1	2	?	?	2	1	?	?	0	2	2

(*continued*)

Table II. (*Continued*)

	Character	Hominid	*Pan*	*Gorilla*	*Pongo*	Hylo
22	shape of distal tibial facet (0 = square, 1 = short AP)	0	1	1	1	1
23	medial malleolar projection (0 = distal, 1 = flared)	1	1	1	1	1
24	fibular robusticity (0 = thin, 1 = robust)	1	1	1	1	0
25	lateral malleolus (0 = small, 1 = intermediate, 2 = large)	0	2	2	2	2
26	talar trochlea depth (0 = deep, 1 = intermediate, 2 = shallow)	2	2	2	2	2
27	talar trochlea (0 = symmetrical, 1 = intermediate, 2 = asymmetrical)	2	2	2	2	2
28	talar neck angle (0 = absent, 1 = present)	0	1	1	1	1
29	dorsoplantar height of talus (0 = tall, 1 = short)	1	1	1	1	1
30	distal calcaneus (0 = long, 1 = intermediate, 2 = short)	2	2	2	2	1
31	flexor hallucis longus grooves (0 = small, 1 = intermediate, 2 = large)	1	2	2	2	1
32	posterior calcaneal contact in gait (0 = heel-off, 1 = heel-down)	1	1	1	1	0
33	posterior talar facet long axis (0 = aligned, 1 = angled)	1	1	1	0	0
34	anterior talar facet (0 = curved, 1 = intermediate, 2 = flat)	2	2	2	2	2
35	plantar calcaneal tubercle (0 = small, 1 = large)	1	1	1	1	0
36	calcaneo-navicular facet (0 = large, 1 = small)	1	1	1	1	1
37	largest area of calcaneocuboid joint (0 = medial, 1 = even, 2 = lateral)	0	0	0	2	0
38	cuboid peg (0 = small, 1 = intermediate, 2 = large)	1	1	1	2	1
39	wedging of cuboid (0 = slight, 1 = stronger)	1	1	1	1	0
40	length of cuboid (0 = long, 1 = short)	0	1	1	1	1
41	entocuneiform facet on navicular (0 = dorsal, 1 = intermediate, 2 = distal)	2	0	0	0	?
42	MT 1 joint on entocuneiform (0 = distal, 1 = medial)	0	1	1	1	1
43	length of cuneiforms (0 = long, 1 = short)	0	1	1	0	0
44	MT 1 size (0 = gracile, 1 = intermediate, 2 = robust)	2	2	2	2	1
45	sesamoid grooves on MT 1 head (0 = small, 1 = large)	1	0	0	0	0
46	shape of MT 1 head (0 = symmetrical, 1 = asymmetrical)	0	1	1	1	1
47	position of MT 1 head (0 = aligned, 1 = twisted)	0	1	1	1	1

Table II. (*Continued*)

	Prop	*Proc*	*Afro*	*Rang*	*Keny*	*Turk*	Moro	*Dryo*	*Plio*	*Siva*	*Oreo*
22	0	0	?	0	0	?	?	1	0	?	?
23	0	0	?	?	0	?	?	0	0	0	1
24	0	1	1	?	1	?	?	?	?	?	1
25	0	1	1	?	1	?	?	?	?	?	?
26	0	1	1	1	1	1	?	1	2	1	2
27	0	1	1	0	1	1	?	1	1	1	2
28	0	1	1	1	1	?	?	1	0	?	1
29	0	1	1	1	1	1	?	1	1	1	1
30	0	0	?	0	?	?	?	?	0	1	2
31	0	1	?	1	1	?	?	1	1	2	2
32	0	0	?	0	?	?	?	?	0	0	0
33	0	0	?	0	0	?	?	?	0	0	0
34	0	1	?	2	1	?	?	?	2	2	2
35	0	0	?	0	0	?	?	?	1	1	1
36	0	0	?	0	0	?	?	?	0	1	1
37	1	1	1	1	1	?	?	?	2	1	1
38	0	2	2	2	?	2	?	?	0	2	2
39	0	0	0	?	?	0	?	?	?	0	1
40	0	0	0	?	?	0	?	?	?	1	1
41	1	1	?	1	?	?	?	1	1	?	0
42	0	1	?	?	0	?	?	1	0	?	1
43	0	0	?	?	?	?	?	0	0	0	?
44	0	1	?	1	1	?	?	?	1	2	?
45	0	1	1	?	1	?	?	?	?	0	?
46	0	1	1	?	1	?	?	?	1	1	?
47	0	1	1	1	?	?	?	?	?	?	?

(*continued*)

Table II. (*Continued*)

	Character	Hominid	*Pan*	*Gorilla*	*Pongo*	Hylo
48	MT 1 length (0 = long, 1 = short)	0	1	1	1	1
49	prehallux facet on MT 1 (0 = present, 1 = absent)	1	1	1	1	0
50	MT 2–5 robusticity (0 = gracile, 1 = robust)	1	1	1	1	0
51	transverse arch in foot (0 = present, 1 = absent)	0	1	1	1	?
52	axis of foot (0 = through digit 3, 1 = through digit 2)	1	1	1	1	0
53	phalangeal robusticity (0 = gracile, 1 = robust)	1	1	1	1	0
54	phalangeal curvature (0 = straight, 1 = curved)	0	1	1	1	1
55	phalangeal flexor ridges (0 = weak, 1 = intermediate, 2 = strong)	2	2	2	2	2

tebral accessory processes and vertebral body proportions of the Moroto vertebra places this hominoid within the large hominoid clade, probably at or near node 9. No characters beyond those identifiable on the vertebra are available for Moroto. However, given consistent correlations between torso shape (including characters of the clavicle, scapula, humerus, and pelvis) and vertebral anatomy among all anthropoids, these characters suggest that Moroto would have shared more characters with hominoids were more of the skeleton known. Thus, its placement at node 9 (Fig. 1) is considered to be reasonable. When more is known about this animal, perhaps from new scapular and femoral fossils from Moroto sites I and II recently reported by MacClatchy *et al.* (1995), its position relative to the other members of its clade may be resolved.

Oreopithecus shares numerous derived features with great apes, reflecting advanced adaptations to below-branch arboreality and strong pedal grasping (Harrison, 1986; Sarmiento, 1987), suggesting a close relationship with extant great apes. Its retention of five lumbar and sacral vertebrae places it as the sister taxon to extant great apes and humans.

Evolutionary Implications

The phylogenetic scheme based on the trunk and hindlimb differs from most other recently proposed hypotheses in a number of important ways. First, trunk and hindlimb characters cannot distinguish *Proconsul, Afropithecus, Kenyapithecus,* and *Rangwapithecus.* Because the functional pattern exhib-

Table II. (*Continued*)

	Prop	*Proc*	*Afro*	*Rang*	*Keny*	*Turk*	Moro	*Dryo*	*Plio*	*Siva*	*Oreo*
48	0	0	0	?	0	?	?	?	0	0	0
49	0	0	0	?	1	?	?	?	0	1	1
50	0	0	0	?	0	?	?	1	0	1	1
51	0	0	0	?	?	?	?	?	?	0	1
52	0	0	0	?	0	?	?	?	0	?	1
53	0	0	0	?	0	?	?	1	0	1	1
54	0	0	0	?	0	?	?	1	1	1	1
55	0	1	1	?	1	?	?	1	2	1	2

ited by the hindlimb and torso skeletons of these taxa appears to be virtually ubiquitous in the early hominoid fossil record, and because these taxa share so many features with propliopithecids and ceboids, it is reasonable to assume that this morphology represents the primitive hominoid condition, as suggested by numerous authors (e.g., Andrews, 1992; Andrews and Martin, 1991; Conroy and Fleagle, 1972; Harrison, 1993; Rose, 1993; Walker and Teaford, 1989).

Early Miocene apes have been shown to have distinctive craniodental and mandibular anatomy, suggesting a diversity of ecological and dietary adaptations among these taxa. *Proconsul* and *Afropithecus* are clearly distinguishable based on craniofacial and dental morphology (Andrews, 1992; Andrews and Martin, 1991; Leakey and Leakey, 1986; Leakey *et al.*, 1988), yet their trunk and hindlimb skeletons are strikingly similar. *Afropithecus* (Leakey and Walker, this volume) was probably a hard-object feeder, while *Proconsul* was a more generalized frugivore (Walker, this volume), suggesting that dietary specialization need not be accompanied by postcranial diversification. The *Rangwapithecus* dentition suggests folivory (Kay, 1977), and shares certain characters with *Oreopithecus* (Harrison, 1986), demonstrating that dietary similarities are not always reflected in the postcranium. It may be that the generalized nature of the early Miocene ape skeleton permitted these animals to exploit a variety of niches, facilitating the ecological and dietary specializations evident in their craniodental anatomy, and leading in part to the broad radiation of hominoids early in the Miocene.

Kenyapithecus is characterized by thick enamel, strong anterior dentition, and a strong inferior mandibular torus, implying tough foods in the diet and perhaps nut-cracking (McCrossin and Benefit, 1993, 1994). Based on these features, *Kenyapithecus* has been suggested to be closely related to the great ape

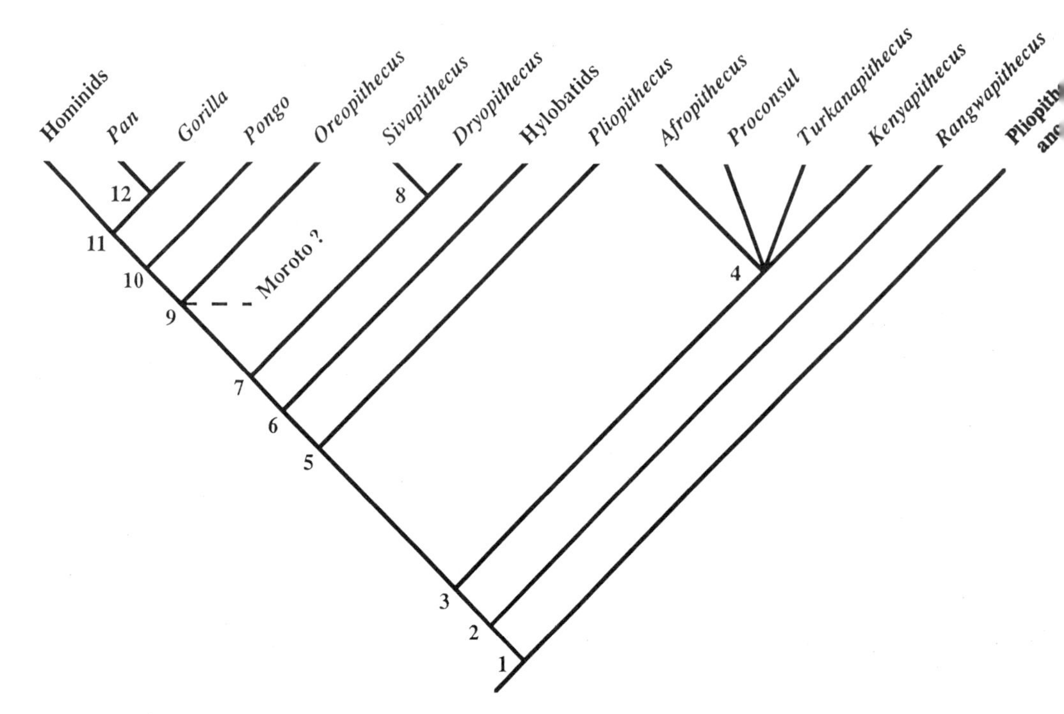

Fig. 1. Character tree depicting results of phylogenetic analysis of 55 trunk and hindlimb characters. Hypothesized position of Moroto hominoid is represented by the dashed line. Moroto was not included in the numeric analysis beause it is known from too few skeletal elements. Characters are listed in Table II.

and human clade on the basis of mandibular and dental morphology (Andrews and Martin, 1987, 1991; Brown and Ward, 1988; Greenfield, 1979; Martin, 1986; McCrossin, 1994a; McCrossin and Benefit, 1993, 1994, this volume). However, *Kenyapithecus* exhibits no derived hominoidlike traits of the trunk and hindlimb (McCrossin, 1994a,b; McCrossin and Benefit, this volume). Begun and Kordos (Chapter 14, this volume) suggests than thick enamel may be convergent among several hominoid lineages as a result of selection for a hard-food diet. Thus, the phylogenetic status of *Kenyapithecus* remains equivocal.

This analysis places *Pliopithecus* as the sister taxon to extant hominoids. It has been widely recognized that *Pliopithecus* and hylobatids share numerous postcranial features, but that many of its postcranial features are similar to platyrrhines, and therefore primitive (Conroy and Rose, 1983; Fleagle, 1983; McHenry and Corruccini, 1976; Rose, 1993). *Pliopithecus* retains many features that are primitive for catarrhines, such as a humeral entepicondylar foramen, hinged carpometacarpal joint, tubular external auditory meatus, and a tail (Fleagle, 1983; Harrison, 1987; Napier, 1961, 1962). These features have led many authors to conclude that *Pliopithecus* represents a hominoid (e.g., Fleagle, 1983; Szalay and Delson, 1979), and even a catarrhine (e.g., Begun, 1992b; Harrison, 1987) sister taxon.

Another distinctive difference between the character tree presented here and others is that this tree (fig. 1) proposes no special relationship between *Sivapithecus* and *Pongo. Sivapithecus* is generally considered to represent the sister taxon of *Pongo* based on a suite of craniofacial characters (Andrews and Cronin, 1982; Andrews and Martin, 1991; Brown and Ward, 1988; Greenfield, 1979; Martin, 1986; Pilbeam, 1982; Ward and Kimbel, 1983) (Fig. 1). However, *Sivapithecus* and *Pongo* share no uniquely derived features of the trunk or hindlimb. *Sivapithecus* is considerably more primitive than *Pongo* in torso shape and pedal characters, lacking many synapomorphies of extant great apes and hylobatids. If *Sivapithecus* does represent the sister taxon of *Pongo,* extensive homoplasy must have occurred in hominoid postcranial evolution (Andrews, 1992; Pilbeam *et al.,* 1990).

Three hypotheses can be proposed to explain the distribution of characters in *Sivapithecus* and *Pongo.* According to hypothesis 1, it is possible that *Sivapithecus* and *Pongo* lack a close phyletic relationship. If so, numerous craniofacial and dental characters seen in both *Sivapithecus* and *Pongo* must be homoplasies (Andrews and Cronin, 1982; Andrews, 1992; Andrews and Martin, 1991; Brown and Ward, 1988; Greenfield, 1979; Martin, 1986; Pilbeam, 1982; Schwartz, 1984; Ward and Pilbeam, 1983). This hypothesis is considered unlikely by several researchers given the number of craniodental characteristics representing multiple functional complexes shared by *Sivapithecus* and *Pongo* alone (see discussion in S. Ward, this volume).

Hypotheses 2 and 3 assume that *Sivapithecus* and *Pongo* belong to a monophyletic clade. According to hypothesis 2, *Sivapithecus* may have reverted to the primitive condition in torso shape, tarsal length, and certain aspects of pedal morphology from a more derived, *Pongo*-like configuration. These anatomical changes may have resulted from selection for pronograde quadrupedalism. It is notable that primitive morphologies that reevolved in *Sivapithecus* according to this scenario are quite similar to those found in the earliest hominoids. So, although this second hypothesis requires the fewest evolutionary steps, it is not obviously superior to the other alternatives given the amount of information currently available.

According to hypothesis 3, which also assumes *Sivapithecus–Pongo* monophyly, derived hominoidlike morphologies of the trunk and foot seen in extant hominoids may have evolved independently in hylobatids, *Pongo,* and the African ape–human clade, with *Sivapithecus* retaining the primitive condition. This hypothesis of homoplasy among extant apes requires more steps than hypothesis 2, but functional arguments can be made to support it.

Numerous hominoid postcranial apomorphies have evolved independently in atelines (34 out of 56 included in the present analysis). These characters represent adaptations to similar loading regimes during locomotion, principally related to an emphasis on forelimb abduction–adduction and hindlimb grasping during arboreality (Benton, 1969, 1974; Kelley, 1986;

Rose, 1983, 1988; Sarmiento, 1987; Ward, 1993). The appearance of this suite of characters in two independent lineages provides strong support for a functional link with forelimb-dominated arboreality. Further evidence is suggested by subtle anatomical differences found in *Pongo* versus the African apes, such as the retention of a ventral median keel on the vertebrae in *Pongo* but the lack of a keel in African apes, and a comparatively broad iliac tuberosity in *Pongo*. All three hypotheses presented here are reasonable. However, further fossil evidence should provide more reliable data with which to exclude one or more of these hypotheses.

Similarly, *Dryopithecus* has been suggested to be a member of the large hominoid clade, recently either as a sister taxon of *Pongo* (Moyà-Solà and Köhler, 1993) or the *Pan–Homo* clade (Begun, 1992b, 1993; Begun and Kordos, Chapter 14, this volume). However, it lacks many of the postcranial apomorphies of these taxa. The phyletic position of *Dryopithecus* could provide critical information allowing us to resolve our interpretation of the evolution of many apparent hominoid synapomorphies. If *Dryopithecus* proves to have a narrow, monkeylike torso and relatively long pedal skeleton, yet belongs within the great ape–human clade, it would support the hypothesis that hylobatids, *Pongo,* and African apes evolved many of their similarities in parallel. On the other hand, retention of primitive morphologies could suggest that *Dryopithecus* shared no unique ancestry with modern hominoids, and instead represents a more primitive form, despite certain derived craniofacial characters (Begun, 1992b; Begun and Kordos, Chapter 14, this volume). Alternatively, if *Dryopithecus* had a great ape-like postcranium, combined with the cranial morphology would present convincing evidence of an association between *Dryopithecus* and extant hominoids.

The Moroto hominoid fits into this tree at or near node 9. The Moroto fossils have been referred to both *Proconsul* (Allbrook and Bishop, 1963; Andrews, 1978; Pilbeam, 1969) and *Afropithecus* (Leakey *et al.,* 1988). Figure 1 illustrates the difference in level of postcranial adaptation between Moroto and the early Miocene hominoids, supporting the conclusion that it did not belong to either of these taxa (Ward, 1991, 1993).

The phyletic position of *Oreopithecus* also has important implications for hominoid locomotor evolution. Some workers have suggested that *Oreopithecus* belongs with the Cercopithecoidea (see also Gervais, 1872; Rosenberger and Delson, 1985; Szalay and Delson, 1979). If they are correct, *Oreopithecus* would represent another case of striking convergence in the postcranial skeleton (Harrison, 1986; Sarmiento, 1987). However, most workers now agree that *Oreopithecus* is closely related to extant hominoids (Harrison, 1986; Hürzeler, 1958; Rose, this volume; Sarmiento, 1987; Schultz, 1960; Szalay and Langdon, 1985; Straus, 1963; this chapter), and its morphologies represent synapomorphies with extant hominoids. Although it has many great ape characters, *Oreopithecus* retains five lumbar and five sacral vertebrae, suggesting that it represents the sister taxon to extant great apes and humans (node 9, Fig. 1).

The pattern of phyletic relationships and locomotor anatomy in Miocene hominoids has critical implications for the interpretation of hominid origins. Until phyletic relationships of Miocene hominoids are resolved, synapomorphies and homoplasies cannot reliably be identified. Hominids lacks several derived characters shared by great apes, and exhibit many primitive features. *Australopithecus* and early *Homo* had six lumbar vertebrae, and modern humans have five. All hominids have five sacral vertebrae, craniocaudally short pelvises with wide iliac tuberosities and blades that face dorsolaterally. Hominids also share with many nonhominoids plantar tubercles on the calcaneus, aspects of the navicular and cuneiform joints, aspects of hallucal morphology, and metacarpal facets on the proximal phalanges that face dorsoproximally, rather than proximally as in other extant hominoids. These hominid morphologies are adaptations to habitual bipedality (Lovejoy *et al.*, 1973; Lovejoy, 1975, 1989; Latimer *et al.*, 1987; Latimer and Lovejoy, 1989). Pedal morphologies shared with nonhominoids may reflect habitual terrestriality, and an absence of powerful hallucal grasping. Primitive characters can also be found in the hominid upper limb, e.g., relatively long thumbs and short fingers (Napier, 1961, 1993).

Two explanations for this character distribution are possible. Either (1) hominids reverted to these primitive features from a more derived, African apelike morphology as a result of selection for habitual bipedality or (2) these characters are primitive features retained by selection for bipedality. Hypothesis 1 is most parsimonious, especially if molecular data continue to support a chimp–human clade (Ruvulo *et al.*, 1991; Sibley and Ahlquist, 1987; Sibley *et al.*, 1990). Hypothesis 1 is also supported by shared derived features of hominids, *Pan*, and *Gorilla* such as the fusion of the os centrale in the wrist. Hypothesis 2 assumes that *Pan* and *Gorilla* are convergent on *Pongo* in many postcranial characters, as one hypothesis suggests based on *Sivapithecus* morphology. Hypothesis 2 is more reasonable if a *Pan–Gorilla* clade is assumed (Marks, 1992, 1993; Smouse and Li, 1987; Templeton, 1983). Although it is less parsimonious, the second hypothesis implies fewer evolutionary reversals in the hominoid lineage.

Choosing between these competing hypotheses depends entirely on interpretation of hominoid relationships, and affects polarity determination for early australopithecines. New *Ardipithecus ramidis* and *Australopithecus anamensis* fossils from the early Pliocene may clarify relations among middle and late Miocene hominoids, and help to determine character polarity among hominids.

Summary: Locomotor Evolution within the Hominoidea

Regardless of the ambiguities that plague interpretation of hominoid phylogeny, certain generalizations about locomotor evolution can be made. If the trunk and hindlimb data are considered on their own merits, it appears as if hominoids became progressively more specialized for forelimb-dominated climbing and bridging locomotion. Early hominoids were generalized arboreal quadrupeds, and probably reflect the primitive hominoid condition, given their similarities to platyrrhines and *Aegyptopithecus* (Conroy and Rose, 1983; Harrison, 1987). Subsequent hominoid evolution produced an increased emphasis on pedal grasping and perhaps climbing or hanging behaviors to some extent in *Pliopithecus*, and more so in *Sivapithecus* and *Dryopithecus*. Not until the appearance of *Oreopithecus* and perhaps the Moroto hominoid, did a broad torso, short spine, shoulders adapted for effective abduction–adduction, and a foot axis through the second digit appear, reflecting habitual below-branch arboreality.

If trunk and hindlimb data are integrated with information from skulls, teeth, and upper limbs, it becomes apparent that this behavioral and morphological transition occurred to some extent in at least one and perhaps up to six lineages (hylobatids, atelines, *Pliopithecus, Oreopithecus, Pongo, Pan, Gorilla*) within anthropoids. Independent evolution of this suite of characters in this many lineages would be striking. However, homoplasy has certainly occurred at least in atelines, implying a strong functional link between forelimb-dominated arboreality and this suite of postcranial characters in anthropoids.

These changes in hominoid locomotor anatomy have been independent of dietary evolution. Some hominoid taxa, particularly those at node 4 (Fig. 1) with distinctly different dental, mandibular, and cranial anatomy have functionally similar postcrania. In other cases, taxa sharing craniodental characters have distinctly different postcrania (e.g., *Oreopithecus, Rangwapithecus*, and cercopithecids). Among hominoids, cranial and postcranial evolution are frequently uncoupled. It is for this reason that great care must be taken to consider all aspects of anatomy when reconstructing hominoid phylogeny.

Conclusions

In conclusion, there are many uncertainties in hominoid phylogeny that obscure our ability to accurately interpret the origins and evolution of hominoid taxa. The phylogeny presented here differs in important ways from those proposed on the basis of dental, mandibular, and cranial anatomy. In several accounts, it is probably unlikely to accurately reflect hominoid relationships because it was constructed using only a partial data set, namely, trunk and hindlimb characters. However, several important conclusions drawn from this analysis can be summarized as follows:

1. Early Miocene "hominoids" *Proconsul, Afropithecus, Turkanapithecus*, and *Kenyapithecus* fit within the stem-based clade Hominoidea.
2. These four taxa cannot be distinguished on the basis of trunk and hindlimb anatomy, and differ from *Rangwapithecus* in only one feature. In most ways, their skeletons probably reflect the primitive hominoid condition.

3. The trunk and hindlimb of *Pliopithecus* shares many similarities with hylobatids, but these may represent homoplasy.
4. *Sivapithecus* shares derived features of the pedal skeleton with extant hominoids, but probably retained a primitive, narrow thorax. Based on extensive craniofacial similarities, *Sivapithecus* probably represent the sister taxon of *Pongo*. So, the derived torso morphology and certain aspects of the hindlimb of *Pongo* suggest either that *Pongo*, hylobatids, and African apes acquired these characters independently, or that *Sivapithecus* reverted to the primitive condition.
5. *Dryopithecus* may or may not have retained a primitive shoulder joint. If it did, inclusion of *Dryopithecus* in the African ape and human clade would support a hypothesis of postcranial homoplasy among extant hominoids.
6. The Moroto hominoid is closely allied functionally with great apes, humans, and *Oreopithecus*. Although this conclusion is based on vertebral data alone, there are sufficient correlations between these vertebral characters and the overall pattern of torso morphology characterizing large hominoids to make this conclusion reasonable at the present time. Moroto provides the earliest evidence for a derived extant hominoidlike postcranium among hominoids.
7. *Oreopithecus* shared numerous postcranial features with great apes and humans, indicating that it may represent the sister taxon of this group. However, the autapomorphic craniodental anatomy of *Oreopithecus* complicates interpretation of its phyletic relations.
8. Hominids retain several primitive characters of the trunk and hindlimb that represent habitual terrestrial bipedality which are either primitive retentions or reversions to the primitive condition.

Because it is based on a limited data set, the character tree presented here should not be interpreted as representing results of a complete phyletic analysis of all possible data. However, this analysis of trunk and hindlimb data reveals important aspects of hominoid evolution. These conclusions can be compared and integrated with other schemes depicting hominoid evolution constructed using other data sets. Such comparison reveals that whichever hypothesis is chosen, homoplasy must have occurred in several hominoid lineages, and probably in several parts of the skeleton. Until we can accurately determine hominoid relationships, we cannot correctly identify these homoplasies. Only by combining all possible data can we resolve hominoid evolutionary history.

References

Allbrook, D., and Bishop, W. W. 1963. New fossil hominoid material from Uganda. *Nature* **197:**1187–1190.

Andrews, P. 1978. A revision of the Miocene Hominoidea of east Africa. *Bull. Br. Mus. Nat. Hist Geol.* **30:**85–224.

Andrews, P. 1992. Evolution and environment in the Hominoidea. *Nature* **360:**641–646.

Andrews, P., and Cronin, J. 1982. The relationships of *Sivapithecus* and *Ramapithecus* and the evolution of the orangutan. *Nature* **297:**541–546.

Andrews, P., and Martin, L. 1987. The phyletic position of the Ad Dabtiyah hominoid. *Bull. Brit. Mus. Nat. Hist. Geol.* **41:**383–393.

Andrews, P., and Martin, L. 1991. Hominoid dietary evolution. *Philos. Trans. R. Soc. London Ser. B* **334:**199–209.

Begun, D. R. 1992a. Phyletic diversity and locomotion in primitive European hominids. *Am. J. Phys. Anthropol.* **87:**311–340.

Begun, D. R. 1992b. Miocene fossil hominids and the chimp–human clade. *Science* **257:**1929–1933.

Begun, D. R. 1993. Phyletic affinities and functional convergence in European Miocene hominoids and living hominids. *Am. J. Phys. Anthropol. Suppl.* **16:**54.

Begun, D. R. 1994. Relations among the great apes and humans: New interpretations based on the fossil great ape *Dryopithecus. Yearb. Phys. Anthropol.* **37:**11–63.

Benton, R. 1969. Morphological evidence for adaptations within the epaxial region of the primates. In H. Vagtborg (ed.), *The Baboon in Medical Research,* pp. 10–20. University of Texas Press, Houston.

Benton, R. S. 1974. Structural patterns in the Pongidae and Cercopithecidae. *Yearb. Phys. Anthropol.* **18:**65–88.

Brown, B., and Ward, S. C. 1988. Basicranial and facial topography in *Pongo* and *Sivapithecus.* In: J. H. Schwartz (ed.), *Orang-utan Biology,* pp. 247–260. Oxford University Press, London.

Clark, W. E. L., and Leakey, L. S. B. 1951. The Miocene Hominoidea of East Africa. *Br. Mus. Nat. Hist. Fossil Mamm. Afr.* **1:**1–117.

Conroy, G. C. 1976. Primate postcranial remains from the Oligocene of Egypt. *Contrib. Primatol.* **8:**1–134.

Conroy, G. C., and Fleagle, J. G. 1972. Locomotor behavior in living and fossil pongids. *Nature* **237:**103–104.

Conroy, G. C., and Rose, M. D. 1983. Evolution of the primate foot from the earliest primates to the Miocene hominoids. *Foot Ankle* **3:**342–364.

Evans, F. G., and Krahl, V. E. 1945. The torsion of the humerus: A phylogenetic study from fish to man. *Am. J. Anat.* **76:**303–337.

Fleagle, J. G. 1975. Ape limb bone from the Oligocene of Egypt. *Science* **189:**135–137.

Fleagle, J. G. 1983. Locomotor adaptations of Oligocene and Miocene hominoids and their phyletic implications. In: R. L. Ciochon and R. S. Corruccini (eds.), *New Interpretations of Ape and Human Ancestry,* pp. 301–324. Plenum Press, New York.

Fleagle, J. G. 1988. *Primate Adaptation and Evolution.* New York: Academic Press.

Fleagle, J. G., and Simons, E. L. 1982. The humerus of *Aegyptopithecus zeuxis:* A primitive anthropoid. *Am. J. Phys. Anthropol.* **59:**175–193.

Ford, S. M. 1988. Postcranial adaptations of the earliest platyrrhine. *J. Hum. Evol.* **17:**155–192.

Gebo, D. L. 1992. Plantigrady and foot adaptation in African apes: Implications for hominid origins. *Am. J. Phys. Anthropol.* **89:**29–58.

Gebo, D. L. 1993. Postcranial anatomy and locomotor adaptation in early African anthropoids. In: D. Gebo (ed.), *Postcranial Adaptation in Nonhuman Primates,* pp. 220–234, Northern Illinois University Press, De Kalb.

Gebo, D. L., and Simons, E. L. 1987. Morphology and locomotor adaptation of the foot in early Oligocene anthropoids. *Am. J. Phys. Anthropol.* **74:**83–101.

Gervais, P. 1872. Sur singe fossile, d'espece non encore décrite, qui a été découvert au Monte Bamboli. *C. R. Acad. Sci.* **74:**1217–1223.

Greenfield, L. O. 1979. On the adaptive pattern of "*Ramapithecus.*" *Am. J. Phys. Anthropol.* **50:**527–548.

Harrison, T. 1986. A reassessment of the phylogenetic relationships of *Oreopithecus bambolii* Gervais. *J. Hum. Evol.* **15:** 541–583.

Harrison, T. 1987. The phylogenetic relationships of the early catarrhine primates: A review of the current evidence. *J. Hum. Evol.* **16:**41–80.

Harrison, T. 1989. New postcranial specimens of *Victoriapithecus* from the middle Miocene of Kenya. *J. Hum. Evol.* **18:**537–557.

Harrison, T. 1992. A reassessment of the taxonomic and phylogenetic affinities of the fossil catarrhines from Fort Ternan, Kenya. *Primates* **33(4):**501–522.

Harrison, T. 1993. Cladistic concepts and the species problem in hominoid evolution. In: W. H. Kimbel and L. B. Martin (eds.), *Species, Species Concepts and Primate Evolution,* pp. 345–371. Plenum Press, New York.

Hürzeler, J. 1949. Neubieschreibung von *Oreopithecus bambolii* Gervais. *Schweiz. Palaeontol. Abh.* **66:**1–20.

Hürzeler, J. 1958. *Oreopithecus bambolii* Gervais: A preliminary report. *Verh. Naturforsch. Ges. Basel* **69:**1–48.

Hürzeler, J. 1960. The significance of *Oreopithecus* in the genealogy of man. *Triangle* **4:**164–174.

Hürzeler, J. 1968. Questions et réflections sur l'histoire des anthropomorphes. *Ann. Paleontol.* **54(2):**13–41.

Kay, R. F. 1977. Diets of early Miocene African hominoids. *Nature* **268:**628–630.

Kelley, J. 1986. Species recognition and sexual dimorphism in *Proconsul* and *Rangwapithecus*. *J. Hum. Evol.* **15:**461–495.

Kelley, J. 1988. A new large species of *Sivapithecus* from the Siwaliks of Pakistan. *J. Hum. Evol.* **17:**305–324.

Kelley, J., and Pilbeam, D. 1986. The dryopithecines: Taxonomy, comparative anatomy and phylogeny of Miocene large hominoids. In: D. R. Swindler and J. Irwin (eds.), *Comparative Primate Biology, vol. 1: Systematics, Evolution and Anatomy,* pp. 361–411. Liss, New York.

Langdon, J. H. 1986. Functional morphology of the Miocene hominoid foot. *Contrib. Primatol.* **22:**1–255.

Lartet, E. 1856. Note sure un grand singe fossile qui se rattache au groupe des singes superieurs. *C. R. Acad. Sci.* **43:**219–223.

Latimer, B., and Lovejoy, C. O. 1989. The calcaneus of *Australopithecus afarensis* and its implications for the evolution of bipedality. *Am. J. Phys. Anthropol.* **78:**369–386.

Latimer, B., Ohman, J. C., and Lovejoy, C. O. 1987. Talocrural joint in African hominoids: Implications for Australopithecus afarensis. *Am. J. Phys. Anthropol.* **74:**155–175.

Leakey, L. S. B. 1963. East African fossil Hominoidea and the classification within this superfamily. In: S. L. Washburn (ed.), *Classification and Human Evolution,* pp. 32–49. Methuen, London.

Leakey, R. E., and Leakey, M. G. 1986. A new Miocene hominoid from Kenya. *Nature* **324:**143–145.

Leakey, R. E., Leakey, M. G., and Walker, A. 1988. Morphology of *Afropithecus turkanensis* from Kenya. *Am. J. Phys. Anthropol.* **76:**289–307.

Lovejoy, C. O. 1975. Biomechanical perspectives on the lower limb of early hominids. In: R. H. Tuttle (ed.), *Primate Functional Morphology and Evolution,* pp. 291–326. Mouton, Paris.

Lovejoy, C. O. 1988. Evolution of human walking. *Sci. Am.* **259:**118–175.

Lovejoy, C. O. 1990. Review of Possible Animal Models for Kinematic/Utility Testing of the Pfizer Anterior Cruciate Ligament Prosthesis (PPCLP). Unpublished report written for Howmedica Division, Pfizer Pharmaceutical Corp.

Lovejoy, C. O., Heiple, K. G., and Burstein, A. H. 1973. The gait of *Australopithecus*. *Am. J. Phys. Anthropol.* **38:**757–780.

MacClatchy, L. M. 1995. *A Three-Dimensional Analysis of the Functional Morphology of the Primate Hip Joint.* Ph.D. dissertation, Harvard University.

MacClatchy, L. M., Gebo, D. L., and Pilbeam, D. R. 1995. New primate fossils from the Lower Miocene of northeast Uganda. *Am. J. Phys. Anthropol. Suppl.* **20:**139.

McCrossin, M. L. 1994a. *The Phylogenetic Relationships, Adaptations, and Ecology of Kenyapithecus.* Ph.D. dissertation, University of California, Berkeley.

McCrossin, M. L. 1994b. Semi-terrestrial adaptations of *Kenyapithecus*. *Am. J. Phys. Anthropol. Suppl.* **18:**142–143.

McCrossin, M. L., and Benefit, B. R. 1992. Comparative assessment of the ischial morphology of *Victoriapithecus macinnesi*. *Am. J. Phys. Anthropol.* **87:**277–290.

McCrossin, M. L., and Benefit, B. R. 1993. Recently discovered *Kenyapithecus* mandible and its implications for great ape and human origins. *Proc. Natl. Acad. Sci. USA* **90:**1962–1966.
McCrossin, M. L., and Benefit, B. R. 1994. Maboko Island and the evolutionary history of Old World monkeys and apes. In: R. S. Corruccini and R. L. Ciochon (eds.), *Integrative Paths to the Past: Paleoanthropological Advances in Honor of F. Clark Howell,* pp. 95–124. Prentice–Hall, Englewood Cliffs, NJ.
McHenry, H. M., and Corruccini, R. S. 1976. Affinities of Tertiary hominoid femora. *Folia Primatol.* **26:**136–150.
Maddison and Maddison. 1992. *MacClade 3.0: Analysis of phylogeny and character evolution.* Sinauer Associates, Sunderland, MA.
Marks, J. 1992. Beads and string: The genome in evolutionary theory. In: E. J. Devor (ed.), *Molecular Applications in Biological Anthropology.* Cambridge University Press, London.
Marks, J. 1993. Hominoid heterochromatin: Terminal C-bands as a complex genetic trait linking chimpanzee and gorilla. *Am. J. Phys. Anthropol.* **90:**237–246.
Martin, L. 1981. New specimens of *Proconsul* from Koru, Kenya. *J. Hum. Evol.* **10:**139–150.
Martin, L. 1986. Relationships among extant and extinct great apes and humans. In: B. Wood, L. Martin, and P. Andrews (eds.), *Major Topics in Primate and Human Evolution,* pp. 161–187. Cambridge University Press, London.
Miller, E. R., and Gunnell, G. F. 1992. Femoral morphology of *Smilodectes mcgrewi. Am. J. Phys. Anthropol.* **14:**123–124.
Morbeck, M. E. 1983. Miocene hominoid discoveries from Rudabánya: implications from the postcranial skeleton. In R. L. Ciochon and R. S. Corruccini (eds.), *New Interpretations of Ape and Human Ancestry.* New York: Plenum Press, pp. 369–404.
Moyà-Solà, S., and Köhler, M. 1993. Recent discoveries of *Dryopithecus* shed new light on evolution of great apes. *Nature* **365:**543–545.
Napier, J. 1961. Prehensility and opposability in the hands of primates. *Symp. Zool. Soc. London* **134:**647–657.
Napier, J. 1962. Fossil hand bones form Olduvai Gorge. *Nature* **196:**409–411.
Napier, J. 1993. *Hands.* Revised by R. H. Tuttle. Princeton University Press, Princeton, NJ.
Pilbeam, D. R. 1969. Tertiary Pongidae of east Africa: Evolutionary relationships and taxonomy. *Peabody Mus. Nat. Hist. Bull.* **31:**1–185.
Pilbeam, D. 1982. New hominoid skull material from the Miocene of Pakistan. *Nature* **295:**232–234.
Pilbeam, D., Rose, M. D., Badgley, C., and Lipschutz, B. 1980. Miocene hominoids from Pakistan. *Postilla* **181:**1–94.
Pilbeam, D., Rose, M. D., Barry, J. C., and Shah, S. M. I. 1990. New *Sivapithecus* humeri from Pakistan and the relationship of *Sivapithecus* and *Pongo. Nature* **348:**237–239.
Preuschoft, H. 1973. Body posture and locomotion in some east African Miocene Dryopithecinae. In: M. H. Day (ed.), *Human Evolution (Symposium of the Society for the Study of Human Biology, Vol. XI),* pp. 13–46. Taylor & Francis, London.
Rafferty, K. L., Walker, A., Ruff, C. B., Rose, M. D., and Andrews, P. J. 1995. Postcranial estimates of body weight in *Proconsul,* with a note on a distal tibia of *P. major* from Napak, Uganda. *Am. J. Phys. Anthropol.* **97:**391–402.
Rose, M. D. 1974. Ischial tuberosities and ischial callosities. *Am. J. Phys. Anthropol.* **40:**375–384.
Rose, M. D. 1983. Miocene hominoid postcranial morphology: Monkey-like, ape-like, neither, or both? In: R. L. Ciochon and R. S. Corruccini (eds.), *New Interpretations of Ape and Human Ancestry,* pp. 405–417. Plenum Press, New York.
Rose, M. D. 1986. Further hominoid postcranial specimens from the late Miocene Nagri Formation of Pakistan. *J. Hum. Evol.* **15:**333–367.
Rose, M. D. 1988. Another look at the anthropoid elbow. *J. Hum. Evol.* **17:**193–224.
Rose, M. D. 1993. Locomotor anatomy of Miocene hominoids. In: D. Gebo (ed.), *Postcranial Adaptation in Nonhuman Primates,* pp. 252–272. Northern Illinois University Press, De Kalb.
Rose, M. D. 1994. Quadrupedalism in some Miocene catarrhines. *J. Hum. Evol.* **26:**387–411.

Rose, M. D., and Walker, A. 1968. Fossil hominoid vertebra from the Miocene of Uganda. *Nature* **217:**980–981.

Rose, M. D., Ishida, H., and Nakano, Y. *Kenyapithecus* postcranial specimens from Nachola, Kenya. (in preparation).

Rosenberger, A. L., and Delson, E. 1985. The dentition of *Oreopithecus bambolii:* Systematic and paleobiological implications. *Am. J. Phys. Anthropol.* **66:**222–223.

Ruvulo, M., Disotell, T. R., Allard, M. W., Brown, W. M., and Honeycutt, R. L. 1991. Resolution of the African hominoid trichotomy by use of a mitochondrial gene sequence. *Proc. Natl. Acad. Sci. USA* **88:**1570–1574.

Sanders, W. J., and Bodenbender, B. E. 1994. Morphometric analysis of lumbar vertebra UMP 67-28: Implications for spinal function and phylogeny of the Miocene Moroto hominoid. *J. Hum. Evol.* **26:**203–238.

Sarmiento, E. E. 1983. The significance of the heel process in anthropoids. *Int. J. Primatol.* **4:**127–152.

Sarmiento, E. E. 1987. The phylogenetic position of *Oreopithecus* and its significance in the origin of the Hominoidea. *Am. Mus. Novit.* **2881:**1–44.

Schultz, A. H. 1960. Einege Beobachtungen und Masse am Skelett von *Oreopithecus* im vergleich mit anderem catarrhinen Primaten. *Z. Morphol. Anthropol.* **50:**136–149.

Schultz, A. H. 1961. Vertebral column and thorax. *Primatologia* **4:**1–66.

Schultz, A. H. 1963. Relations between the lengths of the main parts of the foot skeleton in primates. *Folia Primatol.* **1:**150–171.

Schwartz, J. H. 1984. Hominoid evolution: A review and a reassessment. *Curr. Anthropol.* **25:**655–672.

Sibley, C. G., and Ahlquist, J. E. 1987. DNA hybridisation evidence of hominoid phylogeny. *J. Mol. Evol.* **26:**99–121.

Sibley, C. G., Comstock, J. Q., and Ahlquist, J. E. 1990. DNA hybridization evidence for hominoid phylogeny: A re-analysis of the data. *J. Mol. Evol.* **30:**202–236.

Smouse, P. E., and Li, W.-H. 1987. Likelihood analysis of mitochondrial restriction-cleavage patterns for the human–chimpanzee–gorilla trichotomy. *Evolution* **41:**1162–1176.

Strasser, E. 1992. Hindlimb proportions, allometry, and biomechanics in Old World monkeys (Primates, Cercopithecidae). *Am. J. Phys. Anthropol.* **87:**187–213.

Straus, W. L. 1961. Primate taxonomy and *Oreopithecus. Science* **133:**760–761.

Straus, W. L. 1963. The classification of *Oreopithecus.* In: S. L. Washburn (ed.), *Classification and Human Evolution,* pp. 146–177. Viking Press, New York.

Swofford, D. 1993. PAUP manual.

Szalay, F. S., and Delson, E. 1979. *Evolutionary History of the Primates.* Academic Press, New York.

Szalay, F. S., and Langdon, J. H. 1985. Evolutionary morphology of the foot in Oreopithecus. *Am. J. Phys. Anthropol.* **66:**237.

Teaford, M. F., and Walker, A. 1993. Proconsul: Function and phylogeny. *Am. J. Phys. Anthropol. Suppl.* **16:**194.

Templeton, A. R. 1983. Phylogenetic inference from restriction endonuclease cleavage site maps with particular reference to the evolution of humans and the apes. *Evolution* **37:**221–224.

Walker, A., and Pickford, M. 1983. New postcranial fossils of *Proconsul africanus* and *Proconsul nyanzae.* In: R. Ciochon and R. Corruccini (eds.), *New Interpretations of Ape and Human Ancestry,* pp. 325–351. Plenum Press, New York.

Walker, A., and Teaford, M. F. 1988. The Kaswanga Primate Site: An Early Miocene hominoid site on Rusinga Island, Kenya. *J. Hum. Evol.* **17:**539–544.

Walker, A., and Teaford, M. F. 1989. The hunt for *Proconsul. Sci. Am.* **260:**76–82.

Ward, C. V. 1991. *The Functional Anatomy of the Lower Back and Pelvis of the Miocene Hominoid* Proconsul nyanzae *from Mfangano Island, Kenya.* Ph.D. thesis, The Johns Hopkins University.

Ward, C. V. 1992. Hip joints of *Proconsul nyanzae* and *P. africanus. Am. J. Phys. Anthropol. Suppl.* **14:**171.

Ward, C. V. 1993. Torso morphology and locomotion in catarrhines: Implications for the positional behavior of *Proconsul nyanzae. Am. J. Phys. Anthropol.* **92:**291–328.

Ward, C. V., Walker, A., Teaford, M. F., and Odhiambo, I. 1989. *Proconsul nyanzae* innominate from the early Miocene of Mfangano Island, Kenya. *Am. J. Phys. Anthropol.* **78:**319–320.

Ward, C. V., Walker, A., and Teaford, M. F. 1991. *Proconsul* did not have a tail. *J. Hum. Evol.* **21:**215–220.

Ward, C. V., Walker, A., Teaford, M. F., and Odhiambo, I. 1993. A partial skeleton of *Proconsul nyanzae* from Mfangano Island, Kenya. *Am. J. Phys. Anthropol.* **90:**77–111.

Ward, C. V., Ruff, C. B., Walker, A., Teaford, M. F., Rose, M. D., and Nengo, I. O. 1995. Functional morphology of *Proconsul* patellas from Rusinga Island, Kenya, with implications for other Miocene–Pliocene catarrhines. *J. Hum. Evol.* **29:**1–19.

Ward, S. C., and Kimbel, W. H. 1983. Subnasal alveolar morphology and the systematic position of *Sivapithecus. Am. J. Phys. Anthropol.* **61:**157–151.

Ward, S. C., and Pilbeam, D. R. 1983. Maxillofacial morphology of Miocene hominoids from Africa and Indo-Pakistan. In: R. L. Ciochon and R. S. Corruccini (eds.), *New Interpretations of Ape and Human Ancestry,* pp. 211–238. Plenum Press, New York.

Williams, B. A., and Kay, R. F. 1995. The taxon Anthropoidea and the crown clade concept. *Evol. Anthropol.* **3:**188–190.

7

Dental Evidence for Diet in Some Miocene Catarrhines with Comments on the Effects of Phylogeny on the Interpretation of Adaptation

RICHARD F. KAY and PETER S. UNGAR

Introduction

Studies of the dental anatomy of Miocene catarrhines have concentrated on either phylogenetic or adaptive interpretations. Most investigations of systematics or dental function have been considered more mutually exclusive than reciprocally illuminating. In this chapter, we attempt to develop a balanced view of the two together. Data presented here suggest that functional inferences require consideration of the phylogenetic affinities of the groups

---

RICHARD F. KAY • Department of Biological Anthropology and Anatomy, Duke University Medical Center, Durham, North Carolina 27710. PETER S. UNGAR • Department of Anthropology, University of Arkansas, Fayetteville, Arkansas 72701.

Function, Phylogeny, and Fossils: Miocene Hominoid Evolution and Adaptations, edited by Begun *et al.* Plenum Press, New York, 1997.

being compared. While the way a character functions is independent of its phylogenetic polarity, the way in which that character manifests itself may depend, in part, on phylogenetic considerations.

As a case study in the interpretation of adaptation in the context of phylogeny, we analyze two taxonomically diverse clusters of African early Miocene and European middle to late Miocene catarrhines. All of these taxa have in common a very similar dental pattern consisting of spatulate incisors, projecting canines, two premolars with the anterior lower one wearing against the upper canine, and four-cusped upper molars associated with lower molars with well-developed hypoconulids. This pattern was established by the early Oligocene in Egyptian *Aegyptopithecus* and *Propliopithecus* (Kay *et al.*, 1981; Fleagle and Kay, 1983). Specifically excluded from consideration here are the Ceropithecoidea with their derived bilophodont molar pattern that represents a distinct departure in functional terms derived from the more primitive pattern seen in other catarrhines. For simplicity in what follows we use the term *ape* to encompass all or part of this paraphyletic collection of Miocene–Recent nonceropithecoid catarrhines. Thus, when we refer to European Miocene apes, we mean all nonceropithecoid catarrhines.

Our studies indicate that despite the overall similarity of the molars of early Miocene apes to those of modern apes, there was a subtle morphological shift of this pattern in relation to diet. It appears that the molars of early Miocene apes were less well adapted for cutting up food than those of their extant ape relatives but still exhibited a similar range of diets from more folivorous to more frugivorous, as indicated from their tooth wear. In contrast, the middle to later Miocene European apes show molar shearing and patterns of wear that are similar to or exceed the range of modern apes. Moreover, this similarity cross-cuts phylogeny, suggesting that time, rather than propinquity, may be a more important determinant of molar structure. All of this is interpreted to be an example of the Red Queen hypothesis of Van Valen (1973), which states that because competing species are evolving all the time, there will be pressure for adaptation just to maintain the same level of adaptedness.

Phylogenetic Background

Early Miocene Catarrhines from Africa

Early Miocene apes from Africa are quite diverse. There are two size clusters, the smaller-sized taxa being *Micropithecus, Limnopithecus, Simiolus, Kalepithecus,* and *Dendropithecus* and the larger genera including *Proconsul, Rangwapithecus, Nyanzapithecus, Afropithecus,* and *Turkanapithecus.* Most workers agree that the two size clusters cross-cut phylogenetic boundaries. The dental anatomy of these primates (and extant hominoids) greatly resembles that of

Oligocene Egyptian Propliopithecidae and is probably primitive for catarrhines as a whole (Andrews, 1978; Fleagle and Kay, 1983). Recovery of postcranial materials shows that various early Miocene species were collectively quite different from any living hominoid but a consensus is lacking as to what this might mean about the phyletic placement of early Miocene apes with respect to major extant catarrhine groups.

The recent history of opinions about the phylogenetic position of *Proconsul,* the best known and most completely studied of the early Miocene primates, provides an example (Fig. 1A). Some have argued (e.g., Harrison, 1987) that *Proconsul* possesses plesiomorphic traits expected of a basal catarrhine, so that it preceded the cladogenesis of Old World monkeys and hominoids. Others (e.g., Andrews, 1985, 1992; Andrews and Martin, 1987) take the view that *Proconsul* is a sister taxon to extant hominoids. Finally, still others (e.g., Walker and Teaford, 1989; Rae, this volume) suggest that *Proconsul* has phyletic affinities with the great apes alone.

Study of the relationships *among* early Miocene catarrhines has been undertaken by Harrison and Rae. Harrison (1982) linked *Limnopithecus, Dendropithecus, Proconsul,* and *Rangwapithecus* as a clade separate from *Micropithecus.* He argues further that *Limnopithecus* and *Dendropithecus* are sister taxa, as are *Proconsul* and *Rangwapithecus.* Rae (1993, this volume) recognizes three groups: (1) *Micropithecus,* (2) *Proconsul, Dendropithecus, Rangwapithecus,* and *Turkanapithecus,* and (3) *Afropithecus, Kalepithecus,* and *Simiolus. Micropithecus* is considered a stem hominoid, whereas the second and third groups are considered progressively more derived in the direction of extant great apes. Harrison has recently backed away from his earlier conclusions by suggesting that it may be "simply not possible, given the quality of the information available, to determine the nature of the relationships among the genera of fossil catarrhines from the Miocene of East Africa" (Harrison, 1993, p. 359). Harrison now suspects that the relationships described in his earlier works more closely reflect common adaptive patterns due to similar diet than degree of phyletic affinity.

Middle and Late Miocene Catarrhines from Europe

Middle and late Miocene apes of Europe can be divided into two groups: Pliopithecidae and Hominoidea. Pliopithecids are late surviving primitive catarrhines that lack several derived features common to hominoids and cercopithecoids (Fig. 1B) (e.g., Begun, 1989; Fleagle, 1988; Harrison, 1987). Three hominoid genera are recognized from the late Miocene of Europe: *Oreopithecus, Dryopithecus,* and *Ouranopithecus.* Most now consider *Oreopithecus* a hominoid, but its phylogenetic position within the superfamily remains obscured by a suite of autapomorphic craniodental traits (Harrison, 1986). *Dryopithecus* and *Ouranopithecus* are cladistically great apes, but their positions within that clade remain unclear (Fig. 1B). Some argue that *Dryopithecus* is a

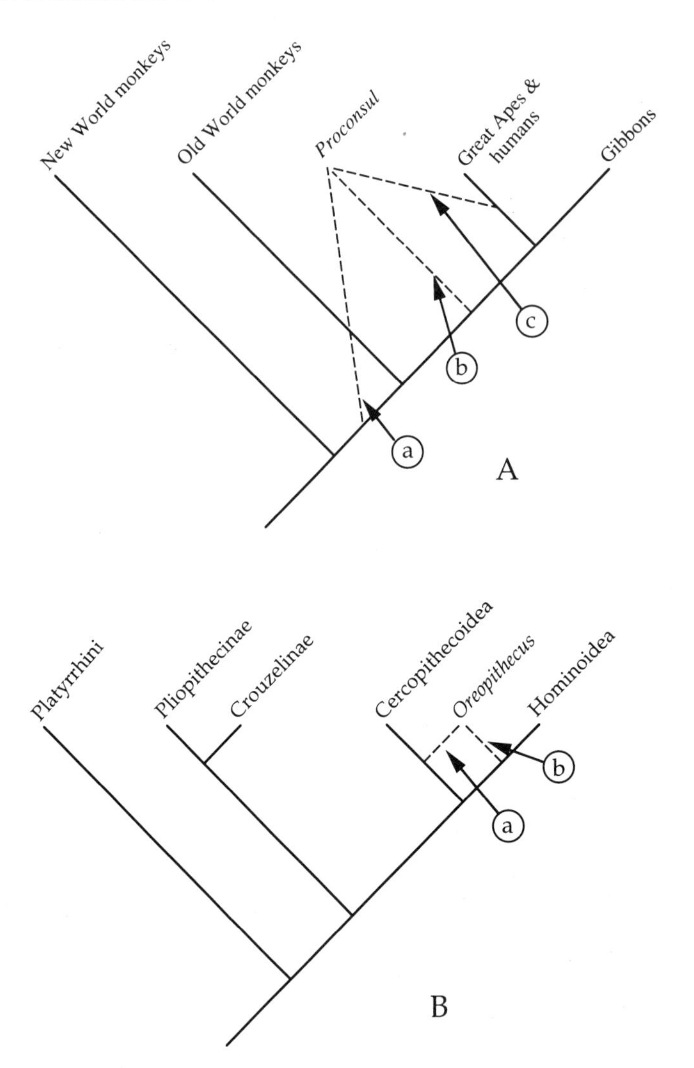

Fig. 1. Various opinions about the phylogenetic affinities of the catarrhines mentioned in this study. (A) Cladogram depicting views of early Miocene African taxa as a sister (a) to extant catarrhines (Harrison, 1987); (b) to hominoids (Andrews, 1992); or (c) to great apes and humans (Walker and Teaford, 1989). (B) Cladogram depicting views of the phylogeny of European Miocene pliopithecids as a sister to extant catarrhines (Harrison, 1987) and the uncertain affinities of *Oreopithecus* as a relative of (b) Hominoidea (Harrison, 1986) or (a) Cercopithecoidea (Szalay and Delson, 1979). (C) Cladogram depicting views of the phylogeny of European Miocene *Dryopithecus* as a sister to (a) *Pongo* (Moyà-Solà and Köhler, 1993; Schwartz, 1990); (b) to African great apes and humans (Begun, 1994; Dean and Delson, 1992); or (c) to all hominoids, Asian and African (Andrews, 1992; Begun, 1992). (D) Cladogram depicting views of the phylogeny of European Miocene *Ouranopithecus* as a sister (a) to Asian great apes (Moyà-Solà and Köhler, 1993; Schwartz, 1990); (b) to African hominoids (Begun, 1994; Dean and Delson, 1992); or (c) to hominids (Koufos and Bonis, 1993, 1994).

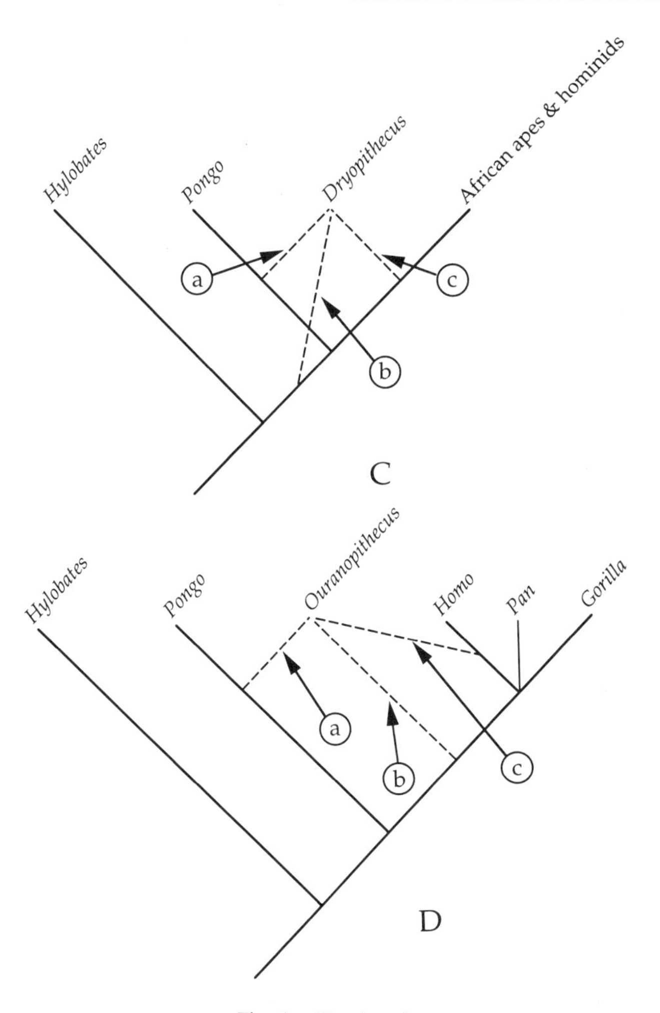

Fig. 1 (*Continued*)

primitive sister group to all extant great apes (Andrews, 1992) or forms an unresolved trichotomy with them (Begun, 1992). Others believe *Dryopithecus* to be as a sister taxon of African apes and humans (e.g., Dean and Delson, 1992; Begun, 1994) or orangutans (Moyà-Solà and Köhler, 1993; Schwartz, 1990). Likewise, while some consider *Ouranopithecus* as an early "African" ape (Andrews, 1992; Begun, 1994; Dean and Delson, 1992), others have suggested phylogenetic affinities with *Australopithecus* (Bonis and Koufos, 1993; Bonis *et al.*, 1990; Koufos, 1993) or *Pongo* (Moyà-Solà and Köhler, 1993; Schwartz, 1990).

The relationships of Miocene apes to one another have also been considered. For example, Harrison (1986) proposed phyletic affinities between *Oreopithecus*, *Nyanzapithecus*, and possibly *Rangwapithecus* from the early and middle Miocene of East Africa. Further, Begun (1994) has recently suggested that *Ouranopithecus* and *Dryopithecus* may form a clade distinct from living African apes and humans. Still, he acknowledges an unresolved trichotomy, and that either of these Miocene hominoids may be more closely related to extant African apes than to the other fossil taxon.

Ape Dental Evolution and Adaptation

Early Miocene Catarrhines from Africa

Harrison (1982) suggested dietary adaptations of several African Miocene apes, based principally on relative incisor size, given a tendency for living frugivorous anthropoids to have relatively larger incisors than their more folivorous close relatives (Hylander, 1975; Kay and Hylander, 1978). He found that the incisors of *Limnopithecus, Dendropithecus, Proconsul,* and *Rangwapithecus* are all small relative to the molars and fall within the range of extant folivores, whereas the incisors of *Micropithecus* are proportionally larger and fall among living frugivores. Harrison's (1993) recent work indicates comparable dietary reconstructions, except that *Limnopithecus* is now considered a frugivore.

Kay (1977) also investigated African early Miocene primate diets. He analyzed molar shearing-crest development, given an association between relative crest length and degree of folivory in extant hominoids. When Kay compared the fossil taxa with an extant baseline series, he found that *Limnopithecus, Dendropithecus,* and *Proconsul* had moderate shearing-crest development suggesting frugivory, whereas *Rangwapithecus gordoni* molars more closely resembled the more folivorous extant apes. Thus, Kay's reconstruction suggested that *Proconsul* and *Dendropithecus* were more frugivorous than suggested by Harrison.

One aspect of Harrison's analysis differs from that of Hylander (1975) and Kay and Hylander (1978), making comparison of their results unclear. Hylander's analysis looked at incisor size relative to body weight whereas Harrison used molar size as a surrogate for size. Thus, Harrison's data could be the result of some taxa having relatively large molars, not small incisors, for their body weight. Moreover, we believe that molar morphology may better reflect trophic adaptations than does incisor size. In fact, incisor size probably relates more to degree of tooth use in ingestion than to diet *per se*. For example, *Hylobates lar* has much smaller incisors (relative to body size) than *Pongo,* though orangutans have been reported to spend more than four times as much of their feeding time eating leaves than do sympatric lar gibbons (Ungar, 1994). Incisor size differences between orangutans and gibbons evidently reflect the greater use by orangutans of their incisors during ingestion than gibbons for most food items (Ungar, 1994). We would therefore argue that one should approach reconstructions of diet based principally on incisor size with caution, particularly when referring to specific, possibly distantly related fossil taxa. As Eaglen (1984) has demonstrated in comparisons of platyrrhines and catarrhines, incisor size differences among distantly related anthropoids in part reflect phylogenetic history.

Middle and Late Miocene Catarrhines from Europe

Some have proposed that the relatively sharp shearing blades and rather small incisors of pliopithecids indicate that they were, as a group, relatively folivorous. Szalay and Delson (1979), for example, suggested that this would have permitted "noncompetitive sympatry" with more frugivorous hominoids. Others, such as Ginsburg and Mein (1980), have stressed dietary variation within the Pliopithecidae, and distinguished subfamilies in part on the basis of their feeding adaptations. These authors wrote that crouzelines (including *Anapithecus*) had sharply crested molars similar to those of extant *Alouatta,* suggesting a more folivorous diet. In contrast, Ginsburg and Mein indicated that *Pliopithecus* had blunter cusps, suggestive of a more "omnivorous" diet, including seeds, berries, and perhaps even meat.

European Miocene hominoids have been suggested to have a range of dietary adaptations. The high-crowned, cuspidate and cristodont cheek teeth and relatively small incisors of *Oreopithecus* led Szalay and Delson (1979) to consider these apes to be folivorous. In contrast, the thick dental enamel and relatively flat occlusal morphology of *Ouranopithecus,* in combination with a reconstructed open environment, led Bonis and Koufos (1994) to suggest that the Greek apes regularly consumed foods such as roots, tubers, seeds, or nuts. Finally, the combination of high-crowned but narrow anterior tooth row and molars with thin enamel and tall, buccolingually constricted cusps has led Begun (1994) to suggest that *Dryopithecus* probably ate leaves and soft fruits.

The only comprehensive study of European Miocene catarrhine diets is that of Ungar and Kay (1995). This work extends Kay's earlier analysis of early Miocene ape molar shearing-crest development and indicates that the European apes had a broad range of molar specializations, suggesting a considerable diversity of dietary adaptations. Pliopithecid species show a spectrum of crest development comparable to that of earlier African catarrhines. *Pliopithecus* sp. (from Castell de Barbera, Spain) had slightly more and *Pliopithecus platydon* slightly less crest development than any extant hominoid, suggesting more folivorous and frugivorous diets, respectively. *Anapithecus* was intermediate, but its crest lengths fell within the extant frugivorous hylobatid range. Among the hominoids, crest length values suggest that *Oreopithecus* was extremely folivorous, *Ouranopithecus* was probably an obligate hard-object feeder, and *Dryopithecus* ate softer fruits, lacking specializations for either folivory or hard-object feeding.

Phylogeny and Adaptation

Since both phylogeny and adaptation play some role in explaining anatomy, it would be useful to assess the relative effects of each on the dentition, and the interplay of the two in determining the morphology observed.

The effects of phylogeny on characters generally considered to have functional significance are well documented. For example, as Hylander (1975), Eaglen (1984), and others have noted among both catarrhines and platyrrhines, there appears to be a relationship between incisor size and degree of incisal use in fruit processing. Nevertheless, when dietary factors are controlled for, platyrrhines tend to have smaller incisors than catarrhines (Eaglen, 1984). Likewise, Kay and Covert (1984) have shown that molar anatomy accurately tracks diet in either living hominoids or cercopithecoids, but that cercopithecoids have better-developed molar shearing than hominoids when diet is controlled for. In both cases, phylogenetic effects come into play when one attempts to infer diet from dental evidence.

These analyses are essentially ahistorical: An inference is made about the extinct species based on the overall similarity of that species with a living model. The analysis relies on inferred overall functional similarity. The phylogenetic effects referred to above may be simply that functional differences in the dental anatomy between the living and extinct groups render specific predictions inexact but allow inferences to be made about the dietary *spectrum*. We therefore hypothesize that for Miocene taxa, those species with larger molar shearing crests were probably more folivorous than other early Miocene taxa with shorter shearing crests and smaller incisors.

A recent review of hominoid dietary evolution by Andrews and Martin (1991) offers a different perspective on reconstructing dietary evolution. Andrews and Martin (1991) accept the view that the primary source of information for inference of diet is the morphology/diet correlation observed in living primates. However, mirroring Coddington (1988) they suggest that it is important to distinguish

> what might be called primary adaptation from heritage characters, the former being a direct response to environmental conditions, improving the fitness of individuals relative to others lacking the adaptation, whereas the latter are retained unchanged from an ancestor, and may not be directly linked with changing diet. (Andrews and Martin, 1991, p. 39)

We regard this distinction as meaningful if the morphological question being posed is, "What is the adaptive explanation for the origin of this morphological structure?", but less useful if the question is, "What is the current adaptedness of this structure?" And it is the latter question that we are asking with respect to the diets of early Miocene catarrhines. An example of this difference is the following: An enlarged hypocone may have evolved as a response to selection favoring a frugivorous diet. However, the presence of an enlarged hypocone cannot be interpreted as evidence that its possessor was a frugivore because some primates with large hypocones are not frugivores. We would argue that only a one-to-one correspondence between the structure and a particular behavior provides evidence about current adaptation. Further, rather than just its *derived* features, the whole digestive system is adapted to the way it makes use of foods. Still, phylogenetic inertia or baggage is bound to play an important role in how such adaptations manifest themselves, as evidenced by the work of Eaglen (1984) and Kay and Covert (1984) cited above. Thus, while the function of traits may be considered independent of character polarity, the way such traits manifest themselves may be influenced by phylogenetic considerations.

Let us now reexamine the dietary adaptations of African and European Miocene apes in the context of their temporal occurrence and phylogenetic affiliations. This will allow us to assess the relative effects of each on the dentition, and the interplay of these in determining the morphology observed. Separate analyses of early Miocene African and later Miocene European ape shearing crests provide examples of how differences among groups being examined can affect functional inferences made using comparative anatomy.

Materials and Methods

For the purposes of reconstructing the diets of Miocene apes, we assembled a data base to express the degree of development of the molar crests in living hominoids that evince a considerable range of diets. If this group follows the pattern seen for many other kinds of living primates, we expect that more folivorous apes will have better-developed shearing crests than their more frugivorous counterparts (i.e., folivorous gorillas and siamang should have better-developed molar crests than frugivorous chimpanzees, gibbons, and, to a lesser degree, orangutans).

Table I lists second-molar mesiodistal occlusal-surface diameter and the summed lengths of eight shearing crests on the M_2s of extant hominoids. Figure 2 illustrates the crests measured. A least-squares regression line is fit to the more frugivorous species [*Pan* spp., *Pongo* and *Hylobates* spp. (except *Hylobates syndactylus*)] with mean log10 M_2 length as the independent variable and mean log10 total crest length as the dependent variable. Variables are log-transformed to maintain linearity of functions. We used frugivorous species alone to fit this line because frugivores span practically the entire range of size for the extant hominoids and because there are more frugivores than folivores. Using a frugivore model allows us to control for allometric effects of crest-length difference within a dietary category. This regression line can be used to predict shearing crest lengths expected of a frugivorous ape with a given mesiodistal M_2 length. The difference between observed and expected crest length [expressed by the formula $100 \times (\text{observed} - \text{expected})/\text{expected}$]

Table I. Shearing-Crest Descriptive Statistics for Extant Taxa[a]

	M_2 length			T_{shear}		
Species	*N*	Mean	SD	Mean	SD	Residual % (SQ)
Pongo pygmaeus	5	12.77	0.92	28.98	1.92	+3.58
Pan troglodytes	10	11.68	1.05	24.74	1.84	−3.68
Pan paniscus	11	9.90	0.35	21.50	1.06	−1.91
Gorilla gorilla	14	18.25	—	41.95	—	+6.46
Hylobates lar	5	5.85	0.34	13.64	1.60	+2.23
Hylobates agilis	5	6.53	0.19	14.72	1.55	+0.09
Hylobates klossi	5	5.85	0.34	12.96	0.71	−2.07
Hylobates moloch	5	6.32	0.63	14.06	1.14	−1.35
Hylobates hoolock	5	7.70	0.32	17.71	1.32	+2.82
Hylobates syndactylus	6	8.71	0.63	20.49	1.30	+5.70

[a]Means and standard deviations provided for M_2 occlusal surface mesiodistal diameter and summed lengths of shearing crests 1–8 (see text for details). Residual from extant frugivorous regression model is expressed as a percentage.

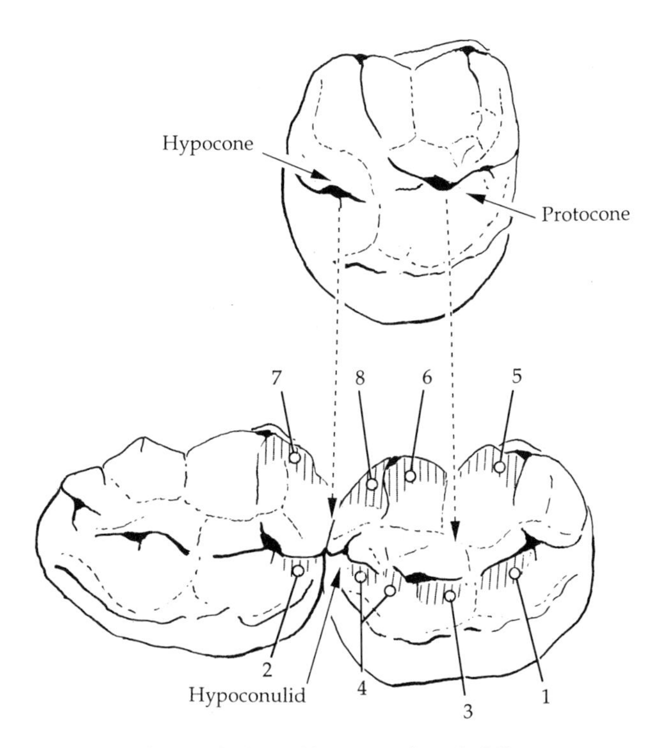

Fig. 2. Right M_{2-3} and left M^2 of *Limnopithecus evansi*, early Miocene, representative of the ground plan of the molars of noncercopithecoid catarrhines. The occlusal relationships of the crests measured are noted. Dashed arrows correspond to the direction taken by the protocone and hypocone to reach centric occlusion.

is called the shearing quotient (SQ), and indicates the magnitude and direction of residuals. The SQ values are therefore a measure of relative degree of crest development with respect to that expected of an extant fruit-eating hominoid.

Miocene fossil specimens used in this study are listed in Table II. The African species include *Dendropithecus macinnesi, Proconsul nyanzae, P. heseloni,* and *Limnopithecus legetet* from Rusinga Island, *Rangwapithecus gordoni, Proconsul major,* and *Limnopithecus evansi* from Songhor, and one specimen each of *Dendropithecus macinnesi* and *Limnopithecus legetet* from Mfangano Island and Koru, respectively. European fossil specimens used in this study include *Anapithecus hernyaki* from Rudabánya, *Pliopithecus platydon* from Göriach, *Pliopithecus* sp. from Castell de Barbera, *Dryopithecus fontani* from St. Gaudens, *Dryopithecus laietanus* from Can Llobateres and La Tarumba, *Oreopithecus bambolii* from Monte Bamboli and Baccinello, and *Ouranopithecus macedoniensis* from Ravin de la Pluie. Taxonomic designations follow Crusafont-Pairo and Golpe-Posse (1981), Harrison (1988), Harrison *et al.* (1991), and Walker *et al.* (1993).

Table II. Fossil Specimens Used in This Analysis

Dendropithecus macinnesi	*Proconsul nyanzae*	*Proconsul heseloni*
RU 2015a	RU 2087	RU 2036a
RU 1849	RU 1710	RU 1706
RU 1850b	RU 1678	RU 1823
RU 1893	RU 1695	RU 1959
RU 2003	RU 1676	RU 1945
RU 1992	RU 1734	RU 7290
MW 53		
Rangwapithecus gordoni	*Proconsul major*	*Limnopithecus evansi*
SO 908	SO 396	SO 424
SO 909	SO 415	SO 444
SO 906	SO 914	SO 911
SO 374		
SO 420		
SO 486		
Limnopithecus legetet	*Dryopithecus laietanus*	*Dryopithecus fontani*
KO 8	IPS 1782	TYPE SPECIMEN
RU 1708	IPS 1796	
	IPS 1797	
	IPS 1808	
	IPS 9001	
Oreopithecus bambolii	*Ouranopithecus macedoniensis*	*Pliopithecus platydon*
IGF 4335	RPL 54	OE 303
IGF 11778	RPL 75	
Pliopithecus sp. (Castell de Barbera)	*Anapithecus hernyaki*	
IPS #1	RUD 89	RUD 106
IPS #2	RUD 91	RUD 108
	RUD 98	RUD 122
	RUD 100	RUD 128

Results

Extant Hominoids

The relationship between total shear and M_2 length for extant frugivorous hominoids is

$$\log10 \text{ total shear} = (\log10\ M_2 \text{ length} \times 0.959) + 0.386$$

or

$$\text{total shear} = 2.432\ M_2\ \text{length}^{0.959}$$

with an R^2 of 0.992. Using this equation, the "expected" total shear is calculated for each living species from its M_2 length. The difference from the observed total shear is expressed as a residual percentage (SQ value).

The range of values for all species is between +6.5% and −4.0% (Table I, Fig. 3). As expected from the way this model is constructed, the values of percentage difference cluster around zero for the frugivorous species. The most folivorous species, gorilla and siamang, having better-developed crests, exhibit the largest positive values for the residual percentage. An important limitation from this sample of extant species is that we do not have sufficient

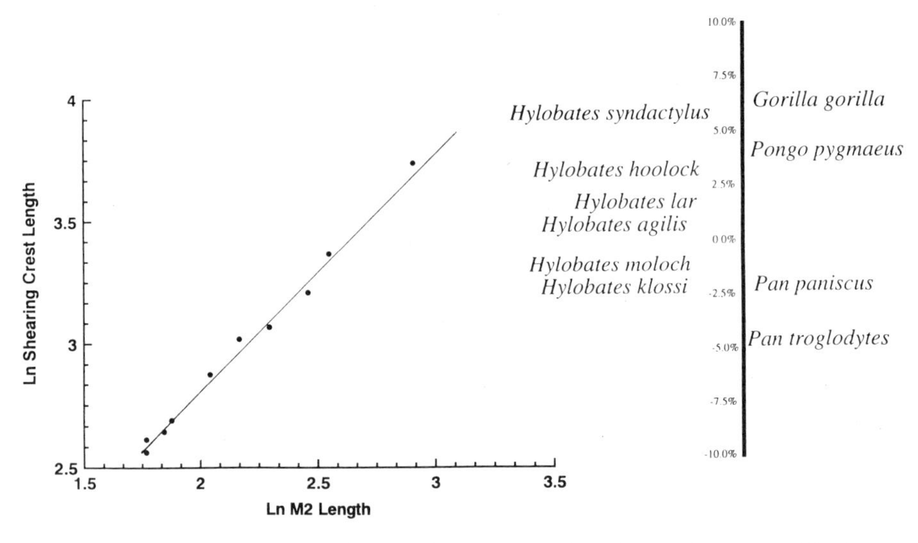

Fig. 3. Relative shearing-crest development in extant apes. This figure illustrates residuals from the frugivorous hominoid regression comparing summed shearing-crest length with M_2 length. Higher values indicate better-developed shearing crests. The more folivorous extant apes show better-developed shearing crests than the frugivores.

taxa to determine whether species that eat hard objects (nuts, tough fruits, seeds) are distinguishable from those that avoid hard objects and pass seeds through their digestive tracts in an unmodified state. An analysis of New World monkeys suggests that the two may be distinguishable and that hard-object feeders like the pitheciines and *Cebus* may have more poorly developed molar shearing than do soft-fruit feeders like *Ateles* (Anthony and Kay, 1993; Fleagle *et al.*, 1997). One additional morphological trait that might be informative is enamel thickness since many hard-object feeders tend to have thick enamel (e.g., Kay, 1981).

Early Miocene Catarrhines from Africa

Descriptive statistics for molar length and summed shearing-crest length for the early Miocene taxa are presented in Table III. Residual percentage values were calculated using the regression line for the living hominoids. The results are summarized graphically in Fig. 4. The range of SQs for these taxa spans a similar degree of morphological distinctness—the largest and smallest values span approximately 12%, compared with 10% for living hominoids.

The taxon with the best-developed molar shearing crests is *Rangwapithecus gordoni;* those with the least-developed crests are *Dendropithecus* and *Proconsul* spp. Shearing quotient values for the taxa range between 0 and −12.0%

Table III. Shearing-Crest Descriptive Statistics for Fossil Taxa[a]

Species		M_2 length		T_{shear}		Residual % (SQ)
	N	Mean	SD	Mean	SD	
Dendropithecus macinnesi	7	7.05	0.20	14.63	0.75	−7.57
Limnopithecus legetet	2	6.00	—	13.09	0.28	−3.46
Limnopithecus evansi	3	5.83	0.16	12.51	0.73	−5.16
Proconsul heseloni	6	9.10	0.43	17.69	0.72	−12.50
Proconsul nyanzae	6	10.94	0.71	21.96	1.86	−8.96
Proconsul major	3	13.29	0.62	26.58	0.72	−8.57
Rangwapithecus gordoni	6	8.69	0.33	19.29	0.33	−0.27
Anapithecus hernyaki	8	8.70	0.34	19.48	0.55	0.60
Dryopithecus fontani	1	11.38	—	25.38	—	1.31
Dryopithecus laietanus	5	10.50	0.39	22.94	0.94	−1.08
Oreopithecus bambolii	2	9.88	—	29.53	—	34.99
Ouranopithecus macedoniensis	2	15.94	—	31.69	—	−8.43
Pliopithecus (Spain)	2	6.00	—	14.53	—	7.16
Pliopithecus platydon	1	7.94	—	16.88	—	−4.84

[a]Means and standard deviations provided for M_2 occlusal-surface mesiodistal diameter and summed lengths of shearing crests 1–8. Residual from extant frugivorous model is expressed as a percentage.

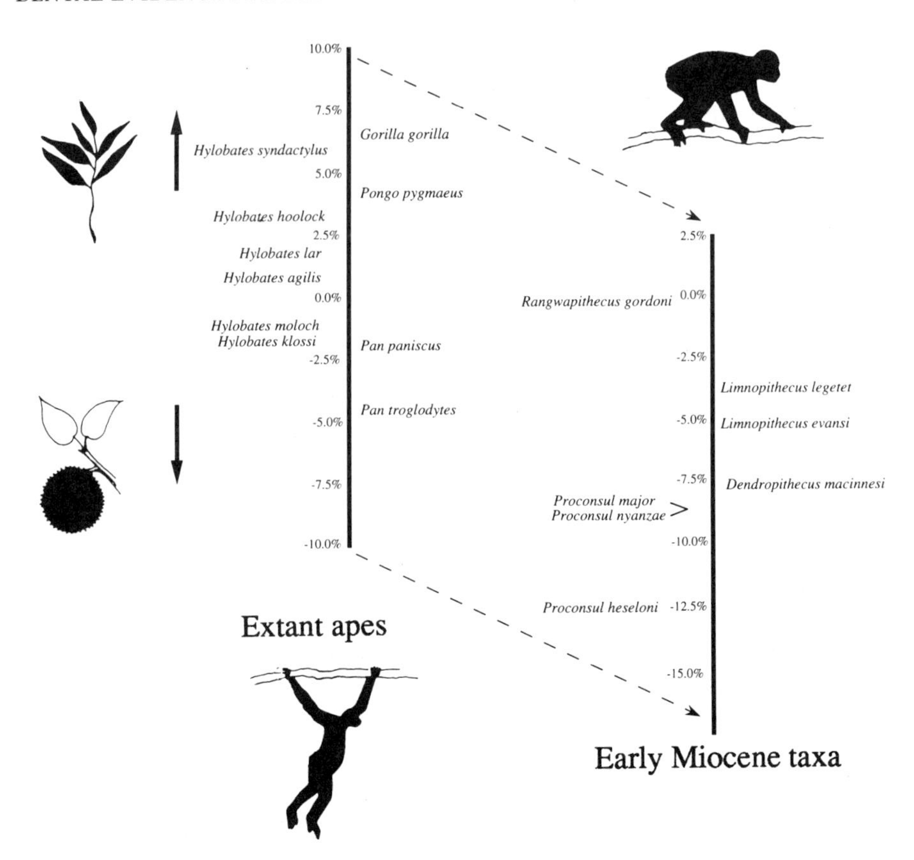

Fig. 4. Relative shearing-crest development in extant apes and early Miocene taxa. Extant hominoids are situated on the bar to the left, and fossil taxa are placed on the bar to the right. The more folivorous extant apes show better-developed shearing crests than the frugivores. Dashed lines indicate a comparable range of residual values (and crest development).

when compared with the frugivorous extant hominoid regression line. Taken at face value, this would suggest that all of the early Miocene taxa examined here were frugivorous. However, it is equally plausible that we are seeing a dietary range from more folivorous (*Rangwapithecus*) to more frugivorous (with or without hard-object feeding) (*Proconsul, Dendropithecus*) and that the whole range is "downshifted": folivorous species were as efficient as they needed to be at that time given the adaptive status of contemporaneous mammalian leaf eaters. We believe the latter scenario is more likely given that all early Miocene taxa show less crest development than highly frugivorous extant gibbons, and most fossil species examined show less crest development than any extant hominoid, even those that feed almost exclusively on soft fruits (e.g., *Hylobates klossi,* Whitten, 1982). An independent source of information that supports our interpretation is found in the patterns of dental microwear. *Rangwapithecus* Phase 2 facets show relatively more scratches than do those of *Proconsul,* suggesting the former are more folivorous, whereas the latter are more frugivorous (Teaford, personal communication; Walker *et al.,* 1994).

Middle and Late Miocene Catarrhines from Europe

As with the African fossil sample, descriptive statistics for molar length and summed shearing-crest length of European apes are presented in Table III. Shearing quotients are presented graphically in Fig. 5. The range of SQs for the pliopithecids is equivalent to that of the early Miocene African hominoids (12%), and slightly exceeds that of living hominoids. This suggests substantial dietary heterogeneity. *Pliopithecus* from Castell de Barbera, Spain has a higher SQ than any extant hominoid, suggesting a more folivorous diet. In contrast, *Pliopithecus platydon* has a lower SQ than any extant hominoid, suggesting a more frugivorous diet. *Anapithecus hernyaki* had an intermediate SQ, suggesting an intermediate diet, though its values fell well within the soft-fruit-eating extant gibbon range.

The range of SQs for the European Miocene hominoids exceeds 43%, a value far greater than that of any other group. The European hominoid with the best-developed crests is *Oreopithecus bambolii.* This taxon has shearing crests so well developed as to extend far down the buccal and lingual surfaces of the molars, substantially below the level of the central basin of the molar (see Fig. 6). This suggests an extremely specialized folivorous diet. In contrast, *Ouranopithecus macedoniensis* has a much lower SQ than any living ape. Lucas *et al.* (1994) have recently demonstrated that blunter, flatter molars require less work to crush hard food items. Indeed, hard-object-feeding primates show lower SQs than closely related soft-fruit specialists (Fleagle *et al.,* 1997). We therefore posit that *O. macedoniensis* habitually consumed harder food items than did other taxa examined in this study. Its remarkably thick enamel is consistent with this hypothesis (Kay, 1981). Finally, *Dryopithecus fontani* and *D. laietanus* have shearing-crest development similar to that of extant frugivorous gibbons and chimpanzees, suggesting a softer-fruit diet lacking specializations for either hard-object feeding or extreme folivory.

Support for the dietary scenarios of pliopithecids and hominoids of Europe comes from dental microwear evidence in progress, wherein *Ouranopithecus* Phase 2 facets are heavily pitted, like modern hard-object-feeding primates; *Oreopithecus* facets show a high incidence of scratches, consistent with a folivorous diet; and *Dryopithecus* molar surfaces appear intermediate in ratios of pits to scratches, suggesting a soft-fruit diet (Ungar, 1996). *Anapithecus hernyaki* and *Pliopithecus platydon* have a combination of scratches and pits suggesting soft-fruit eating while Spanish *Pliopithecus* microwear surfaces are dominated by scratches indicating folivory.

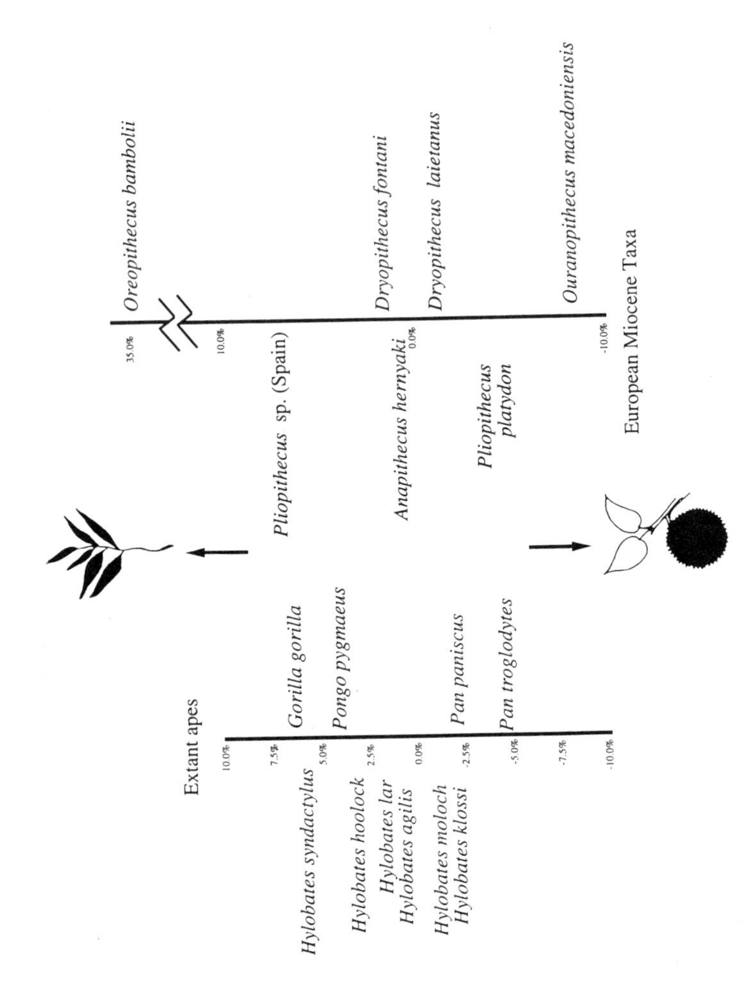

Fig. 5. Relative shearing-crest development in extant apes and European Miocene taxa. See caption of Fig. 4 for more details.

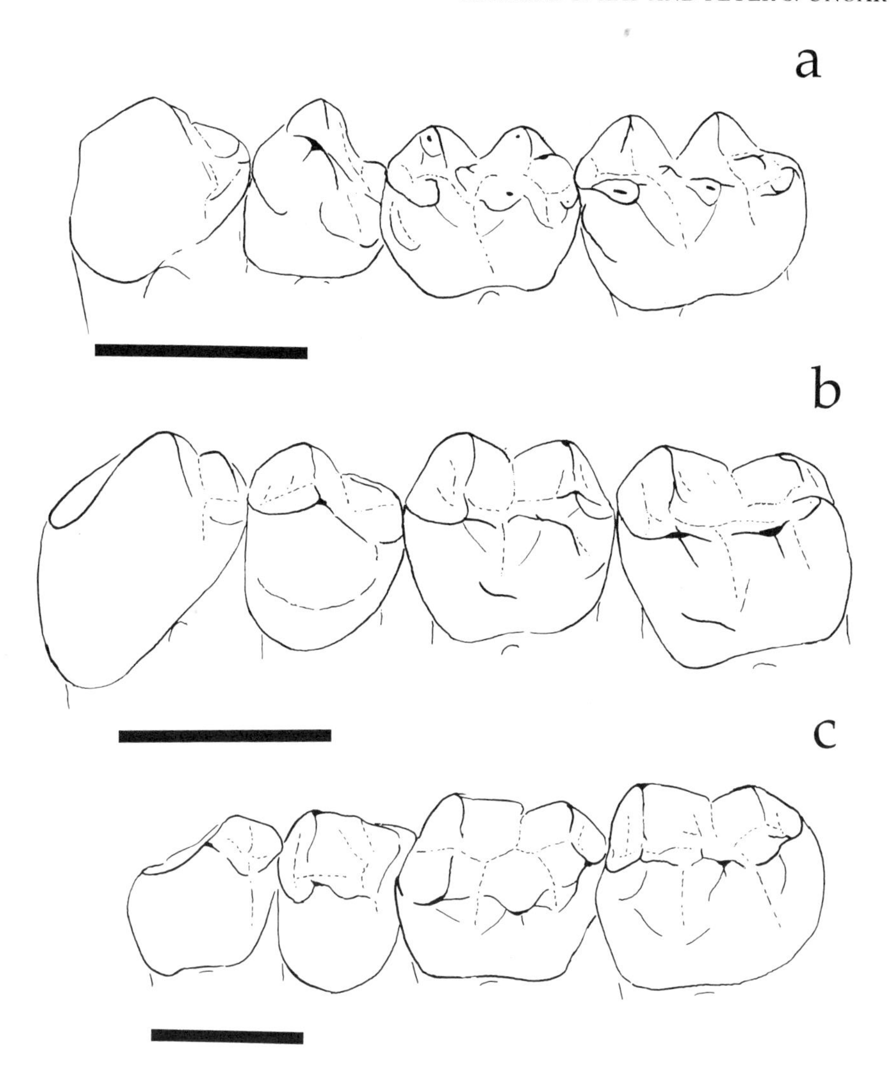

Fig. 6. Occlusolateral views of the right P_3–M_2 of *Oreopithecus bambolii,* IGF 11778, *Dryopithecus fontani* TYPE, and *Ouranopithecus macedoniensis,* RPL 54, exemplifying the differing degree of development of shearing crests, as reflected in the SQs in Table III. Scale bars = 2 cm.

Discussion

The dental anatomy of early Miocene catarrhines appears to have departed to only a small degree from the primitive anatomy of an inferred ancestral hominoid of the Oligocene. Likewise, at first glance, the molar anatomy of extant hominoids also retains many conservative traits laid down in the early Oligocene. However, in our effort to reconstruct the dietary behavior of early Miocene catarrhines we have encountered a peculiar phenomenon: The molars of early Miocene apes demonstrate a substantial range of morphological variation, but this range is downshifted from the modern hominoid range. Many early Miocene taxa actually have less overall molar shearing than their extant counterparts. *Rangwapithecus,* the early Miocene taxon with the most shearing, still has substantially less shearing than folivorous extant apes, gorilla, or siamang. *Proconsul* has the least well-developed shearing of our early Miocene sample, less than any living ape. Tooth-wear studies suggest that *Rangwapithecus* was probably folivorous and *Proconsul* frugivorous. In contrast, our analysis of the molar morphology of middle to later Miocene European apes exhibits no major discordance with the tooth-wear data. Thus, there seems to have been a morphological shift toward enhanced molar shearing between the early and middle Miocene, apparently unaccompanied by a dietary shift.

The shift toward increased molar shearing seems to have occurred at the same time and in several different clades. No early Miocene taxon examined here exhibits a shift even though some early Miocene apes could be a sister clade to extant Hominoidea and others may fall within the hominoid clade. Furthermore, the shift toward increased molar shearing is found in all middle-to-late Miocene taxa examined irrespective of their phylogenetic position: Pliopithecids are sister to early Miocene taxa and hominoids so their development of increase shearing seems to have occurred independently from hominoids. Thus, the morphological shift is temporally constrained and cross-cuts phylogeny.

These phenomena lead us to propose that we are seeing an example of the Red Queen hypothesis of Van Valen (1973). This hypothesis states that because competing species are evolving all the time, there will be pressure for adaptation just to maintain the same level of adaptedness. If this is what is going on, we might ask, what was the driving selection for the change? Two possibilities suggest themselves. One is coevolution of the plant foods to better resist the activities of their primate predators. The overall expansion of shearing crests may have been selected to maintain the ability of hominoids to more finely comminute plant foods of all kinds in response to the plants themselves evolving more resistant structures. Rather than extant hominoids feeding on something different from early Miocene apes, they have to be more efficient to extract the same nutrients from plant foods.

A second scenario is that the increased shearing evolved in response to competition from the Cercopithecoidea that first appear in the fossil record between the early and middle Miocene, after the early Miocene taxa that we discuss above but well before hominoids and pliopithecids are first reported in Europe. Temerin and Cant (1983) argue that the evolution of cheek pouches and a bilophodont dentition in cercopithecoids are traits that enhance the digestive efficiency and harvesting rates giving them a competitive advantage over extant hominoids which in turn may explain why living cercopithecoids are more diverse and abundant than their ape competitors. We suggest something slightly different, namely, that cercopithecoids may have "pushed" European hominoids and pliopithecids to increase their digestive efficiency before they left Africa. Nevertheless, it appears probable that there are structural limitations to how far the apomorphic "apelike" molar design catarrhine pattern of early Miocene catarrhines and living apes can be "improved" by natural selection. *Oreopithecus* has shearing values within the cercopithecoid range. Interesting peculiarities in its molar occlusion suggest a convergence on the bilophodont condition independent from Old World monkeys. Any changes beyond these limits must involve a design breakthrough. If the evolution of the bilophodont molar pattern is such a breakthrough, competition with Old World monkeys might drive changes in the hominoid dental organization only so far. The masticatory system of living apes has been demonstrated to be less efficient than cercopithecoids in chewing up plant foods (Walker and Murray, 1975). The masticatory system of early Miocene apes may have been even less efficient given that they have less molar shearing than extant apes. Some improvement may have allowed hominoids to survive and flourish in some habitats up to the present day.

Our attempt to infer the adaptations of early Miocene noncercopithecid catarrhines emphasizes that functional inferences employing the comparative method must, at the outset, take into consideration the phylogenetic affinities of the groups being compared. In particular, we recommend that clusters of closely related and temporally similar taxa be compared in such analyses. This would enable us to identify patterns of functional shifts and get a sense of (and interpret) variation among contemporaneous taxa. For example, an analysis of just one early Miocene taxon would not have identified the functional shift we were able to identify comparing early Miocene taxa collectively. While the way a character functions is independent of its polarity, the way in which that character manifests itself may depend, in part, on phylogenetic considerations.

Acknowledgments

We thank David Begun, Carol Ward, and Mike Rose for this opportunity to discuss the interaction of adaptation and phylogeny in early Miocene catarrhines. Much of the data on which this work is based were collected by R.F.K. in the 1970s with support from NSF grant 43262 and permission from R. Leakey, P. Andrews, R. Thorington, S. Anderson, and T. Van Den Audenaerde. European data were collected by P.S.U. with support from the L. S. B. Leakey and Andrew Mellon Foundations and permission from L. Kordos, G. Koufos, J. Agusti, S. Moyà-Solà, C. Rosselet, and R. Martin. Also, we especially thank Mark Teaford and Alan Walker for advice and access to specimens.

References

Andrews, P. J. 1978. A revision of the Miocene Hominoidea of East Africa. *Bull. Br. Mus. Nat. Hist. Geol.* **30:**85–224.

Andrews, P. J. 1985. Family group systematics and evolution among catarrhine primates. In: E. Delson (ed.), *Ancestors: The Hard Evidence,* pp. 14–22. Alan R. Liss, New York.

Andrews, P. J. 1992. Evolution and environment in the Hominoidea. *Nature* **360:**641–646.

Andrews, P. J., and Martin, L. B. 1987. Cladistic relationships of extant and fossil hominoids. *J. Hum. Evol.* **16:**101–118.

Andrews, P. J., and Martin, L. B. 1991. Hominoid dietary evolution. *Philos. Trans. R. Soc. London B Ser.* **334:**199–209.

Anthony, M. R. L., and Kay, R. F. 1993. Tooth form and diet in ateline and alouattine primates: Reflections on the comparative method. *Am. J. Sci.* **293-A:**1–26.

Begun, D. R. 1989. A large pliopithecine molar from Germany and some notes on the Pliopithecinae. *Folia primatol.* **52:**156–166.

Begun, D. R. 1992. Miocene fossil hominids and the chimp–human clade. *Science* **257:**1929–1933.

Begun, D. R. 1994. Relations among the great apes and humans: New interpretations based on the fossil great ape *Dryopithecus. Yearb. Phys. Anthropol.* **37:**11–63.

Bonis, L. de, and Koufos, G. 1993. The face and mandible of *Ouranopithecus macedoniensis:* Description of new specimens and comparisons. *J. Hum. Evol.* **24:**469–491.

Bonis, L. de, and Koufos, G. 1994. Our ancestors' ancestor: *Ouranopithecus* is a Greek link in human ancestry. *Evol. Anthropol.* **3:**75–84.

Bonis, L. de, Bouvrain, G., Geraads, D., and Koufos, G. 1990. New hominoid skull material from the late Miocene of Macedonia in northern Greece. *Nature* **345:**712–714.

Coddington, J. A. 1988. Cladistic tests of adaptational hypotheses. *Cladistics* **4:**3–22.

Crusafont-Pairo, M., and Golpe-Posse, J. M. 1981. Estudio de la denticion inferior del premer Pliopithecido hallado en Espana (Vindoboniense termial de Castell de Barbera, Cataluna, Espana). *Bol. Inf. Sabadell.* **13:**25–38.

Dean, D., and Delson, E. 1992. Second gorilla or third chimp? *Nature* **359:**676–677.

Eaglen, R. H. 1984. Incisor size and diet revisited: The view from a platyrrhine perspective. *Am. J. Phys. Anthropol.* **69:**262–275.

Fleagle, J. G. 1988. *Primate Adaptation and Evolution.* Academic Press, New York.

Fleagle, J. G., and Kay, R. F. 1983. New interpretations of the phyletic position of Oligocene hominoids. In: R. L. Ciochon and R. S. Corruccini (eds.), *New Interpretations of Ape and Human Ancestry,* pp. 181–210. Plenum Press, New York.

Fleagle, J. G., Kay, R. F., and Anthony, M. R. L. 1997. Fossil New World monkeys. In: R. F. Kay, R. L. Cifelli, R. H. Madden, and J. J. Flynn (eds.), *Paleobiology of an Extinct Neotropical Fauna.* Smithsonian Institution Press, Washington, DC, pp. 473–495.

Ginsburg, L., and Mein, P. 1980. *Crouzelia rhodanica,* nouvelle espece de Primate catarhinien, et essai sur la position systematique des Pliopithecidae. *Bull. Mus. Natl. Hist. Nat. Paris 4e Ser. Sect. C* pp. 57–85.

Harrison, T. 1982. *Small-Bodied Apes from the Miocene of East Africa.* Ph.D. thesis, University of London.

Harrison, T. 1986. A reassessment of phylogenetic relationships of *Oreopithecus bambolii* Gervais. *J. Hum. Evol.* **15:**541–583.

Harrison, T. 1987. The phylogenetic relationships of the early catarrhine primates: A review of the current evidence. *J. Hum. Evol.* **16:**41–80.

Harrison, T. 1988. A taxonomic revision of the small catarrhine primates from the early Miocene of East Africa. *Folia Primatol.* **50:**59–108.

Harrison, T. 1993. Cladistic concepts and the species problem in hominoid evolution. In: W. H. Kimbel and L. B. Martin (eds.), *Species, Species Concepts and Primate Evolution,* pp. 345–371. Plenum Press, New York.

Harrison, T., Delson, E., and Jian, G. 1991. A new species of *Pliopithecus* from the middle Miocene of China and its implications for early catarrhine zoogeography. *J. Hum. Evol.* **21:**329–361.

Hylander, W. L. 1975. Incisor size and diet in anthropoids with special reference to Cercopithecidea. *Science* **189:**1095–1098.

Kay, R. F. 1977. Diets of early Miocene African hominoids. *Nature* **268:**628–630.

Kay, R. F. 1981. The nut-crackers: A new theory of the adaptations of the Ramapithecinae. *Am. J. Phys. Anthropol.* **55:**141–151.

Kay, R. F., and Covert, H. H. 1984. Anatomy and behavior of extinct primates. In: D. J. Chivers, B. A. Wood, and A. Bilsborough (eds.), *Food Acquisition and Processing in Primates,* pp. 467–508. Plenum Press, New York.

Kay, R. F., and Hylander, W. L. 1978. The dental structure of mammalian folivores with special reference to primates and Phalangeroidea (Marsupialia). In: G. G. Montgomery (ed.), *The Ecology of Arboreal Folivores,* pp. 173–191. Smithsonian Institution Press, Washington, DC.

Kay, R. F., Fleagle, J. G., and Simons, E. L. 1981. A revision of the African Oligocene apes of the Fayum Province, Egypt. *Am. J. Phys. Anthropol.* **55:**293–322.

Koufos, G. D. 1993. Mandible of *Ouranopithecus macedoniensis* (Hominidae, Primates) from a new late Miocene locality of Macedonia (Greece). *Am. J. Phys. Anthropol.* **91:**225–234.

Lucas, P. W., Peters, C. R., and Arrandale, S. R. 1994. Seed-breaking forces exerted by orang-utans with their teeth in captivity and a new technique for estimating forces produced in the wild. *Am. J. Phys. Anthropol.* **94:**365–378.

Moyà-Solà, S., and Köhler, M. 1993. Recent discoveries of *Dryopithecus* shed new light on evolution of great apes. *Nature* **365:**543–545.

Rae, T. 1993. *Phylogenetic Analysis of Proconsulid Facial Morphology.* Ph.D. dissertation, State University of New York at Stony Brook.

Schwartz, J. 1990. *Lufengpithecus* and its potential relationship to an orang-utan clade. *J. Hum. Evol.* **19:**591–605.

Szalay, F. S., and Delson, E. 1979. *Evolutionary History of the Primates.* Academic Press, New York.

Temerin, L. A., and Cant, J. G. H. 1983. The evolutionary divergence of Old World monkeys and apes. *Am. Nat.* **122:**335–351.

Ungar, P. S. 1994. Patterns of ingestive behavior and anterior tooth use differences in sympatric anthropoid primates. *Am. J. Phys. Anthropol.* **95:**197–219.

Ungar, P. S. 1995. Fruit preferences of four sympatric primate species at Ketambe, northern Sumatra, Indonesia. *Int. J. Primatol.* **16:**221–245.

Ungar, P. S. 1996. Dental microwear of European Miocene catarrhines: Evidence for diets and tooth use. *J. Hum. Evol.* **31:**335–366.

Ungar, P. S., and Kay, R. F. 1995. The dietary adaptations of European Miocene catarrhines. *Proc. Natl. Acad. Sci. USA* **92:**5479–5481.

Van Valen, L. 1973. A new evolutionary law. *Evol. Theory* **1:**1–30.

Walker, A., and Teaford, M. 1989. The hunt for *Proconsul. Sci. Am.* **260:**76–82.

Walker, A., Teaford, M. F., Martin, L., and Andrews, P. 1993. A new species of *Proconsul* from the early Miocene of Rusinga/Mfangano Islands, Kenya. *J. Hum. Evol.* **25:**43–56.

Walker, A., Teaford, M. F., and Ungar, P. S. 1994. Enamel microwear differences between species of Proconsul from the early Miocene of Kenya. *Am. J. Phys. Anthropol. Suppl.* **18:**202–203.

Walker, P., and Murray, P. 1975. An assessment of masticatory efficiency in a series of anthropoid primates with special reference to the Colobinae and Cercopithecinae. In: R. H. Tuttle (ed.), *Primate Functional Morphology and Evolution,* pp. 135–150. Mouton, The Hague.

Whitten, A. J. 1982. Diet and feeding behaviour of Kloss gibbons on Siberut Island, Indonesia. *Folia Primatol.* **37:**177–208.

Miocene Hominoid Mandibles

8

Functional and Phylogenetic Perspectives

BARBARA BROWN

Introduction

It is axiomatic in vertebrate paleontology that mandibles and mandibular fragments are among the most frequently recovered specimens from fossil localities. The dense, cohesive structure of the mandibular corpus and embedded tooth roots improve the likelihood that they will survive, in some form, the vagaries of deposition and postdepositional diagenetic processes. This is as true in the hominoid fossil record as it is in other areas of paleontology, a fact that mandates continuing analysis of the factors that mediate mandibular size and shape. As is true for all mammals, the morphology of primate mandibles is influenced in many ways, including patterns of food procurement, masticatory mechanics, cranial growth trajectories, and body size. Such influences can be analyzed to reconstruct functional/behavioral patterns, and, to a more limited extent, provide data useful in phylogenetic assessments.

BARBARA BROWN • Department of Anatomy, Northeastern Ohio University College of Medicine, Rootstown, Ohio 44272-0095.

Function, Phylogeny, and Fossils: Miocene Hominoid Evolution and Adaptations, edited by Begun *et al.* Plenum Press, New York, 1997.

Functional Perspectives

Function in Symphyseal Shape

Despite its complex, anisotropic structure (Smith, 1983; Brown, 1989; Daegling *et al.*, 1992), the mandible has been extensively analyzed in terms of masticatory function and stress distribution. Hylander and colleagues (Hylander, 1979, 1984, 1985, 1988; Daegling, 1989; Ravosa, 1991; Daegling *et al.*, 1992) have proposed a number of functional correlates of primate mandibular morphology based on *in vivo* experimental stress analysis. These studies on the mandibular corpus and symphysis have addressed variation in morphology as possible adaptations to counter bending, torsional, and dorsoventral shearing stresses. The fused mandibular symphysis of higher primates resists a variety of stresses generated by a number of different loading patterns (Hylander, 1984, 1985). The corpora can either be forced medially, pulled laterally as in a wishbone, twisted unilaterally or twisted bilaterally depending on the actions of the masticatory musculature (Hylander, 1984). It therefore seems probable that variations in the sectional anatomy of the symphysis can be used as first-order reconstructions of the symphyseal loading environment.

Variation in mandibular symphyseal morphology, particularly the superior and inferior transverse tori, is known to be extensive in Miocene and extant hominoids (Fig. 1). Within this general range of toral variation, the following four basic morphologies are present and a combination of these may exist within a taxon: (1) superior and inferior transverse tori may have equal prominence, (2) the superior transverse torus may be more prominent than the inferior, (3) the inferior transverse torus may be more prominent than the superior, or (4) neither torus is prominent.

The Early Miocene hominoid genus *Proconsul* is known from deposits primarily from Kenya and eastern Uganda. *Proconsul* mandibles have a dominant superior transverse torus and an inferior transverse torus which is either minimal or completely lacking (Andrews, 1978; Brown, 1989). (The most prominent superior transverse tori are characteristic of the largest species, *Proconsul major*). In *P. major,* the planum alveolar is relatively long extending from the symphysis inferiorly to about midcorpus at a shallow angle to the alveolar plane. In the smaller proconsulid species of *P. heseloni, P. africanus,* and *Rangwapithecus,* there is a corresponding reduction in superior transverse torus prominence and lingual buttressing.

Another Kenyan taxon, *Afropithecus,* is from later Lower Miocene deposits dated between 16 and 18 Ma (Leakey and Walker, 1985; Leakey *et al.,* 1988). The *Afropithecus* mandibles which preserve relatively complete symphyses appear to be male (Leakey *et al.,* 1988). They are characterized by tall narrow corpora, and their symphyses are tall with superior and inferior transverse tori separated by a relatively shallow genial fossa. The inferior transverse

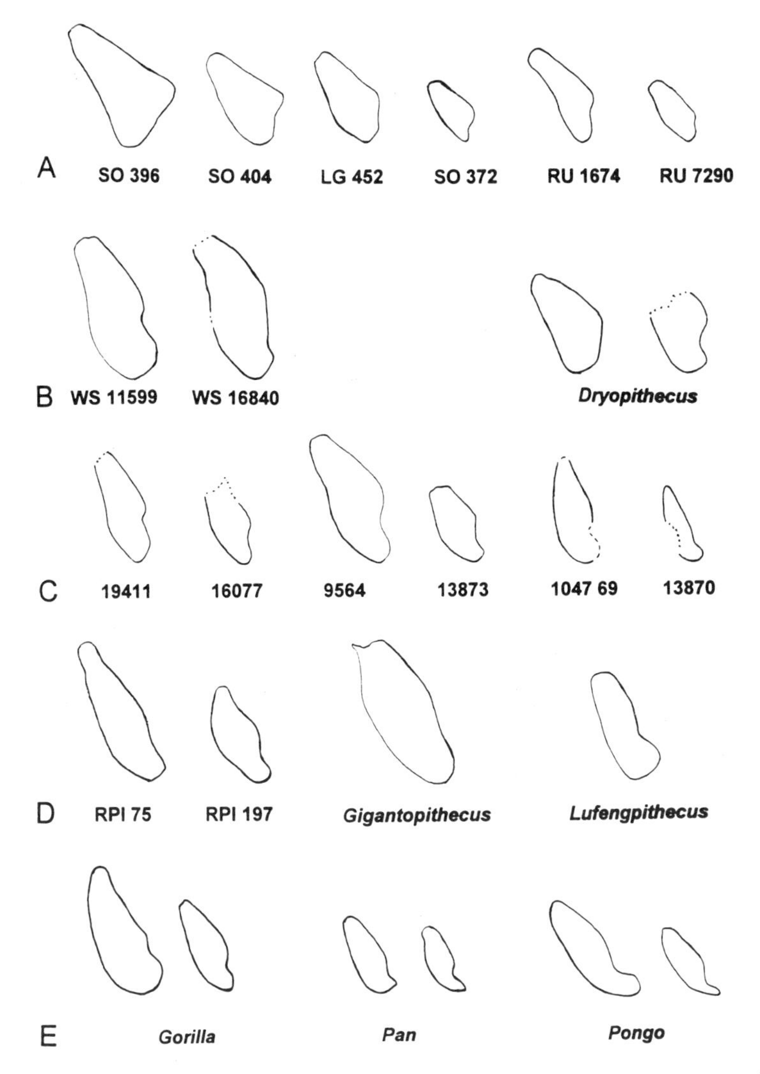

Fig. 1. Symphyseal sections of Miocene and extant hominoids. (A) *Proconsul* specimens as labeled. (B) *Afropithecus* specimens as labeled and *Dryopithecus* specimens. (C) *Sivapithecus* specimens as labeled. (D) *Ouranopithecus* specimens as labeled, *Gigantopithecus*, and *Lufengpithecus*. (E) Male and female specimens of *Gorilla*, *Pan*, and *Pongo*.

torus is robust and slightly more prominent in KNM-WS 11599, and neither torus is particularly dominant in KNM-WS 16840 (Leakey *et al.*, 1988). If there is phylogenetic change in mandibular form between these early Miocene genera, it may be toward the development of an inferior transverse torus (McCrossin and Benefit, 1993); however, this may also be a response to an increase in mandibular size.

Kenyapithecus has been recovered at a number of Middle Miocene localities in Kenya. The most complete adult mandibular specimen is from Kaloma, western Kenya (Pickford, 1982). The Kaloma mandible has been crushed along its occlusal surface, and the symphysis is damaged. Most of the symphyseal damage has excavated the alveolar plane, resulting in a relatively deep concavity. The most distal aspect of the damaged planum alveolar preserves the posterior part of a superior transverse torus that borders a relatively deep genial fossa superior to an inferior transverse torus. Other mandibles attributable to *Kenyapithecus* have been recovered from various localities in Kenya (Pickford, 1982, 1985; Ishida *et al.,* 1984; McCrossin and Benefit, 1993; Brown *et al.,* 1995), but are either undescribed or poorly preserved. Thus, for the moment, there are disagreements concerning the lingual topography of the *Kenyapithecus* symphysis (Pickford, 1982, 1985; Ishida *et al.,* 1984; McCrossin and Benefit, 1993).

McCrossin and Benefit (1993) have recently described a complete juvenile *Kenyapithecus* mandible from Maboko (KNM-MB 20573). This specimen is undistorted, and based on symphyseal contours, McCrossin and Benefit have posited certain functional changes in the hominoid symphysis through time. They regard the mandibular form of *Kenyapithecus* as an intermediate stage between that of *Proconsul* and the extant apes. In contrast to *Proconsul,* extant apes possess an inferior transverse torus that is more prominent than the superior transverse torus, and, in some cases, the superior transverse torus may also be absent (Brown, 1989).

Later Miocene hominoids have been recovered from localities throughout much of the Old World. These include *Sivapithecus,* which spans the greatest time range, with *Dryopithecus, Oreopithecus, Ouranopithecus, Gigantopithecus,* and *Lufengpithecus* existing within that range. In its symphyseal morphology, some *Dryopithecus* specimens are similar to *Proconsul.* Among the larger specimens of *Dryopithecus,* two have more prominent superior transverse tori with minimally expressed inferior transverse tori, and so are similar to *Proconsul.* Another has both superior and inferior transverse tori with a distinct and relatively deep genial fossa.

Sivapithecus, from Middle–Late Miocene localities in the Siwalik Hills of Indo-Pakistan, is characterized by superior and inferior transverse tori of relatively equal prominence, although there is considerable variation (Brown, 1989). The *Sivapithecus* mandibular symphysis, AMNH 19411, for example, has symphyseal contours similar to *Afropithecus* (KNM-WS 11599). AMNH 19411 is from deposits dated between 10 and 11 Ma (Barry *et al.,* unpublished data). One of the large *Sivapithecus* mandibles, GSI 18039, also has a symphyseal shape like a male *Afropithecus.* Both have relatively tall corpora with more prominent inferior transverse tori. GSI 18039 is from one of the youngest Siwalik localities at Hari Talyangar dated between 7 and 8 Ma (Barry *et al.,* unpublished data).

Shape differences seen in fossil collections of *Sivapithecus* and other Miocene hominoid mandibles have been commonly considered genus-specific.

However, *Ouranopithecus*, from late Miocene deposits in Greece, incorporates mandibles of two size classes which have been identified as either males or females on the basis of canine morphology (Bonis and Melentis, 1977, 1980; Bonis *et al.*, 1981; Bonis and Koufos, 1993; Koufos, 1993). The larger male *Ouranopithecus* mandible is quite tall and narrow with a relatively long planum alveolar extending below midcorpus just superior to a shallow genial fossa. In contrast, the smaller female *Ouranopithecus* symphysis is like that of similarly sized *Sivapithecus*, having relatively shallow, robust corpora with deep equally prominent transverse tori. One of the *Lufengpithecus* mandibles from Chinese deposits has a preserved symphysis which is also relatively narrow labiolingually. The inferior transverse torus is more prominent than the relatively indistinct superior transverse torus.

Gorilla symphyses are similar in shape to those of *Pan*. Both have relatively robust superior transverse tori often with deep genial fossae and narrow inferior transverse tori. A more prominent inferior transverse torus which extends further posteriorly than the more robust superior transverse torus is commonly described as a "simian shelf." This is typical of *Pongo*. In most hominoid mandibles the inferior transverse torus extends posteriorly to P3 or P4. In *Pongo* it is often at P4 or M1. Some large male *Gorilla* and *Pongo* symphyses have massive transverse tori, shallow genial fossae, and inferior transverse tori which extend posteriorly to M1. This is especially evident in the massive mandible of *Gigantopithecus* from Pakistan.

Pongo is unique among hominoids (extant and extinct) because an anterior diagastric muscle is absent. Although the posterior digastric muscle achieves a similar function in opening the mouth, the absence of the anterior digastric muscle has certainly affected the *Pongo* mandible, especially the area of the inferior transverse torus. Other than the platysma, there are no muscle fibers that pull at a common anterior digastric attachment site on the *Pongo* symphyseal base. A narrow "simian shelf" or increase in the posterior extent of the inferior transverse torus in *Pongo* may be associated with anterior digastric absence.

If the symphysis is modeled as an oval or an ellipse, an increase in the minor axis increases the polar moment of inertia to counteract torsional stress at the symphysis (Hylander, 1984, 1985). Medial transverse bending normally occurs during jaw opening which results in tensile stress along the labial surface. Lateral transverse bending occurs during the power stroke of mastication. This results in tensile stress along the lingual surface. An increase in the transverse dimension (labiolingual thickness) of the symphysis increases the area moment of inertia and counters tensile stress. Therefore, the superior transverse tori of *Proconsul* and some *Dryopithecus* mandibles would have compensated for a high degree of torsional and bending stresses at the symphysis caused by either medial or lateral transverse bending of the mandible.

Hylander (1984) also proposes that an adaptive response to bending stresses during long-axis corporal twisting can be accommodated in two different ways. The first is to increase symphyseal height, the second is to asymmetrically add bone to the lingual surface of the symphyseal base. The first strategy describes a difference often seen among sexually dimorphic hominoids where male mandibles are simply larger overall than female mandibles. Many of the largest male hominoid mandibles lack pronounced transverse tori evident in smaller male and female forms (compare *Afropithecus, Ouranopithecus, Gigantopithecus, Gorilla,* and *Pongo*). Their symphyses are taller and larger thus increasing the area moment of inertia as a response not only to greater size but also to greater muscle mass.

The second arrangement involves an inferior transverse torus, providing a twofold advantage. First, an enlarged inferior transverse torus bridges the lower borders of the mandibular corpora at the symphysis. This distribution of bone would counter mandibular eversion at the junction of the symphyseal base and corpora. Second, it forms an asymmetrical symphyseal section which also decreases tensile stress at the symphyseal base along the lingual border. Asymmetrical shape promotes differentiality in material strength. A greater reduction in tensile stress over compressive stress is critical for material like bone which fails more readily under tension than compression (Hylander, 1984). The functional inference is that symphyses with large inferior transverse tori, like those of extant apes and possibly *Lufengpithecus,* may be reflecting an adaptation to greater eversion on the corpus. Either strategy mentioned above would increase the area moment of inertia and minimize stress at the symphyseal section. However, an increase in symphyseal base dimension eventually becomes limited near M1 because of soft structure attachments. Most large hominoid symphyses, therefore, demonstrate an increased symphyseal height rather than an increased inferior transverse toral length.

Structure and Function in Corpus Shape

Serial coronal sections effectively demonstrate differences in anterior-to-posterior buttressing patterns in hominoid mandibles (Fig. 2). Those taken at P4 and M3 are the most informative. Anterior coronal sections are generally elliptical. Posteriorly a pronounced lateral eminence increases the transverse dimension buccally at M3. A thickened corpus at M2 also results from an anterior extension of the lateral eminence. The coronal sections of *Proconsul* manifest minimum contour change through the length of the corpus. Their most significant feature exists at P4 sections where the superior transverse torus of *Proconsul* is very prominent. This is especially exaggerated in *P. major* specimens, KNM-SO 396 and KNM-SO 404. Generally, *Proconsul,* like the extant hominoids, shows ellipsoid sections with little asymmetrical change distal to P4. Sections at P4 may be affected by either an enlarged superior transverse torus, as in *Proconsul,* or a more prominent inferior transverse torus, as in *Pongo.* At M1 the superior transverse torus of *Proconsul* may blend with a mylohyoid ridge. Coronal sections at M3 demonstrate a narrow triangular shape which shows the formation of the lateral eminence. The buccina-

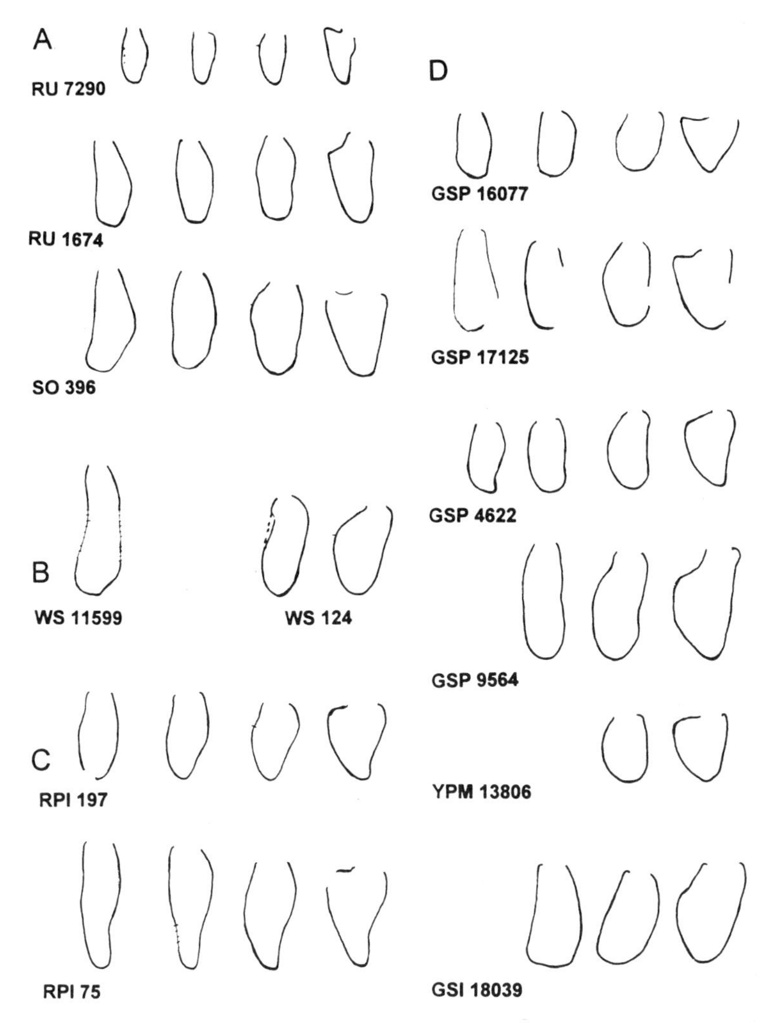

Fig. 2. Coronal sections of Miocene hominoids. Each row of coronal sections begins with the section for P4 on the left through to the section for M3 specimens on the right, when present. (A) *Proconsul* specimens as labeled. (B) *Afropithecus* specimens as labeled. (C) *Ouranopithecus* specimens as labeled. (D) *Sivapithecus* specimens as labeled.

tor groove is shallow and the oblique line develops relatively high on the corpus.

The coronal section at P4 in male *Afropithecus* is tall and narrow. Lingually the base has a depression for the anterior digastric muscle that extends through the section at M1. At M3 the lateral eminence is relatively high along the alveolar plane. The more shallow and robust form of KNM-WS 124 has a coronal section at M3 which is similar to the late Miocene forms of *Sivapithecus*. It shows the development of the lateral eminence just above midacorpus.

Most *Dryopithecus* coronal sections show areas of damage. Because of their relatively small size and narrow corpora, they are similar to *Proconsul* of a comparable size. The larger *Dryopithecus* mandibles with their shallow robust corpora have M3 sections more similar to those seen in comparably sized *Sivapithecus*. Some evidence for a submandibular fossa is present.

Sivapithecus coronal sections are quite distinctive. They clearly demonstrate the extreme sculptural relief of the *Sivapithecus* mandibular corpus. Sections at P4 are generally elliptical to oval, but there is most often an intertoral sulcus which is expressed as a concavity near the corporal lingual base. This sulcus may extend posteriorly to M2. Sections at M3 are more like rounded triangles. Both the large and smaller forms have a very prominent lateral eminence which continues inferiorly to the base. Among other hominoids, especially *Pongo*, the lateral eminence is often merely a bony swelling above midcorpus. Larger *Sivapithecus* specimens show a lateral eminence which is relatively low along the corpus, inferior to a deep buccinator groove. The largest *Sivapithecus* species, *S. parvada* (Kelley, 1988), dated around 10 Ma (Barry *et al.*, unpublished data), shows an inferior transverse torus which extends posteriorly through the M1 level. This is often seen in some of the larger male *Gorilla* and *Pongo*. The inferior transverse torus in *Sivapithecus* forms the inferior border of an intertoral sulcus which is evident as a concavity along the lingual surface. An intertoral sulcus is also present along the lingual border of most other *Sivapithecus* P4 coronal sections.

Female *Ouranopithecus* coronal sections are most like those of *Sivapithecus*, especially the section at M3. Male *Ouranopithecus* sections are taller than the female sections and more narrow or columnar than the *Sivapithecus* sections. An intertoral sulcus is quite distinct at P4 and it continues posteriorly where it blends with the submandibular fossa at M2. The buccinator groove is shallow with the lateral eminence forming high on the corpus along the alveolar plane.

Lufengpithecus has damaged corpora. However, they are also narrow and columnar with little sculptural distribution. The sections for *Gigantopithecus* from Pakistan demonstrate the extreme posterior extension of the lingual symphysis. A part of the inferior transverse torus is still present at M1. A submandibular sulcus is present at M2 and continues posteriorly toward the ramus. In this single *Gigantopithecus* specimen the lateral eminence is quite extensive at M3 with a relatively shallow buccinator groove at the alveolar plane.

Both coronal sections and transverse sections allow one to compare lingual versus buccal buttressing patterns along the corpus. Because of the extensive superior transverse torus in *Proconsul*, most buttressing is along the anterolingual portion of the corpus. Comparatively, posterobuccal buttressing is small. The opposite is true for *Sivapithecus:* posterobuccal buttressing is more extensive than anterolingual. The buttressing for *Afropithecus* is similar along both sides of the occlusal line. Although variable, buttressing in *Dryopithecus* seems to be more prominent on the distobuccal surface. In

Ouranopithecus the corpora are quite narrow both anteriorly and posteriorly, and the buttressing is relatively equal. *Lufengpithecus* has little buttressing.

Coronal sections of modern ape mandibles show generally elliptical sections and a sexually dimorphic size difference, primarily in *Gorilla* and *Pongo*. In both *Gorilla* and *Pan* the sections at P4 display an anterior digastric fossa at the base, the swelling of the inferior transverse torus sometimes extending to M1, and a slight intertoral sulcus. The M3 section shows lateral eminence formation relatively high on the corpus generally with a shallow buccinator groove. Coronal sections of *Pongo* are more narrow and columnar. As mentioned above, *Pongo* coronal sections at P4 display more extensive inferior transverse tori that extend posteriorly as a true simian shelf. M3 sections are narrow with a relatively gracile lateral eminence. Although *Pongo* has distinctive sections at P4 and M3, *Ouranopithecus* and the larger *Afropithecus* have coronal sections similar to those of some male *Pongo*. Unlike the Miocene hominoids, coronal sections of *Pongo* at M3 are quite gracile and narrow. The lateral eminence either has minimal expression or appears as a swelling on the superior half of the corpus which is countered by a narrow base with a relatively deep submandibular fossa.

Body Size and Sexual Dimorphism

A complicating factor with respect to the relationship between corpus sectional geometry and functional reconstruction is body size. Among hominoids, males are typically larger in body size than females. Most hominoid taxa are currently recognized as sexually dimorphic, but there is always a question of how dimorphic a fossil really is. Sexual dimorphism is difficult to assess in Miocene hominoid samples, especially those of *Proconsul* and *Sivapithecus*, because multiple species may overlap within variable stratigraphic ranges. *Proconsul* exhibits differential morphologies associated with a large size diversity among species including those that are clearly sexually dimorphic. Among the *Sivapithecus* mandibular sample there are three different size morphologies within one taxon.

The smallest size class of both *Proconsul* and *Sivapithecus* has columnar and ellipsoid coronal sections. They lack the robusticity or characteristic heavy buttressing evident in many of the larger adult hominoid mandibles. Smith (1983) reports that a greater corpus robusticity occurs with increases in body size in a number of female primates. Intermediate to large *Sivapithecus* mandibles do exhibit greater variation in overall corpus buttressing. The intermediate-sized specimens of *Sivapithecus* present the classic "ramamorph" shape characterized by robust, shallow, and broad corpora. This morphology tends toward a triangular coronal section posteriorly at M3 where the lateral eminence is most prominent. It typifies intermediate-sized *Sivapithecus*, *Ouranopithecus* females, and some *Dryopithecus*, and is highly effective in resisting torsion. Coronal sections taken anterior to M_2 of the intermediate-sized

Sivapithecus and smaller *Ouranopithecus* are elliptical tending toward a more circular shape.

Larger *Sivapithecus* mandibles are both tall and transversely broad. Tall, broad corpora with a relatively large cross-sectional area are assumed to be adapted to withstand large sagittal bending loads on the balancing side, and large torsional and shearing loads incorporated on the working side during unilateral mastication (Hylander, 1979). For example, Kelley and Pilbeam (1986) submit that the corpus of GSP 9564, a large *Sivapithecus* mandible, withstood significant twisting moments resulting from great molar biting and masticatory force. Large twisting forces could also be inferred from the oblique position of the tooth axes and corpora such that the basal portion of the corpus is angled farther buccally than the alveolar portion. Powerful incisal biting may also influence corpus shape by producing twisting moments along the long axis of the corpus. However, Kelley and Pilbeam (1986) propose incisal biting as secondary to more powerful and consistent forces during mastication and molar biting. Male *Afropithecus* and *Ouranopithecus* mandibles do not have such extensive lateral buttressing. Thus, using Hylander's model, the tall narrow mandibles of *Afropithecus* and *Ouranopithecus* are more adapted for sagittal bending than large torsional loads.

Differences in shape between taller versus shallower corpora are mechanically analogous to shape differences among other skeletal features. Since hollow tubular sections are designed to resist the greatest torsional stress (Frankel and Burstein, 1965, 1970), one can make inferences concerning the strength of the mandibular corpus if it is modeled as a tube (Hylander, 1988; Daegling, 1989). A section of bone that is columnar, or mediolaterally flattened, versus one that is more triangular with a greater mediolateral breadth is assumed to withstand greater bending strain in an anterior–posterior direction as well as greater torsional strain. However, this morphological difference results in a decrease in resistance if subjected to mediolateral bending (Lovejoy *et al.*, 1976). A triangular shape is most efficient in resisting both torsion and bending stress from a variety of directions rather than a circular section which resists torsion from constant directions (Frankel and Burstein, 1965, 1970). This condition is quite analogous for some coronal sections at M2 and most coronal sections at M3 for the intermediate-sized *Sivapithecus* and smaller *Ouranopithecus*. Coronal sections at M3 correspond to the junction between the mandibular corpus and ascending ramus. At this junction the torsional and bending stress come from a variety of directions since most masticatory muscle forces and resultant occlusal forces occur within this region.

If mandibles are shaped and remodeled primarily as an adaptation to distinct loading regimes, are we to infer that male and female mandibles within a species are loaded differently? Based on their sectional geometries, male mandibles of *Pongo, Gorilla,* and *Ouranopithecus* seem to be adapted for higher bending stress during unilateral biting and, as in some primates, powerful incisor biting (Hylander, 1979). On the other hand, female mandibles from these taxa appear to be adapted to withstand greater torsional stress. If variation in mandibular depth does reflect specific functional responses, then it becomes necessary to reconcile differing biomechanical parameters that characterize males and females in highly dimorphic species. In biomechanical analyses among *Gorilla, Pongo,* and *Pan,* cross-sectional shape reveals no significant difference in mandibular mechanical design following correction for size (Daegling, 1989). The transverse dimension of the mandibular corpora in both male and female *Ouranopithecus* specimens is relatively equal. Thus, there is simply an increase in male corpus height, rather than the corresponding structural change as is seen in most other hominoids.

Selection pressures on large size in male hominoids are most often assumed to aid in predator defense and reproductive success. However, a greater body size usually results in a longer growth period. Some changes in male mandibular morphology not seen among females are caused by the increased growth of the male canine. In a number of primates including modern *Homo sapiens,* a large male canine grows steadily in length after puberty (Walker and Kowalski, 1972; Shea, 1985, 1986). A larger muscle mass and deeper corpora are also associated with the increased depth of the male canine root (Smith, 1983). Although there does not appear to be a significant difference in length of molar and premolar roots, there is a great difference in canine crown size among males and females. A male mandible must allocate an increased alveolar process volume and consequently greater anterior corpus depth to accommodate the larger canine. In addition, the male corpus must have a greater muscle mass to accommodate an increase in both corporal and symphyseal mass.

Allometric scaling may play a role in the observed differences in size between male and female corpora. The increased size of male mandibular corpora may be a response to the greater muscle mass needed for chewing fibrous foods over longer periods. For example, orangutans normally eat seasonal fruits and seeds, but male orangutans were observed to consume selected barks and leaves only when fruit was not abundant (Rodman, 1988). Reported differences in observed male/female food preference have been contradictory, however (Rodman, 1977, 1978; Galdikas, 1978). Also, little difference is seen in male/female diet or behavior among mountain gorillas that also have dimorphic mandibles (Remis, 1994). Differences in food variety, feeding bouts, and food volumes are dependent on body size as well as substrate use, seasonability, and female lactation periods (Remis, 1994). Clearly, the interdependence of each of these factors plus basic physiological constraints are highly complex.

Phylogenetic Perspectives

Subocclusal Anatomy

Intergeneric variation in mandibular morphology includes many features that are often unaccounted for or unseen. Assessment of subocclusal relationships can provide evidence for differences in growth and development affecting root structure and form. These differences may be broadly related to phylogenetic assessments insofar as root structure can be used for character analysis. The angulation and structure of mandibular root systems can only be visualized with the use of lateral and symphyseal X rays of the alveolar processes, and coronal CT imaging. In specimens of extant hominoids, radiographs demonstrate aspects of foot morphology such as root angulation, length, extent of pulp chamber, and spatial relationships associated with corpus height, and the inferior alveolar canal (Fig. 3).

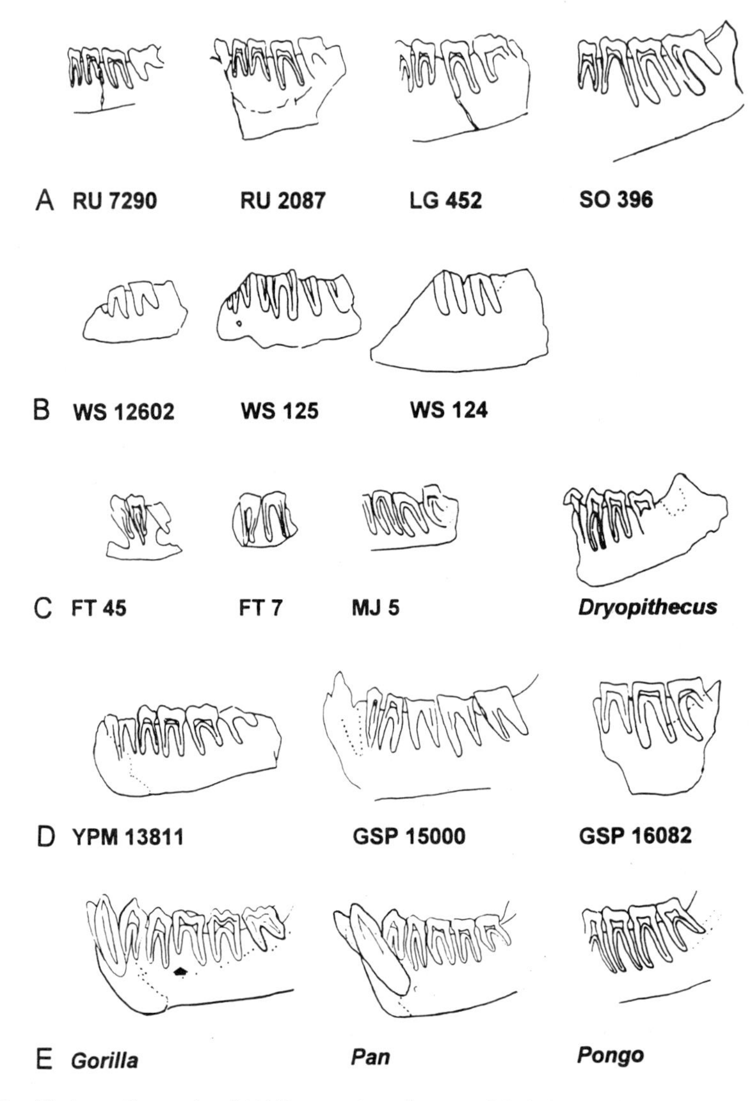

Fig. 3. Mandibular radiographs of (A) *Proconsul* specimens as labeled, (B) *Afropithecus* specimens as labeled, (C) *Kenyapithecus* specimens as labeled and *Dryopithecus*, (D) *Sivapithecus* specimens as labeled, (E) *Gorilla* female (note short mesial M1 root), *Pan* male, and *Pongo* female.

Among the extant hominoids, *Gorilla* is noted for having the mesial M1 root shorter than the distal M1 and the roots of both P4 and M2. Although variation exists, both *Pan* and *Pongo* have premolar and molar root lengths that are relatively constant from P4 to M3. Premolar and molar root lengths also show little variation with changes in hominoid corpus height. The subocclusal anatomy of *Pongo* features long and narrow postcanine roots, which is similar to that of *Sivapithecus*. Many Miocene hominoid mandibles are also amenable to standard radiography and it is possible to compare aspects of root morphologies (Fig. 3).

Proconsul exhibits great variability in root length. Three specimens from Songhor (KNM-SO 391, KNM-SO 396, and KNM-SO 404) have short mesial M1 roots with the distal M1 root being the same length as that of the M2. The P4 root lengths are similar to those of the mesial M1. This form is reminiscent of *Gorilla* root structure. However, the P4 root length for the Songhor specimens is relatively comparable to that of the mesial M1 root rather than being as long as the distal M1 and M2 roots as in *Gorilla*. The older specimen, KNM-LG 452, considered by some to be the same species as KNM-SO 396 and KNM-SO 404, has a morphology in which both roots of M1 are shorter than those of M2. Generally, specimens from Rusinga have roots of approximately the same length, except that the larger specimens, KNM-RU 1674 and KNM-RU 2087, differ in root length. KNM-RU 1674 displays shorter roots for P4 and M1 than for M2 and mesial M3 like that of KNM-LG 452. This is similar to the general root pattern of *Sivapithecus*. KNM-RU 2087 has M1 roots shorter than M2 with an additionally shortened mesial M1 root which also seems apparent in KNM-RU 1678 and KNM-RU 1728. *Rangwapithecus* specimens have the same root morphology as most *Proconsul* specimens of their size class.

Although fragmentary, *Afropithecus* and *Kenyapithecus* specimens preserve roots that have been broken along the alveolar margin or just below it. Few pulp chambers are complete. An *Afropithecus* mandible from Buluk, KNM-WS 125, has an M1 mesial root that appears slightly shorter than its neighboring molar roots. The smaller specimen, KNM-WS 12602, has relatively short, thick roots that extend to midcorpus. The small *Kenyapithecus* sample is also of interest. The M1 mesial roots of *Kenyapithecus* KNM-FT 7 and KNM-FT 45 are of a similar length to the M2 roots of KNM-FT 7 and the P4 roots of KNM-FT 45, respectively, The Kaloma mandible, KNM-MJ 5, has roots like those from Ft. Ternan. They are relatively long, narrow, and of equal length. They extend beyond midcorpus level close to the base.

Radiographs of *Dryopithecus*, published by Simons and Meinel (1983), display long narrow roots for P4–M2. Lateral mandibular radiographs of *Sivapithecus* demonstrate that most of the largest specimens have M1 roots that are distinguished by being shorter in length than those of the M2 and M3 roots, while the intermediate-sized specimens have molar roots of similar length. The general root pattern of *Sivapithecus* is like that seen in the proconsulids KNM-RU 1674 and KNM-LG 452. A size factor including sexual dimorphism may also pertain to this observed similarity among the larger

Rusinga specimens and those of intermediate-sized *Sivapithecus,* for example, or other comparably sized hominoids for which data are lacking.

Life-History Strategies

It may be necessary to consider additional factors besides biomechanics, diet, and positional behavior when constructing taxonomic exercises on fossil samples. In the case of Miocene hominoid mandibles, elements of life history may also influence mandibular form beyond basic jaw mechanics. For example, a complex social structure that includes selection for dimorphism could affect mandibular form. A case in point is the life-history pattern of male orangutans. Morphological and behavioral differences occur in the two subspecies of *Pongo, P p. pygmaeus,* and *P. p. abelii* from Borneo and Sumatra, respectively. Mandibles of both subspecies are differentiated primarily on the basis of the anterior corpus which is deeper and broader anterioposteriorly in *P. p. pygmaeus* and *P. p. abelii* (Brown, 1989). Bornean orangutan mandibles exhibit symphyseal sections that are generally larger, thicker, and more bulbous than those from Sumatra (Fig. 4). Those of Sumatran mandibles are long

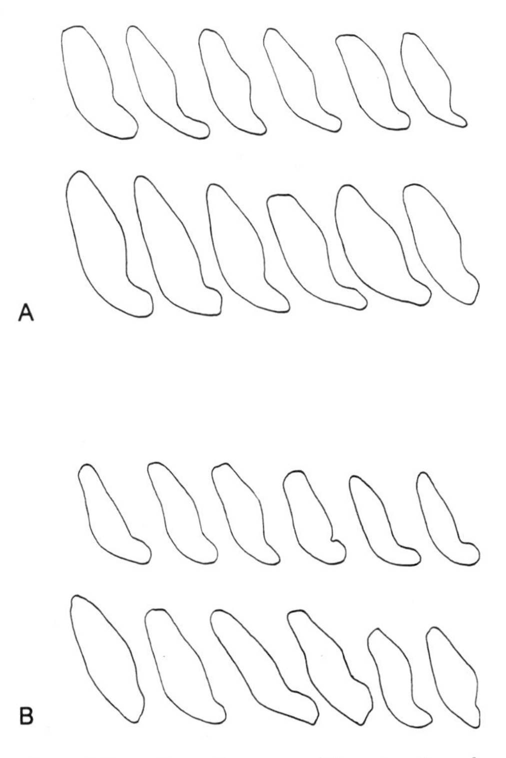

Fig. 4. Symphyseal sections of *Pongo* from Borneo and Sumatra. Females are shown in the top row, males in the bottom row. (A) *Pongo pygmaeus pygmaeus,* (B) *Pongo pygmaeus abelii.*

and narrow with extensive inferior transverse tori. Despite these shape differences, the symphyseal surface areas are generally equal among Bornean and Sumatran orangutans. As mentioned previously, an increase in transverse dimension at the symphysis, by increasing either the superior transverse torus or the inferior transverse torus, reduces bending moments at the symphysis (Hylander, 1979, 1984, 1985). Consequently, little biomechanical difference exists among the symphyses of the two *Pongo* subspecies. A relatively thin symphysis with a minimal superior transverse torus and extensive inferior transverse torus achieves the identical function as a robust superior transverse torus with minimal inferior transverse torus, assuming a similarity in symphyseal depth. Subspecific differences in mandibular structure are therefore influenced by factors such as feeding preferences or behavior involving epigamic display. Morphological differences characterizing *Pan troglodytes* subspecies have also been reported (Shea, 1985).

In *Pongo pygmaeus* (Brown, 1989) there are male mandibles that have an overall structure similar to that of females. Apart from their definitively male canines, mandibles with associated crania can be easily mistaken as female. Not only is there extreme variability in the corpus size and structure among orangutan males and females, but there is also variability among the males within this single species. The small male specimens have fully occluded canines, M3s with slight wear, greater wear on the M2s, and buccally exposed secondary dentine on the M1s. In most normal circumstances their crania are classified as adult.

Rijksen (1978) describes two variants of *P. p. abelii;* a dark-haired, long-fingered type makes up approximately 25% of the population and its features could indicate a recessive trait. The presence of a recessive trait for a smaller form of Sumatran orangutan may therefore represent three male morphologies. Either smaller males are small because of a recessive gene as in the dark-haired, long-fingered *P. p. abelii* labeled "sneakers" as seen in sunfish (Gross, 1979; Gross and Charnov, 1980, as cited in Daly and Wilson, 1983), or they are repressed hormonally because of subordinate status. MacKinnon (1974) and others (Maple, 1980; Rodman and Mitani, 1987; Rodman, 1988) suggest an inhibition mechanism may occur in the subadult male maturity process among captive orangutans. The speculation that it is a result of testosterone suppression was tested by Kingsley (1982) who documented a correlation between male dominance, testosterone suppression, and secondary sexual character growth. Similar behavioral/hormonal effects are also known to occur among adult male rhesus macaques (Bernstein *et al.*, 1974) and other primates including humans (Kingsley, 1982).

A combined sample of orangutans from both Borneo and Sumatra could manifest four to six types of male skeletal morphologies. These are: Bornean and Sumatran dominant males, Bornean and/or Sumatran subordinant males, smaller because of testosterone suppression, and Bornean and/or Sumatran subordinant males, smaller because of a recessive gene (Brown, 1989). It is also possible that smaller male mandibles with associated gracile crania represent males that did not fully develop into potential alpha males. A sample of male mandibles with tall, robust corpora are not exclusive to mature males. Also, small mandibles with male canines and a morphology similar to females are not exclusive to *P. p. abelii*. Whatever the cause, either developmental or genetic, variation in male corpora is extensive within this single species.

It will still be quite some time before we fully understand the effects of intraspecies body size dimorphism on primate life history patterns among extant species, let alone our hypotheses on those few skeletal differences evident in fossil samples. With respect to feeding behavior, we are becoming more knowledgeable in identifying life history parameters. For example, we now know that in free-ranging orangutans, subadult males are sequestered from other males and females. The observed morphological differences among extant orangutan populations may mirror the variability that we see in *Sivapithecus* mandibular samples, for example. Therefore, reproductive behavior could influence morphology, thus complicating our efforts to reconstruct phylogenies.

The range of variation seen among extant *Pongo pygmaeus* could occur in fossil populations and must be accounted for when species-level or generic attributions are contemplated. The few mandibles that do not fall neatly within a cluster of points expressing mandibular shape or structure may be indicative of life history patterns that strongly influence maturation of the individual and skeletal morphology. However, one of the principal observations that one can make for small male mandibles is to see if they are fully adult. Cranial material is most often classified as adult when the M3s are in occlusion. Often M3s are in occlusion, but the male canine root has not completed development. If mandibles are radiographed, an assessment of the degree of root apex closure, especially of the canines, is evident. The canine root of *Otavipithecus namibiensis* (Conroy *et al.*, 1992), for example, is not yet fully fused (Conroy *et al.*, 1992). Therefore, the Otavi mandible may have developed a deeper mandibular corpus later in life.

Sensitivity to the process of hominoid mandibular growth trajectories to maturity is clearly essential in taxonomic analysis of both extant and extinct specimens. It is imperative that the status of canine development and third-molar occlusion be evaluated for proper taxonomic assignment for most fossil hominoid mandibles. The symphysis with the canine or canine root should be present, although we are often not that fortunate.

Conclusion

There is still considerable uncertainty about the relationships between skeletal anatomy and behavior. Despite extensive experimental work and application of engineering principles to the study of primate mandibular form and function, mandibles alone cannot adequately answer questions concerning function and masticatory mechanics. However, they do provide information on tooth angulation, buttressing patterns of surrounding alveolar tissue, the angulation of the ascending ramus, and inferior cranial attachment sites for masticatory musculature. These are all useful features in functional reconstructions and taxonomic exercises. I have shown here that across the entire span of the Miocene, hominoid mandibles vary enormously in size and shape. I have also attempted to illuminate problems associated with interpreting this variation by noting components of social and reproductive behavior that can demonstrably affect anatomical and metrical attributes of hominoid mandibles. There never will be a single useful formula whereby a given mandible can be associated with a specific diet, habitat, or species. However, when mandibles are considered within the larger context of the masticatory apparatus, including occlusal and subocclusal dental anatomy, muscle attachments, and internal architecture, much can be learned. Hopefully, new experimental approaches to mandibular forms (Chen and Povirk, 1996) will promote a better understanding of how primate mandibles function, what shape differences mean biomechanically, and how intraspecific variations in size and shape reflect behavioral patterns. We can assume that mandibles and mandibular fragments will continue to be prevalent in the Miocene hominoid fossil record, and continual effort must be extended to interpret their functional biology.

References

Andrews, P. J. 1978. A revision of the Miocene Hominoidea of East Africa. *Bull. Br. Mus. Nat. Hist. Geol.* **30:**85–224.

Bernstein, I. S., Rose, R. M., and Gordon, T. P. 1974. Behavioral and environmental events influencing primate testosterone levels. *J. Hum. Evol.* **3:**517–525.

Bonis, L. de, and Koufos, G. D. 1993. The face and the mandible of *Ouranopithecus macedoniensis:* Description of new specimens and comparisons. *J. Hum. Evol.* **24:**469–491.

Bonis, L. de, and Melentis, J. 1977. Les primates hominoides du Ballesien de Macedoine (Greece). Etude de la Machoire inferieure. *Geobios* **10:**849–885.

Bonis, L. de, and Melentis, J. 1980. Nouveles remarquest sur l'anatomie d'un Primate hominoide du Miocene: *Ouranopithecus macedoniensis.* Implications sur la phylogenie de Hominides. *C. R. Acad. Sci.* **290:**755–758.

Bonis, L. de, Johanson, D., Melentis, J. and White, T. 1981. Variations metriques de la denture chez les Hominides primitifs: Comparaison entre *Australopithecus afarensis* et *Ouranopithecus macedoniensis. C. R. Acad. Sci.* **292:**373–376.

Brown, B. 1989. *The Mandibles of Sivapithecus.* Ph.D. dissertation, Kent State University, Kent, Ohio.

Brown, B., Ward, S. C., and Hill, A. 1995. New *Kenyapithecus* partial skeleton from the Tugen Hills, Baringo District, Kenya. *Am. J. Phys. Anthropol. Suppl.* **20:**69.

Chen, X., and Povirk, G. 1996. Assessing errors introduced by modeling the anisotropic human mandible isotropically with finite element method. *Am. J. Phys. Anthropol. Suppl.* **22:**83.

Conroy, G. C., Pickford, M., Senut, B., van Couvering, J., and Mein, P. 1992. *Otavipithecus namibiensis,* first Miocene hominoid from southern Africa. *Nature* **356:**144–148.

Daegling, D. J. 1989. Biomechanics of cross-sectional size and shape in the hominoid mandibular corpus. *Am. J. Phys. Anthropol.* **80:**91–106.

Daegling, D. J., Ravosa, M. J., Johnson, K. R., and Hylander, W. 1992. Influence of teeth, alveoli, and periodontal ligaments on torsional rigidity in human mandibles. *Am. J. Phys. Anthropol.* **89:**59–72.

Daly, M., and Wilson, M. 1983. *Sex, Evolution, and Behavior.* Willard Grant Press, Boston.

Frankel, V. H., and Burstein, A. H. 1965. Load capacity of tubular bone. In: *Biomechanics and Related Bio-Engineering Topics.* Pergamon, London, p. 382.

Frankel, V. H., and Burstein, A. H. 1970. *Orthopaedic Biomechanics.* Lea & Febiger, Philadelphia.

Galdikas, B. M. F. 1978. *Orangutan Adaptation at Tanjung Puting Reserve, Central Borneo.* Ph.D. dissertation, University of California, Los Angeles.

Gross, M. R. 1979. Cuckoldry in sunfishes. (*Lepomis:* Centrarchidae). *J. Zool.* **57:**1507–1509.

Gross, M. R., and Charnov, E. L. 1980. Alternative male life histories in bluegill sunfish. *Proc. Natl. Acad. Sci. USA* **77:**6937–6940.

Hylander, W. L. 1979. The functional significance of primate mandibular form. *J. Morphol.* **160:**223–239.

Hylander, W. L. 1984. Stress and strain in the mandibular symphysis of primates: A test of competing hypotheses. *Am. J. Phys. Anthropol.* **64:**1–46.

Hylander, W. L. 1985. Mandibular function and biomechanical stress and scaling. *Am. Zool.* **25:**315–330.

Hylander, W. L. 1988. Implications of *in vivo* experiments for interpreting the functional significance of "robust" australopithecine jaws. In: F. Grine (ed.), *Evolutionary History of the "Robust" Australopithecines,* pp. 55–83. Aldine De Gruyter, New York.

Ishida, H., Pickford, M., Nakaya, H., and Nakano, Y. 1984. Fossil anthropoids from Nachola and Samburu Hills, Samburu District, Kenya. *Afr. Stud. Monogr. Suppl.* **2:**73–85.

Kelley, J. 1988. A new large species of *Sivapithecus* from the Siwaliks of Pakistan. *J. Hum. Evol.* **17:**305–324.

Kelley, J., and Pilbeam, D. 1986. The dryopithecines: Taxonomy, comparative anatomy, and phylogeny of Miocene large hominoids. In: D. R. Swindler and J. Irwin (eds.), *Comparative Primate Biology, Vol. I: Systematics, Evolution and Anatomy,* pp. 361–411. Liss, New York.

Kingsley, S. 1982. Causes of non-breeding and the development of the secondary sexual characteristics in the male orang utan: A hormonal study. In: L. E. M. de Boer (ed.), *The Orangutan. Its Biology and Conservation,* pp. 215–229. Junk, The Hague.

Koufos, G. D. 1993. Mandible of *Ouranopithecus macedoniensis* (Hominidae, Primates) from a new Late Miocene locality of Macedonia (Greece). *Am. J. Phys. Anthropol.* **91:**225–234.

Leakey, R. E., and Walker, A. C. 1985. New higher primates from the Miocene of Buluk, Kenya. *Nature* **318:**173–175.

Leakey, R. E., Leakey, M. G., and Walker, A. C. 1988. Morphology of *Afropithecus turkanensis.*

Lovejoy, C. O., Burstein, A. H., and Heiple, K. G. 1976. The biomechanical analysis of bone strength: A method and its application to platycnemia. *Am. J. Phys. Anthropol.* **44:**489–506.

McCrossin, M., and Benefit, B. R. 1993. Recently recovered *Kenyapithecus* mandible and its implications for great ape and human origins. *Proc. Natl. Acad. Sci. USA* **90:**1962–1966.

MacKinnon, J. 1974. The behavior and ecology of wild orangutan populations. In: D. A. Hamburg and E. R. McCown (eds.), *The Great Apes,* pp. 257–274. Benjamin/Cummings, Menlo Park, CA.

Maple, T. L. 1980. *Orang-utan Behavior.* Litton Educational Publishing, New York.

Pickford, M. 1982. New higher primate fossils from the middle Miocene deposits at Majiwa and Kaloma, western Kenya. *Am. J. Phys. Anthropol.* **58:**1–19.

Pickford, M. 1985. A new look at *Kenyapithecus* based on recent discoveries in western Kenya. *J. Hum. Evol.* **14:**113–143.

Ravosa, M. J. 1991. Structural allometry of the prosimian mandibular corpus and symphysis. *J. Hum. Evol.* **20:**3–20.

Remis, M. 1994. *Feeding Ecology and Positional Behavior of Western Lowland Gorillas (Gorilla gorilla gorilla) in the Central African Republic.* Ph.D. dissertation, Yale University.

Rijksen, H. D. 1978. A field study on Sumatran orang utans (*Pongo pygmaeus abelii* Lesson 1827): *Ecology, behavior, and conservation.* H. Veenman & Zonen B. V. Wageningen, The Netherlands.

Rodman, P. S. 1977. Feeding behavior of orangutans of the Kutai Nature Reserve, East Kalimantan. In: T. H. Clutton-Brock (ed.), *Primate Ecology: Studies of Feeding and Ranging Behavior in Lemurs, Monkeys, and Apes.* pp. 383–413. Academic Press, New York.

Rodman, P. S. 1978. Diets, densities, and distributions of Bornean primates. In: G. G. Montgomery (ed.), *The Ecology of Arboreal Folivores,* pp. 465–478. Smithsonian Institution Press, Washington, DC.

Rodman, P. S. 1988. Diversity and consistency in ecology and behavior. In: J. H. Schwartz (ed.), *Orang-utan Biology,* pp. 21–51. Oxford University Press, London.

Rodman, P. S., and Mitani, J. C. 1987. Orangutans: Sexual dimorphism in a solitary species. In: B. B. Smuts, D. L. Cheney, R. M. Seyfarth, R. W. Wrangham, and T. T. Struhsaker (eds.), *Primate Societies,* pp. 146–154. University of Chicago Press, Chicago.

Shea, B. T. 1985. The ontogeny of sexual dimorphism in the African apes. *Am. J. Primatol.* **8:**183–188.

Shea, B. T. 1986. Ontogenetic approaches to sexual dimorphism. *Hum. Evol.* **1:**97–110.

Simons, E. L., and Meinel, W. 1983. Mandibular ontogeny in the Miocene great ape *Dryopithecus. Int. J. Primatol.* **4:**331–337.

Smith, R. J. 1983. The mandibular corpus of female primates: Taxonomic, dietary, and allometric correlates of interspecific variations in size and shape. *Am. J. Phys. Anthropol.* **61:**315–330.

Walker, G. F., and Kowalski, C. J. 1972. On the growth of the mandible. *Am. J. Phys. Anthropol.* **36:**111–118.

Paleobiological and Phylogenetic Significance of Life History in Miocene Hominoids

9

JAY KELLEY

Features by which extant hominoids are distinguishable from cercopithecoids can be readily identified. Anyone schooled in primate biology could generate a comprehensive list of these features; a less comprehensive although still generally accurate list could probably be generated by any observant individual. From such a list it would not be too difficult to come up with a short characterization of hominoids that unified the various features into a few more generalized traits, abilities, and so forth. I suspect there would be sufficient conformity in these characterizations such that there would be few disagreements as to how to define a hominoid. It is surely the case, however, that character lists would become shorter and the entire exercise more challenging the further back one went into the hominoid record.

In fact, as is well known, this kind of exercise has been carried out repeatedly, both before there was any catarrhine fossil record to speak of and since that record has expanded to an unanticipated degree. It seems, though, that at least the more recent efforts along these lines have been less concerned with

JAY KELLEY • Department of Oral Biology, College of Dentistry, University of Illinois at Chicago, Chicago, Illinois 60612.

Function, Phylogeny, and Fossils: Miocene Hominoid Evolution and Adaptations, edited by Begun *et al.* Plenum Press, New York, 1997.

what a hominoid is in a broad biological sense than with whether or not particular genera should be considered hominoids, the same duality that has also enlivened the debate over primate origins during the last two decades. It is the former question, however, that to my mind is the more intriguing, all the more so since the latter question seems to have turned many away from broader biological considerations of the groups whose phylogenies they are attempting to work out. It is my perception that one of the themes of this volume is to reverse this tendency and to encourage a return to the underlying biology of morphology to enlighten phylogenetic studies. It is also my objective to emphasize that morphology can inform about other kinds of biological traits, traits that can also be discussed in terms of function and that have a place in considerations of phylogeny.

What Is a Hominoid?

Traditional Anatomical Definitions

This question can be addressed from a strictly phylogenetic (cladistic) perspective or from a more traditional adaptationist perspective. A cladistic definition is a matter of recognition and inclusivity, that which allows one to recognize an animal as a member of the hominoid clade regardless of the biological significance of the feature(s) used to make the determination. An adaptive definition is a matter of characterization, a description of the attributes that confer biological (e.g., behavioral, ecological, demographic) distinctiveness on hominoids as a group. Although there is a clear distinction between the two definitions, they are not necessarily mutually exclusive. Rather, they have different points of emphasis. In principle, a synapomorphy of a group can be anything from an important biological attribute to a simple anatomical oddity; cladogenesis does not necessarily have to result in a fundamental divergence in biology even in what ultimately become higher taxa (e.g., divergence through vicariance). The thrust of this contribution is not so much about defining the inclusivity of the Hominoidea, as about attempting to identify unifying adaptive attributes of those animals that most of us can comfortably place within the Hominoidea.

More adaptively oriented definitions of the Hominoidea have traditionally centered on the anatomy and behaviors related to suspensory posture and locomotion, usually with a more specific link to "brachiation" (Keith, 1923). Again, this does not mean that phylogeny was not a consideration. Earlier discussions prior to the introduction of the more formalized constructs of cladistic thinking were in fact often implicitly or explicitly phylogenetic; it is simply that the thrust was often more adaptive conceptually. Historically, however, brachiation is a much-abused and often ill-defined term, embodying very different sets of behaviors for different workers (see summaries and references in Hunt, 1992, and Tuttle, 1969, 1974).* The anatomy and behaviors relating to strictly forelimb suspension arguably constitute a compelling adaptive characterization of the Hominoidea if we restrict the analysis to extant genera (Hunt, 1992). Admittedly, many of the anatomical features associated with suspension can also be plausibly interpreted with respect to other positional behaviors, most notably "climbing," especially vertical climbing (Fleagle *et al.*, 1981; Stern, 1971; Stern *et al.*, 1977), or cautious, deliberate quadrupedalism with extended postures (Cartmill and Milton, 1977). But as Hunt (1992) has observed, these constellations of behaviors are neither well defined nor particularly important for all large hominoids. Moreover, they are routinely utilized by many primates with a variety of postcranial morphologies. Finally, they seemingly would not confer the same adaptive benefits as suspension (Hunt, 1992; see below).

With an expanding fossil record, it quickly became apparent that earlier genera of presumed hominoids could not be encompassed by a definition that is based on suspensory positional behavior. Such a definition would certainly exclude *Proconsul* and its relatives, all of which were clearly unspecialized pronograde quadrupeds (Beard *et al.*, 1986; Begun *et al.*, 1994; Rose, 1993, 1994; Walker and Pickford, 1983; Ward, 1993; Ward *et al.*, 1993). For some, though (e.g., Harrison, 1987), the few presumed synapomorphies linking proconsulids (*Proconsul, Afropithecus, Turkanapithecus?*) with extant hominoids are questionable, making the phylogenetic position of these genera equivocal at best (but see Rose, 1992, this volume).

Regardless, a definition based on suspension would now apparently exclude *Sivapithecus* as well, which is linked to the extant orangutan by a suite of uniquely shared cranial features (Andrews, 1992; Andrews and Cronin, 1982; Ward and Brown, 1986; Ward and Pilbeam, 1983). Not only are there few features within the known *Sivapithecus* postcranium that suggest habitual use of suspensory behaviors (Rose, 1986, 1993, 1994), more recent finds of the upper limb suggest instead a somewhat surprising commitment to pronograde quadrupedality, and perhaps terrestriality (Pilbeam *et al.*, 1990; Spoor *et al.*, 1991). In certain behaviorly diagnostic features of the humerus (Pilbeam *et al.*, 1990), phalanges (Rose, 1986), carpus (Spoor *et al.*, 1991), and tarsus (Rose, 1994), *Sivapithecus* is strikingly reminiscent of *Proconsul* and both are generally similar to more primitive catarrhines, all presumably sharing something close to the primitive anthropoid conditions. It must be noted that

*This matter touches on the rich and often colorful literature on the phylogenetic relationships of humans within Primates (e.g., Straus, 1949). Virtually all protagonists in this century-long debate, regardless of differing opinions regarding the place of humans, viewed features relating to suspensory posture and locomotion ("brachiation") as the defining features of apes, implicitly or explicitly understood as derived from common descent (see Straus, 1949; Tuttle, 1969, 1974).

most of these specimens represent a species, *S. parvada,* that is substantially larger than other species of *Sivapithecus* (Kelley, 1988). They may not, therefore, be representative of the genus as a whole, but there is as yet nothing in the more limited postcranial remains of other *Sivapithecus* species to suggest that suspensory positional behaviors were habitually used. If *Sivapithecus* is in fact on the *Pongo* lineage, then not only are suspensory adaptations and behavior not defining characteristics of the Hominoidea, they would not even be synapomorphies of the great ape and human clade (Andrews, 1992, Pilbeam *et al.*, 1990). They would have had to be acquired in parallel in the African and Asian great ape lineages (unless *Sivapithecus* has secondarily reacquired pronograde posture). The implications of the *Sivapithecus* postcranium for attempts to define the Hominoidea are unchanged even if *Sivapithecus* is not on the orangutan lineage since it is still surely a hominoid based on the possession of numerous hominoid synapomorphies of the cranium (Andrews and Martin, 1987) and postcranium (Rose, this volume).

Interpreting the Anatomy of Early Hominoids

If the Hominoidea cannot be defined with regard to suspensory positional behavior, are there clues within hominoid anatomy that might suggest other biologically distinctive attributes by which they can be defined? As a beginning we might look to a list of suggested hominoid synapomorphies since, among catarrhines, these are presumably both exclusive and common (unless further modified) to all early as well as later hominoids. The list is limited to extant hominoid synapomorphies also present in *Proconsul* (and in *Afropithecus* and perhaps *Turkanapithecus* to the extent that they are known) but not in other early Miocene or earlier taxa. These are ennumerated in Table I together with the sources from which they were compiled.

The dental traits might indicate a dietary shift from that of the hominoid–cercopithecoid common ancestor but it is difficult to envision what fundamental change in diet would produce these seemingly disparate and, in some cases, relatively trivial changes. Further, the more important shift in diet would appear to have taken place in the cercopithecoid lineage, associated with the advent of bilophodonty (Andrews, 1981; Andrews and Aiello, 1984). The issue of brain size increase is interesting and, relative to postcranial skeletal features, has received little consideration in discussions of hominoid (as opposed to extant great ape and human) adaptations, perhaps for good reason since information on fossil hominoid brain sizes has been all but unavailable. I will return to the issue of brain size below.

By contrast, postcranial features, including some of those listed in Table I, have been the focus of much attention in considerations of *Proconsul* in particular, or of hominoids in general. This is especially true of forelimb features. The features listed in Table I can be associated with a short list of functional attributes (Rose, this volume, personal communication), which can

Table I. Presumed Hominoid Synapomorphies[a]

Character	Source
Craniodental	
1 Low crowned P3	1
2 Upper premolar cusp heteromorphy reduced	1
3 Reduced upper molar breadth	1
4 Relatively broad upper central incisor	1
5 Relative increase in brain size	1
Humerus	
6 Deep olecranon fossa	2
7 Lateral epicondyle extends proximal to capitulum	2
8 Narrow zona conoidea	2
9 Deep zona conoidea	2
10 Capitulum extends onto distolateral surface of humerus	2
11 Capitular tail absent	2
Radius	
12 Head beveled	2
13 Neck approximates circularity in cross section	2
Trapezium	
14 Dorsal tubercle present	2
15 Surface for first metacarpal strongly trochleiform	2
Capitate	
16 Surface for third metacarpal dorsopalmarly elongate	2
Metacarpal I	
17 Proximal articular surface strongly curved mediolaterally	2
Femur	
18 Posterior buttress to the trochanteric fossa	3
Fibula	
19 Talar facet faces mediodistally	3
Metatarsals	
20 Degree of torsioning in metatarsals I and II	3
Axial skeleton	
21 Sacrum narrow	3, 4
22 Sacroiliac joint small and linear	3, 4
23 Absence of a tail	5

[a]Sources: 1, Andrews (1993); 2, Rose (this volume); 3, Rose (1993); 4, Ward (1991); 5, Ward *et al.* (1991).

in turn be viewed in terms of a lesser number of more general anatomical transformations or behavioral abilities. The posited functional attributes and associated characters are:

- Increased stability of the humeronulnar and especially humeroradial joints in all positions (6, 8, 9, 12)
- More extensive forearm pronation/supination (10, 11, 13)
- More highly mobile and fully opposable thumb (14, 15, 17)
- Some capacity for abduction/adduction at the talocrural joint (19)
- Increased mobility of the hip joint (18)
- Enhanced grasping ability in the foot, particularly of the hallux (20)

All of the above can be related either to increased joint mobility in the limbs, either through enhanced mobility or increased stability over greater ranges of motion (particularly in the forelimb), or to enhanced grasping ability in both the hand and foot. Two of the remaining features from Table I not listed above (21, 22) can be related to reorganization of the caudal musculature as a consequence of loss of the tail (23) (Ward, 1991). In addition, *Proconsul* has an apparently autapomorphous condition in the articulation of the ulnar styloid process with the proximal carpal row, and in the conformation of the hamate, that permit greater ulnar deviation, further promoting forelimb mobility (Rose, 1993, this volume).

Opinions have differed as to what some or all of these abilities signify in *Proconsul.* To some, the forelimb features promoting increased mobility reflect incipient development (in a still basically quadrupedal animal) of features related to increased usage of suspensory positional behaviors (Aiello, 1981; Feldesman, 1982; Fleagle, 1983). To others, fore- and hindlimb mobility in hominoids more generally (as well as a number of other extant hominoid "suspensory" traits) can instead be linked to cautious quadrupedal locomotion such as practiced by lorises. This also requires grasping with the hand in a variety of positions and entails frequent bridging or transferring within the tree canopy (Cartmill and Milton, 1977). Cartmill and Milton expressly relate this mode of locomotion to taillessness, noting that animals that practice this type of locomotion are either basically tailless or have prehensile tails operating as a fifth limb. In their view, this type of locomotor pattern permits taillessness and, implicitly, precedes loss of the tail. However, although tailless (Ward *et al.,* 1991), *Proconsul* lacks most of the other features noted by Cartmill and Milton that are homoplastic for extant hominoids and lorises and that form the basis of their hypothesis regarding postcranial evolution in the apes. They briefly discuss *Proconsul* and do note certain resemblances to *Alouatta,* another cautious pronograde quadruped. However, it is now known that *Proconsul* also lacked at least two of the three anatomical specializations [a shortened lumbar spine and relatively long forelimbs (Ward, 1993; Ward *et al.,* 1993)] mentioned by Cartmill and Milton as being shared by *Alouatta* and hominoids. Thus, increased limb mobility and enhanced grasping in *Proconsul* are present within an anatomical complex that otherwise suggests neither suspension nor lorislike or *Alouatta*-like modes of cautious arboreal quadrupedalism. These capabilities would therefore seem to require another, or at least a fuller explanation.

Rose (1993, this volume) interprets the increased limb mobility and grasping abilities in *Proconsul* as adaptations for climbing, thus placing *Proconsul* within the more general framework of positional behavioral evolution in the Hominoidea advocated by others (see above). Based on a study of the greatly expanded collection of phalanges from Rusinga, Begun *et al.* (1994) also conclude that climbing was an important element in the positional behavior of *Proconsul.* They describe a number of phalangeal features related to powerfully grasping hands and feet, more fully developed than in most cercopithecids or cebids but not quite as well developed as in the extant hominoids. They see *Proconsul* specifically as a powerfully grasping vertical climber, but one lacking digital specializations related to suspensory grasping. They further see in *Proconsul* only incipient development of the pronounced hand/foot differentiation characteristic of the forelimb-dominant extant hominoids.

Begun *et al.* view enhanced manual and pedal grasping, and the inferred locomotor pattern, largely as a response to increasing body size in an arboreal pronograde animal. They view the limitations imposed by taillessness mostly in the context of a lack of rapid, frequent leaping. Like others before, they interpret locomotion in *Proconsul* as having been slow and deliberate, in keeping with previous assessments of overall skeletal robusticity (Langdon, 1986; Rose, 1983; Walker and Pickford, 1983). Thus, by a different route, they arrive at similar elements of positional behavior to those proposed by Cartmill and Milton in their lorisid model for the basal hominoid locomotor adaptation. Begun *et al.* even note similarities in manual ray proportions between *Proconsul* and *Perodicticus,* which they relate to increasing the span of the hand and therefore the range of branch sizes that could be effectively utilized. However, while admitting that *Proconsul* must have been fundamentally a pronograde animal (Ward, 1993; Ward *et al.,* 1993), Begun *et al.* are compelled to explain the highly developed grasping abilities of *Proconsul.* To them, *Proconsul* was still a pronograde quadruped but one that kept contact with the substrate in an unusual manner compared to most extant quadrupedal primates, and, as a presumed vertical climber, one that frequently adopted more orthograde postures. There is, however, no supporting anatomical evidence for this kind of behavior.

An Alternative Interpretation of the Proconsul Postcranium

To me, the two real novelties of the character list in Table I are the absence of a tail and a completely opposable thumb. The functional implications of taillessness in *Proconsul* were not discussed by Ward *et al.* (1991). Taillessness was only briefly addressed by both Begun *et al.* (1994) and Cartmill and Milton (1977). I would suggest, however, that taillessness and an opposable thumb are functionally related and are in fact the keys to a comprehensive understanding of the *Proconsul* postcranium. The enhanced capacity for both hand and foot grasping generally, and the fully opposable thumb in particular, might be best interpreted with respect to taillessness. Although taillessness has been one of the most frequently cited synapomorphies of the hominoid clade, predating cladistic thinking and terminology, the full behavioral implications of this trait have been less frequently considered.

The importance of a tail in maintaining balance in an arboreal setting has been established for a variety of mammals. Again, both Cartmill and Milton (1977) and Begun *et al.* (1994) comment on the matter of taillessness in *Proconsul* or in hominoids generally; for each, it is related to the common element of their respective interpretations of early hominoid locomotion, deliberate movements through the canopy. However, the degree to which a large, above-branch quadruped is surely compromised by the loss of a tail seems not to have been fully appreciated. As typical palmigrade or digitigrade quadrupeds, large-bodied *Proconsul* species would have been in constant jeopardy of falling, particularly from smaller supports. For a tailless animal to maintain balance during above-branch progression, especially a large animal, hands and feet adapted for powerful grasping, as opposed to a more strictly palmigrade or digitigrade set, would seemingly be a necessary concomitant to, or compensation for, loss of a tail. In *Proconsul,* the powerful hand grasp was further augmented by a first carpometacarpal joint that was essentially modern hominoidlike, resulting in a truly opposable thumb and "more sophisticated manipulative and grasping use of the hand" (Rose, 1992). Both Rose (1993) and Andrews (1992) have stressed the enhanced manipulative abilities, the latter having also raised the possibility of tool use in *Proconsul.* Begun *et al.* (1994) instead interpreted the increased range of polliceal abduction, as well as the previously noted pattern of digital length, in terms of further enhancement of manual grasping, an interpretation that I favor as well.

As noted by Cartmill and Milton (1977), the need to grasp in turn requires substantial limb mobility, especially in the forelimb, to accommodate ever-altering orientations of the substrate. Enhanced limb mobility also would have further refined a balancing mechanism dependent on maneuvering the torso over hand and foot holds, as opposed to positioning the counterweight of the tail. These abilities would seemingly be necessary consequences of combining large body size and taillessness regardless of the particular locomotor pattern or overall positional behavior. Parenthetically, this combination of traits might also provide an explanation for the perplexing association in *Sivapithecus* (Rose, 1994), and incipiently in *Proconsul,* of an essentially modern-hominoid distal humerus with a very cercopithecoid-like humeral diaphysis. The latter clearly reflects continuing habitual pronograde posture while the former might reflect the concomitant need for substantial forelimb mobility and stabilization of the elbow joint in a variety of positions.

This is not to say that *Proconsul* was incapable of or did not engage in the behaviors suggested by Cartmill and Milton or Begun *et al.* Indeed, cautious or deliberate progression would appear to have been inescapable for *Proconsul,* and most arboreal primates engage in climbing of some sort. Rather, it is to say that there has not necessarily been selection for these particular modes of locomotion *per se,* either lorislike or *Alouatta*-like quadrupedalism or (orthograde) climbing. Nor is there a need to infer habitual positional behaviors involving anything other than the generalized quadrupedalism that is so clearly revealed by the *Proconsul* postcranium as a whole (Rose, 1994; Ward, 1993; Ward *et al.,* 1993). Those anatomical features that seem somewhat odd in this context may simply represent the cumulative compensatory mechanisms by which these large arboreal animals maintained balance without being able to rely on the almost universal balancing organ among arboreal animals both large and small, a tail.

Adaptive Considerations

Regardless of particular interpretations of early hominoid anatomy, are any of the hypothesized positional modes indicative of some fundamental adaptation that permits a particular exploitation of the environment? All are in a sense similar in that they view key features of early hominoid anatomy as largely compensatory for other early hominoid attributes, in particular the first evidence of large body size among catarrhines, or large body size in association with loss of the tail. None suggests any particular adaptive advantage. These hypothesized postural and locomotor patterns differ in this respect from suspension, which to a degree is also compensatory for large body size, but which in principle also opens up additional opportunities for efficient foraging among the abundant foods located in the periphery of the tree canopy (Avis, 1962; Grand, 1972, 1984; Ripley, 1979) or, more accurately, allows large-bodied animals to *retain* access to the resources located there, which are readily available to most smaller-bodied animals (Hunt, 1992). The attraction of suspension as the defining element of the Hominoidea was as much in its ecological implications as in the anatomy itself.

Thus, there appears to have been little that was anatomically distinctive about the earliest hominoids compared to earlier, more primitive catarrhines. Postcranially, those features that are distinctive do not suggest any obvious advantageous ecological or behavioral innovations, and can instead be interpreted as compensatory for the loss of the tail. It would further appear that, in many respects, little had changed postcranially by the late Miocene in at least one of the extant hominoid lineages, represented by *Sivapithecus.* Was there in fact no distinctive biological change associated with the appearance of hominoids? Were the first hominoids nothing more than large, arboreally compromised catarrhines?

A Definition of the Hominoidea Based on Life History

Given the apparent lack of an adaptively significant anatomical definition of the Hominoidea, a more fruitful realm of inquiry into the question "what is a hominoid?" might instead lie in aspects of biology other than those anatomical. In particular we might look to areas of biology represented by life history. This is an aspect of biology that has been almost completely overlooked in anatomically biased comparisons of apes and monkeys, especially cladistic analyses.

Life-history traits are among the most fundamental attributes of the biology of a species. They describe the rate and pattern of maturation of a species literally from conception until death. Principal among these attributes are gestation period, age at weaning, female age at sexual maturity and first breeding, interbirth interval, and longevity. Living apes and Old World monkeys are more distinctly different in their life-history profiles than they are in most other aspects of their biology (Smuts *et al.*, 1987). This is evident in a comparison of monkeys, gibbons, and great apes for certain of these traits (Fig. 1). These data can be summarized by saying that apes have greatly prolonged maturation relative to monkeys. Since life-history traits are highly correlated with body size (Calder, 1984; Harvey *et al.*, 1987; Peters, 1983; see below), the differences between apes and monkeys are most apparent in the comparison of gibbons with the similarly sized or much larger monkeys.

Since life history is so fundamental to a species's overall biology, it is important to know when the differences between monkeys and apes became manifest. I have elsewhere suggested that a life-history profile characterized by proloned maturation is characteristic (apomorphic) for the great ape and human clade, and perhaps for the Hominoidea as a whole (Kelley, 1992, 1993). Given similarly prolonged life histories in both gibbons and great apes,

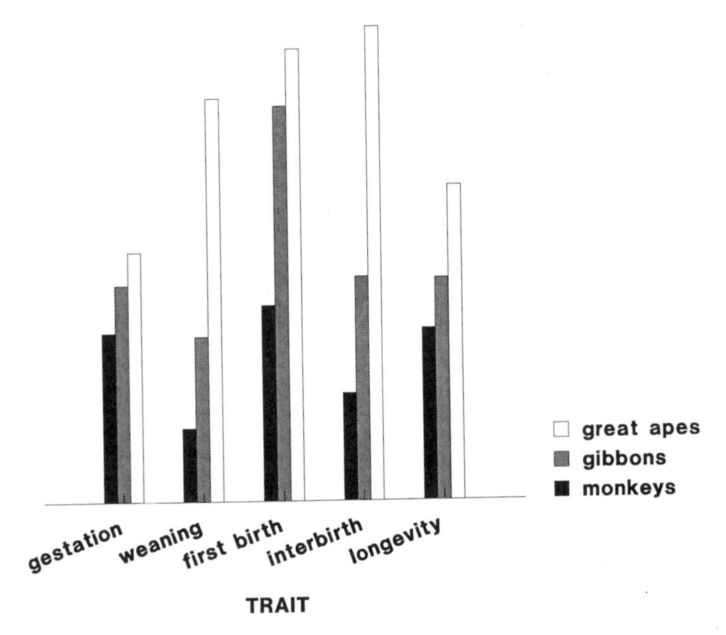

Fig. 1. Relative timing of life-history traits in cercopithecoid monkeys, gibbons, and great apes. Values are proportional among taxa for each trait but the absolute scale varies among traits (e.g., interbirth interval expressed in months, gestation in weeks). Values are means of included genus means. Data from Harvey *et al.* (1987) with adjustments to great ape values as explained in the text.

I further suggested that this might be an important defining "trait" of the Hominoidea, in an adaptive sense as well as a strictly cladistic sense. It is possible, however, that gibbons and great apes, or even different lineages among the great apes, have acquired their maturational profiles in parallel. Therefore, these phylogenetic hypotheses need to be tested in the fossil record.

A means for inferring the life-history profiles of fossil primates has been advanced by Smith (1989, 1991). She has demonstrated that among living primates, individual life-history traits are strongly correlated with the time of eruption of the first molar (M1). Thus, if the average age of eruption of M1 can be determined, the overall maturational profile of a species can be inferred and the timing of individual life-history traits can be calculated within reasonable limits. Calculating the age of eruption of M1 requires knowing the age at death of individuals that were in the process of, or had recently completed, erupting their first molars. Over the last several years methods have been developed that permit fairly precise estimates of age at death in juvenile individuals using incremental growth lines of the teeth (Beynon *et al.*, 1991; Bromage and Dean, 1985; Dean, 1987a,b; Dean *et al.*, 1993; see below). These methods have been applied to a number of individuals representing *Australopithecus, Paranthropus,* and early *Homo*. Downward revisions of the ages at death for these individuals from previous estimates (Mann, 1975) have produced earlier ages of M1 eruption. From these it can be concluded that maturational rates in all early hominids were more or less equivalent to those of modern great apes and unlike those of humans (Beynon and Dean, 1987; Bromage and Dean, 1985; Dean *et al.*, 1993).

To date, no such analysis has been carried out for any fossil ape. Thus, we presently have no direct evidence of life-history evolution within the Hominoidea as a whole. However, juvenile specimens of a species of *Sivapithecus* from the late Miocene of Pakistan that are at the appropriate developmental stage, and that preserve the necessary morphology to determine the age at first molar eruption, have recently been recovered. For some time *Sivapithecus* has been widely considered to be the sister taxon of *Pongo* (e.g., Andrews, 1992; Andrews and Cronin, 1982; Andrews and Martin, 1987; Pilbeam, 1982; Ward and Brown, 1986). However, as described above, it has recently been shown to be surprisingly primitive in important features of the postcranium. This has led some to question what had become the most secure phylogenetic position of any fossil ape (Pilbeam *et al.*, 1990; Pilbeam, this volume). Rather than question the phylogenetic position of *Sivapithecus*, others have instead interpreted this new evidence as revealing the parallel evolution of suspensory positional behaviors among the extant great apes (Andrews, 1992). If *Sivapithecus* is the sister taxon of *Pongo*, then any other characters in this genus that are similarly primitive with respect to derived characters common to the extant great apes reveal further cases of homoplasy among the latter. It is therefore of some importance to determine if the life-history profile of *Sivapithecus* is like those of the extant great apes, which are more or less equivalently prolonged with respect to monkeys (Harvey *et al.*, 1987). Anything substantially more monkeylike would indicate homoplasy in maturational biology among the extant apes and would eliminate prolonged maturation as both a synapomorphy and a possible defining attribute of the Hominoidea as a whole.

Life History in Sivapithecus

Juvenile Specimens of S. parvada

Several years ago a juvenile mandibular fragment of *Sivapithecus* with indications of a recently erupted first molar was recovered from locality Y311 ca. 10 Ma) in the upper Nagri Formation of the Pakistan Siwaliks (Pilbeam *et al.*, 1980). *Sivapithecus* specimens from Y311 have been assigned to *S. parvada*, which is most notable for its very large size (Kelley, 1988). The specimen, GSP 11536, is a left hemimandible preserving the deciduous premolars, the germs of the permanent I_1–P_4 within their crypts, and an isolated M2 germ, freed from its crypt through postfossilization breakage (Fig. 2). The first molar is not preserved but it can be seen from the capacious, matrix-filled alveolus that it had undergone alveolar emergence and had fallen out prior to fossilization (Fig. 2). The entire mesial margin of the I_1 is exposed from the incisive edge to the developing cervix by breakage at the symphysis (Fig. 3), while the I_2 is clearly discernible in radiographs. Crown formation was nearly complete in both teeth; on the I_1, root formation had begun mesially but not labially or lingually (Fig. 3).

More recently, another juvenile specimen that almost certainly represents the same individual as GSP 11536 was recovered during excavation at Y311 (Kelley *et al.*, 1995). The specimen, GSP 46460, is a premaxillary fragment preserving the unerupted but nearly complete crowns of the right central and lateral incisors (Fig. 4). That the two juvenile specimens represent the same individual is based on several lines of evidence. First, the premaxillary fragment is at precisely the same stage of development as the mandible, with incisor crown formation nearing completion. Like the mandibular I_1, the maxillary I^1 shows the beginnings of root formation mesially but not labially or lingually. In all apes, upper and lower incisor homologues develop nearly in synchrony (Beynon *et al.*, 1991; Kuykendall, 1992). Further, GSP 11536 and 46460 were recovered from the same stratigraphic horizon, and from the same small section of outcrop, in deposits in which hominoid fossils are rare. Finally, the taphonomy of the site suggests two brief periods of bone accumulation, largely through predation and scavenging with little or no fluvial transport (Behrensmeyer, 1987; Kelley, unpublished data), and there are other likely instances of multiple bones from single individuals. It cannot be demonstrated unequivocally that the two specimens represent the same individual, but given the circumstantial developmental evidence, the spatial prox-

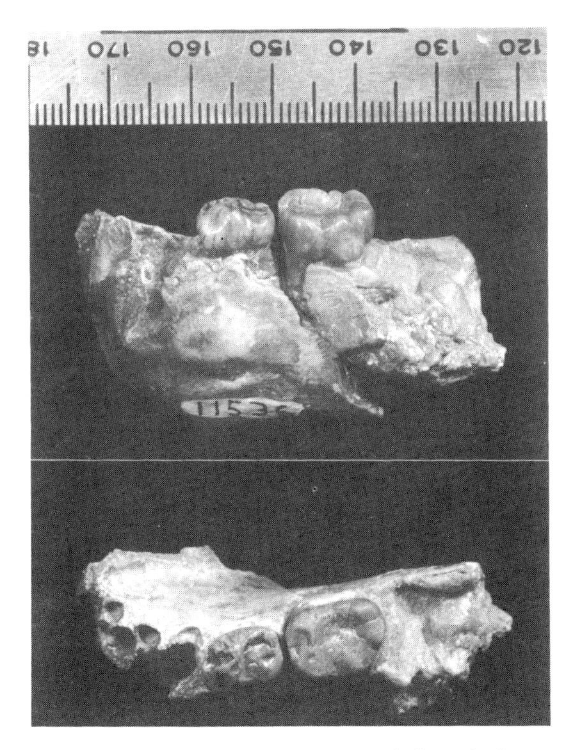

Fig. 2. *Sivapithecus parvada* juvenile mandible, GSP 11536, in lateral (above) and occlusal (below) views. The specimen is a left hemimandible bearing the dP_3 and dP_4, and preserving the damaged alveoli of the dC and the dI_2 and the mesial portion of the matrix-filled alveolus of the M1. Note the interradicular crest in the M_1 alveolus indicating the point of root bifurcation. Radiographs reveal all of the antemolar teeth in their crypts at various stages of development.

imity and temporal equivalence of the two specimens, the rarity of hominoids even at this relatively rich locality, and the inferred conditions under which bones accumulated, it is very likely.

To estimate the age at eruption of the first molar in the individual represented by GSP 11536 and GSP 46460, it is necessary to know (1) how much time elapsed between the emergence of the first molar and the death of the individual and (2) the absolute age of the individual at the time of death. The first can be determined from the mandible while the second can be calculated from the maxillary incisors.

M1 Emergence in *S. parvada*

First Molar Development and Eruption. Several lines of evidence suggest that the amount of time between the emergence of M1 and death in the individual represented by GSP 11536 was less than 6 months, and perhaps closer to 4 months. These largely rely on radiographic and actual physical inspection of

Fig. 3. GSP 11536 naturally fractured symphyseal section showing the left I_1 germ within its crypt; scale bar equals 10 mm. Arrow indicates the cementoenamel junction mesially; there is no evidence of root formation inferior to either the labial or lingual enamel extensions.

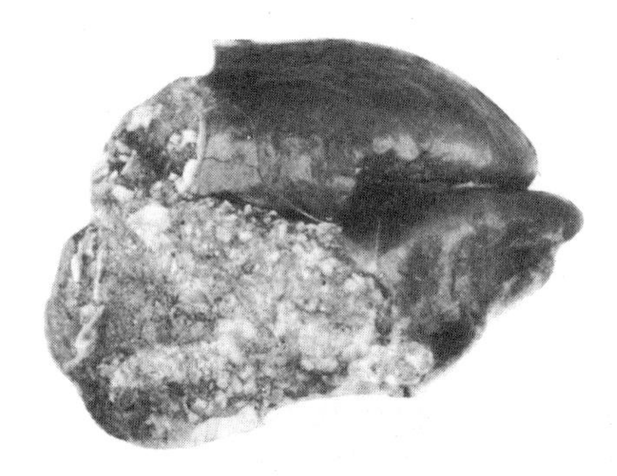

Fig. 4. *Sivapithecus parvada* juvenile premaxillary fragment, GSP 46460, in labial (left) and distal (right) views. The specimen is a fragment of premaxillary bone preserving the germs of the unerupted, permanent right I^1 and I^2. The specimen appears to be undistorted with the teeth in true developmental position. The labial side of the I^1 crown is broken cervically, but removal of the tooth from the remainder of the specimen revealed the lingual surface to be intact cervically, terminating in a hypomineralized region of developing enamel. Scale bar equals 10 mm.

the progress of root development in the first molar, the condition of the dP_4, and the stage of development of the M_2 in this individual. Full details of the various methods by which first molar development was assessed in GSP 11536 are given elsewhere (Kelley, in preparation); the relevant observations and conclusions will only be summarized here.

Among extant great apes, despite some interspecific differences in the overall duration of first molar development and in the percentage of total developmental time devoted to particular growth stages (Beynon *et al.*, 1991), the stage of tooth development at which emergence occurs, as well as the progression and duration of eruption (emergence to occlusion), appear to be roughly similar and, with respect to eruption, to encompass a relatively short amount of time (see below). Among great apes, therefore, chronologies specifically of the eruption process are relatively uniform and can be used with some confidence to assess developmental stages in fossil hominoids.

Chronologies of dental development and eruption in apes have been most thoroughly investigated in chimpanzees (Anemone *et al.*, 1991; Anemone and Siegel, 1993; Chandrasekera *et al.*, 1993; Conroy and Mahoney, 1991; Kuykendall *et al.*, 1992). That the relative timing of molar development and emergence in chimpanzees is an appropriate standard for *Sivapithecus*, and that the *Sivapithecus* juvenile died during the period of M1 eruption, is supported by the degree of development of the M2 germ relative to M1 in GSP 11536. M2 crown formation was nearly complete mesially and distally; in fact, this tooth would probably be scored as crown-complete in a lateral radiograph (see discussions in Beynon and Dean, 1988, Beynon *et al.*, 1991, Chandrasekera *et al.*, 1993, and Dean *et al.*, 1993, regarding the discrepancy between radiographic and histological determinations of crown completion). In chimpanzees, on average, the developmental stage representing M2 crown completion (radiographic) corresponds closely to the time of M1 emergence (Anemone *et al.*, 1991; Anemone and Siegel, 1993). Thus, the *relative* timing of molar development as a whole appears to be roughly the same in the two taxa. Importantly, it is approximately the same in cercopithecoids as well (Swindler and Beynon, 1993).

That the M1 of GSP 11536 was still undergoing eruption is supported by two further pieces of evidence. The first is the absence of even the beginning of an interproximal facet on the distal surface of the dP_4, indicating that the M1 had not or was just beginning to come into occlusion. The second is that the M1 fell out of the mandible prior to the onset of fossilization while the two deciduous premolars remained in place, indicating that the M1 root–periodontal ligament complex was still in a relatively early formative stage.

This assessment is further supported by data on root development in relation to molar eruption in chimpanzees and gorillas, and equivalent data from GSP 11536 and other *Sivapithecus* individuals. In the two great apes, the erupting M1 begins to approximate the dental occlusal plane when the root length measured from the point of bifurcation is approximately one-third of its fully formed length. In chimpanzees at least, full occlusion has been achieved when this length is approximately two-thirds to three-fourths of the fully formed length (S. Simpson, unpublished data). There are at present no data on fully formed M1 root lengths for *S. parvada.* However, M1 root lengths of the largest adult individuals of other *Sivapithcus* species, measured from bifurcation, range between about 12 and 15 mm (S. Ward, unpublished data). These large individuals were selected because tooth size, and therefore probably root length, is somewhat greater in *S. parvada* than in all other *Sivapithecus* species (see Kelley, 1988). The distal root length of the GSP 11536 M1 measured from bifurcation is approximately 5–6 mm. If we assume that the average length of fully formed M1 roots in *S. parvada* was approximately equal to the maximum root lengths found in smaller *Sivapithecus* species, then the roots of the GSP 11536 M1 were roughly one-third of the fully formed length, suggesting a developmental stage somewhere between emergence and full occlusion.

In chimpanzees, the M1 reaches full occlusion approximately 4 months after alveolar emergence (Zuckerman, 1928). Since the M1 in GSP 11536 had undergone alveolar emergence but had very likely not yet achieved full occlusion, it is reasonable to conclude that in this individual death followed M1 emergence by no more than a few months.

All evidence then is consistent with the interpretation that the GSP 11536 individual died within months of the emergence of the first molar. For the purpose of calculating the age at emergence of M1, the interval between emergence and death will be estimated to be 4 months.

Age at M1 Emergence. It now remains to establish the absolute age at death of the GSP 11536/46460 individual. This can be done with some precision using incremental growth lines on the teeth. The use of growth lines for establishing ages at death in juvenile fossil specimens is now well established (Bromage and Dean, 1985; Dean, 1987b; Dean and Beynon, 1991; Dean *et al.*, 1986, 1993). The most commonly used technique utilizes counts of perikymata, the surface manifestations of periodic growth lines known as the brown striae of Retzius that extend from the dentine–enamel interface to the surface of the tooth. Although there is both intraspecific and interspecific variation in the periodicity of Retzius lines, within any particular individual the period appears to be constant both within and between teeth (Beynon *et al.*, 1991; Dean, 1987a,b, 1989; Dean and Beynon, 1991; Dean *et al.*, 1993). Thus, if the periodicity is known, perikymata become accurate chronometers of dental development.

Incisor perikymata are very often expressed over the entire interval from the cusp apex (incisal edge) to the cervix, although more frequently in humans than in extant apes (Dean, 1987a). In both humans and apes incisor calcification begins near the time of birth, anywhere from 1 to several months postnatally (Beynon *et al.*, 1991; Dean, 1987a; Dean *et al.*, 1986; Dean and Beynon, 1991). Importantly, in most taxa incisor crown development is still in progress when the first molar has completed eruption, and often continues for some time afterward (Beynon and Dean, 1988; Beynon *et al.*, 1991). Thus, with knowledge of the duration of postnatal delay in the onset of incisor calcification, the growth record of the incisors can be used to evaluate that of the first molar, even subsequent to first molar eruption (Beynon *et al.*, 1991; Bromage and Dean, 1985).

As noted earlier, the GSP 11536 mandible preserves both left permanent incisors in their crypts, with the central incisor exposed over its entire mesial surface (Fig. 2 and 3). Perikymata counts would logically be carried out on this tooth. Unfortunately, GSP 11536 is not presently in the collections of the Geological Survey of Pakistan and is presumed to have been lost subsequent to the production of the radiographic and photographic documentation presented here. With the recovery, however, of the GSP 46460 premaxillary fragment with its incisors, and the numerous indications that it and GSP 11536 represent the same individual, it has again become possible to attempt to establish an age at death for this individual and, from this, the age at M1 emergence.

Both the central and lateral incisors of GSP 46460 bear exquisitely preserved perikymata running from near the incisive edge all the way to the base of the crown (Fig. 5). The very base of the labial surface of the central incisor is damaged but the lingual surface is intact and tapers to a hypomineralized region showing the typical cracking pattern of immature enamel. Thus, the

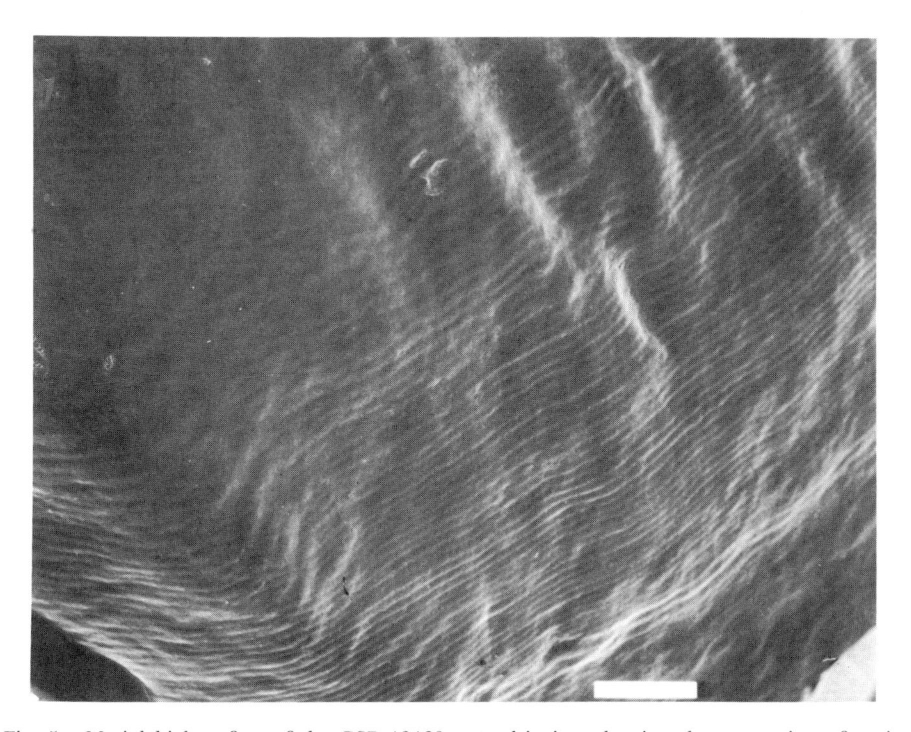

Fig. 5. Mesiolabial surface of the GSP 46460 central incisor showing the expression of perikymata; scale bar equals 1 mm. Incisal edge is toward the upper left.

lingual surface faithfully records the development of the tooth right up to the death of the individual.

The total number of perikymata on the GSP 46460 central incisor is 116. This value was increased by 17 to compensate for the lack of perikymata over the approximately first 2.5 mm of the crown from the incisive edge cervically, the number being based on the density of perikymata adjacent to the nonexpressed region. This brought the total number of perikymata to 133.

There are three variables for which values must be determined to translate perikymata counts into an estimate of age at death: (1) the number of daily increments (cross-striations, varicosities) between adjacent perikymata (the periodicity), (2) the duration of appositional enamel formation (the first-formed enamel over the dentine apex, in which Retzius lines are buried under subsequently formed enamel rather than being expressed as surface perikymata), and (3) the duration of the postnatal delay in the inception of incisor calcification (Beynon *et al.*, 1991; Bromage and Dean, 1985; Dean *et al.*, 1993). Since information on these growth parameters has been reported for only one fossil primate individual, a South African australopithecine (Dean *et al.*, 1993), ranges of values were tabulated from the literature on extant ape and human tooth development (Kelley, in preparation). From these data, maximum, minimum, and approximate modal values were determined for each of the three variables. These values were combined with the perikymata count from the GSP 46460 central incisor to produce maximum, minimum, and modal estimates for the age at death of the *Sivapithecus* juvenile (Table II). The modal age at death is between 44 and 45 months. The minimum and maximum values are 30.1 and 57.5 months, respectively.

It might be argued that this result is biased by assuming ape–human values for the three growth parameters, thereby ensuring ages at death and M1 emergence that are closer to those of apes than to monkeys. However, it appears form the very limited available data that living large monkeys and great apes differ little in either cross-striation repeat number, duration of appositional enamel formation, or postnatal delay in the onset of incisor

Table II. Estimated Age at Eruption of the First Molar in GSP 11536, *Sivapithecus parvada*[a]

	Perikymata count	No. of cross-striations	Duration of appositional enamel growth	Length of I1 postnatal delay	Age at death	Age at M_1 emergence
Minimum	133	6	4	0	30.1	26.1
Modal	133	8	6	4	44.7	40.7
Maximum	133	10	8	6	57.5	53.5

[a]All temporal measurements are in months. Age at M1 eruption set at 4 months prior to the death of the individual. Sources for cross-striation number, duration of appositional enamel growth, and length of I1 postnatal delay: Beynon and Reid (1987), Beynon *et al.* (1991), Boyde (1990), Bromage and Dean (1985), Dean (1987a), Dean and Beynon (1991), Dean *et al.* (1986, 1993).

calcification. Averages or ranges for these variables in several *Theropithecus* individuals are, respectively, 7 (constant), 4.6–6.5 months, and approximately 3–4 months (Swindler and Beynon, 1993). Not only do these values fit comfortably within the ape–human ranges of Table II, they are, in fact, quite close to the ape–human modal values.

Four months was subtracted from the estimates of age at death to produce maximum, minimum, and modal estimates for the age at M1 emergence (Table II). The modal value for the age at emergence of M1, at around 40–41 months, is comparable to the mean ages of emergence of M1 in chimpanzees and gorillas, at approximately 39 and 42 months, respectively (Conroy and Mahoney, 1991; Kuykendall *et al.*, 1992; Nissen and Riesen, 1964; Willoughby, 1978). The minimum and maximum estimates are coincidentally close to the chimpanzee extremes (Table III).

The modal age at M1 emergence in *S. parvada* is substantially greater than those of the cercopithecoids for which data are available (Table III). Included in this list are two species, *Papio cynocephalus* and *P. anubis,* that, because of their size, would be expected to have among the longest periods of maturation and, consequently, the latest ages for M1 emergence among all cercopithecoids. Moreover, the minimum estimate for M1 emergence in *S. parvada,* which incorporates the recorded lows in each of the three developmental variables for I1, is beyond the ranges of variation of the cercopithecoid

Table III. Age at First Molar Emergence in *Sivapithecus parvada, Pan troglodytes,* and Seven Old World Monkeys[a]

	Mean	Minimum	Maximum
Sivapithecus parvada	40.7	—	—
Pan troglodytes	38.9	25.7	48.0
Macaca mulatta	16.2	12.5	22.6
M. fascicularis	16.8	14	20
M. nemestrina	16.4	—	18.6+
M. fuscata	18.0	—	<24
Cercopithecus aethiops	9.9	7.9	12.0
Papio cynocephalus	20.0	—	—
P. anubis	20.0	>16	<25

[a]All ages are in months. Sources: *Pan troglodytes* (Kuykendall *et al.*, 1992), *Macaca mulatta, Cercopithecus aethiops* (Hurme and van Wagenen, 1961), *Macaca fascicularis* (Bowen and Koch, 1970), *Macaca nemestrina* (Swindler, 1985; B. H. Smith, personal communication—based on Swindler data), *Macaca fuscata* (Smith *et al.*, 1994; B. H. Smith, personal communication—based on data in Iwamoto *et al.*, 1987), *Papio cynocephalus* (Smith *et al.*, 1994—based on Reed in Phillips-Conroy and Jolly, 1988), *Papio anubis* (Smith *et al.*, 1994; B. H. Smith, personal communication—based on data in Kahumbu and Eley, 1991). Reliability of range data varies considerably depending on methodology and sample size.

species for which range data are available, well beyond those of the three species for which the data are most reliable (Table III).

Life History in S. parvada

I conclude from these results that the age at eruption of the first molar in *S. parvada* was within the ranges of the extant great apes, and may well have been close to the great-ape mean values. First molar emergence was clearly much later than in extant cercopithecoid monkeys. It follows, therefore, that *S. parvada* would have had a maturational profile, and individual life-history values, that also approached those of modern apes (Fig. 1).

For a species presumably on the lineage of one of the extant great apes, the postcranial anatomy of *S. parvada* is surprisingly primitive, yet it probably possessed a modern great-ape life-history profile. We might therefore conclude that, unlike adaptations for suspensory locomotion, the ontogenetic and reproductive features that collectively constitute the maturational or life-history profile do in fact represent a synapomorphy(ies) of at least the great ape and human clade. Equally or more prolonged maturation in relation to body size in gibbons further suggests that this might also be true for more distantly related members of the hominoid clade (see below).

Life History in a Functional and Phylogenetic Perspective

Appropriate to the theme of this volume, and before embarking on a discussion of life-history evolution in Old World anthropoids, I will present a brief review of recent theoretical and empirical studies of life-history biology in mammals generally to provide a "functional" context for investigating life-history variation in catarrhines.

Life Histories: Scaling Phenomena or Adaptive Strategies of Reproduction?

Mammalian life histories can be ordered on a fast–slow continuum (Promislow and Harvey, 1990; Read and Harvey, 1989). Such a continuum is an expression of the fact that life-history variables are more highly correlated with one another than they are even with body size or brain size. Thus, life histories can be described as distinct suites or packages of traits that represent, at one extreme, rapid development and maturation, early and prolific reproduction and early demise, and, at the other extreme, the converse. As expected, the fast–slow continuum is highly correlated with body size. Historically, the clear relationship between life-history traits and body size had led to a general perception among many workers that life history is constrained by body size in much the same way as many physiological and ecological traits. However, unlike the latter two categories of variables, there was no clearly demonstrated mechanism to explain the scaling relationship. For many, in fact, explanatory mechanisms other than those related to body size itself were considered to be unnecessary (see Pagel and Harvey, 1990). Therefore, the allometric relationship became the "explanation" although, as discussed by Boyce (1988), Harvey *et al.* (1989a), and others, in the absence of hypotheses about causation, such allometric relationships are really nothing more than "restatements of the empirical facts they are purported to explain" (Harvey *et al.*, 1989a, p. 15).

Others meanwhile had generated a variety of causal hypotheses relating life-history variation to metabolic rates, brain size and development, or ecology. In a series of recent publications, however, Harvey and colleagues have demonstrated that, in every case, when the effects of body size are removed, there is either no correlation between the residual variance in life-history variables and variance in the given explanatory variable, or correlations are insignificant (Harvey, 1990; Harvey *et al.*, 1989a,b, 1991; Pagel and Harvey, 1990; Partridge and Harvey, 1988; Read and Harvey, 1989); the purported explanations are in effect simply surrogate variables for body size. These authors argue that it is the residual variance after the removal of body size that must be the focus of attempts to explain life-history variation. Explanations derived from consideration of the residual variance must then also make comprehensible the variation that *is* concordant with body size, that is, variance in the explanatory variable must also correlate with body size variation. Somewhat surprisingly, most of this residual variance within mammalian life histories is found at higher taxonomic levels, especially among orders (Read and Harvey, 1989). Certain orders or families within orders display dramatic departures from the general body size/life-history relationship. For example, animals such as suids and large carnivores (canids, felids, ursids) are large and have relatively fast life histories, while chiropterans are small and relatively very slow (see also Martin and MacLarnon, 1990). There are less dramatic but equally important departures between body size and life history at lower taxonomic levels as well. Therefore, the fast–slow continuum is not entirely concordant with body size. What might explain these departures and at the same time provide a mechanism to explain the overriding association between life history and body size?

It has long been recognized among theoretical ecologists that life history is driven in some fashion by demographic necessity, that components of fecundity must be matched by components of mortality (see Boyce, 1988; Sutherland *et al.*, 1986). Thus, age-specific mortality schedules, along with measures of fecundity and age at sexual maturity, have been key variables in optimality models of life-history evolution which seek to identify the trade-offs that define particular life-history strategies (e.g., Caswell, 1982; Charlesworth, 1980; Charnov, 1991, 1993; Stearns, 1992; see Hill, 1993, for a very accessible summary). Harvey and co-workers have tested these models by examining the relationships between mortality, body size, and life-history variables at various taxonomic levels (Harvey and Zammuto, 1985; Harvey *et al.*, 1989a,b; Pagel and Harvey, 1990; Promislow and Harvey, 1990, 1991; Read and Harvey, 1989; Sutherland *et al.*, 1986). They found that, unlike other presumed explanatory variables, mortality rates are highly correlated with life-history parameters after the effects of body size have been removed. In terms of the fast–slow continuum, animals that suffer high nonreproductive mortality for their size grow faster, reproduce earlier, and have larger litters with less postnatal parental investment than similarly sized animals with lower levels of mortality. Citing several lines of evidence, Promislow and Harvey (1990, 1991) have forcefully argued that the relationships between mortality rates and the various elements of fecundity encompassed in life history are in fact evolved characteristics of taxa and not just more proximate expressions of the necessity for demographic equilibrium of local populations. These authors acknowledge, however, that questions about the directionality of cause and effect in these relationships are not resolved by their comparative methods. Nevertheless, patterns of mortality do provide an explanatory mechanism for life-history variation that is both consistent with empirical findings and grounded in more general theory of life-history evolution, which is not true of other purported explanations.

Harvey and colleagues (in particular, Harvey *et al.*, 1989a, Pagel and Harvey, 1990, and Promislow and Harvey, 1990, 1991) have made two further general observations regarding mortality with implications for life-history evolution. First, mortality is also highly correlated with body size so that there is size-dependent variation in nonreproductive mortality. Thus, mortality rates offer an explanation both for the relationship between life history and body size, and for life-history variation that is independent of body size. Second, it is variation in *age-specific* mortality that defines the fast–slow life-history continuum. With the effects of body size removed, animals that suffer low rates of adult mortality in relation to juvenile mortality delay maturity and reproduction and in general occupy the slow end of the fast–slow life-history continuum (see Horn, 1978). The explanation offered in optimality models of life history is that reproductive efficiency increases with age, or the costs of reproduction decrease, at least through the prime reproductive years. Therefore, if mortality is relatively low once past infancy (where it tends to be absolutely high for all mammals), then there may be a benefit to delaying reproduction (see Hill, 1993). Promislow and Harvey (1990) note that there is in fact evidence from a variety of animals that reproductive efficiency does increase with age. Data on age and reproductive efficiency in primates strongly support this supposition. Long-term studies of macaques (Drickamer, 1974; Sade, 1990), langurs (Harley, 1990), and chimpanzees (Nishida *et al.*, 1990) have all found that the rates of unsuccessful pregnancies and/or infant mortality are significantly higher among the youngest primiparous females. Data from a second study of chimpanzees are less reliable but show a similar tendency (Sugiyama, 1994), as do more anecdotal data for the Gombe chimpanzees (Goodall, 1986, p. 85).

Thus, the most coherent picture of life-history traits is one that views them as the integrated elements of adaptive strategies for optimizing lifetime reproductive effort. In this context, the major focus in life history is not maturation *per se* but, rather, maturation as a reflection of reproductive scheduling and the allocation of reproductive effort and resources, largely in response to age-specific patterns of mortality (see Hill, 1993). To the extent that there are constraints imposed by body size, they are perhaps more in the nature of "constraints of optimization" in which "size imposes boundaries on the genetic variance on which selection operates" (Read and Harvey, 1989). Given the noted departures from body size trends that characterize some families or entire orders, the genetic variance is surely considerable. To some extent, size may itself be simply another variable responding to the same selective pressures as other life-history variables (Partridge and Harvey, 1988), causally linked to other variables by growth rates (Hill, 1993). Other types of "constraints" are certainly possible, e.g., constraints resulting from developmental canalization such as those suspected to be operative on morphology (Alberch, 1990). Constraints of this kind are suggested by Martin and MacLarnon (1990) and Sade (1990) to be operating in the realm of life history as well. These possibilities are acknowledged by Harvey and co-workers, as are other possible factors influencing life-history evolution (Promislow and Harvey, 1991; see also Pereira, 1994), but they also point out that it is the burden of the proponents of constraints to provide plausible mechanisms by which they function, in addition to demonstrations of their effects. Optimality models based on fecundity and mortality have the present advantage of having been substantiated by demonstrations of predicted patterns.

The conception of life histories as adaptive reproductive strategies leads logically to questions about the nature of the selective forces by which they are shaped. If life histories are adaptive, then they are so in response to environmental factors, and attempts have been made to rationalize life-history strategies with respect to ecology. To the extent that ecology is equated with traditional habitat classifications (e.g., Boyce, 1988), there has been little success in finding associations with various elements of the fast–slow life-history continuum, suggesting greater complexity than is afforded by these classifications (Harvey and Read, 1988; Harvey *et al.*, 1989a; Wootton, 1987). Attempts to link life-history strategies with ecology have been reviewed by Southwood (1988). He defined two more general habitat variables, frequency of disturbances and the overall level of adversity or harshness, that appear to unify different approaches regarding the influence of ecology in life-history evolution. Viewed in terms of these characterizations, however, the parameter values of particular habitats cease to be absolute. To a great extent they are instead relative with respect to the body sizes and longevities of the animals occupying them; environmental grain and stability are not "perceived" uniformly (see also Partridge and Harvey, 1988). Thus, as pointed out by Southwood, there is likely more than one stable strategy for a given environment. In light of such complexity, Partridge and Harvey (1988) have questioned whether there are in fact any general associations to be made, and have suggested that, if there are, they might be generalizable only within rather than among clades. Interestingly, one apparent success with this kind of more restrictive application has been with primates. Ross (1987, 1991, 1992) found that primates living in tropical rain forests have significantly lower potential reproductive rates (r_{max}) than those living in other habitats, that is, they are slower on the fast–slow continuum. Ross's interpretation of these findings based on r- and K-selection theory (itself highly controversial; see Boyce, 1984) has been questioned by Promislow and Harvey (1990) and Harvey *et al.* (1989b), who speculated that the phenomenon might instead be related to lower rates of mortality at any given body size in tropical rain forests.

Life History in Extant Apes and Monkeys: Comparison with Predictions

Primates as a whole have protracted life histories with respect to many other mammals (Richard, 1985); in the terms of Harvey and co-workers, they are somewhat on the slow side of the fast–slow life-history continuum in absolute measures of life history (Read and Harvey, 1989). For their body sizes, however, they have among the most protracted life histories of all mammals, exceeded only by chiropterans for some variables (Harvey, 1991; Harvey *et al.*, 1989a; Read and Harvey, 1989). Richard (1985) was among the first to characterize primate life histories as a whole in terms of optimization models centered on age-specific rates of mortality, but there has been comparatively little discussion of life-history variation *within* primates in this theoretical context. Of particular interest here is a comparison of the life histories of monkeys, particularly cercopithecoids, and apes.

As noted, earlier, in absolute terms apes have greatly prolonged maturation (delayed reproduction) relative to monkeys (Fig. 1). The comparison between monkeys and gibbons suggests that ape life histories are also relatively prolonged when the effects of body size are taken into account. A more in-depth examination of primate life-history data reinforces this view and confirms that great apes also have prolonged life histories for their sizes, although perhaps not so relatively prolonged as those of gibbons (Figs. 6 and 7). It is also apparent that, for their body sizes, cercopithecoid monkeys have accelerated maturation/reproduction compared with most other higher primate groups. These conclusions are based on an analysis of the data in Harvey *et al.* (1987), with some of the great ape and ateline data adjusted to incorporate more recent information (Galdikas and Wood, 1990; Sugiyama, 1994; K. B. Strier, personal communication). The quality and completeness of the data for many of the remaining taxa is an issue. However, the analysis was weighted toward two life-history parameters, gestation length and interbirth

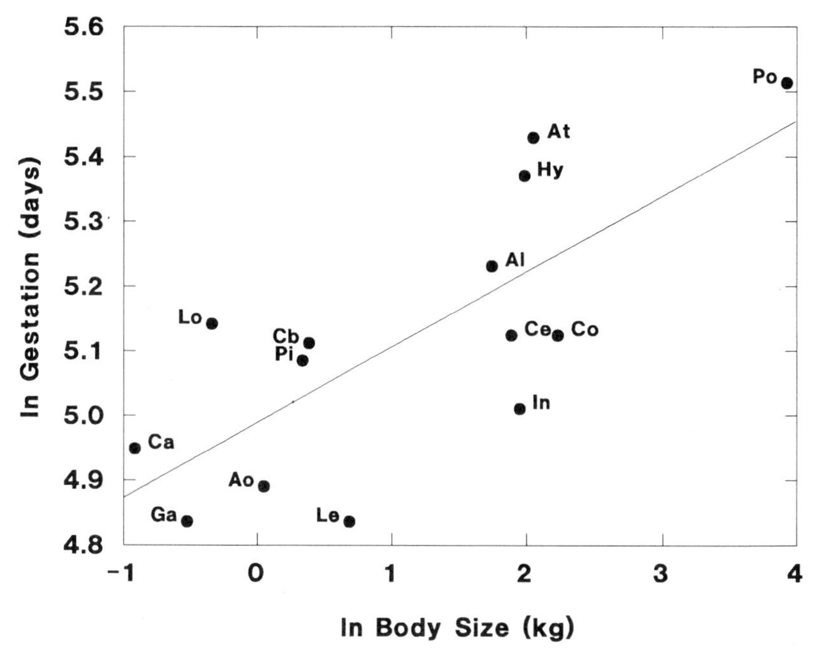

Fig. 6. Least-squares regression of gestation length on body size (natural logarithms) in primate higher taxa. Symbols are as follows: Lemuridae (Le), Inriidae (In), Lorisinae (Lo), Galaginae (Ga), Callitrichidae (Ca), Cebinae (Cb), Alouattinae (Al), Atelinae (At), Aotinae (Ao), Pithiciinae (Pi), Cercopithecinae (Ce), Colobinae (Co), Hylobatidae (Hy), great apes (Po). Values are means of included genus means. Body weights based on included species only. Data from Harvey *et al.* (1987) with adjustments to ateline and great ape values as explained in the text.

interval, for which the data were most complete and presumably more reliable because they require observation for shorter durations than many other parameters, at least for nonapes. Nevertheless, the tendencies observed in these two variables were present to a greater or lesser extent in other variables as well, e.g., the critical variable of age at sexual maturity. Further, a tendency toward early reproduction in cercopithecoids has been noted in other studies with more restricted scope and, consequently, greater reliability in the data (e.g., Watts, 1990).

How might these departures from body size expectations be interpreted? The theory of life-history evolution outlined above predicts that adult mortality among cercopithecoid monkeys should be relatively high compared with that of apes. Data are limited, especially for apes, but there is at least the suggestion that this is indeed the case. Mortality–survivorship curves for both apes and monkeys have the characteristics expected for animals with slow or protracted life histories, very high infant–juvenile mortality in relation to adult mortality (Dunbar, 1987; Goodall, 1986). However, data on chimpanzees (Goodall, 1986; Nishida *et al.*, 1990) appear to show a more pronounced drop in mortality after the infant–juvenile stage, and uniformly

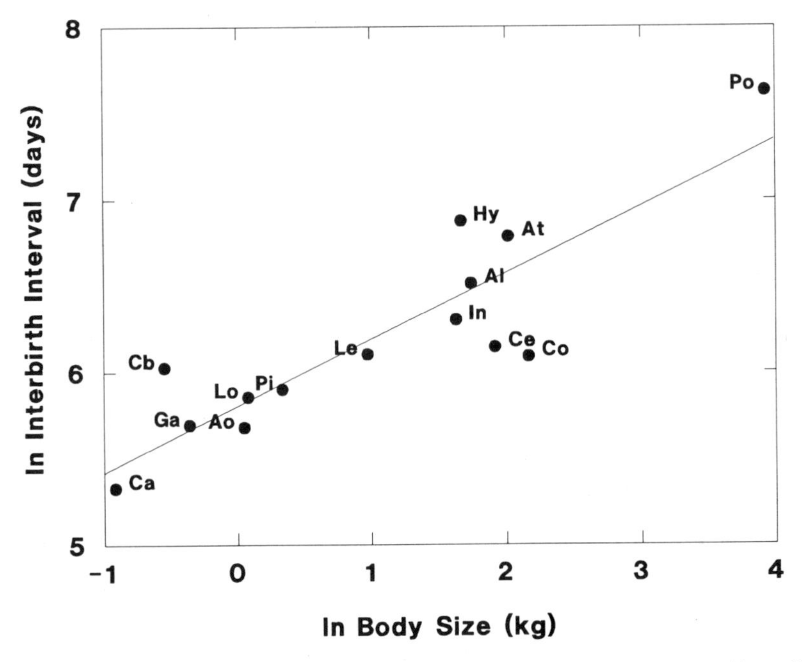

Fig. 7. Least-squares regression of interbirth interval on body size (natural logarithms) in primate higher taxa. Symbols and explanation as in Fig. 6. Body sizes may differ somewhat from those in Fig. 6 because of changes in the species composition of given samples, dependent on the availability of data.

lower rates of mortality during the prime adult years, than is often the case for monkeys (Dunbar, 1987; Gage and Dyke, 1988). This may be less true for the chimpanzee population studied by Boesch (1991), but the age-class structure of the mortality data presented by him is not sufficiently complete to be fully comparable with the data from the other studies.

Another way to approach this question is by looking at data on one aspect of extrinsic mortality only, predation. We might expect data on predation to be informative since predation should be an especially important component of adult mortality in species in which adult mortality from all causes is relatively low (Boesch, 1991; Condit and Smith, 1994). The quality of the data relating to predation is inconsistent and, as might be expected, opinions differ about the importance of predation as a cause of adult mortality among higher primates; contrast for example Goodall (1986) and Boesch (1991) for chimpanzees and Schaller (1972) and Condit and Smith (1994) for baboons. One reason for this may be that variation in predation pressure on primates is substantial, even within species, and might be linked to habitat-dependent differences in predator presence and abundance (Boesch, 1991; Isbell, 1994). Nevertheless, the absence of definitive evidence for predation on apes led Cheney and Wrangham (1987) to exclude apes from a compilation that, by contrast, reported observed predations in 15 populations of cercopithecoid monkeys representing eight species. More recent observations substantiate that predation on cercopithecoids can be high, even devastating in the short term (Boesch and Boesch, 1989; Isbell, 1990; Stanford *et al.*, 1994; Struhsaker and Leakey, 1990). Reports by Boesch (1991) and Tsukahara (1993) reveal that the evidence for predation on apes is no longer so one-sided but the imbalance in the data on incidence persists. Thus, the scanty data on predation are at least compatible with the notion that apes have lower rates of adult mortality than do monkeys.

It appears, therefore, that the limited demographic data from apes and monkeys are consistent with their relative positions along the fast–slow continuum, in accordance with predictions of life-history models based on fecundity and mortality. This information from living catarrhines can be used in conjunction with other data on cercopithecoid and hominoid biology and paleontology to suggest a possible adaptive scenario of life-history evolution in the two groups.

Life-History Divergence in Apes and Monkeys

One notable feature of the life-history data presented by Harvey *et al.* (1987) is the general consistency in values among cercopithecoid monkeys across the range of body sizes represented, which is again most evident for gestation length, interbirth interval, and age at sexual maturity. As with the hominoids, this suggests the establishment of a fundamental pattern at the initial radiation of at least the extant taxa and perhaps earlier. Martin and MacLarnon (1990) have speculated that suites of life-history parameters typically may be laid down early in the history of major lineages and may then be subject to limited subsequent variation, particularly in relation to body-size variation. This is supported by the findings of Harvey *et al.* (1989a), Promislow and Harvey (1990), and Read and Harvey (1989) that a disproportionate amount of mammalian life-history variance is found at higher taxonomic levels, especially among orders. These data together suggest that divergence in life-history biology might have accompanied the phyletic divergence of Old World monkeys and apes and, more importantly, might have been related to the ecology of this divergence.

Although, as noted earlier, attempts to link life-history evolution with specific ecological parameters have largely proved futile, Ross's (1987, 1991, 1992) findings that primates inhabiting tropical forests generally have lower values of r_{max} (slower life histories) than those occupying other types of habitats provide a basis for interpreting the ape–monkey divergence in terms of life-history evolution. There is now considerable evidence to support the inferences that the earliest cercopithecoids were cursorial terrestrial or semiterrestrial species occupying woodland–savanna habitats, whereas early hominoids were primarily or strictly arboreal species occupying tropical forests.

The evidence derives from analyses of postcranial anatomy of both extant and fossil species, as well as from interpretations of the paleoecology of early and middle Miocene fossil primate sites (Andrews, 1982, 1992; Andrews and Aiello, 1984; Andrews *et al.*, this volume; Harrison, 1989; Pickford and Senut, 1988; Strasser, 1988). These inferences are further supported by the fact that in the late Miocene of Eurasia, monkeys everywhere succeed rather than supplant the previously extinct hominoids, usually in the context of a transition from forest to more open-country faunas (Bonis *et al.*, 1986, 1988; Jablonski, 1993; Kelley, 1994; Pan and Jablonski, 1987; Szalay and Delson, 1979). Interpreted in light of Ross's analysis and life-history models, these patterns suggest that from the time of the divergence and initial radiations of the two groups, they may have been characterized by divergent life histories, achieved through accelerated maturation and reproduction in monkeys and prolonged maturation and delayed reproduction in apes. This presumably would have been in response to different rates of adult mortality, associated with their different habitats and body sizes. This scenario is entirely compatible with the inferred life-history characteristics of Siwalik *Sivapithecus* from 10 Ma.

At present we have no knowledge of dental maturation in earlier hominoids, including *Proconsul* and its relatives, and can therefore draw no inferences about life histories from this source. However, given the strong correlations between life-history variables and body size (Harvey *et al.*, 1987), the large body sizes of proconsulids compared with earlier catarrhines and contemporaneous nonhominoids (Fleagle, 1988) are perhaps indicative of the beginnings of prolonged life history.

There is also a reasonable estimate of brain size for one individual of *Proconsul heseloni*, KNM-RU 7290 (Walker *et al.*, 1983). The very strong correlation across primates between brain size and age at M1 eruption demonstrated by Smith (1989, 1991) and Smith *et al.* (1995) permits at least an indirect approximation of this important maturational marker for this species. Inserting the ln-transformed mean and 95% confidence intervals for the estimated brain size of KNM-RU 7290 (Walker *et al.*, 1983) into a regression equation generated from the 23 primate species for which there are data for both age at M1 eruption (Smith, 1989; Smith *et al.*, 1994) and brain size (Harvey *et al.*, 1987), resulted in an estimated mean age at M1 eruption in *P. heseloni* of 20.6 months, with an approximate range for the mean (based on the confidence interval for brain size rather than the error of the estimate) of between 19.6 and 21.6 months. The mean is just beyond the upper end of the range of means of the seven cercopithecoid species included in the comparative sample (Table III). Since this small sample contains two of the larger-brained cercopithecids, *Papio cynocephalus* and *P. anubis*, the results suggest that the average age at M1 eruption in *P. heseloni* was at the upper end of the range of M1 eruption ages for all extant nonhominoid catarrhines. This part of the cercopithecoid range is occupied by species that are all considerably larger on average than *P. heseloni* (Ruff *et al.*, 1989; Walker *et al.*, 1993). These results suggest a more prolonged life history for *P. heseloni* than would be expected for an early Miocene catarrhine of this size.

Two important qualifiers need to be attached to the foregoing discussion of *P. heseloni.* First, as with the determination of age at M1 eruption in *Sivapithecus,* the result is based on one individual. We can only presume that brain size in the *P. heseloni* individual, and whatever its actual age at M1 eruption might have been, were both near the average for the species. In actuality, there is probably substantial individual variation in the relationship between brain size and M1 eruption; departures from species averages in the two variables in any single individual may not necessarily be concordant nor of equal magnitude. Because the differences in dental maturation between *Proconsul* and extant cercopithecoids are likely to be relatively small, the issue of individual variation is more problematic in this instance than it was in the analysis of *Sivapithecus,* for which the distinction to be made was simply whether its life-history profile was more modern apelike or monkeylike.

Second, the KNM-RU 7290 individual is a female (Kelley, 1986, 1995). Therefore, its absolute brain size might actually slightly underestimate the average for the species as a whole. On the positive side, it is at least the case that intraspecific variation in brain size is less than that known for many other features, including body size (Economos, 1980).

In light of these caveats and given the error factors in the estimation of brain size in KNM-RU 7290 and in the brain size/age at M1 eruption regression, the estimated age at eruption of M1 in *P. heseloni* presented here should serve as nothing more than an approximation. Nevertheless, with the notable exceptions described earlier, both brain and body size are reasonably good predictors of overall rates of maturation, particularly within higher taxonomic groups. Thus, whether or not the above estimate of M1 eruption in *Proconsul* is indeed accurate, brain size and especially body size estimates still suggest at a more *qualitative* level that proconsulids had started down the path toward prolonged maturation. This can serve as a working hypothesis to be tested through direct measurement of dental maturation in *Proconsul* and other proconsulids.

Conclusions

Viewed as "optimal" reproductive strategies, ape life histories are adaptive responses to a particular set of environmental conditions. Larger body size and protracted life history might have been the most adaptively important features of apes at the time of their divergence from monkeys. As such, they might constitute the best attributes by which to construct a unifying biological definition of the Hominoidea. As a package, they would have placed hominoids on a biological course that was different from that of both cercopithecoid monkeys and the catarrhine antecedents of both groups. In this interpretation, rather than being a competitive liability with respect to the "demographically superior" monkeys (Lovejoy, 1981), the life histories of hominoids would have been instrumental to their success until the forested environments to which they were adapted disappeared over large areas of the Old World. If the earliest hominoids are ultimately shown to have had protracted life histories in comparison with their nonhominoid contemporaries and extant cercopithecoid monkeys, then this would also constitute an important synapomorphy of the hominoid clade.

Acknowledgments

I would first like to thank the editors of this volume, David Begun, Carol Ward, and Mike Rose, for inviting me to participate in the original symposium at the AAPA meeting in Toronto. My foray into life-history evolution has benefited greatly from discussion and correspondence with Paul Harvey and Holly Smith. I would also like to thank Holly, Robert Anemone, Scott Simpson, Karen Strier, and Steve Ward, for either procuring information or sharing unpublished data with me. The section on the evolution of hominoid positional behavior was greatly improved by comments from and discussion with Mike Rose. I thank Jim Drummond for assistance with scanning electron microscopy and William Wynn for photography. I would also like to express my gratitude to the permanent Y311 field crew, Mohammad Anwar, Ahmed Khan, Sandy Madar, and Melanie McCollum, and to occasional interlopers Steve Ward, Stephanie Pangas, and Bobbie Brown; all have the scarred hands to show their mettle. Finally, I thank Dr. S. Mahmood Raza and the Directors General, Geological Survey of Pakistan, for their continued support of field work in Pakistan. This work was funded by NSF grants BNS-9196211 and INT-9296063.

References

Aiello, L. C. 1981. Locomotion in the Miocene Hominoidea. In: C. B. Stringer (ed.), *Aspects of Human Evolution,* pp. 63–98. Taylor & Francis, London.

Alberch, P. 1990. Natural selection and developmental constraints: External versus internal determinants of order in nature. In: C. J. DeRousseau (ed.), *Primate Life History and Evolution,* pp. 15–35. Wiley–Liss, New York.

Andrews, P. 1981. Species diversity and diet in monkeys and apes during the Miocene. In: C. B. Stringer (ed.), *Aspects of Human Evolution,* pp. 25–61. Taylor & Francis, London.

Andrews, P. 1982. Ecological polarity in primate evolution. *Zool. J. Linn. Soc.* **74:**233–244.

Andrews, P. 1992. Evolution and environment in the Hominoidea. *Nature* **360:**641–646.

Andrews, P., and Aiello, L. 1984. An evolutionary model for feeding and positional behavior. In: D. J. Chivers, B. A. Wood, and A. Bilsborough (eds.), *Food Acquisition and Processing in Primates,* pp. 429–466. Plenum Press, New York.

Andrews, P., and Cronin, J. E. 1982. The relationship of *Sivapithecus* and *Ramapithecus* and the evolution of the orang-utan. *Nature* **297:**541–546.

Andrews, P., and Martin, L. 1987. Cladistic relationships of extant and fossil hominoids. *J. Hum. Evol.* **16:**101–118.

Anemone, R. L., and Siegel, M. I. 1993. A longitudinal study of molar development in chimpanzees. *Am. J. Phys. Anthropol. Suppl.* **16:**49 (abstr.).

Anemone, R. L., Watts, E. S., and Swindler, D. R. 1991. Dental development of known-age chimpanzees, *Pan troglodytes. Am. J. Phys. Anthropol.* **86:**229–241.

Avis, V. 1962. Brachiation: The crucial issue for man's ancestry. *Southwest. J. Anthropol.* **18:**119–148.

Beard, K. C., Teaford, M. F., and Walker, A. 1986. New wrist bones of *Proconsul africanus* and *P. nyanzae* from Rusinga Island, Kenya. *Folia Primatol.* **47:**97–118.

Begun, D. R., Teaford, M. F., and Walker, A. 1994. Comparative and functional anatomy of *Proconsul* phalanges from the Kaswanga Primate Site, Rusinga Island, Kenya. *J. Hum. Evol.* **26:**89–165.

Behrensmeyer, A. K. 1987. Miocene fluvial facies and vertebrate taphonomy in northern Pakistan. *Soc. Econ. Paleontol. Min. Spec. Publ.* **39:**169–176.

Beynon, A. D., and Dean, M. C. 1987. Crown formation time of a fossil hominid premolar tooth. *Arch. Oral Biol.* **32:**773–790.

Beynon, A. D., and Dean, M. C. 1988. Distinct dental development patterns in early fossil hominids. *Nature* **335:**509–514.

Beynon, A. D., and Reid, D. J. 1987. Relationships between perikymata counts and crown formation times in the human permanent dentition. *J. Dent. Res.* **66:**889.

Beynon, A. D., Dean, M. C., and Reid, D. J. 1991. Histological study on the chronology of the developing dentition in gorilla and orangutan. *Am. J. Phys. Anthropol.* **86:**189–204.

Boesch, C. 1991. The effects of leopard predation on grouping patterns in forest chimpanzees. *Behaviour* **117:**220–242.

Boesch, C., and Boesch, H. 1989. Hunting behavior of wild chimpanzees in the Tai National Park. *Am. J. Phys. Anthropol.* **78:**547–573.

Bonis, L. de, Bouvrain, G., Koufos, G., and Melentis, J. 1986. Succession and dating of the late Miocene primates of Macedonia. In: J. G. Else and P. C. Lee (eds.), *Primate Evolution,* pp. 107–114. Cambridge University Press, London.

Bonis, L. de, Bouvrain, G., and Koufos, G. D. 1988. Late Miocene mammal localities of the lower Axios Valley (Macedonia, Greece) and their stratigraphic significance. *Mod. Geol.* **13:**141–147.

Bowen, W. H., and Koch, G. 1970. Determination of age in monkeys (*Macaca irus*) on the basis of dental development. *Lab. Anim.* **4:**113–123.

Boyce, M. S. 1984. Restitution of r- and K-selection as a model of density dependent natural selection. *Annu. Rev. Ecol. Syst.* **15:**427–448.

Boyce, M. S. 1988. Evolution of life histories: Theory and patterns from mammals. In: M. S. Boyce (ed.), *Evolution of Life Histories of Mammals,* pp. 3–30. Yale University Press, New Haven.

Boyde, A. 1990. Developmental interpretations of dental microstructure. In: C. J. DeRousseau (ed.), *Primate Life History and Evolution,* pp. 229–267. Wiley–Liss, New York.

Bromage, T. G., and Dean, M. C. 1985. Re-evaluation of the age at death of Plio-Pleistocene fossil hominids. *Nature* **317:**525–528.

Calder, W. A. 1984. *Size, Function and Life History.* Harvard University Press, Cambridge, MA.

Cartmill, M., and Milton, K. 1977. The lorisiform wrist joint and the evolution of "brachiating" adaptations in the Hominoidea. *Am. J. Phys. Anthropol.* **47:**249–272.

Caswell, H. 1982. Optimal life histories and the age specific costs of reproduction. *J. Theor. Biol.* **98:**519–529.

Chandrasekera, M. S., Reid, D. J., and Beynon, A. D. 1993. Dental chronology in chimpanzee (*Pan troglodytes*). *J. Dent. Res.* **72:**729 (abstr.).

Charlesworth, B. 1980. *Evolution in Age-Structured Populations.* Cambridge University Press, London.

Charnov, E. L. 1991. Evolution of life history variation among female mammals. *Proc. Natl. Acad. Sci. USA* **88:**1134–1137.

Charnov, E. L. 1993. *Life-History Invariants.* Oxford University Press, London.

Cheney, D. L., and Wrangham, R. W. 1987. Predation. In: B. B. Smuts, D. L. Cheney, R. M. Seyfarth, R. W. Wrangham, and T. T. Struhsaker (eds.), *Primate Societies,* pp. 227–239. University of Chicago Press, Chicago.

Condit, V. K., and Smith, E. O. 1994. Predation on a yellow baboon (*Papio cynocephalus cynocephalus*) by a lioness in the Tana River National Primate Reserve, Kenya. *Am. J. Primatol.* **33:**57–64.

Conroy, G. C., and Mahoney, C. J. 1991. Mixed longitudinal study of dental emergence in the chimpanzee, *Pan troglodytes* (Primates, Pongidae). *Am. J. Phys. Anthropol.* **86:**243–254.

Dean, M. C. 1987a. Growth layers and incremental markings in hard tissues: A review of the literature and some preliminary observations about enamel structure in *Paranthropus boisei. J. Hum. Evol.* **16:**157–172.

Dean, M. C. 1987b. The dental developmental status of six juvenile fossil hominids from Koobi Fora and Olduvai Gorge. *J. Hum. Evol.* **16:**197–213.

Dean, M. C. 1989. The developing dentition and tooth structure in hominoids. *Folia Primatol.* **42:**160–177.

Dean, M. C., and Beynon, A. D. 1991. Histological reconstruction of crown formation times and initial root formation times in a modern human child. *Am. J. Phys. Anthropol.* **86:**215–228.

Dean, M. C., Stringer, C. B., and Bromage, T. G. 1986. A new age at death for the Neanderthal child from the Devil's Tower, Gilbraltar and the implications for studies of general growth and development in Neanderthals. *Am. J. Phys. Anthropol.* **70:**301–309.

Dean, M. C., Beynon, A. D., Thackeray, J. F., and Macho, G. A. 1993. Histological reconstruction of dental development and age at death of a juvenile *Paranthropus robustus* specimen, SK 63, from Swartkrans, South Africa. *Am. J. Phys. Anthropol.* **91:**401–419.

Drickamer, L. C. 1974. A ten year summary of reproductive data for free-ranging *Macaca mulatta. Folia Primatol.* **21:**61–80.

Dunbar, R. I. M. 1987. Demography and reproduction. In: B. B. Smuts, D. L. Cheney, R. M. Seyfarth, R. W. Wrangham, and T. T. Struhsaker (eds.), *Primate Societies,* pp. 240–249. University of Chicago Press, Chicago.

Economos, A. C. 1980. Brain–lifespan conjecture: A re-evaluation of the evidence. *Gerontology* **26:**82–89.

Feldesman, M. R. 1982. Morphometric analysis of the distal humerus of some Cenozoic catarrhines. *Am. J. Phys. Anthropol.* **59:**73–95.

Fleagle, J. G. 1983. Locomotor adaptations of Oligocene and Miocene hominoids and their phyletic implications. In: R. L. Ciochon and R. S. Corruccini (eds.), *New Interpretations of Ape and Human Ancestry,* pp. 301–324. Plenum Press, New York.

Fleagle, J. G. 1988. *Primate Adaptation and Evolution.* Academic Press, New York.

Fleagle, J. G., Stern, J. T., Jungers, W. L., Susman, R. L., Vangor, A. K., and Wells, J. P. 1981. Climbing: A biomechanical link with brachiation and with bipedalism. *Symp. Zool. Soc. London* **48:**359–375.

Gage, T. B., and Dyke, B. 1988. Model life tables for the larger Old World monkeys. *Am. J. Primatol.* **16:**305–320.

Galdikas, B. M. F., and Wood, J. W. 1990. Birth spacing patterns in humans and apes. *Am. J. Phys. Anthropol.* **83:**185–191.

Goodall, J. 1986. *The Chimpanzees of Gombe: Patterns of Behavior.* Harvard University Press, Cambridge, MA.

Grand, T. I. 1972. A mechanical interpretation of terminal branch feeding. *J. Mammal.* **53:**198–201.

Grand, T. I. 1984. Motion economy within the canopy: Four strategies for mobility. In: P. S.

Rodman and J. G. H. Cant (eds.), *Adaptations for Foraging in Nonhuman Primates: Contributions to an Organismal Biology of Prosimians, Monkeys and Apes*, pp. 54–72. Columbia University Press, New York.

Harley, D. 1990. Aging and reproductive performance in Langur monkeys (*Presbytis entellus*). *Am. J. Phys. Anthropol.* **83:**253–261.

Harrison, T. 1987. The phylogenetic relationships of the early catarrhine primates: A review of the current evidence. *J. Hum. Evol.* **16:**41–80.

Harrison, T. 1989. New postcranial remains of *Victoriapithecus* from the middle Miocene of Kenya. *J. Hum. Evol.* **18:**3–54.

Harvey, P. H. 1990. Life-history variation: Size and mortality patterns. In: C. J. DeRousseau (ed.), *Primate Life History and Evolution*, pp. 81–88. Wiley–Liss, New York.

Harvey, P. H. 1991. Comparing life histories. In: S. Osawa and T. Honjo (eds.), *Evolution of Life*, pp. 215–228. Springer-Verlag, Berlin.

Harvey, P. H., and Read, A. F. 1988. How and why do mammalian life histories vary? In: M. S. Boyce (ed.), *Evolution of Life Histories of Mammals: Theory and Pattern*, pp. 213–232. Yale University Press, New Haven, CT.

Harvey, P. H., and Zammuto, R. M. 1985. Patterns of mortality and age at first reproduction in natural populations of mammals. *Nature* **315:**319–320.

Harvey, P. H., Martin, R. D., and Clutton-Brock, T. H. 1987. Life histories in comparative perspective. In: B. B. Smuts, D. L. Cheney, R. M. Seyfarth, R. W. Wrangham, and T. T. Struhsaker (eds.), *Primate Societies*, pp. 181–196. University of Chicago Press, Chicago.

Harvey, P. H., Read, A. F., and Promislow, D. E. L. 1989a. Life history variation in placental mammals: Unifying the data with the theory. *Oxford Surv. Evol. Biol.* **6:**13–31.

Harvey, P. H., Promislow, D. E. L., and Read, A. F. 1989b. Causes and correlates of life history differences among mammals. In: R. Foley and V. Standen (eds.), *Comparative Socioecology: The Behavioural Ecology of Humans and Other Mammals*, pp. 305–318. Blackwell, Oxford.

Harvey, P. H., Pagel, M. D., and Rees, J. A. 1991. Mammalian metabolism and life histories. *Am. Nat.* **137:**556–566.

Hill, K. 1993. Life history theory and evolutionary anthropology. *Evol. Anthropol.* **2:**78–88.

Horn, H. S. 1978. Optimal tactics of reproduction and life history. In: J. R. Krebs and N. B. Davies (eds.), *Behavioural Ecology: An Evolutionary Approach*, pp. 272–294. Blackwell, Oxford.

Hunt, K. D. 1992. Positional behavior of *Pan troglodytes* in the Mahale Mountains and Gombe Stream National Parks, Tanzania. *Am. J. Phys. Anthropol.* **87:**83–105.

Hurme, V. O., and van Wagenen, G. 1961. Basic data on the emergence of permanent teeth in the rhesus monkey (*Macaca mulatta*). *Proc. Am. Philos. Soc.* **105:**105–140.

Isbell, L. A. 1990. Sudden short-term increase in mortality of vervet monkeys (*Cercopithecus aethiops*) due to leopard predation in Amboseli National Park, Kenya. *Am. J. Primatol.* **21:**41–52.

Isbell, L. A. 1994. Predation on primates: Ecological patterns and evolutionary consequences. *Evol. Anthropol.* **3:**61–71.

Iwamoto, M., Watanabe, T., and Hamada, Y. 1987. Eruption of permanent teeth in Japanese monkeys (*Macaca fuscata*). *Primate Res.* **3:**18–28.

Jablonski, N. G. 1993. Quaternary environments and the evolution of primates in East Asia, with notes on two new species of fossil Cercopithecoidea from China. *Folia Primatol.* **60:**118–132.

Kahumbu, P., and Eley, R. M. 1991. Teeth emergence in wild live baboons in Kenya and formulation of a dental schedule for aging wild baboon populations. *Am. J. Primatol.* **23:**1–9.

Keith, A. 1923. Man's posture: Its evolution and disorder. *Br. Med. J.* **1:**451–454, 499–502, 545–548, 587–590, 624–626, 669–672.

Kelley, J. 1986. Species recognition and sexual dimorphism in *Proconsul* and *Rangwapithecus*. *J. Hum. Evol.* **15:**461–495.

Kelley, J. 1988. A new large species of *Sivapithecus* from the Siwaliks of Pakistan. *J. Hum. Evol.* **17:**305–324.

Kelley, J. 1992. Evolution of apes, In: S. Jones, R. D. Martin, and D. Pilbeam (eds.), *The Cambridge Encyclopedia of Human Evolution,* pp. 223–230. Cambridge University Press, London.

Kelley, J. 1993. Life history profile of *Sivapithecus. Am. J. Phys. Anthropol. Suppl.* **16:**123 (abstr.).

Kelley, J. 1994. A biological hypothesis of ape species density. In: B. Thierry, J. R. Anderson, J. J. Roeder, and N. Herrenschmidt (eds.), *Current Primatology, Vol. 1: Ecology and Evolution,* pp. 11–18. Université Louis Pasteur, Strasbourg.

Kelley, J. 1995. Sex determination in Miocene catarrhine primates. *Am. J. Phys. Anthropol.* **96:**390–417.

Kelley, J., Anwar, M., McCollum, M., and Ward, S. C. 1995. The anterior dentition of *Sivapithecus parvada* with comments on the phylogenetic significance of incisor heteromorphy in hominoids. *J. Hum. Evol.* **28:**503–517.

Kuykendall, K. L. 1992. *Dental Development in Chimpanzees (Pan troglodytes) and Implications for Dental Development Patterns in Fossil Hominids.* Ph.D. dissertation, Washington University.

Kuykendall, K. L., Mahoney, C. J., and Conroy, G. C. 1992. Probit and survival analysis of tooth emergence ages in a mixed-longitudinal sample of chimpanzees (*Pan troglodytes*). *Am. J. Phys. Anthropol.* **89:**379–399.

Langdon, J. H. 1986. Functional morphology of the Miocene hominoid foot. *Contrib. Primatol.* **22:**1–225.

Lovejoy, C. O. 1981. The origin of man. *Science* **211:**341–350.

Mann, A. E. 1975. *Some Paleodemographic Aspects of the South African Australopithecines.* University of Pennsylvania Press, Philadelphia.

Martin, R. D., and MacLarnon, A. M. 1990. Reproductive patterns in primates and other mammals: The dichotomy between altricial and precocial offspring. In: C. J. DeRousseau (ed.), *Primate Life History and Evolution,* pp. 47–79. Wiley–Liss, New York.

Nishida, T., Takasaki, H., and Takahata, Y. 1990. Demography and reproductive profiles. In: T. Nishida (ed.), *The Chimpanzees of the Mahale Mountains: Sexual and Life History Strategies,* pp. 63–97. University of Tokyo Press, Tokyo.

Nissen, H. W., and Riesen, A. H. 1964. The eruption of the permanent dentition of chimpanzee. *Am. J. Phys. Anthropol.* **22:**285–294.

Pagel, M. D., and Harvey, P. H. 1990. Diversity in the brain sizes of newborn mammals. *Bioscience* **40:**116–122.

Pan, Y.-R., and Jablonski, N. G. 1987. The age and geographical distribution of fossil cercopithecids in China. *Hum. Evol.* **2:**59–69.

Partridge, L., and Harvey, P. H. 1988. The ecological context of life history evolution. *Science* **241:**1449–1455.

Pereira, M. E. 1994. Life history and social cognition in gregarious primates: Robotic monkeys and clever lemurs? *Folia Primatol.* **62:**220 (abstr.).

Peters, R. H. 1983. *The Ecological Implications of Body Size.* Cambridge University Press, London.

Phillips-Conroy, J. E., and Jolly, C. J. 1988. Dental eruption schedules of wild and captive baboons. *Am. J. Primatol.* **15:**17–29.

Pickford, M., and Senut, B. 1988. Habitat and locomotion in Middle Miocene cercopithecoids. In: F. Bourliere, J.-P. Gautier, and A. Gautier-Hion (eds.), *Evolutionary Biology of the African Guenons,* pp. 35–52. Cambridge University Press, London.

Pilbeam, D. 1982. New hominoid skull material from the Miocene of Pakistan. *Nature* **295:**232–234.

Pilbeam, D., Rose, M. D., Badgley, C., and Lipschutz, B. 1980. Miocene hominoids from Pakistan. *Postilla* **181:**1–94.

Pilbeam, D., Rose, M. D., Barry, J. C., and Shah, S. M. I. 1990. New *Sivapithecus* humeri from Pakistan and the relationship of *Sivapithecus* and *Pongo. Nature* **348:**237–239.

Promislow, D. E. L., and Harvey, P. H. 1990. Living fast and dying young: A comparative analysis of life-history variation among mammals. *J. Zool.* **220:**417–437.

Promislow, D. E. L., and Harvey, P. H. 1991. Mortality rates and the evolution of mammal life histories. *Acta Oecol.* **12:**119–137.

Read, A. F., and Harvey, P. H. 1989. Life history differences among the eutherian radiations. *J. Zool.* **219:**329–353.

Richard, A. F. 1985. *Primates in Nature.* Freeman, San Francisco.

Ripley, S. 1979. Environmental grain, niche diversification, and positional behavior in Neogene primates: An evolutionary hypothesis. In: M. E. Morbeck, H. Preuschoft, and N. Gomberg (eds.), *Environment, Behavior and Morphology: Dynamic Interactions in Primates,* pp. 91–104. Smithsonian Institution Press, Washington, DC.

Rose, M. D. 1983. Miocene hominoid postcranial morphology: Monkey-like, ape-like, neither, or both? In: R. L. Ciochon and R. S. Corruccini (eds.), *New Interpretations of Ape and Human Ancestry,* pp. 405–417. Plenum Press, New York.

Rose, M. D. 1986. further hominoid postcranial specimens from the Late Miocene Nagri Formation of Pakistan. *J. Hum. Evol.* **15:**333–367.

Rose, M. D. 1992. Kinematics of the trapezium–1st metacarpal joint in extant anthropoids and Miocene hominoids. *J. Hum. Evol.* **22:**255–266.

Rose, M. D. 1993. Locomotor anatomy of Miocene hominoids. In: D. Gebo (ed.), *Postcranial Adaptations in Nonhuman Primates,* pp. 252–272. Northern Illinois University Press, De Kalb.

Rose, M. D. 1994. Quadrupedalism in some Miocene catarrhines. *J. Hum. Evol.* **26:**387–411.

Ross, C. R. 1987. The intrinsic rate of natural increase and reproductive effort in primates. *J. Zool.* **214:**199–220.

Ross, C. R. 1991. Life history patterns of New World monkeys. *Int. J. Primatol.* **12:**481–502.

Ross, C. R. 1992. Environmental correlates of the intrinsic rate of natural increase in primates. *Oecologia* **90:**383–390.

Ruff, C., Walker, A., and Teaford, M. F. 1989. Body mass, sexual dimorphism and femoral proportions of *Proconsul* from Rusinga and Mfangano Islands, Kenya. *J. Hum. Evol.* **18:**515–536.

Sade, D. S. 1990. Intrapopulational variation in life-history parameters. In: C. J. DeRousseau (ed.), *Primate Life History and Evolution,* pp. 181–194. Wiley–Liss, New York.

Schaller, G. 1972. *The Serengeti Lion.* University of Chicago Press, Chicago.

Smith, B. H. 1989. Dental development as a measure of life history in primates. *Evolution* **43:**683–688.

Smith, B. H. 1991. Dental development and the evolution of life history in Hominidae. *Am. J. Phys. Anthropol.* **86:**157–174.

Smith, B. H., Crummet, T. L., and Brandt, K. L. 1994. Age of eruption of primate teeth: A compendium for aging individuals and comparing life histories. *Yearb. Phys. Anthropol.* **37:**177–231.

Smith, R. J., Gannon, P. J., and Smith, B. H. 1995. Ontogeny of australopithecines and early *Homo:* evidence from cranial capacity and dental eruption. *J. Hum. Evol.* **29:**155–168.

Smuts, B. B., Cheney, D. L., Seyfarth, R. M., Wrangham, R. W., and Struhsaker, T. T. (eds.). 1987. *Primate Societies.* University of Chicago Press, Chicago.

Southwood, T. R. E. 1988. Tactics, strategies and templates. *Oikos* **52:**3–18.

Spoor, C. F., Sondaar, P. Y., and Hussain, S. T. 1991. A hominoid hamate and first metacarpal from the Late Miocene Nagri Formation of Pakistan. *J. Hum. Evol.* **21:**413–424.

Stanford, C. B., Wallis, J., Matama, H., and Goodall, J. 1994. Patterns of predation by chimpanzees on red colobus monkeys in Gombe National Park, 1982–1991. *Am. J. Phys. Anthropol.* **94:**213–228.

Stearns, S. 1992. *The Evolution of Life Histories.* Oxford University Press, London.

Stern, J. T. 1971. Functional myology of the hip and thigh of cebid monkeys and its implications for the evolution of erect posture. *Bibl. Primatol.* **14:**1–318.

Stern, J. T., Wells, J. P., Vangor, A. K., and Fleagle, J. G. 1977. Electromyography of some muscles of the upper limb in *Ateles* and *Lagothrix. Yearb. Phys. Anthropol.* **20:**498–507.

Strasser, E. 1988. Pedal evidence for the origin and diversification of cercopithecid clades. *J. Hum. Evol.* **17:**225–245.

Straus, W. L. Jr. 1949. The riddle of man's ancestry. *Q. Rev. Biol.* **24:**200–223.

Struhsaker, T. T., and Leakey, M. 1990. Prey selection by crowned hawk-eagles on monkeys in the Kibale Forest, Uganda. *Behav. Ecol. Sociobiol.* **26:**435–443.

Sugiyama, Y. 1994. Age-specific birth rate and lifetime reproductive success of chimpanzees at Bossou, Guinea. *Am. J. Primatol.* **32:**311–318.

Sutherland, W. J., Grafen, A., and Harvey, P. H. 1986. Life history correlations and demography. *Nature* **320:**88.

Swindler, D. R. 1985. Nonhuman primate dental development and its relationship to human dental development. In: E. S. Watts (ed.), *Nonhuman Primate Models for Human Growth and Development,* pp. 67–94. Liss, New York.

Swindler, D. R., and Beynon, D. 1993. The development and microstructure of the dentition of *Theropithecus.* In: N. G. Jablonski (ed.), *Theropithecus: The Rise and Fall of a Primate Genus,* pp. 351–381. Cambridge University Press, London.

Szalay, F. S., and Delson, E. 1979. *Evolutionary History of the Primates.* Academic Press, New York.

Tsukahara, T. 1993. Lions eat chimpanzees: The first evidence of predation by lions on wild chimpanzees. *Am. J. Primatol.* **29:**1–11.

Tuttle, R. H. 1969. Knuckle walking and the problem of human origins. *Science* **166:**953–961.

Tuttle, R. H. 1974. Darwin's apes, dental apes and the descent of man: Normal science in evolutionary anthropology. *Curr. Anthropol.* **15:**389–426.

Walker, A. C., and Pickford, M. 1983. New postcranial fossils of *Proconsul africanus* and *Proconsul nyanzae.* In: R. L. Ciochon and R. S. Corruccini (eds.), *New Interpretations of Ape and Human Ancestry,* pp. 325–351. Plenum Press, New York.

Walker, A. C., Falk, D., Smith, R., and Pickford, M. 1983. The skull of *Proconsul africanus:* Reconstruction and cranial capacity. *Nature* **305:**525–527.

Walker, A., Teaford, M. F., Martin, L. B., and Andrews, P. 1993. A new species of *Proconsul* from the Early Miocene of Rusinga/Mfangano Islands, Kenya. *J. Hum. Evol.* **25:**43–56.

Ward, C. V. 1991. *The Functional Anatomy of the Lower Back and Pelvis of the Miocene Hominoid Proconsul nyanzae from Mfangano Island, Kenya.* Ph.D. dissertation, The Johns Hopkins University.

Ward, C. V. 1993. Torso morphology and locomotion in *Proconsul nyanzae. Am. J. Phys. Anthropol.* **92:**291–328.

Ward, C. V., Walker, A., and Teaford, M. F. 1991. *Proconsul* did not have a tail. *J. Hum. Evol.* **21:**215–220.

Ward, C. V., Walker, A., Teaford, M. F., and Odhiambo, I. 1993. A partial skeleton of *Proconsul nyanzae* from Mfangano Island, Kenya. *Am. J. Phys. Anthropol.* **90:**77–111.

Ward, S. C., and Brown, B. 1986. Facial anatomy of Miocene hominoids. In: D. R. Swindler and J. Erwin (eds.), *Comparative Primate Biology, Vol. 1: Systematics, Evolution, and Anatomy,* pp. 413–452. Liss, New York.

Ward, S. C., and Pilbeam, D. 1983. Maxillofacial morphology of Miocene hominoids from Africa and Indo-Pakistan. In: R. L. Ciochon and R. S. Corruccini (eds.), *New Interpretations of Ape and Human Ancestry,* pp. 211–238. Plenum Press, New York.

Watts, E. S. 1990. Evolutionary trends in primate growth and development. In: C. J. DeRousseau (ed.), *Primate Life History and Evolution,* pp. 89–104. Wiley–Liss, New York.

Willoughby, D. P. 1978. *All About Gorillas.* Barnes, South Brunswick, NJ.

Wootton, J. T. 1987. The effects of body mass, phylogeny, habitat and trophic level on mammalian age at first reproduction. *Evolution* **41:**732–749.

Zuckerman, S. 1928. Age changes in the chimpanzee. *Proc. Zool. Soc. London* **Pt. 1:**1–42.

Proconsul

Function and Phylogeny

10

ALAN WALKER

Introduction

The genus *Proconsul* was recognized over 60 years ago as the first Miocene anthropoid from sub-Saharan Africa (Hopwood, 1933). The collections of *Proconsul* fossils have grown steadily since then, so that it is now probably the best-known Miocene primate. We have hundreds of fossils of several species from many localities in Kenya and Uganda. Nearly every body part is now represented and much is known about sexual dimorphism, body proportions, growth and development, and paleoecology. Because of this, the functional anatomy of the genus is relatively well known.

Phylogenetically, *Proconsul* is thought to be either the sister group of all catarrhines [e.g., Harrison, 1982, 1987, 1993 (as one possibility)], the sister group of all apes and humans, including lesser apes [e.g., Andrews, 1985; Andrews and Martin, 1987a; Harrison, 1993 (as one possibility); Martin, 1986; Szalay and Delson, 1979], the sister group of just great apes and humans (e.g., Walker and Teaford, 1989; Rae, 1993, this volume). It is a puzzle that such a well-known genus can be placed in such different positions in a simple phylogenetic scheme of catarrhine primates. Harrison (1993) has given the following four reasons why the relationships between closely related species might not be easily sorted out: (1) there is not enough morphological

---

ALAN WALKER • Departments of Anthropology and Biology, The Pennsylvania State University, University Park, Pennsylvania 16802.

Function, Phylogeny, and Fossils: Miocene Hominoid Evolution and Adaptations, edited by Begun *et al.* Plenum Press, New York, 1997.

information in the fossils, (2) it is difficult to determine the polarities of traits, (3) stem groups are difficult to define because the members retain a high frequency of primitive characters, and (4) closely related species often converge functionally. The same reasons might apply when trying to sort out higher levels of relationship. An outsider to this debate might wonder how much can be said about any fossil primate if the huge collection of *Proconsul* fossils is inadequate for phylogenetic reconstruction.

The Species of Proconsul

There are four species of *Proconsul* that are recognized by most workers. Two of these are from Tinderet localities in Kenya and two are from Kisingiri localities in western Kenya. The larger Tinderet species is also found at Napak in Uganda. *P. africanus* is a medium-sized (cf. siamang) species known from the type locality of Koru (Hopwood, 1933) and other Tinderet localities, *P. heseloni,* a medium-sized species from the type locality of R114, Rusinga Island (Walker *et al.,* 1993) and other Kisingiri localities, *P. major,* a large species from the type site of Songhor (Clark and Leakey, 1951), other Tinderet localities, and Napak, and *P. nyanzae,* a large species from the type locality R1-3, Rusinga Island (Clark and Leakey, 1951), and other localities on Rusinga/Mfangano Islands.

Specimens claimed to be *Proconsul* from other Kenyan and Ugandan localities have either been shown to be of other hominoids or are too fragmentary to be certain. Andrews and Walker (1976) placed a few isolated teeth from the middle Miocene site of Fort Ternan in *Dryopithecus cf. africanus* and *D. cf. nyanzae* (for which read *Proconsul cf. africanus* and *P. cf. nyanzae*). Andrews later (1978) seemed more certain that these were *P. africanus* and *P. nyanzae* and referred them to those species. Harrison (1992) has recently reassessed all of the Fort Ternan material and shows that at least two isolated teeth are closest in their size and morphology to *P. africanus,* but prefers to consider them as belonging to an indeterminate species of *Proconsul.* As he notes, they are clearly from a medium-sized proconsulid, and in view of the way that many fragmentary specimens that are originally named with certainty have to be renamed in the light of further discoveries, they are probably better left as such, rather than placed definitely in the genus. Andrews (1978) also referred some specimens from the Lothidok Formation in northern Kenya to *Proconsul major.* These are now known either to be from a new genus of large Oligocene primate *Kamoyapithecus* (Leakey *et al.,* 1995) or from early Miocene *Afropithecus* (see Leakey and Walker, this volume), or from indeterminate catarrhine species from the late Oligocene and early Miocene. Pilbeam (1969) reported *P. major* from Napak and Moroto in Uganda. The Napak specimens seem, on the parts known, to be indistinguishable from *P. major* from Tinderet, but the Moroto taxon now has no certain taxonomic home.

Andrews and Martin (1987b) and Leakey *et al.* (1988) see its affinities with *Afropithecus* and *Heliopithecus*.

The Hypodigm

No attempt is undertaken here to list the complete hypodigm. A summary of the known body parts is given instead.

Cranial Material

Only one reasonably preserved skull is known, KNM-RU 7029 (Clark and Leakey, 1951); several vertebrae are also associated with the type of *P. heseloni* (Walker and Pickford, 1983; Walker *et al.*, 1983). Several isolated skull parts, mostly of juveniles, are known from the Kaswanga Primate Site (Walker and Teaford, 1988). A few palatal specimens are known that provide some information about the facial skeleton, the most complete ones being the type of *P. nyanzae* BM(NH) M 16647 (Whybrow and Andrews, 1978) and KNM-RU 16000 (Teaford *et al.*, 1988) that either belongs to *P. heseloni* or *P. nyanzae*. Several complete or nearly complete mandibles are known of all species, with *P. africanus* again being the poorest known. Hundreds of teeth representing all species are in mandibles and maxillas or pieces of them or as isolated teeth.

Postcranial Material

Thoracic and lumbar vertebrae are known for *P. nyanzae* (Ward *et al.*, 1993) and many, mostly juvenile, from the Kaswanga Primate site, including half of an atlas. One isolated, large lumbar vertebral body is also known from Rusinga. A last sacral vertebra from the Kaswanga site shows that this species did not have a tail (Ward *et al.*, 1991). Rib fragments have been found at Kaswanga and the skeleton KNM-RU 2036, the type of *P. heseloni,* has rib pieces. Several unpublished sternebrae come from the Kaswanga Primate Site. They are broad and flat, like those of hylobatids, not thin and rodlike as in cercopithecoids. But caution is advised when taking this observation at face value. This is because their absolute size, which is at the small end of the hominoid range and the large end of the cercopithecoid range, might mask hidden allometric effects. Isolated elements of pelvic bones have also been found at Kaswanga, but the best specimen, which is a practically complete pelvic bone, is from Mfangano (Ward *et al.*, 1993).

Upper limb material is best seen in KNM-RU 2036, which has part of a clavicle, part of a scapula, and large parts of both arms and hands (Napier and

Davis, 1959; Walker and Pickford, 1983; Walker *et al.,* 1985). A proximal humerus (Gebo *et al.,* 1988) might belong to *P. heseloni.* Several forelimb bones are known from the Kaswanga Primate Site, including complete and partial hand skeletons (Beard *et al.,* 1986; Begun *et al.,* 1993). The clavicles reported by Clark and Leakey (1951) are really crocodile femurs. Only small parts of the forelimb are known for both larger species. Isolated forelimb pieces of *P. nyanzae* include a distal humerus (Senut, 1986), a proximal ulna (Clark, 1952), and several hand bones (Beard *et al.,* 1986). A piece of a humerus (Rafferty *et al.,* 1995) must belong to *P. major,* but a distal ulna fragment (Nengo and Rae, 1992) must be taken with caution because it does not fit with the rest of what we know about this genus and it has not been firmly established that the bone is, in fact, a distal ulna and/or primate. Some isolated hand bones are known for all species.

The hindlimb skeleton is represented in KNM-RU 2036 by parts of both femurs, tibias, and foot skeletons. Other *P. heseloni* hindlimb bones include nearly complete hindlimb skeletons of adult and juvenile individuals from the Kaswanga Primate Site (Walker and Teaford, 1988). Two partial *P. nyanzae* hindlimbs are known. One, KNM-RU 5872 from Rusinga, has tibia, fibula, and much of a foot skeleton (Walker and Pickford, 1983). It might have been associated with the type of *P. nyanzae* as it is preserved in the same way and comes from the site. The other, KNM-MW 13412, comes from Mfangano and has femurs, tibia, fibula, and two foot bones associated with pelvis and vertebrae (Ward *et al.,* 1993). Only a tibia (Rafferty *et al.,* 1995) and a talus and calcaneum (Clark and Leakey, 1951) are published of the hindlimb of *P. major,* but there are unpublished femoral pieces. Patellae of both Kisingiri species have been described (Ward *et al.,* 1995). They are broad, flat, and thin, a condition found in all fossil and extant hominoids. As with the forelimb, numerous isolated bones of the hindlimb and foot are known for all species.

Body Size in Proconsul

Several attempts have been made to determine the body size of the species of *Proconsul.* The most recent, and the ones based on postcranial bones and therefore more reliable, are by Ruff *et al.* (1989), Sanders and Bodenbender (1994), and Rafferty *et al.* (1995). There are no postcranial bones associated with teeth of *P. africanus,* which is probably a little smaller than *P. heseloni* in tooth size. Following Rafferty *et al.* (1995), whose values are based on the largest sample of individuals, *P. heseloni* has an estimated mean weight of 10.9 kg ($n = 6$), *P. nyanzae,* 35.6 kg ($n = 12$), and *P. major,* 75.1 kg ($n = 3$). Thus, the smaller Kisingiri species is about the size of a siamang and the larger about the size of a pygmy chimpanzee, while the largest species, *P. major,* is about midway in average size between male orangutans and female mountain gorillas. Sanders and Bodenbender (1994) give several reasons why their estimates of body weight based on vertebral body surface area are so much smaller than those of Ruff *et al.*, which were based on hindlimb joint or long-bone cross-sectional area. Their estimates are also much smaller than those of Rafferty *et al.* based on ankle joint surface properties. Sanders and Bodenbender give estimates of 23.4 kg (95% confidence limits from 22.6 to 29.4 kg) for the *P. nyanzae* individual KNM-MW 13142, for which Ruff *et al.* obtained values ranging from 34.1 to 36.3 kg from femoral shaft cross-sectional properties and 31.2 to 36.0 kg from femoral head size. Rafferty *et al.* give an estimate of 41.1 kg from talar joint size. Sanders and Bodenbender also give a rough estimate of 5–8 kg for the *P. heseloni* individual KNM-RU 2036, which Ruff *et al.* estimate weighed between 8.2 and 10.4 kg based on femoral shaft cross sections and which Rafferty *et al.* estimate to have weighed 10.6 kg from talar and tibial dimensions and 9.4 kg from humeral dimensions. It is true, as Sanders and Bodenbender point out, that regression analyses of measurements from different parts of the skeleton and using extant species as the main data base often give different body weight retrodictions for fossil animals. In this case, however, the estimates using vertebral body area give answers that are much lower than those retrodicted by using long-bone shaft properties or hindlimb joint properties. This is clearly a cause for concern and may simply indicate that body proportions and patterns of body use in locomotor and postural activities in *Proconsul* were not the same as those in extant anthropoids. It should be noted, however, that while there is no problem with homologous element and function when using ankle bones or femurs, there is such a problem when using vertebrae. Sanders and Bodenbender, for instance, considered the 6th lumbar of *Macaca,* the 4th lumbar of *Homo,* and the 3rd of *Pan* as equivalent in their analyses.

Sexual Dimorphism

The issue of tooth and body size sexual dimorphism has been an interesting one. Ruff *et al.* (1989), Teaford *et al.* (1993), and Walker *et al.* (1993) review some of this history and Rafferty *et al.* (1995) give new data. The position taken here is that the amount of male/female body weight dimorphism in both *P. heseloni* and *P. nyanzae* is about 1.3:1. It cannot be determined for the other species, although we know that the three individuals of *P. major* for which weight can be determined range from 64.4 to 86 kg (a ratio of 1.34:1). A ratio of 1.3:1 is a moderate amount of body size dimorphism in primates, near that of many cercopithecoids and the common chimpanzee. Canine dimorphism is fairly strong in *Proconsul,* with females having small, conical canines that hone only a little and males having large, bladelike canines that are honed down with age.

Function

This is only a superficial account of function in *Proconsul* for three main reasons apart from space considerations. These are: (1) practically every part of the skeleton is known and much has been written about it; (2) only *P. heseloni* and *P. nyanzae* are well known, the other two species are really known only from fragments; and (3) body size varies by nearly an order of magnitude in the species and this is bound to have had some consequences for function in this genus.

Social System

Based on knowledge of present-day anthropoids, Clutton-Brock *et al.* (1977) showed that body size sexual dimorphism was not inevitably correlated positively with species body weight. Harvey *et al.* (1978) then went on to examine sexual dimorphism in primate teeth. They concluded that both sexual selection and predator defense are important selective forces. Terrestrial species had significantly larger male upper canines than arboreal species as well as being larger bodied. They also claimed that monogamous species tended to have nondimorphic canines and that having considerable canine sexual dimorphism was usually the rule in polygynous species. This last idea has not gone unchallenged, however (Rowell and Chism, 1986). Fleagle *et al.* (1980), taking their lead from Harvey and colleagues, looked for dimorphism in early anthropoids from the Fayum. They found that there was considerable canine dimorphism in three genera and concluded that this showed they were polygynous rather than monogamous. Taking the evidence without Rowell and Chism's cautions, we can look at *Proconsul.* In all species there is moderate canine sexual dimorphism at the level of that seen in chimpanzees, but nothing like the level found in baboons. We can conclude that, probably, these species lived in polygynous single- or multi-male groups. The body weight dimorphism of about 1.3:1 that is probably true for all species, predicts a socionomic sex ratio similar to some macaque and howling monkey species (Clutton-Brock *et al.*, 1977).

Relative Brain Size

This is possible for only one individual and even then we have to estimate the body weight from other specimens. Walker *et al.* (1983) were able to estimate the cranial capacity for the *P. heseloni* skull, KNM-RU 7029. They were able to show that this species was as relatively large-brained as modern anthropoids. Leaf-eating primates have smaller brains than closely related frugivorous primates, a relationship that seems to hold for most mammalian groups (Martin, 1983). Clutton-Brock and Harvey (1980) have proposed that leaf-eating primates have smaller brains than their fruit-eating relatives because the central nervous system processing necessary to find dispersed fruit is more than that required to find leaves, which are present on all trees whether fruiting or not. This idea did not take into account the fact that many trees produce leaf toxins to discourage leaf-eating and that selecting those leaves without toxins might be as difficult if not more difficult than finding fruits (Martin, 1983). Further, McNab (1980) has shown that mammalian folivores generally have a lower basal metabolic rate than their frugivorous counterparts and that this effect alone would produce small brains in a leaf-eating group through the action of maternal metabolic rate on the fetus. Whatever the merits of either hypothesis, it seems that *P. heseloni,* at least, had a "normal" anthropoid basal metabolic rate and because of this and the size of its body was, most likely, a frugivore.

Diet

The conformation of the dentition, the length of the shearing crests (Kay, 1977), and the molar microwear (Walker *et al.*, 1994) all show that *Proconsul* species were frugivorous. Reference to enamel thickness in *Proconsul* is anecdotal to date, but individual teeth from the Kaswanga Primate Site are now being sectioned and examined by A. D. Beynon and M. C. Dean.

Locomotion

A lot has been written about the locomotion of this genus, starting with the speculations of Clark and Leakey (1951) based on very few specimens, some of which were not even mammalian. The associated forelimb skeleton described by Napier and Davis (1959) was, for a long time, practically the only decent fossil evidence, but conclusions concerning hand and arm function were not unanimous. This was related in part to distortion in the original fossils, poor study casts, the fact that important parts were missing or damaged, and the like. Today we have ample evidence from most body parts to put together a fairly complete picture of the joint capabilities of these primates, their body proportions, and body weight. This account will progress by anatomical region.

Body Proportions

The intermembral index of *P. heseloni* is 86.9 (Walker and Pickford, 1983). This value is close to that of some anthropoids. They include some terrestrial cercopithecines, some arboreal atelines, and the colobine *Nasalis*

(Napier and Napier, 1967). The brachial index of 96.4 is in the ranges of some atelines and cebines as well as both subfamilies of Old World monkey and *Pan*. It is worth noting that it lies at the lower end of the range for *Pongo* although it is far lower than the extremes for gibbons. The crural index is not very variable in anthropoid primates, and the index of 91.8 lies within the ranges of a great variety of locomotor types. In robusticity, the long bones of *Proconsul* resemble more those of cercopithecoids, chimpanzees, or ceboids with less elongated limbs. They are not very long and gracile like most atelines or the lesser apes (Ruff *et al.*, 1989).

Several ribs and vertebrae of *Proconsul* are known and the configuration of the iliac blades also serves to tell us what the torso morphology was like. Ward (1993) has convincingly shown that the torso was mediolaterally narrow and dorsoventrally deep and that the epaxial muscles were thin and straplike like those of most monkeys. The torso shape is typical of primates that engage in pronograde quadrupedalism. There were probably six lumbar vertebrae, a condition like that in some cercopithecoids and some gibbons. No evidence is found, therefore, of the lumbar reduction and stiffening that are so characteristic of the great apes. *Proconsul* had a relatively long, flexible lumbar region.

The scapula most resembles large platyrrhines and colobines (Rose, 1993). The acromial end of a clavicle from KNM-RU 2036 (Walker *et al.*, 1985) is not very instructive, but is flat and does have a strong anteroposterior curvature. The acromial articular surface is relatively small. The proximal humerus (Gebo *et al.*, 1988) that probably belongs to this genus is most similar to those of *Alouatta seniculus*. The humeral head seems to lack the medial torsion typical of living apes. Medial torsion has been reconstructed in KNM-RU 2036 (Napier and Davis, 1959; Walker and Pickford, 1983), but both humeri are distorted plastically and a skeptic could easily make the argument that there was no torsion in that individual. The bicipital groove is shallow and the tuberosities not very developed. All of these features and more indicate a relatively mobile shoulder joint with limited arm-swinging abilities and no special adaptation for bimanual suspension, cursorial quadrupedalism or brachiation.

The distal humerus (see Rose, 1993, for review) is similar to those of small apes, but the lateral keel of the trochlea is more salient. There was a zona conoidea and this is confirmed by the reciprocal articulation on the radial head. This condition confers stability of the radial head during pronation and supination. The olecranon is directed slightly posteroproximally which might indicate some terrestriality (Harrison, 1982) or, alternatively, overhead use of the arm in the trees (Rose, 1983). The posterior border of the ulna is relatively straight, like most anthropoids except apes.

The wrist has an extensive literature pertaining to it (see Beard *et al.*, 1986, for a review). It is of a nonape anthropoid type, in that the ulnar styloid articulates with the pisiform and triquetrum (Napier and Davis, 1959; Beard *et al.*, 1986). There is a centrale present. A couple of features resemble those of hominoid wrists: These are a midcarpal joint that permits marked ulnar deviation and a relatively mobile carpometacarpal joint for digit 1 (Rose, 1992). Although there is a clear articulation of the ulna with the wrist, it is constructed in such a way as to permit more ulnar deviation than in cercopithecoids. This is surely related to its climbing capabilities and probably is part of the primarily arboreal adaptations of this genus, rather than arboreal adaptations built on a more terrestrial anatomical substrate as in the cercopithecoids (see Harrison, 1989). The hand proportions are rather like those of certain macaque species. The phalanges (Begun *et al.*, 1994) are relatively strongly built and their midshaft curvatures and relatively poor development of secondary shaft features indicate that *Proconsul* was primarily an above-branch quadruped. The joint surfaces suggest that the hand might have had slightly hyperextended metacarpophalangeal joints on large branches.

Pieces of pelvis are known for adults and juvenile *P. heseloni* from the Kaswanga Primate Site. An almost complete hipbone is known for a *P. nyanzae* adult (Ward *et al.*, 1993). The iliac blades are narrow relative to their length, like those of other catarrhines except apes. It has a relatively short pubis and long ischia with no flaring tuberosities. Thus, the pelvis supports the notion that these animals practiced pronograde quadrupedal locomotion and they did not have the ischial callosities that are seen in Old World monkeys and gibbons. The acetabulum is shallow, like hominoids and most ceboids, yet the lunate surface is not expanded cranially as it is in hominoids and the more suspensory ceboids. *Proconsul* shows, then, no evidence from the pelvis of an emphasis on upright posture.

The hindlimb skeleton is known completely from several individuals, and these include juveniles. The femoral heads are uniform partial spheres. The fovea capitis is situated in a position that indicates frequently abducted hip postures during weight-bearing activities. The ratio between the head and neck diameters also shows this, as does the high neck–shaft angle of 135°, which is in the range of great apes and suspensory platyrrhines.

The femoral shafts (Ruff *et al.*, 1989) are nearly circular in cross section and relatively straight. Primates that are very cursorial or that leap often, have anteroposteriorly elongated cross sections, so this signifies little emphasis on these two activities in this genus. The patellar notch is shallow and the patellae themselves and flat, broad and thin (Ward *et al.*, 1995). This is the condition in all living and fossil hominoids and indicates, once again, the *Proconsul* had no specializations for leaping or cursoriality. The femoral condyles are relatively small and convex in both major directions. There are strong collateral ligament impressions and a very deep groove and insertion pit for the popliteus muscle. The tibia is a stout bone, oval in section with the long axis anteroposteriorly directed. The femoral articular surfaces are about equal in area and the cnemial crest extends quite distally. The latter, together with the patellar shape and popliteus development, suggests a knee held by muscular activity in many varied positions including some with a lot of rotation, but not one in which full and rapid extension was needed. The fibula is a strong element in *Proconsul.* This is like the condition in hominoids and not at all like that in cercopithecoids where it is reduced to a thin splintlike bone. The large fibula is related to the relatively large hallux in *Proconsul,* for the long hallucial flexor has its origin on the fibula. Old World monkeys have had a period of terrestrial existence in their ancestry (Harrison, 1989) during which both the hallux and the fibula were reduced.

The tibia has a stout malleolus with an articular surface laterally and anteriorly that fits into a discrete cotylar fossa on the talus. This fossa appears to be a kind of articular stop at the limit of dorsiflexion. Stability in dorsiflexion is also seen in the wedging of the talus, whereby the trochlear surface becomes increasingly wider anteriorly. As the ankle moved into dorsiflexion, this wedging would tend to spread the two malleoli apart and tighten the interosseus ligaments and stretch the interosseus membrane. This in turn would have the effect of tightening the grip of the malleoli on the talus. In plantar flexion, however, ankle stability would be compromised and would have to be maintained by muscular activity. In its overall proportions the foot is rather like that of certain colobines (Strasser, 1993). The similarity to colobines lies in the relative length of the tarsus and the triceps surae load arm/lever arm ratio, and the relative lengths of parts of the third digital ray. It is more similar to apes, particularly hylobatids, in the relative length of the hallux. The hallux was strong and there was considerable motion possible at the midtarsal joint. There is only a hint, and then only in some specimens, of a plantar tubercle for the origin of flexor digitorum brevis (Sarmiento, 1983). Sarmiento correlates this process with slow climbers or those species descended from slow climbers. The absence or near absence in *Proconsul* is a puzzle if one accepts this interpretation, as practically all other parts of the postcranial skeleton evince no signs of adaptations for cursorial, suspensory, or leaping behavior but yet show many related to slow arborealism. Much has been written about the function of the foot in *Proconsul* and the reader is referred to Langdon (1986), Lewis (1989), and Rose (1993) and references therein.

All of the phalanges of the foot are known for several individuals (Begun *et al.,* 1994). The foot phalanges, particularly those of the hallux, strongly support the notion that *Proconsul* species had powerful grasping and climbing abilities that were adapted to arboreal substrates of relatively small diameter.

In summary, the postcranial skeleton of this genus is most reminiscent of that of certain platyrrhines and nonterrestrial cercopithecines. Taking the bony features overall, these arboreal quadrupeds most probably had a varied positional repertoire and indulged in relatively slow climbing. There are no overt signs of bony adaptations to fore- or hindlimb suspension, cursorial quadrupedalism, or leaping behavior. This is the conclusion reached by Rose (1983, 1993) and Walker and Pickford (1983). This is somewhat different from the picture originally put forward by Napier (1963) who considered *Proconsul* to have been a more active, leaping genus as a "semibrachiator." He meant by this term animals that were essentially quadrupedal but which used upper limb extension to suspend the body, propel it through space and check momentum at the end of a leap. It now seems, with our vastly improved collections and a more complete knowledge of primate locomotion, that there are few signs of forelimb suspensory behavior in *Proconsul.*

Phylogeny

It was earlier thought that *Proconsul* had special relationships with living hominoids. Hopwood (1933), for instance, thought it was ancestral to the chimpanzee and Pilbeam (1969) thought that *P. major* was probably ancestral to the *G. gorilla* and *P. africanus* possibly ancestral to *P. troglodytes.* However, it is now the consensus of nearly all workers that *Proconsul* is a primitive catarrhine. The question really boils down to whether the genus represents the sister group of all catarrhines, all hominoids, or just great apes and humans. Harrison has argued most forcibly that there are hardly any shared derived features of the hominoids to be seen in *Proconsul.* He admits, however, that there are a few possible synapomorphies in the postcranial skeleton that might show that the genus is a basal hominoid.

Clark and Leakey (1951) reported that *Proconsul* had a frontal sinus. This fact seems to have been forgotten, perhaps because until recently the post-fossilization defect in the upper orbital margin had been filled with plaster. This has now been removed and there is a clear and extensive frontal sinus present that excavates most of the frontal squame. Andrews (1992) referring to this and knowing that a frontal sinus is also present in *Afropithecus* and *Dryopithecus,* concluded that the presence of a frontoethmoid sinus is "almost certainly an ancestral hominoid character." This would mean, of course, that hylobatids as well as orangutans have lost this primitive condition. Since Old World monkeys do not have these sinuses, and they are the sister group to the hominoids, it would be more parsimonious to consider a sinus in *Proconsul* as a derived feature shared with humans and great apes. The loss would then be restricted to orangutans and their fossil relatives. Deciding the polarity of this trait is complicated, though, by the fact that several New World monkey genera have frontal sinuses (Hershkowitz, 1977). Shea (1985) has linked frontal sinus development with marked supraorbital tori, but *P. heseloni* has a large sinus and no large tori.

Note that Rae (this volume) has undertaken a cladistic analysis of the facial skeleton of Miocene hominoids and reports that *P. heseloni* and *P. nyanzae* have different faces. He says that *P. heseloni* was excluded from his analysis because it (and other taxa) had insufficient derived features that would unequivocally place them with any catarrhine clades without severely diminishing resolution between other taxa. But his results place *P. nyanzae* (with others) at the base of the great ape crown group. This is surprising in view of the long debate over whether or not there was only one species of

Proconsul at the Kisingiri sites (see Teaford *et al.,* 1993, for a review), and the recent suggestion by Rettalack *et al.* (1995) that these are time-successive species in a single lineage.

There is also a derived feature of the postcranial skeleton that also links gibbons and Old World monkeys, namely, the flat, flared ischial tuberosities that form the bony substrate for ischial callosities. Here the polarity is more easily determined, since prosimians and platyrrhines lack the condition. *Proconsul* did not have callosities and, like great apes and humans, retained the primitive condition. If this is a shared derived condition in gibbons and cercopithecoids, then it argues either for *Proconsul* being the sister group of all catarrhines or of great apes and hominoids. *Proconsul* differs from monkeys and resembles hominoids in that it probably had six lumbar vertebrae rather than seven and had a narrow sacrum with a small sacroiliac joint (Ward, 1993; Ward *et al.,* 1993). *Proconsul* did not have a tail (Ward *et al.,* 1991) and this condition is a derived one for all hominoids. It has occurred independently in several genera of catarrhines, but the ancestral catarrhine condition must surely have been to have a tail. Rose (1983) has argued for the condition of the distal humerus in *Proconsul* to be the derived hominoid one. Rose (1992) also makes a case for the morphology and kinematics of the trapezium–first metacarpal joint in *Proconsul, Afropithecus,* and large living hominoids to be a shared derived complex. Beard *et al.* (1986) also consider that the increased ulnar deviation that the *Proconsul* wrist would have allowed is derived for catarrhines and, while not equivalent to the hominoid condition, foreshadows it. Begun *et al.* (1994) suggest that several characters of the phalanges may be derived features shared with living apes. These include relatively long thumbs and halluces, broader hallucial terminal phalanx articular surface, very powerful hallucial and pollical flexors, relatively short, robust intermediate phalanges, and differentiated foot and hand phalanges. And finally, Rae (1993, this volume) has claimed that several metric and nonmetric characters of the face link several; Miocene taxa (*Proconsul* among them) with the living hominoids and, in particular, with the great apes.

Suggestions that *Proconsul* had postcrania close to the ancestral catarrhine morphotype (e.g., Harrison, 1993) must be reconciled with the configuration of the few postcranial bones attributed to *Aegyptopithecus* and *Propliopithecus* (see Gebo, 1993, and Rose, 1993, for reviews). This is because this Oligocene catarrhine has postcranial bones that several important differences from those of the Miocene ones. These include those listed in Table I. These early catarrhines apparently were more stockily built and had less joint mobility and/or excursions at the shoulder, elbow, and ankle than the Miocene forms. Unlike the latter, in which leaping was, at the most, a minor part of the locomotor repertoire, these had some leaping capabilities (Gebo, 1993). However, it is clear that both the Oligocene and Miocene groups were arboreal quadrupeds, and that neither had any clear morphological adaptations for terrestrial, arm-swinging, or suspensory locomotion. My conclusion at present, then, is that *Proconsul* mostly has the morphology of a primitive homi-

Table I. Some Differences betwen Postcranial Bones of *Aegyptopithecus* and Early Miocene Large Hominoids

Aegyptopithecus	Miocene forms
Humeral head faces posteriorly	Humeral head faces posterosuperiorly
Humeral head small relative to the tuberosities	Head bigger and elevated above the tuberosities
Humeral lesser tuberosity less anteriorly positioned	Lesser tuberosity more anteriorly positioned
Humeral shaft not elongated	Shaft elongated
The olecranon fossa is wide and shallow	Fossa deep and confined
Great lateral extension of brachialis flange	Flange moderately developed
Zona conoidea flat	Zona conoidea bevelled
Coronoid process low	Coronoid process higher
Olecranon process long	Olecranon process short
Posterior border of ulnar shaft convex proximally, straight distally	Posterior border of shaft sigmoidal
Ulnar shaft short	Ulnar shaft elongated
Tibia short and robust (for *Propliopithecus*)	Tibia elongated slender
Talar trochlea parallel-sided	Trochlea tapers posteriorly

noid in the sense of Rose's "Miocene hominoid-like" (1983), but that it also has several derived features of living hominoids and just a few features that link it with living great apes and humans. More detailed analysis may reveal more derived features linking this genus with hominoids or even great apes and humans, but they are unlikely to change the overall picture of this genus as a primitive hominoid one.

References

Andrews, P. J. 1978. A revision of the Miocene Hominoidea of East Africa. *Bull. Br. Mus. Nat. Hist. Geol.* **30:**85–225.

Andrews, P. J. 1985. Family group systematics and evolution among catarrhine primates. In: E. Delson (ed.), *Ancestors: The Hard Evidence,* pp. 14–22. Liss, New York.

Andrews, P. 1992. Evolution and environment in the Hominoidea. *Nature* **360:**641–646.

Andrews, P., and Martin, L. 1987a. Cladistic relationships of extant and fossil hominoids. *J. Hum. Evol.* **16:**101–118.

Andrews, P. J., and Martin, L. 1987b. The phyletic position of the Ad Dabtiyah hominoid. *Bull. Br. Mus. Nat. Hist. Geol.* **41:**383–393.

Andrews, P., and Walker, A. 1976. The primate and other fauna from Fort Ternan, Kenya. In: G. Isaac and E. R. McCown (eds.), *Human Origins: Louis Leakey and the East African Evidence,* pp. 279–304, Benjamin, Menlo Park, CA.

Beard, K. C., Teaford, M. F., and Walker, A. 1986. New wrist bones of *Proconsul africanus* and *nyanzae* from Rusinga Island, Kenya. *Folia Primatol.* **47:**97–118.

Begun, D. R., Teaford, M. F., and Walker, A. 1993. Comparative and functional anatomy of *Proconsul* phalanges from the Kaswanga Primate Site, Rusinga Island, Kenya. *J. Hum. Evol.* **26:**89–165.

Clark, W. E. L. 1952. Report on fossil hominoid material collected by the British–Kenya Miocene expedition. *Q. J. Geol. Soc. London* **105:**260–262.

Clark, W. E. L., and Leakey, L. S. B. 1951. The Miocene Hominoidea of East Africa. *Br. Mus. Nat. Hist. Fossil Mamm. Afr.* **1:**1–117.

Clutton-Brock, T. H., and Harvey, P. H. 1980. Primates, brains and ecology. *J. Zool.* **190:**309–323.

Clutton-Brock, T. H., Harvey, P. H., and Rudder, B. 1977. Sexual dimorphism, socionomic sex ratio and body weight in primates. *Nature* **269:**797–800.

Fleagle, J. G. 1988. *Primate Adaptation and Evolution.* Academic Press, New York.

Fleagle, J. G., Kay, R. F., and Simons, E. L. 1980. Sexual dimorphism in early anthropoids. *Nature* **287:**328–330.

Gebo, D. L. 1993. Postcranial anatomy and locomotor adaptations in early African anthropoids. In: D. L. Gebo (ed.), *Postcranial Adaptation in Nonhuman Primates,* pp. 220–234. Northern Illinois University Press, De Kalb.

Gebo, D. L., Beard, K. C., Teaford, M. F., Walker, A., Larson, S. G., Jungers, W. L., and Fleagle, J. G. 1988. A hominoid proximal humerus from the Early Miocene of Rusinga Island, Kenya. *J. Hum. Evol.* **17:**393–401.

Harrison, T. 1982. *Small-Bodied Apes from the Miocene of East Africa.* Ph.D. thesis, University of London.

Harrison, T. 1987. The phylogenetic relationships of the early catarrhine primates: A review of the current evidence. *J. Hum. Evol.* **16:**41–80.

Harrison, T. 1989. New postcranial remains of Victoriapithecus from the Middle Miocene of Kenya. *J. Hum. Evol.* **18:**3–54.

Harrison, T. 1992. A reassessment of the taxonomic and phylogenetic affinities of the fossil catarrhines from Fort Ternan, Kenya. *Primates* **33:**501–522.

Harrison, T. 1993. Cladistic concepts and the species problem in hominoid evolution. In: W. H. Kimbel and L. B. Martin (eds.), *Species, Species Concepts, and Primate Evolution,* pp. 345–371. Plenum Press, New York.

Harvey, P. H., Kavanagh, M., and Clutton-Brock, T. H. 1978. Sexual dimorphism in primate teeth. *J. Zool.* **186:**475–485.

Hershkowitz, P. 1977. *Living New World Monkeys (Platyrrhini),* Vol. 1. University of Chicago Press, Chicago.

Hopwood, A. T. 1933. Miocene primates from British East Africa. *Ann. Mag. Nat. Hist.* **11:**96–98.

Kay, R. F. 1977. Diets of early Miocene African hominoids. *Nature* **268:**628–630.

Langdon, J. H. 1986. Functional morphology of the Miocene hominoid foot. *Contrib. Primatol.* **22:**1–225.

Leakey, M. G., Ungar, P. S., and Walker, A. 1995. A new genus of large primate from the late Oligocene of Lothidok, Turkana District, Kenya. *J. Hum. Evol.* **28:**519–531.

Leakey, R. E., Leakey, M. G., and Walker, A. C. 1988. Morphology of *Afropithecus turkanensis* from Kenya. *Am. J. Phys. Anthropol.* **76:**289–307.

Lewis, O. J. 1989. *Functional Morphology of the Evolving Hand and Foot.* Clarendon Press, Oxford.

McNab, B. K. 1980. Food habits, energetics and the populations biology of mammals. *Am. Nat.* **116:**106–124.

Martin, L. 1986. Relationships among extant and extinct great apes and humans. In B. Wood, L. Martin, and P. Andrews (eds.), *Major Topics in Primate and Human Evolution,* pp. 161–187, Cambridge University Press, London.

Martin, R. D. 1983. Human brain evolution in an ecological context. 52nd James Arthur Lecture, American Museum of Natural History, New York.

Napier, J. R. 1963. Brachiation and brachiators. *Symp. Zool. Soc. Lond.* **10:**183–195.

Napier, J. R., and Davis, P. R. 1959. The forelimb and associated remains of *Proconsul africanus. Br. Mus. Nat. Hist. Fossil Mamm. Afr.* **16:**1–69.

Napier, J. R., and Napier, P. H. 1967. *A Handbook of Living Primates.* Academic Press, New York.

Nengo, I. O., and Rae, T. C. 1992. New hominoid fossils from the early Miocene site of Songhor, Kenya. *J. Hum. Evol.* **23:**423–429.

Pilbeam, D. R. 1969. Tertiary Pongidae of East Africa: Evolutionary relationships and taxonomy. *Bull. Peabody Mus. Nat. Hist.* **31:**1–185.

Rae, T. C. 1993. Early Miocene hominoid evolution: Phylogenetic considerations. *Am. J. Phys. Anthropol. Suppl.* **16:**161.

Rafferty, K. L., Walker, A., Ruff, C. B., Rose, M. D., and Andrews, P. J. 1995. Postcranial estimates of body weights in *Proconsul,* with a note on a distal tibia of *Proconsul major* from Napak, Uganda. *Am. J. Phys. Anthropol.* **97:**391–402.

Rettalack, G. J., Bestland, E. A., and Dugas, D. P. 1995. Miocene paleosols and habitats of *Proconsul* on Rusinga Island, Kenya. *J. Hum. Evol.* **29:**53–91.

Rose, M. D. 1983. Miocene hominoid postcranial morphology: Monkey-like, ape-like, neither, or both? In R. L. Ciochon and R. S. Corruccini (eds.), *New Interpretations of Ape and Human Ancestry,* pp. 405–417. Plenum Press, New York.

Rose, M. D. 1992. Kinematics of the trapezium–1st metacarpal joint in extant anthropoids and Miocene hominoids. *J. Hum. Evol.* **22:**255–266.

Rose, M. D. 1993. Locomotor anatomy of Miocene hominoids. In: D. L. Gebo (ed.), *Postcranial Adaptation in Nonhuman Primates,* pp. 252–272. Northern Illinois University Press, De Kalb.

Rowell, T. E., and Chism, J. 1986. Sexual dimorphism and mating systems: Jumping to conclusions. In: M. Pickford and B. Chiarelli, eds., *Sexual Dimorphism in Living and Fossil Primates,* pp. 107–111. Il Sedicesimo, Florence.

Ruff, C. B., Walker, A., and Teaford, M. F. 1989. Body mass, sexual dimorphism and femoral proportions of *Proconsul* from Rusinga and Mfangano Islands, Kenya. *J. Hum. Evol.* **18:**515–536.

Sanders, W. J., and Bodenbender, B. E. 1994. Morphological analysis of lumbar vertebra UMP 67-28: Implications for spinal function and phylogeny of the Moroto hominoid. *J. Hum. Evol.* **26:**203–237.

Sarmiento, E. 1983. The significance of the heel process in anthropoids. *Int. J. Primatol.* **4:**127–152.

Senut, B. 1986. New data on Miocene hominoid humeri from Pakistan and Kenya. In: J. Else and P. Lee (eds.), *Primate Evolution,* pp. 151–161. Cambridge University Press, London.

Shea, B. T. 1985. On aspects of skull form in African apes and orangutans, with implications for hominoid evolution. *Am. J. Phys. Anthropol.* **68:**329–342.

Strasser, E. 1993. Kasawanga *Proconsul* foot proportions. *Am. J. Phys. Anthropol. Suppl.* **16:**191.

Szalay, F. S., and Delson, E. 1979. *Evolutionary History of the Primates.* Academic Press, New York.

Teaford, M. F., Beard, K. C., Leakey, R. E., and Walker, A. 1988. New hominoid facial skeleton from the Early Miocene of Rusinga Island, Kenya, and its bearing on the relationship between *Proconsul nyanzae* and *Proconsul africanus. J. Hum. Evol.* **17:**461–477.

Teaford, M. F., Walker, A., and Mugaisi, G. S. 1993. Species discrimination in *Proconsul* from Rusinga and Mfangano Islands, Kenya. In: W. H. Kimbel and L. B. Martin (eds.), *Species, Species Concepts, and Primate Evolution,* pp. 373–392. Plenum Press, New York.

Walker, A., and Pickford, M. 1983. New postcranial fossils of *Proconsul africanus* and *Proconsul nyanzae.* In: R. L. Ciochon and R. S. Corruccini (eds.), *New Interpretations of Ape and Human Ancestry,* pp. 325–351. Plenum Press, New York.

Walker, A., and Teaford, M. 1988. The Kaswanga Primate Site: An early Miocene hominoid site on Rusinga Island, Kenya. *J. Hum. Evol.* **17:**539–544.

Walker, A., and Teaford, M. 1989. The hunt for *Proconsul. Sci. Am.* **260:**76–82.

Walker, A., Falk, D., Smith, R., and Pickford, M. 1988. The skull of *Proconsul africanus. Nature* **305:**525–527.

Walker, A., Teaford, M. F., and Leakey, R. E. 1985. New information regarding the R114 *Proconsul* site, Rusinga Island, Kenya. In: J. Else and P. Lee (eds.), *Primate Evolution,* pp. 143–149. Cambridge University Press, London.

Walker, A., Teaford, M. F., Martin, L., and Andrews, P. 1993. A new species of *Proconsul* from the early Miocene of Rusinga/Mfangano Islands, Kenya. *J. Hum. Evol.* **25:**43–56.

Walker, A., Teaford, M. F., and Ungar, P. S. 1994. Enamel microwear differences between species of *Proconsul* from the Early Miocene of Kenya. *Am. J. Phys. Anthropol. Suppl.* **18:**202–203.

Ward, C. V. 1993. Torso morphology and locomotion in *Proconsul nyanzae. Am. J. Phys. Anthropol.* **92:**291–328.

Ward, C. V., Walker, A., and Teaford, M. F. 1991. *Proconsul* did not have a tail. *J. Hum. Evol.* **21:**215–220.

Ward, C. V., Walker, A., Teaford, M. F., and Odhiambo, I. 1993. A partial skeleton of *Proconsul nyanzae* from Mfangano Island, Kenya. *Am. J. Phys. Anthropol.* **90:**77–112.

Ward, C. V., Ruff, C. B., Walker, A., Rose, M. D., Teaford, M. F., and Nengo, I. O. 1995. Functional morphology of *Proconsul* patellas from Rusinga Island, Kenya, with implications for other Miocene–Pliocene catarrhines. *J. Hum. Evol.* **29:**1–19.

Whybrow, P. J., and Andrews, P. J. 1978. Restoration of the holotype of *Proconsul nyanzae. Folia Primatol.* **30:**115–125.

Afropithecus Function and Phylogeny

11

MEAVE LEAKEY and ALAN WALKER

Introduction

The genus *Afropithecus* Leakey & Leakey, 1986 contains fossil large hominoids probably of a single species, *A. turkanensis,* from four early Miocene sites east and west of Lake Turkana, Kenya. The type (KNM-WK 16999) is a palate, facial skeleton, and anterior part of the cranium of an adult, presumed male, individual. Apart from the type, there are many specimens of upper and lower teeth and several postcranial elements. Nearly all of these have now been described (Leakey and Leakey, 1986; Leakey *et al.*, 1988; Leakey and Walker, 1985). There have been several discussions about the relationships of this genus and the tribe Afropithecini has been named for it together with *Heliopithecus* and *Otavipithecus* (Andrews, 1992). Not much has been written about the postcranial anatomy, but it has been favorably compared with *Proconsul nyanzae* (Leakey *et al.*, 1988) and several functional analyses indicate that *P. nyanzae* was an arboreal, relatively slow-moving quadruped (Ward *et al.*, 1993). It is likely that *Afropithecus,* like *Proconsul,* had a postcranial skeleton that is very close to the primitive hominoid condition.

Often grouped with *Afropithecus* is the Miocene hominoid, *Heliopithecus,* from Ad Dabtiyah, Saudi Arabia, named by Andrews and Martin (1987a) in

MEAVE LEAKEY • Division of Palaeontology, National Museums of Kenya, Nairobi, Kenya. ALAN WALKER • Departments of Anthropology and Biology, The Pennsylvania State University, University Park, Pennsylvania 16802.

Function, Phylogeny, and Fossils: Miocene Hominoid Evolution and Adaptations, edited by Begun *et al.* Plenum Press, New York, 1997.

an announcement that was submitted for publication before the naming of *Afropithecus,* but published afterwards. In a note added in proof (p. 391) they doubted whether the generic distinction between the two was justified. Thus, although there are clear resemblances between this Saudi Arabian hominoid and the known hypodigm of *Afropithecus,* the detailed relationship is less certain (Leakey and Leakey, 1986; Leakey *et al.,* 1988).

The Moroto specimens provide another complication in this account. They have been placed previously in *Proconsul major* (Pilbeam, 1969; Andrews, 1978), but are now thought to be close to *Afropithecus* (Andrews, 1992; Andrews and Martin, 1987b; Kelley and Pilbeam, 1986; Leakey, 1963; Leakey and Leakey, 1986; Leakey *et al.,* 1988; Madden, 1980; Martin, 1981). Although the facial specimen from Moroto might possibly be accommodated in *Afropithecus,* the vertebrae that are probably associated with it differ strongly from those of *P. nyanzae,* which in other postcranial bones resembles *Afropithecus* closely. This is a point that has been taken up by Ward (1993) to mean that the Moroto specimens cannot belong in *Afropithecus.* Her conclusion has not, however, gone unchallenged (Sanders and Bodenbender, 1994). It is worth pointing out, though, that the conformation of the incisive canal, which is often taken to be a reliable character (Ward and Pilbeam, 1983), is very different in the two. The Moroto palate has an open fossa, like those seen in *Proconsul* (Ward and Pilbeam, 1983), while *Afropithecus* has a narrow, inclined canal. This is a greater difference than found in modern catarrhine genera.

Afropithecus is known from cranial and dental specimens from Kalodirr, Moruorot, and Buluk, two isolated teeth from Locherangan, and postcranial fossils from Kalodirr and Buluk. All sites are dated to between about 17 and 17.5 Ma (Boschetto *et al.,* 1992; McDougall and Watkins, 1985; Anyonge, 1991).

The Hypodigm of Afropithecus turkanensis

A nearly complete list of specimens from Kalodirr that can be attributed to *Afropithecus* is given in Leakey *et al.* (1988). An undescribed specimen from Kalodirr includes both maxillas and a nearly complete mandible and has almost all of its tooth crowns relatively unworn. The only definite *Afropithecus* specimen from Moruorot is a juvenile mandible, previously referred to *Proconsul major* by Andrews (1978). The specimens from Buluk are listed in Leakey and Walker (1985). A complete list of the hypodigm is given in Table I.

Size

The face of the type specimen is about the size of that of a male common chimpanzee. The maxillary teeth in another specimen, KNM-WK 17012, are much smaller, and the comparison invites the commonly asked question: is

Table I. Hypodigm of *Afropithecus turkanensis*

From Kalodirr	
Cranial and associated postcranial	
KNM-WK 16840	Mandible (lt. I_{1-2}, P_4–M_2, rt. roots and/or broken crowns I_1–M_3)
KNM-WK 16901	Lower rt. C, root lt. C, rt. fibula, prox. rt. m/t III, rt. m/t IV, dist. m/t I
KNM-WK 16959	Weathered rt. and lt. mandible frags.
KNM-WK 16962	Lt. maxilla and premaxilla (partial roots I^2, C, and P^3–M^3), rt. maxilla frag. (roots M^3), rt. frontal frag., isolated rt. C
KNM-WK 16999	Facial and frontal region of a cranium with dentition (type)
KNM-WK 17010	Associated lower teeth: frag. lt. I_1, I_2 crown, tip crown C, P_4, M_{2-3}, frag. rt. I_1 crown, I_2–P_4, and frags. M
KNM-WK 17011	Mandible frag. (root rt. C and I_2)
KNM-WK 17012	Associated upper teeth: rt. P^3 (lingual frag.), P^4 (buccal frag.), M^{1-3}, lt. P^4, M^1, and M^3
KNM-WK 17013	Rt. I^1
KNM-WK 17014	Mandible frags. (rt. C, P_4, M_{2-3}, lt. P_4 and M_2)
KNM-WK 17015	Fragment of distal humerus
KNM-WK 17016	Associated lower teeth: lt. I_1–M_3, rt. I_1–P_4, M_{2-3}, partial shaft rt. ulna
KNM-WK 17021	Rt. mandibular body, ramus and associated teeth, lt. I_1–C, P_4 and broken P_3, rt. I_{1-2}, M_{1-3}, roots P_{3-4}
KNM-WK 17023[a]	Frag. lt. lower C crown
KNM-WK 17024[a]	Lt. M_2 crown
KNM-WK 17025[a]	Lt. M_1 frag.
KNM-WK 17026[a]	Frag. lower I crown
KNM-WK 17037[a]	Lt. P_3 crown
KNM-WK 17162	I_1 and frag. rt. P_3
KNM-WK 17168[a]	Frag. lower I crown
KNM-WK 24300	Associated maxilla and mandible body with associated part rt. P^3 and P^4–M^3, lt. I^2, C root, and P^3–M^3, rt. I_2, C root, P_3–M_3, lt. I_{1-2}, C root, part P_3, P_4–M_3, and piece of frontal
Unassociated postcranial	
KNM-WK 16961	Proximal phalanx lacking base
KNM-WK 17008	Frag. distal lt. humerus
KNM-WK 17027[a]	Lt. trapezium
KNM-WK 17028[a]	Prox. frag. lt. m/c II
KNM-WK 17029[a]	Dist. frag. proximal phalanx
KNM-WK 17030[a]	Dist. frag. proximal phalanx
KNM-WK 17031[a]	Dist. frag. middle phalanx
KNM-WK 17032[a]	Terminal pollical phalanx
KNM-WK 17033[a]	Prox. frag. proximal phalanx
KNM-WK 17034[a]	Prox. rt. m/c II
KNM-WK 17035	Dist. frag. proximal phalanx
KNM-WK 17038[a]	Lt. lunate
KNM-WK 17040[a]	Prox. frag. middle phalanx

[a]Specimens recovered from one locality. Some may be associated.

(continued)

Table I. *(Continued)*

KNM-WK 18119	Prox. frag. rt. m/t I
KNM-WK 18120[a]	Lt. talus
KNM-WK 18121[a]	Proximal phalanx
KNM-WK 18356	Proximal and middle phalanges
KNM-WK 18364[a]	Lt. cuboid
KNM-WK 18365[a]	Lt. capitate
KNM-WK 18366[a]	Prox. lt. m/c I
KNM-WK 18367[a]	Prox. frag. proximal phalanx
KNM-WK 18368[a]	Middle phalanx lacking base
KNM-WK 18369[a]	Prox. frag. middle phalanx
KNM-WK 18370[a]	Prox. frag. lt. m/c IV
KNM-WK 18372[a]	Frag. lt. scaphoid
KNM-WK 18395[a]	Frag. prox. lt. ulna
KNM-WK 18397[a]	Prox. frag. middle phalanx
From Buluk	
KNM-WS 124	Rt. mandible
KNM-WS 125	Lt. mandible
KNM-WS 11599	Associated mandible and lt. maxilla with lower rt. C, upper C, P^3, M^2
KNM-WS 12601	Rt. mandible with I and M_2 tooth germs
KNM-WS 12606	Lt. M_3
KNM-WS 12608	Proximal phalanx
From Moruorot	
KNM-MO 26	Rt. mandible with dp_4 and M_1 with I_2 and C in alveolus
From Locherangan	
KNM-LC 17590	Lt. M^3
KNM-LC 18405	Rt. P^3

there one very variable species present or two very similar less-variable ones? Leakey *et al.* (1988) compared variability in *Afropithecus* with that in extant sexually dimorphic catarrhines. They showed that, while the two specimens differ more in size than the means of these extant male and female catarrhines, they could nonetheless be accommodated within the observed male and female ranges of those catarrhines. It appears, then, that there is considerable size dimorphism in *Afropithecus*, but there is a distinct possibility that two similar species are sampled at Kalodirr.

A quantitative assessment of the body size of *Afropithecus* individuals can be made by using the regression equations of Conroy (1987). Using his all-anthropoid equation, the only complete M_1s from an *Afropithecus turkanensis* individual (which was of medium tooth and jaw size for the species) give an estimate of 34 kg. This is about the size of females of *Pan paniscus* or *Pan troglodytes schweinfurthii* (Jungers, 1985) and is near the middle of *Proconsul nyanzae* size estimates (Ruff *et al.*, 1989). An *Afropithecus* talus, KNM-WK 18120 (Leakey *et al.*, 1988), gives a body weight estimate of about 35 kg when the regression equation derived by Rafferty *et al.* (1995) is used. This is very close to the mean body weight of 35.6 kg reported for 12 *P. nyanzae* individuals based on their ankle joint size (Rafferty *et al.*, 1995).

Function

Cranial remains of *Afropithecus* are limited to the anterior part of the cranium and the teeth and jaws. Even though only the anterior part of the cranium is preserved, it is possible to determine certain proportions of the cranium. Using Finite Element Scaling Analysis, Leakey *et al.* (1991) showed that the overall facial proportions of *Afropithecus* are like *Aegyptopithecus zeuxis* from the early Oligocene Jebel Qatrani Formation of Egypt. They found only minor shape differences between the two species, but there are major size differences. *Aegyptopithecus* is estimated to have weighed about 6–8 kg (Conroy, 1987; Fleagle, 1988), which is about a fifth of the estimated weight of *Afropithecus*.

The facial configuration in these two genera is probably close to the primitive catarrhine condition, an idea discussed in detail by Benefit and McCrossin (1993) following their account of the condition in *Victoriapithecus*. There are several anatomical conditions that were apparently stabilized in this conformation over a considerable time and size range. They include: a frontal placed at relatively steep angle to the facial skeleton and having a frontal trigon; relatively bulky anterior temporalis muscle bundles and/or small frontal lobes of the brain; projecting and anteriorly converging premaxillae with large, procumbent central incisors and smaller more posteriorly placed lateral incisors; moderately long facial skeleton with long nasal bones; relatively small pyriform aperture with lower border grooved between the central incisors; deep zygomatic process of the maxilla with its origin low on the maxilla; large posterior lacrimal crest; bean-shaped orbital outline; supraorbital costae present; interorbital diameter broad; low-crowned molars with strong cingula; premolars with cusp tips much more approximated than the tooth base; and buccolingually broad molars.

In an age series of *Aegyptopithecus* there are strong supraorbital costae and a frontal trigon that decreases in size with age (Simons, 1987). The similarities between *Aegyptopithecus* and *Afropithecus* suggest the latter may show a similar shape change with age. Since the two adult, but not old, specimens of *Afropithecus* have very diminutive frontal trigons, the logical expectation is that, when old individuals are found, their trigons will have been shrunken to slits between temporal crests that meet above glabella. One explanation for the encroachment of the temporal surfaces onto the frontal in early catarrhines is that the brain case is small relative to their temporal musculature. The alternative explanation—that the temporal muscles were relatively large and the brain case of normal anthropoid size—implies that the same condition would always occur in anthropoids with strong teeth and jaws, which it often does, as in, for example, *Gorilla, Chiropotes, Australopithecus (Paranthropus),* and *Cebus*. The only known cranium of the early Miocene catarrhine *Proconsul* has a brain case the size of modern anthropoids (Walker *et al.*, 1983) and no frontal trigon (Clark and Leakey, 1951). It has been suggested that this species (*P. heseloni*) is relatively megadont (Teaford *et al.*, 1993), but it does not have enlarged masticatory musculature. Thus, compared with the only other large early Miocene hominoid skull known, *Afropithecus* seems to have had well-developed masticatory muscles. The anterior teeth of *Afropithecus* are relatively large and very strongly built. The projecting premaxilla bears large-rooted canines and procumbent incisors with the lateral ones smaller and set back from the centrals. These occluded with elongated, procumbent, and labiolingually compressed lower incisors in a way that is not seen in any modern Old World primate, but which is very reminiscent of the pitheciines, a group of New World seed-predators. The configuration of the anterior teeth, in fact, conforms to the picture of pitheciine morphology painted by Kinsey (1992) where he records styliform lower incisors, anteriorly inclined upper and lower incisors, and enormous and thick-calibered, laterally splayed canines. Both upper and lower canines have large, stout roots and rather short, conical crowns. Kinsey points out that the combination of splayed canines and very procumbent incisors in pitheciines means that incisor and canine function can be uncoupled so that the incisors can nip and crop effectively and the canines can be used to puncture large food items and apply high forces to hard objects. In this regard *Afropithecus* appears to be a pitheciine mimic. The details of the resemblance might not be expected to be great, given the different geological age, geographical distribution, and body size, and some of Kinsey's list of shared derived tooth features for the pitheciines only apply to New World monkeys, but the majority of features also occur in *Afropithecus*.

It is likely that *Afropithecus* individuals used their canines in food preparation more than in display or intraspecific aggressive encounters (see review of canine function in Walker, 1984). Practically every stage of canine wear is documented in *Afropithecus*, from slight, in both the type and the best Buluk specimen, to severe in KNM-WK 16901 (and see Leakey *et al.*, 1988, Fig. 4). Honing wear on the upper tooth precedes wear on the lower canine caused by the upper canine. Lower canine wear results in the removal of almost the whole crown, leaving a flat, almost horizontal wear plane. Initially, the canine is honed on the sectorial P_3, but, even when newly honed, the length of the canine blade projects only about 7 mm past the buccal cusps of the cheek teeth. This is in distinct contrast to the condition in most male catarrhines, including *Aegyptopithecus* (Fleagle *et al.*, 1980), in which the canine blade projects far below the tooth row. This relative reduction in canine crown length combined with the relatively large canine roots, and a marked lack of sexual dental dimorphism, makes a dietary use for the canines more likely than an aggressive or agonistic one. Such a condition is found in the pitheciine *Chiropotes* (Swindler, 1976), an animal that uses its canines to break open hard fruit (Kinsey, 1992), and the seeming lack of sexual dimorphism in canine size in *Afropithecus* (Leakey *et al.*, 1988), if confirmed, would further support this assessment.

In an elegant review of canine size in primates, Lucas (1981) presented a model accounting for the empirical observation that the opening angle of the jaw required to clear the canine in males, is inversely related to the scaled height of the jaw joint above the occlusal plane. The partial mandible, KNM-WK 17021, does not have the condyle preserved, but possesses enough of the condylar neck to permit reconstruction of its position fairly accurately. Calculating Lucas's index of height over jaw length yields a value of 0.18, which falls squarely within his baboon and drill ranges. The implication—that *Afropithecus* males would have very long canines, like baboon and drills—is contradicted by the fossils. It might be that *Afropithecus* had yawning displays that showed the big, but short, canines, similar to displays used by baboons to exhibit their long bladelike canines, but it could also be that it does not fit the model because its canines were used for some other function that required a large gape and strong temporalis muscles. The roots of these teeth are extraordinarily thick and long relative to the crowns, suggesting that high stresses were applied to the canine crowns. Perhaps they peeled bark by stripping it with their canines or dealt with extremely large, hard fruit. Tooth microwear studies might shed light on this paradox.

Because the maxillary specimens of *Afropithecus* have not yet been X-rayed, an analysis of facial buttressing like that performed by Ward and Brown (1986) has not been done. However, Ward (personal communication) conducted a laser-scanning study of the facial skeletons of *Sivapithecus* and *Afropithecus* and found the lower faces similar but the upper faces different, with each genus having a different pattern of contact between the two parts of the face.

There are only two specimens in which the presence or absence of a frontal sinus can be determined. Frontal sinuses are clearly seen. Maxillary sinuses are present, but are restricted anteriorly by the large canine roots; they neither excavate deeply between the molar roots nor invaginate the palatine process of the maxilla. The full range of paranasal sinus functions in humans has been debated for nearly a century (see Shea, 1977, for a review of maxilloturbinal sinus function) with little or no resolution. There is, however, no doubt that they must act as sound resonators whatever other functions each of the sinuses may serve. Any air-filled cavity in an animal resonates when excited by a broadband sound spectrum and those in the head that are downstream from the vocal folds must be involved in coloring vocalizations with overtones. The experience of having paranasal sinuses filled with mucus, as when suffering from a common cold, shows the results of removing sinus resonance. The voice is reedy, lacking in color and individuality, and reduced in intensity. It is no surprise that those primates that have loud signaling calls have either large paranasal sinuses or specialized resonators in the hyoid or larynx.

Interpreting the facial skeleton in functional terms is more difficult than interpreting the mandible for several reasons. The primary one is that the facial skeleton is involved in many functions and the mandible few. The facial skeleton shape is influenced by being joined to the brain case and is involved functionally in vision, smell, taste, vomeronasal function, resonance in vocalization, breathing, mastication, and facial expression. The mandible is involved in some of these indirectly, but its mechanical role in food preparation and mastication seems so clear-cut that experimental studies have concentrated on it rather than the maxilla.

The symphysis in *Afropithecus* is long and deep, with moderately developed superior and inferior mandibular tori. The genial pit, however, is low in all jaws in which it can be seen. This, in turn, indicates that the tongue was deep and narrow, especially anteriorly, and probably less mobile than in those species that have a more open anterior part of the jaw. Moderate to strong anterior digastric impressions are seen in *Afropithecus* and there are invariably strong markings for the depressor muscles of the lips and mouth which might be associated with the use of the canines in feeding, as posited before, and the medial frenulum. The mandibular buttresses are very similar to those in *Sivapithecus,* a resemblance that in part led Leakey and Walker (1985) at first to include the Buluk sample in that genus. *Afropithecus* mandibles are characterized by strong basal buttresses and a very strong lateral tubercle where the oblique line meets the body. There is a strong lateral hollow on the buccal surface that is centered above the mental foramen. The latter is bounded anteriorly by the juga caused by enormous canine roots. All of these features, together with tooth rows that are roughly parallel and set close together and a long and narrow post incisive plane are also found in *Sivapithecus.*

Microwear analysis of the molar teeth has yet to be carried out to determine diet. The cusp morphology and lack of large shearing crests suggest a frugivorous regimen, as does the superficial resemblance of the teeth to those of *Proconsul* species whose morphology is consistent with frugivory (Kay, 1977). The incisor teeth are relatively large and also strong, a feature that Kay (1984) has pointed out also relates to frugivory. However, the extremely heavy wear that is seen on much of the Kalodirr sample begs for explanation, as does the extremely thick molar enamel in *Afropithecus* (Lawrence Martin, personal communication).

The pervasive similarities to the pitheciines, the pattern of facial and mandibular buttressing, the massive chewing musculature, the heavy canine wear, and so on, point convincingly to *Afropithecus* being a sclerocarp forager. It is not known at present whether the forests at the sites from which this genus is known were rich in this resource.

The anatomy of *Afropithecus* is known for some postcranial elements (Leakey *et al.,* 1988, and Table I). In practically all parts preserved, it is similar to *Proconsul* spp. Rose (1993) found such similarity between *Proconsul* and *Afropithecus* postcranial bones that he was able to discuss their anatomy as though they were one genus. Most likely the configuration of the postcranial skeleton conforms to the primitive hominoid morphotype with suggestions of a relatively deliberate, even-limbed quadrupedal arboreal animal that lacked overt specializations for arm-swinging or terrestrial locomotion, or leaping.

Phylogeny

Afropithecus is a large hominoid. Comparing it with other large species is relatively easy, but comparing it with smaller species is difficult because we lack scaling protocols. The overall facial shape in *Afropithecus* has been shown to resemble that of a smaller, much earlier species, *Aegyptopithecus zeuxis* (Leakey *et al.*, 1991). However, for two reasons, making phylogenetic sense of this resemblance is a different matter than dealing with discrete morphological traits. First, the Finite Element Scaling method used in that analysis enables one to separate the effects of size from shape. Second, by dealing with overall shape, the effects of the different functional demands of the various parts of the facial skeleton are integrated into a simple whole shape. We can consider canine teeth as an example of how size can complicate morphology. Canine length and breadth do not scale in a simple relationship across a wide size range because of many effects including differential mechanical or social function, tooth shape and tooth tissue strength, and limits on motion at the temporomandibular joint. And this nonisometry will, in turn, affect the premaxilla size and shape, the size and shape of the edges of the nasal aperture, the shape of the P_3, the diastema, and so on. And if, as is most likely the case, there are phylogenetic limits to the body size ranges (there are no very small or very large gibbons and no very small extant great apes, for instance), then analytical problems are exaggerated by phylogenetic bias (Harvey and Pagel, 1991).

Afropithecus is clearly a large catarrhine. In many features it may be close to the condition of the stem catarrhine, but it is highly derived in its feeding apparatus. In his recent review of hominoid relationships, Andrews (1992) has grouped it in a tribe with *Heliopithecus* (Andrews and Martin, 1987a,b) and *Otavipithecus* (Conroy *et al.*, 1992). The former genus has no published postcranial remains, and the latter has only two postcranial parts that also are known for *Afropithecus*. These are a proximal ulna and a middle hand phalanx (Conroy *et al.*, 1993). The ulna of *Afropithecus* has only the posterior border preserved so even here direct comparison is not possible, but the *Otavipithecus* ulna is said to be like that of *Proconsul* (Conroy *et al.*, 1993) and in the parts preserved so is the one from Kalodirr (Leakey *et al.*, 1988). The phalanx is larger than those of the type of *Proconsul heseloni*, but smaller than a third ray middle phalanx from a hand of *P. nyanzae* [KNM-RU 15001 (Beard *et al.*, 1986, 1993)]. Both the Namibian phalanx and those from Kalodirr seem to be very similar in morphology to those of *Proconsul*, and this may reflect the primitive catarrhine condition. Comparisons of *Afropithecus* with *Otavipithecus* and *Heliopithecus* must therefore be limited mainly to the maxilla and mandible, respectively.

As we have seen, *Afropithecus* was almost certainly a committed sclerocarp feeder and its jaws and teeth are specialized for this task. Its relative, *Heliopithecus leakeyi,* is a smaller species than *A. turkanensis,* yet its canine was large, as judged from the remaining alveolar margin, and its lateral incisor root may well have been procumbent. But apart from this, only the widely spread crown of the P^3, which might function to distribute high forces from the cusps across a bigger crown base, gives a hint at this being a hard fruit eater.

In the case of *Otavipithecus,* even less resemblance is seen, with the type and only mandible having relatively small, vertically implanted canine root and incisor alveoli that show that the roots were definitely not procumbent (Conroy *et al.,* 1992). As Begun (1994) points out, mandibles have proven very difficult to assess from a phylogenetic perspective, a point raised much earlier by Leakey and Walker (in Howell and Isaac, 1976, p. 473). This means that one can develop character trees based on mandibles, but that they might not bear much resemblance to trees based on a more extensive anatomical data base (Begun, 1994).

McCrossin and Benefit (1993) described a new mandible of *Kenyapithecus africanus.* They noted that it, too, had the procumbent incisors and robust canines like those of *Chiropotes satanus* and *Pongo pygmaeus,* and suggested on that occasion and later (Benefit and McCrossin, 1995) that it had a nut-cracking adaptation. Now *Kenyapithecus* is not the same for all workers. Some, like McCrossin and Benefit, use the genus both for the type Fort Ternan material and for the Maboko fossils. They caution waiting for a full analysis when considering fossils from Nachola, but hint that those might be in another genus. Pickford (1985) keeps the Fort Ternan and Maboko fossils in *Kenyapithecus,* but notes that it might prove necessary to separate them generically. In Andrews's new scheme, he separates the present genus into two different tribes with the Maboko material in one and *Kenyapithecus* proper (from Fort Ternan) linked with fossils from Turkey and Austria, as also does Begun (1992). Andrews (1992) regards the Maboko material as rather close to *Afropithecus* and a member of the Afropithecini. For the sake of convenience the Maboko material is called "*Kenyapithecus*" here as it is in Andrews's account.

If *Otavipithecus* is removed from the Afropithecini (and transferred to Andrews's Dryopithecini, perhaps), then the clade at present probably has two genera, plus the Moroto taxon, of medium to large pitheciine mimics. These are clearly derived in a number of ways relative to the earlier *Proconsul* group. However, the presence of large primates from the late Oligocene of northern Kenya complicates the matter. Leakey *et al.* (1995) name a new genus, *Kamoyapithecus,* for large Oligocene catarrhines from Lothidok. Not enough is known about this genus to determine even if it is a hominoid, but it does have very large-rooted and stubby-crowned canines and a lower lateral incisor identical to those of *Afropithecus.* The molars do not have thick enamel and the premolars were not, as judged by the remains of the P^3 roots, cusp heteromorphic as in the latter genus. These fossils suggest, then, another sclerocarp foraging primate that might be more than 10 million years earlier than *Afropithecus* and much earlier than the earliest known *Proconsul*. If the latter is related to *Afropithecus*, as the postcranial similarities suggest, then this hard-fruit-eating adaptation probably had at least two separate origins in these early hominoids.

There are several major differences between *Afropithecus* and "*Kenyapithecus*." McCrossin and Benefit (1993) and Pickford (1985) report that the canines from Maboko (as well as Fort Ternan) show considerable sexual dimorphism. This is not the case in *Afropithecus*, and, further, the western Kenya species have canines that are more buccolingually compressed with crowns longer relative to their roots. The premolars are less cusp heteromorphic in "*Kenyapithecus*" and the molars have, apparently, thinner enamel [L. B. Martin (personal communication) reports that *Afropithecus* has the thickest enamel of any African Miocene hominoid]. Just as a caution, however, it is clear that the use of relative thickness of tooth enamel in phylogenetic studies has been poorly controlled. Apart from lack of specimens that can be cut, which means that sample sizes are always likely to be small, claims such as that of Andrews (1992) that members of the Afropithecini have enamel intermediate in thickness between *Proconsul* and *Kenyapithecus* are based on little or no data. In any case, it seems that if the Maboko species had a sclerocarp foraging adaptation, it was much less developed than in *Afropithecus*. Since the newly described jaw of *Kenyapithecus* is from a juvenile (McCrossin and Benefit, 1993), and since tooth procumbency and mandibular morphology change as individuals age, the resemblance to adult *Afropithecus* mandibles might not be as close as it presently seems.

This particular adaptation to deal with hard-shelled fruits is likely to create taxonomic confusion in fossil samples dominated by jaws and teeth, but, taken at its face value, it links *Afropithecus* with the Moroto taxon and probably with *Heliopithecus*. These form a group derived in feeding apparatus relative to *Proconsul* (see Walker, this volume). If the late Oligocene genus shows, with further specimens, to also be in this group, then the split with the *Proconsul* group must have been more ancient than has previously been thought. Finally, it is worth noting that resemblances to *Sivapithecus* have been seen in "*Kenyapithecus*" (Clark and Leakey, 1951) and *Afropithecus* (Leakey and Walker, 1985). Features such as relatively procumbent incisors, thick-enameled teeth, heavy face and jaw buttresses, and deep mandibles with close-set tooth rows can all be seen to be associated, as they are in orangutans, with a diet of hard fruit. The resemblances between the early Miocene sclerocarp feeders and the middle Miocene members of the orangutan clade may be purely convergent ones related to similar feeding adaptations (Benefit and McCrossin, 1995). Or, alternatively, they may attest to a close relationship. If some of the characters we think are present in *Afropithecus* are confirmed, such as the lack of sexual dimorphism in canine shape and size, then that genus at least appears to be too derived to belong to the same clade as *Sivapithecus*.

Suggestions that *Proconsul* (and therefore by extension *Afropithecus*) have postcrania close to the ancestral catarrhine morphotype (e.g., Harrison, 1993) must be reconciled with the configuration of the few postcranial bones attributed to *Aegyptopithecus* and *Propliopithecus* (see Gebo, 1993, and Rose, 1993, for reviews). These Oligocene catarrhines have postcranial bones with several important differences from those of Miocene catarrhines (Table II). Summaries of the locomotor and postural potentials of these two different groups are, however, similar. This may be related to the fact that apart from being arboreal quadrupeds, neither has any clear morphological adaptations for terrestrial, arm-swinging, or suspensory locomotion.

In summary, it is most likely, because of the postcranial skeleton, which is derived relative to that of *Aegyptopithecus* and *Propliopithecus,* that *Afropithecus* and *Proconsul* can be placed in the same early hominoid group. Based on the postcranial anatomy, it would be difficult to make a case that these two early Miocene genera belong in different tribes. This is not the case, apparently, with these genera and *Kenyapithecus,* for the latter has substantial postcranial adaptations for terrestrial locomotion (Benefit and McCrossin, 1995). When it comes to the skull, however, *Afropithecus* retains the primitive condition of relationships of parts of the face that we also see in Oligocene forms, but has overlain on it derived characters, such as enlarged and procumbent incisors and thick, splayed canines, relating to sclerocarp feeding. *Proconsul* lacks these latter characters and has what is probably the primitive condition in incisor and canine morphology and dimorphism, but has a derived facial configuration that foreshadows that of later hominoids (Rae, this volume). Whether

Table II. Some Differences between Postcranial Bones of *Aegyptopithecus* and Early Miocene Large Hominoids

Aegyptopithecus	Miocene forms
Humeral head faces posteriorly	Humeral head faces posterosuperiorly
Humeral head small relative to the tuberosities	Head bigger and elevated above the tuberosities
Humeral lesser tuberosity less anteriorly positioned	Lesser tuberosity more anteriorly positioned
Humeral shaft not elongated	Shaft elongated
The olecranon fossa is wide and shallow	Fossa deep and confined
Great lateral extension of brachialis flange	Flange moderately developed
Zona conoidea flat	Zona conoidea beveled
Coronoid process low	Coronoid process higher
Olecranon process long	Olecranon process short
Posterior border of ulnar shaft convex proximally, straight distally	Posterior border of shaft sigmoidal
Ulnar shaft short	Ulnar shaft elongated
Tibia short and robust (for *Propliopithecus*)	Tibia elongated and slender
Talar trochlea parallel-sided	Trochlea tapers posteriorly

these derived features of the skull in *Afropithecus* are enough to warrant a separate tribe is a matter for debate, especially because the other genera thought to belong to the Afropithecini are either known only from a small part of the anatomy, as is the case with *Heliopithecus*, or else do not have the sclerocarp feeding adaptations unambiguously expressed, as in "*Kenyapithecus.*"

References

Andrews, P. 1978. A revision of the Miocene Hominoidea of East Africa. *Bull. Brt. Mus. Nat. Hist. Geol.* **30:**85–224.

Andrews, P. 1992. Evolution and environment in the Hominoidea. *Nature* **360:**641–646.

Andrews, P., and Martin, L. 1987a. The phyletic position of the Ad Dabtiyah hominoid. *Bull. Brt. Mus. Nat. Hist. Geol.* **41:**383–393.

Andrews, P., and Martin, L. 1987b. Cladistic relationships of extant and fossil hominoids. *J. Hum. Evol.* **16:**101–118.

Anyonge, W. 1991. Fauna from a new lower Miocene locality west of Lake Turkana, Kenya. *J. Vertebr. Paleontol.* **11:**378–390.

Beard, K. C., Teaford, M. F., and Walker, A. 1986. New wrist bones of *Proconsul africanus* and *Proconsul nyanzae* from Rusinga Island, Kenya. *Folia Primatol.* **47:**97–118.

Beard, K. C., Teaford, M. F., and Walker, A. 1993. New hand bones of the early Miocene hominoid *Proconsul* and their implications for the evolution of the hominoid wrist. In: H. Preuschoft and D. J. Chivers (eds.), *Hands of Primates*, pp. 387–403. Springer-Verlag, Berlin.

Begun, D. R. 1992. Phyletic diversity and locomotion in primitive European hominids. *Am. J. Phys. Anthropol.* **87:**311–340.

Begun, D. R. 1994. The significance of *Otavipithecus namibiensis* to interpretations of hominoid evolution. *J. Hum. Evol.* **27:**385–394.

Benefit, B. R., and McCrossin, M. L. 1993. Facial anatomy of *Victoriapithecus* and its relevance to the ancestral cranial morphology of Old World monkeys and apes. *Am. J. Phys. Anthropol.* **92:**329–370.

Benefit, B. R., and McCrossin, M. L. 1995. Miocene hominoids and hominid origins. *Annu. Rev. Biol.* **24:**237–256.

Boschetto, H. B., Brown, F. H., and McDougall, I. 1992. Stratigraphy of the Lothidok Range, northern Kenya, and K/Ar ages of its Miocene primates. *J. Hum. Evol.* **22:**44–71.

Clark, W. E. L., and Leakey, L. S. B. 1951. The Miocene Hominoidea of East Africa. *Br. Mus. Nat. Hist. Fossil Mamm. Afr.* **1:**1–117.

Conroy, G. C. 1987. Problems of body-weight estimation in fossil primates. *Int. J. Primatol.* **8:**115–137.

Conroy, G. C., Pickford, M., Senut, B., and van Couvering, J. 1992. *Otavipithecus namibiensis*, first Miocene hominid from southern Africa. *Nature* **356:**144–148.

Conroy, G. C., Pickford, M., Senut, B., and Mein, P. 1993. Additional Miocene primates from the Otavi Mountain, Namibia. *C. R. Acad. Sci. Ser. II* **317:**987–990.

Fleagle, J. G. 1988. *Primate Adaptation and Evolution.* Academic Press, New York.

Fleagle, J. G., Kay, R. F., and Simons, E. L. 1980. Sexual dimorphism in early anthropoids. *Nature* **287:**328–330.

Gebo, D. L. 1993. Postcranial anatomy and locomotor adaptations in early African anthropoids. In: D. L. Gebo (ed.), *Postcranial Adaptation in Nonhuman Primates*, pp. 220–234. Northern Illinois University Press, De Kalb.

Harrison, T. 1993. Cladistic concepts and the species problem in hominoid evolution. In: W. H. Kimbel and L. B. Martin (eds.), *Species, Species Concepts, and Primate Evolution,* pp. 345–371. Plenum Press, New York.

Harvey, P. H., and Pagel, M. D. 1991. *The Comparative Method in Evolutionary Biology,* Oxford University Press, London.

Howell, F. C., and Isaac, G. L. 1976. Introduction to Part 3, Paleoanthropology. In: Y. Coppens, F. C. Howell, G. L. Isaac, and R. E. F. Leakey (eds.), *Earliest Man and Environments in the Lake Rudolf Basin,* pp. 471–475. University of Chicago Press, Chicago.

Jungers, W. L. 1985. Body size and scaling in limb proportions in Primates. In: W. L. Jungers (ed.), *Size and Scaling in Primate Biology,* pp. 345–381. Plenum Press, New York.

Kay, R. F. 1977. Diets of early Miocene hominoids. *Nature* **268:**628–630.

Kay, R. F. 1984. On the use of anatomical features to infer foraging behavior in extinct primates. In: P. S. Rodman and J. G. H. Cant (eds.), *Adaptations for Foraging in Nonhuman Primates,* pp. 21–53. Columbia University Press, New York.

Kelley, J., and Pilbeam, D. 1986. The dryopithecines: Taxonomy, anatomy and phylogeny of Miocene large hominoids. In: D. R. Swindler and J. Erwin (eds.), *Comparative Primate Biology,* Vol. 1, pp. 361–411. Liss, New York.

Kinsey, W. G. 1992. Dietary and dental adaptations in the Pithecinae. *Am. J. Phys. Anthropol.* **88:**499–514.

Leakey, L. S. B. 1963. East African fossil Hominoidea and the classification within this superfamily. In: S. L. Washburn, (ed.), *Classification and Human Evolution,* pp. 32–49. Aldine, Chicago.

Leakey, M. G., Leakey, R. E., Richtsmeier, J. T., Simons, E. L., and Walker, A. C. 1991. Similarities in *Aegyptopithecus* and *Afropithecus* facial morphology. *Folia Primatol.* **56:**65–85.

Leakey, M. G., Ungar, P. S., and Walker, A. 1995. A new genus of large primate from the late Oligocene of Lothidok, Turkana District, Kenya. *J. Hum. Evol.* **28:**519–531.

Leakey, R. E., and Leakey, M. G. 1986. A new Miocene hominoid from Kenya. *Nature* **324:**143–145.

Leakey, R. E., and Walker, A. C. 1985. New higher primates from the early Miocene of Buluk, Kenya. *Nature* **318:**173–175.

Leakey, R. E., Leakey, M. G., and Walker, A. C. 1988. Morphology of *Afropithecus turkanensis* from Kenya. *Am. J. Phys. Anthropol.* **76:**289–307.

Lucas, P. W. 1981. An analysis of canine size and jaw shape in some Old and New World nonhuman primates. *J. Zool.* **195:**437–448.

McCrossin, M. L., and Benefit, B. R. 1993. Recently recovered *Kenyapithecus* mandible and its implications for great ape and human origins. *Proc. Natl. Acad. Sci. USA* **90:**1962–1967.

McDougall, I., and Watkins, R. 1985. Age of hominoid-bearing sequence at Buluk, northern Kenya. *Nature* **318:**175–178.

Madden, C. 1980. New *Proconsul (Xenopithecus)* from the Miocene of Kenya. *Primates* **21:**241–252.

Martin, L. 1981. New specimens of *Proconsul* from Koru, Kenya. *J. Hum. Evol.* **10:**139–150.

Pickford, M. 1985. A new look at Kenyapithecus based on recent discoveries in western Kenya. *J. Hum. Evol.* **14:**113–143.

Pilbeam, D. R. 1969. Tertiary Pongidae of East Africa: Evolutionary relationships and taxonomy. *Bull. Peabody Mus. Nat. Hist.* **31:**1–185.

Rafferty, K. L., Walker, A., Ruff, C. B., Rose, M. D., and Andrews, P. J. 1995. Postcranial estimates of body weight in *Proconsul,* with a note on a distal tibia of *P. major* from Napak, Uganda. *Am. J. Phys. Anthropol.* **97:**391–402.

Rose, M. D. 1993. Locomotor anatomy of Miocene hominoids. In: D. L. Gebo (ed.), *Postcranial Adaptation in Nonhuman Primates,* pp. 252–272. Northern Illinois University Press, De Kalb.

Ruff, C. B., Walker, A., and Teaford, M. F. 1989. Body mass, sexual dimorphism and femoral proportions of *Proconsul* from Rusinga and Mfangano Islands, Kenya. *J. Hum. Evol.* **18:**515–536.

Sanders, W. J., and Bodenbender, B. E. 1994. Morphological analysis of lumbar vertebra UMP

67-28: Implications for spinal function and phylogeny of the Moroto hominoid. *J. Hum. Evol.* **26:**203–237.
Shea, B. T. 1977. Eskimo craniofacial morphology, cold stress and the maxillary sinus. *Am. J. Phys. Anthropol.* **47:**289–300.
Simons, E. L. 1967. The significance of primate paleontology for anthropological studies. *Am. J. Phys. Anthropol.* **27:**307–332.
Simons, E. L. 1987. New faces of Aegyptopithecus, early human forebear from the Oligocene of Egypt. *J. Hum. Evol.* **16:**273–290.
Swindler, D. R. 1976. *Dentition of Living Primates.* Academic Press, New York.
Teaford, M. F., Walker, A., and Mugaisi, G. S. 1993. Species discrimination in *Proconsul* from Rusinga and Mfangano Islands, Kenya. In: W. H. Kimbel and L. B. Martin (eds.), *Species, Species Concepts, and Primate Evolution,* pp. 373–392. Plenum Press, New York.
Walker, A. 1984. Mechanisms of honing in the male baboon canine. *Am. J. Phys. Anthropol.* **65:**47–60.
Walker, A., Falk, D., Smith, R., and Pickford, M. 1983. The skull of *Proconsul africanus. Nature* **305:**525–527.
Ward, C. V. 1993. Torso morphology and locomotion in *Proconsul nyanzae. Am. J. Phys. Anthropol.* **92:**291–328.
Ward, C. V., Walker, A., Teaford, M. F., and Odhiambo, I. 1993. A partial skeleton of *Proconsul nyanzae* from Mfangano Island, Kenya. *Am. J. Phys. Anthropol.* **90:**77–111.
Ward, S. C., and Brown, B. 1986. The facial skeleton of *Sivapithecus indicus.* In: D. Swindler and J. Erwin (eds.), *Comparative Primate Biology, Vol. 1: Systematics, Evolution, and Anatomy,* pp. 413–452. Liss, New York.
Ward, S. C., and Pilbeam, D. R. 1983. Maxillofacial morphology of Miocene hominoids from Africa and indo-Pakistan. In: R. L. Ciochon and R. S. Corruccini (eds.), *New Interpretations of Ape and Human Ancestry,* pp. 211–238. Plenum Press, New York.
Watkins, R. 1982. Ph.D. thesis, University of London.

On the Relationships and Adaptations of *Kenyapithecus*, a Large-Bodied Hominoid from the Middle Miocene of Eastern Africa

12

MONTE L. McCROSSIN and BRENDA R. BENEFIT

Introduction

The phylogenetic relationships and adaptations of *Kenyapithecus* have been of special interest since Leakey (1962, p. 696) first described the genus as possessing a number of characters exhibiting "a marked tendency in the direction of the Hominidae." Expectations regarding the hominid affinities of *Kenyapithecus* influenced the reconstruction of many functionally and phylogenetically significant aspects of its anatomy in the absence of tangible fossil

MONTE L. McCROSSIN and BRENDA R. BENEFIT • Department of Anthropology, Southern Illinois University, Carbondale, Illinois 62901.

Function, Phylogeny, and Fossils: Miocene Hominoid Evolution and Adaptations, edited by Begun *et al.* Plenum Press, New York, 1997.

evidence. The type species *Kenyapithecus wickeri* is represented by four jaw fragments, 11 isolated teeth, and a distal humerus collected from a single site, Fort Ternan (Pickford, 1985). As of 1985, the hypodigm of the referred species *Kenyapithecus africanus* consisted of the type maxilla together with 46 isolated teeth, four incomplete postcranial pieces, and a poorly preserved mandible collected from the Maboko Formation and an isolated lower molar from Nyakach (Pickford, 1985). Features for which little or no fossil evidence existed, but which *Kenyapithecus* was said to share with hominids, include: small lower incisors relative to cheek tooth size, reduced incisor procumbency, arcuate dental arcade, short rostrum, and a humerus that is longer than the femur (Simons and Pilbeam, 1972). The supposed reduction in upper canine and lower incisor size, and facial abbreviation, in combination with thick-enameled molars in *Kenyapithecus* was interpreted as being related to an australopithecine-like emphasis on molar grinding resulting from the consumption of hard objects (Andrews and Walker, 1976), with incisors being "relatively unimportant in food preparation" (Simons and Pilbeam, 1978, pp. 149, 152).

Of the two species known for *Kenyapithecus,* an advanced hominoid status is generally accepted for *K. wickeri,* but lingering uncertainty exists regarding the phylogenetic relationships of *K. africanus.* Perhaps related to confusion about the provenience of type maxilla BMNH M. 16649 (Clark and Leakey, 1951), jaws and teeth of *Proconsul nyanzae* and *P. major* from Songhor and Rusinga were erroneously attributed to *K. africanus* (Leakey, 1967, 1968). As a result, features uncharacteristic of *Kenyapithecus,* such as a slender mandibular corpus, strong superior transverse torus, and retention of beaded molar cingula, became associated with the referred species. This, and the failure of most cladistic analyses of Miocene hominoids to take into account ranges of variation present in living and extinct taxa, influenced some scholars to assign *K. africanus* to *P. nyanzae* (Simons and Pilbeam, 1978), Afropithecini (a tribe based on *Afropithecus,* Andrews, 1992), and "*Griphopithecus africanus*" (Begun, 1987). At the same time, *K. africanus* was treated by some as conspecific with *K. wickeri* (Kay and Simons, 1983).

In recent years, the sample of *Kenyapithecus* fossils has dramatically increased. The number of *K. africanus* fossils from Maboko Island alone has grown from 43 (Pickford, 1985) to 175, 32 of which are postcranial remains (Benefit and McCrossin, 1993a, 1994; McCrossin and Benefit, 1993a,b, 1994; McCrossin, 1994a,b). Comparison of the expanded collection of *K. africanus* fossils with those from Fort Ternan indicates that differences between them, such as in the height and position of the zygomatic root above the molar row, are minimal, being well within the range of extant anthropoid genera and species (Benefit and McCrossin, 1994; McCrossin, 1994a). They are therefore treated as at least congeneric in this study. This chapter focuses on the contribution made by new and previously described fossil material from Maboko Island and Fort Ternan to an understanding of the phylogenetic position and paleobiology of *Kenyapithecus.*

Craniodental Morphology and Adaptations

Knowledge of the facial morphology of *Kenyapithecus* comes primarily from three maxillary specimens, KNM-FT 46 and 47, which represent one individual of *K. wickeri* (Leakey, 1962) and *K. africanus* maxilla BMNH M. 16649 (Clark and Leakey, 1951). KNM-FT 46 and BMNH M. 16649 are both left maxillary fragments, the former with associated female C^1, P^4–M^2, and the latter, a probably male individual based on the large size of the teeth, with P^3–M^1. Both preserve the distal wall of the C^1 alveolus, portions of the palatal process, and lower portions of the zygomatic root. An additional left premaxilla and maxilla with P^3–M^3 from Baragoi has provisionally been attributed to *K. africanus* (Ishida, 1986). The Baragoi specimen exhibits a subnasal pattern similar to *Dryopithecus* and some individuals of *Gorilla* but unlike *Sivapithecus* and *Pongo,* in that the incisive foramen is somewhat constricted and the premaxilla is obliquely oriented relative to the palatal process of the maxilla (Ishida, 1986, Fig. 3). Consequently, Brown and Ward (1988) have interpreted *K. africanus* as exhibiting a subnasal morphology consistent with the expected condition for the last common ancestor of *Gorilla, Pan,* and *Homo.*

The maxillae of both *K. wickeri* and *K. africanus* are characterized by a relatively anterior position of the root of the zygomatic process (Leakey, 1962). With few exceptions (e.g., some colobinans), the zygomatic root originates above the second (usually) or third molar (more rarely) in platyrrhines, cercopithecoids, *Proconsul, Afropithecus, Dryopithecus, Sivapithecus, Hylobates, Gorilla,* and *Pan.* The relatively anterior position of the root of the zygomatic exhibited by *Kenyapithecus* is also seen, however, in members of the subfamily Pitheciinae, particularly the bearded saki (*Chiropotes*), the uakari (*Cacajao*) (Kinzey, 1992), and a fossil form from the Miocene of Colombia (*Cebupithecia*), as well as in some individuals of the orangutan (*Pongo*).

Relative to M^1 area, the height of the zygomatic process above the alveolar margin of both *K. wickeri* and *K. africanus* is low, barely overlapping the lowest portion of the range observed for extant hominoids and falling well below the regression for modern apes (Fig. 1; McCrossin, 1994a). Similarly low roots of the zygomatic process, as measured relative to facial height, are characteristic of many fossil catarrhines, including *Aegyptopithecus, Victoriapithecus, Rhinocolobus, Proconsul, Afropithecus,* and *Turkanapithecus,* as well as in some fossil and extant platyrrhines such as *Cebupithecia, Pithecia, Cacajao,* and *Alouatta* (Benefit and McCrossin, 1993b). Although a higher origin of the zygomatic process is seen in extant cercopithecoids, *Dryopithecus brancoi, Sivapithecus indicus,* and living hominoids, the lower position shared by some platyrrhines, propliopithecids, fossil cercopithecoids, and Miocene hominoids may be the primitive condition for the last common ancestor of living catarrhines (Benefit and McCrossin, 1993b).

Only one upper central incisor (KNM-FT 49) and one upper lateral incisor (KNM-FT 3637) of *K. wickeri* are known (Pickford, 1985). Contrary to

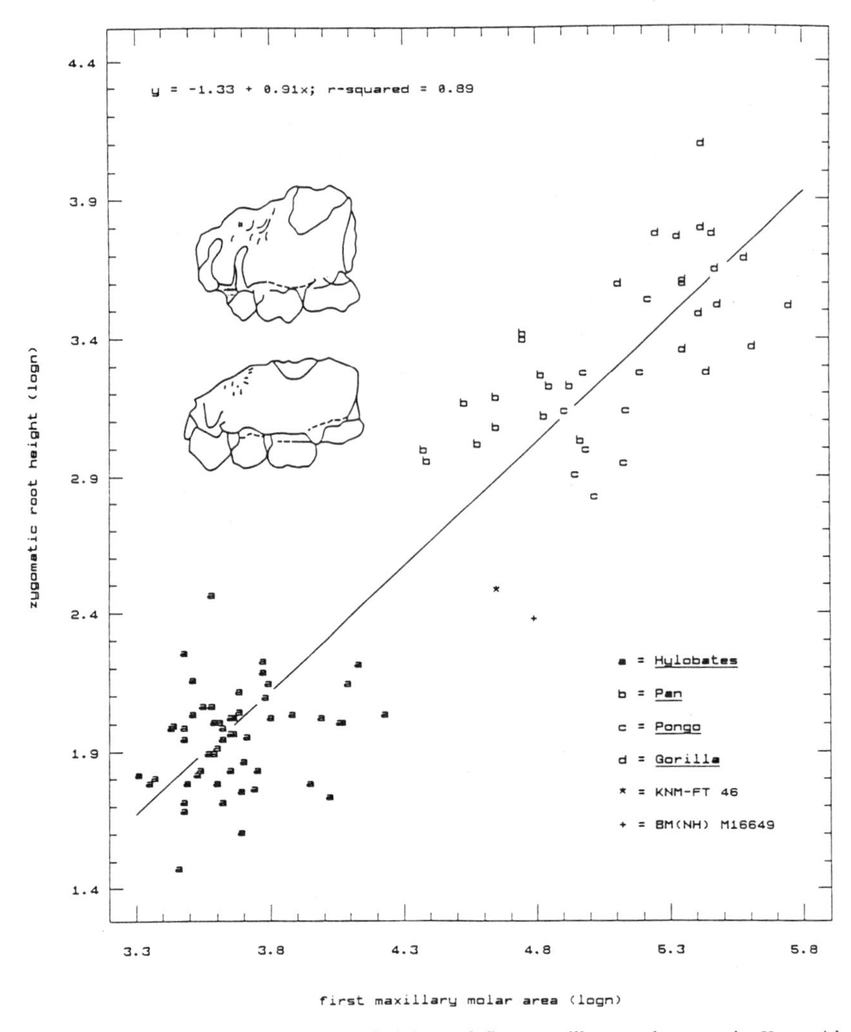

Fig. 1. Bivariate plot of zygomatic root height and first maxillary molar area in *Kenyapithecus wickeri* (KNM-FT 46, upper), *K. africanus* (BMNH M. 16649, lower), and extant hominoids.

Begun (1992, Table 1), their size and shape dimensions overlap the ranges of mesiodistal breadth, labiolingual thickness, and crown height observed for *K. africanus* I^1s ($N = 7$) and I^2s ($N = 5$) currently known from Maboko Island and Majiwa (Benefit and McCrossin, 1994; McCrossin, 1994a). Some of the I^1s recently collected from Maboko match KNM-FT 49 in size, shape, and lingual relief. All *Kenyapithecus* upper central incisors are spatulate, mesiodistally broader than labiolingually thick, moderately high crowned, and lack a strong central pillar. They exhibit a pattern of lingual relief similar to that of *Sivapithecus sivalensis*, with a diffuse basal swelling and a series of vertically disposed crenulations above a clearly defined lingual cingulum (Leakey,

1967). They differ from the more primitive upper central incisors of *Proconsul nyanzae, Griphopithecus alpani, Dryopithecus laietanus,* and *D. brancoi* which manifest a strongly developed central pillar extending uninterrupted from the cervix toward the occlusal margin, and are mesiodistally narrower as well as taller relative to their labiolingual thickness.

Both species of *Kenyapithecus* are characterized by high degrees of upper incisor heteromorphy, mean dimensions of the upper lateral incisor being considerably smaller those of the upper central incisor in terms of mesiodistal length (61%), labiolingual thickness (86%), and crown height (67%). Similarly high degrees of upper incisor heteromorphy are observed for *Chiropotes, Cacajao, Afropithecus, Griphopithecus, Sivapithecus,* and *Pongo* (Leakey and Leakey, 1986; Kinzey, 1992).

Two clear size groups are apparent in the combined sample of *Kenyapithecus* upper canines, small specimens that are inferred to represent females and larger examples probably belonging to males (Pickford, 1985; McCrossin, 1994a). Upper canines of *Kenyapithecus* resemble those of *Proconsul* as well as those of extant great apes in being relatively robust (mesiodistally) and low crowned. In contrast, the upper canines of cercopithecoids and hylobatids are more gracile and relatively high crowned. A similar diversity in upper canine robusticity is observed among platyrrhines. The pitheciines *Cebupithecia, Chiropotes,* and *Cacajao* have very robust, tusklike canines while other New World anthropoids typically have more gracile upper canines (Kinzey, 1992). The male upper canine of *K. wickeri* (KNM-FT 39) is inferred to have been medially inclined and externally rotated and differs in these respects from the upper canines of *Proconsul* and *Dryopithecus.* The medial inclination and external rotation of the *Kenyapithecus* upper canine crown and root is shared with *Afropithecus* (Leakey and Leakey, 1986), *Sivapithecus,* and *Pongo* (Ward and Pilbeam, 1983). The same pattern of medial inclination and lateral rotation is also characteristic of Pitheciinae, especially *Cebupithecia, Chiropotes,* and *Cacajao* (Kinzey, 1992). Only female upper canines are known for both *K. wickeri* and *K. africanus,* but they are virtually indistinguishable in terms of their size and shape.

Upper fourth premolar buccolingual breadth as measured relative to M^1 breadth for both type specimens of *K. wickeri* and *K. africanus* is well within the range shown by extant hominoids (Fig. 2). However, values for both species lie above the hominoid regression line. Similar enlargement of upper premolar area is seen in pitheciines, compared with other New World anthropoids. As stated by Leakey and Leakey (1986), buccal-lingual expansion of the P^4 of *Afropithecus* (including both KNM-WK 16999 from Kalodirr and UMP 62-11 from Moroto) is much more extreme than that observed in any other hominoid, including both *K. africanus* and *K. wickeri* (Fig. 2). Such evidence contradicts claims that *K. africanus* is more closely related to *Afropithecus* than to *K. wickeri* (Andrews, 1992).

Assessment of lingual cingulum development on the upper molars of *K. wickeri* is constrained by the fact that only first and second molars from a

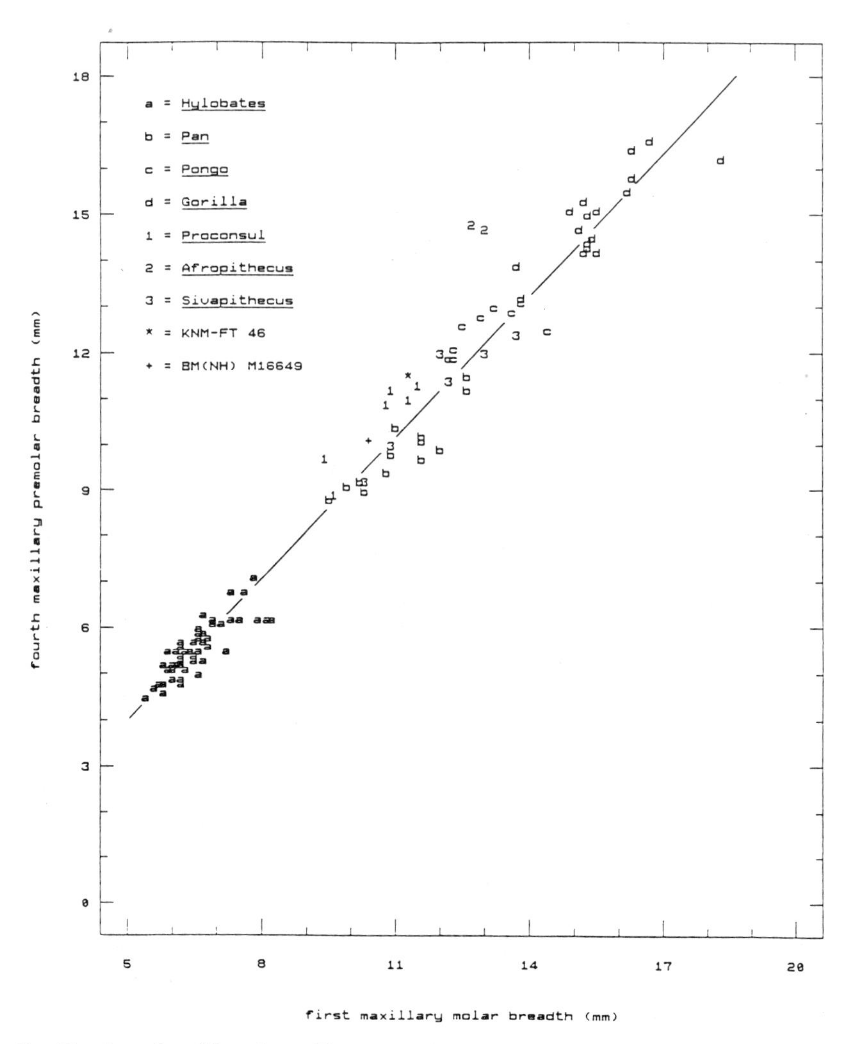

Fig. 2. Bivariate plot of fourth maxillary premolar breadth and first maxillary molar breadth in *Kenyapithecus*, *Proconsul* (BMNH M 14084, and 16647, KNM-RU 1674, 1677, and 7290, KNM-SO 418), *Afropithecus* (KNM-WK 16999, UMP 62-11), *Sivapithecus* (GSI-D 1, 185, and 196, MTA 2125, GSP 15000, YPM 13799), and extant hominoids. Regression line: $y = 0.07 + 0.74x$, r-squared = 0.97.

single individual (KNM-FT 46 and 47) and no third molars are known. Development of a lingual cingulum varies greatly among the 20 upper molars known from Maboko Island, being absent on 85% (17/20), moderately developed on 10% (2/20), and well developed on only one specimen KNM-MB 9728 (Pickford, 1985; McCrossin, 1994a). The lingual cingulum development of *K. wickeri* and *K. africanus* differs from the strongly beaded cingulum seen in early Miocene hominoids such as *Proconsul* and *Afropithecus*.

The mandibular morphology of *K. wickeri* is known from two small fragments, a right mandible with P_4 and broken M_1 KNM-FT 7, and a left mandible with P_{3-4} KNM-FT 45 (Andrews and Walker, 1976). Until discovery of the Kaloma mandible (KNM-MJ 5; Pickford, 1982), the mandibular morphology of *K. africanus* was entirely unknown. Unfortunately, the Kaloma mandible is badly distorted and crushed, particularly in the symphyseal region, and all teeth are too poorly preserved to provide reliable information. A juvenile mandible with LI_2, ldp_{3-4}, LM_{1-2}, rdp_{3-4}, RM_1, and associated RM_2 (KNM-MB 20573) recently recovered from Maboko Island preserves the first intact and undistorted mandibular symphysis and corpus of *Kenyapithecus* (McCrossin and Benefit, 1993a,b, 1994; McCrossin, 1994a).

The mandibular symphyses of *K. wickeri* (KNM-FT 45) and *K. africanus* (KNM-MB 20573) are long, low, and have strong inferior transverse tori which extend farther posteriorly than their weakly developed superior transverse tori (Fig. 3; McCrossin and Benefit, 1993a,b, 1994; McCrossin, 1994a). The inferior transverse torus of KNM-FT 45 extends below the mesial root of the M_1 while that of KNM-MB 20573 extends to below distal dp_4 (McCrossin and Benefit, 1993a). Genioglossal fossae are posteriorly directed on both specimens. *Kenyapithecus* shares these features with extant great apes and several Miocene hominoids, including *Griphopithecus, Dryopithecus, Sivapithecus,* and *Ouranopithecus* (Ward and Brown, 1986; McCrossin and Benefit, 1993a). A weak superior transverse torus, strong inferior torus, and posteriorly directed genioglossal fossa are also seen in *Cebupithecia sarmientoi.* Mandibles of *Proconsul* and Miocene "small-bodied apes" generally differ in combining a larger superior transverse torus with absence or slight development of an inferior transverse torus and an inferiorly directed genioglossal fossa (Clark and Leakey, 1951).

Comparison of extant hominoid juvenile and adult symphyseal cross sections indicates that following eruption of the permanent incisors, immature

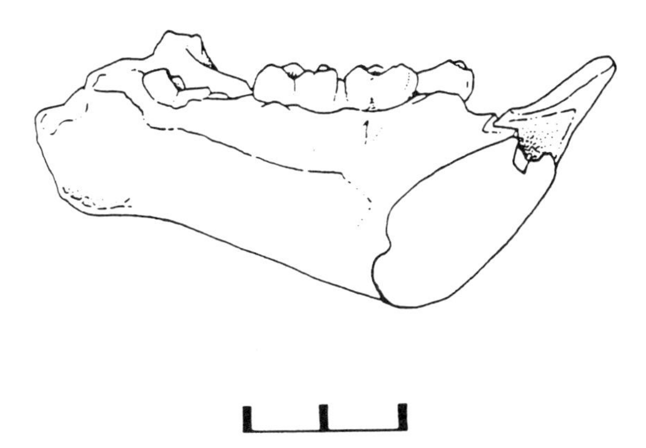

Fig. 3. Cross section of the mandibular symphysis of *Kenyapithecus africanus* specimen KNM-MB 20573, including intact I_2. Each scale segment = 1 cm.

individuals acquire an orientation of the symphyseal long axis that closely resembles that of adults (McCrossin and Benefit, 1993a). *K. wickeri* and *K. africanus* are unique among extinct large-bodied hominoids in having a strongly proclined symphysis, forming an angle of approximately 30–40° to the alveolar margin of the postcanine teeth. A similarly proclined symphyseal axis is present in the fossil pitheciine *Cebupithecia. Proconsul, Griphopithecus, Sivapithecus,* and *Ouranopithecus* have more vertically oriented symphyseal axes (McCrossin and Benefit, 1993a,b, 1994; McCrossin, 1994a). Strong proclination of the symphyseal axis had been suggested previously for *K. wickeri* and *K. africanus* (Pickford, 1982) but was questioned because of the fragmentary and distorted nature of the remains (Simons and Pilbeam, 1978; Kelley and Pilbeam, 1986). Specimen KNM-MB 20573 provides confirmation that *Kenyapithecus* truly has a strongly proclined mandibular symphysis and shows that this condition is one of the most diagnostic features presently known for the genus (McCrossin and Benefit, 1993a,b, 1994; McCrossin, 1994a).

The corpora of KNM-FT 45, KNM-MJ 5, and KNM-MB 20573 are robust, unlike the tall and slender mandibular bodies of *Proconsul* and other early Miocene hominoids (McCrossin and Benefit, 1993a). Like *Griphopithecus, Sivapithecus,* and *Ouranopithecus,* the mandibular corpus of *Kenyapithecus* (KNM-MJ 5, KNM-MB 20573) is thickest (mediolaterally) adjacent to the origin of the ramus, because of pronounced development of the *proeminentia lateralis*.

The unworn left I_2 found in mandible KNM-MB 20573 provides the first fossil evidence about the morphology and implantation of *Kenyapithecus* lower incisors (McCrossin and Benefit, 1993a,b, 1994). Before this discovery, lower incisors of the genus were reconstructed as having been very small, with strong anteroposterior curvature of the root and vertical orientation of the crown (Andrews and Walker, 1976). Based on this reconstruction, tall lateral incisors discovered at Maboko prior to 1987 were removed from the hominoid collections of the National Museums of Kenya and placed among the suids—until the discovery of KNM-MB 20573. The well-preserved lateral incisor in the juvenile *K. africanus* mandible is quite large, resembling *Proconsul, Afropithecus,* and *Dryopithecus* in being very tall relative to the mesiodistal and labiolingual dimensions (McCrossin and Benefit, 1993a). The labiolingual thickness of the *K. africanus* I_2 crown is significantly greater than the mesiodistal dimension, similar to proportions seen in *Proconsul, Afropithecus, Griphopithecus, Dryopithecus, Sivapithecus,* and *Ouranopithecus* (McCrossin and Benefit, 1993a). Extant great apes, in contrast, often exhibit substantially broader (mesiodistally) lower lateral incisor crowns. The strong procumbency of *Kenyapithecus* lower incisors, as shown definitively by mandible KNM-MB 20573, differs from the condition observed in other apes, including *Proconsul, Dryopithecus, Sivapithecus,* and *Ouranopithecus* (McCrossin and Benefit, 1993a,b). However, similarly narrow, high-crowned, and strongly procumbent lower incisors are seen in the pitheciines *Chiropotes* and *Cacajao* (Kinzey, 1992). Four isolated I_1s of *K. africanus* are somewhat smaller in terms of length, thickness, and crown height than all four known I_2s. The distal end of the long and thick lateral incisor roots curve medially, presumably to buttress the shorter and smaller roots of the lateral incisors which they extend below.

Two lower third premolars of *K. wickeri* (KNM-FT 35 and 45) and one P_3 of *K. africanus* are known. Specimens from both sites have a moderately developed mesiolingual beak and a single distally oriented lingual crest connecting the tip of the protoconid with the shorter and much less distinct metaconid. Lower third premolar KNM-MB 9737, previously attributed to *K. africanus* by Pickford (1985), is removed from that species because it is substantially smaller and differs in morphology from others referred here to *Kenyapithecus*. Thus, claimed distinctions of P_3 morphology between *K. wickeri* and *K. africanus* (relative breadth, development of a mesiolingual beak, presence or absence of a metaconid) based on assignment of KNM-MB 9737 to *Kenyapithecus* (Begun, 1992, Table 1) no longer hold true.

Morphologically, *K. africanus* and *K. wickeri* lower fourth premolars are very similar, with distinct entoconids, the cusps of the trigonid being of approximately equivalent size to each other, and the talonid being between about two-thirds and three-fourths the height of the trigonid. No differences of P_4 length and P_4 talonid height between *K. wickeri, K. africanus,* and *Dryopithecus brancoi* are evident (*contra* Martin, 1986; Begun, 1992).

Lower molars are better known for *K. africanus* (N = 45) than for *K. wickeri* (N = 4). First and second lower molar dimensions for *K wickeri* fall within the range of variation of *K. africanus* for these elements. Coefficients of variation for samples combining both species do not exceed those observed for extant hominoids. As for *Gorilla, Pan,* and some individuals of *Proconsul major,* M_2s of *K. africanus* are broader relative to their lengths than are those of *Proconsul africanus, P. nyanzae* and other Miocene apes. Cingulum development is difficult to assess on all Fort Ternan and 20 of the 45 Maboko specimens because of heavy wear, weathering, and damage. Of the remaining 25 lower molars from Maboko and the molars from Fort Ternan, full buccal cingula occur on only two teeth, both of which are M_3s. Buccal cingula are definitely absent on 23 Maboko lower molars, although tiny remnants of cingula restricted to the base of the median buccal cleft or to both the median and distal buccal clefts occur on 11 Maboko specimens, as well as on two teeth from Fort Ternan. A comparable degree of buccal cingulum reduction is seen in *Afropithecus turkanensis, Dryopithecus laietanus, D. brancoi,* and *Sivapithecus sivalensis,* but is usually much stronger in lower molars of *Proconsul, Griphopithecus alpani,* and *Dryopithecus fontani.*

All elements of the deciduous dentition except the lower incisors are known for *K. africanus,* of which the most informative are the dp_3 and dp_4. The dp_3 of KNM-MB 20573 resembles that of *Pongo, Gorilla, Pan, Dryopithecus laietanus* (IPS 1784), and *Ardipithecus ramidus* (White *et al.,* 1994) in being ovoid, having a large protoconid, a distinct metaconid that is closely apposed against the lingual face of the protoconid, and a low and weakly developed hypoconid, as well as a minute entoconid bordering the simple talonid basin

(McCrossin and Benefit, 1993a). The dp_3s of these taxa have a mesiodistally shorter and less steeply inclined preprotocristid and have a reduced extension of enamel onto the mesial root than large-bodied early Miocene hominoids, including *Proconsul. Kenyapithecus, Dryopithecus* and *A. ramidus* dp_3s resemble *Gorilla* and *Pan* in that the premetacristid is short, intersecting the lingual margin a short distance mesial to the tip of the metaconid, and in having a mesiolingually oriented crest from the protoconid that divides the anterior fovea (McCrossin and Benefit, 1993a). In contrast, *Sivapithecus sivalensis* dp_3s (exemplified by GSP 11536) are more molariform because of the larger talonid basin and higher and more strongly developed hypoconid and entoconid, and have an expansive anterior fovea bounded buccally by the preprotocristid and lingually by the premetacristid. *Kenyapithecus* dp_4s are similar to those of *Dryopithecus, Sivapithecus,* and extant hominoids in lacking distinct buccal cingula such as occur on similar teeth of *Proconsul* and *Griphopithecus.*

Postcranial Morphology and Adaptations

Until recently, postcranial remains confidently referable to *Kenyapithecus* consisted of only a right distal humerus (KNM-FT 2751) from Fort Ternan as well as a left humerus shaft (BMNH M 16334), left proximal femur (BMNH M 16331), left femur shaft (BMNH M 16330), and right femur shaft (BMNH M 16332/16333) from Maboko (Morbeck, 1983). Since 1987, our excavations on Maboko Island have led to the recovery of 28 new postcranial elements, including the first specimens of the *Kenyapithecus* shoulder (proximal humerus), forearm (proximal ulna), knuckle (third metacarpal), thumb (pollicial proximal phalanx), knee (distal femur, patella), ankle (astragalus), foot (all but the third metatarsals), and digits (several phalanges). Consequently, more is known of the postcranial morphology of *Kenyapithecus africanus* than for any other middle or late Miocene hominoid except *Oreopithecus.*

The proximal humerus of *K. africanus* that we recovered from disturbed sediment on Maboko in 1992, perfectly conjoins the humerus shaft originally collected by W. E. Owen in 1933 (Benefit and McCrossin, 1993a; McCrossin and Benefit, 1994; McCrossin, 1994a). The distal humerus from Fort Ternan is nearly identical to the proximal humerus and shaft from Maboko in size, degree of anteroposterior compression of the shaft, development of the brachioradialis crest, and proximal contour of the olecranon fossa. By combining the three specimens, it is possible to reconstruct a complete *Kenyapithecus* humerus (Fig. 4; McCrossin, 1994a). Judging from this reconstruction, the humerus would have been approximately 264 mm long, less than its previously estimated length of 280 mm (Clark and Leakey, 1951).

The *Kenyapithecus* humerus is long and slender, unlike the moderately robust humeri of fossil and modern cercopithecines, including *Victoriapithecus.* Comparison of its length to the proximodistal diameter of its head

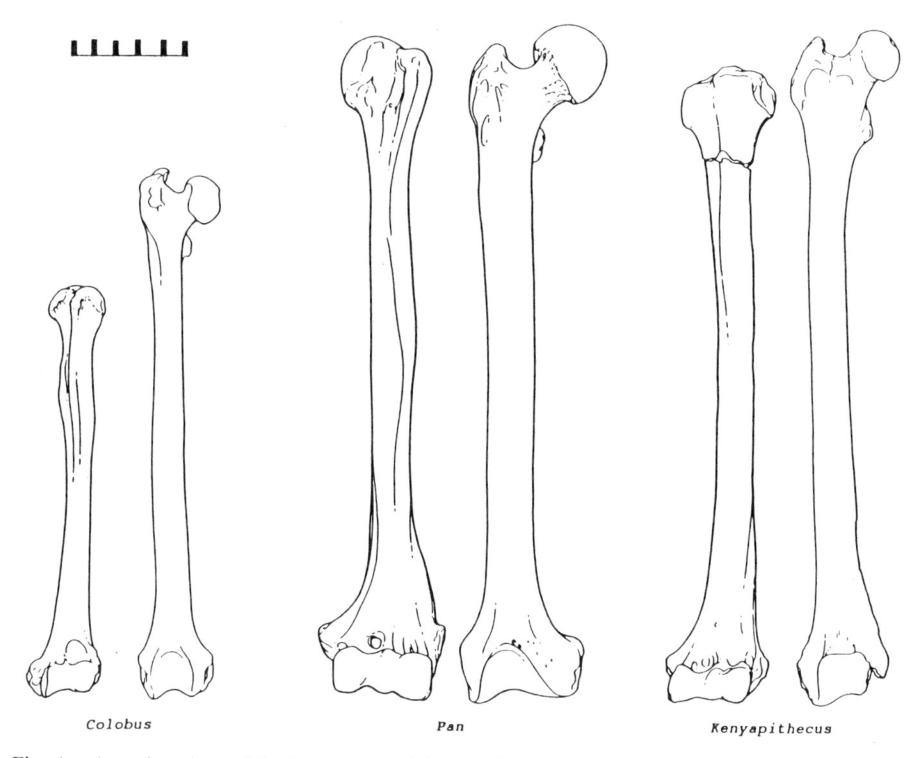

Fig. 4. Anterior view of the humerus and femur of *Colobus guereza* (left), *Pan paniscus* (center), and *Kenyapithecus* reconstruction (right). Each scale segment = 1 cm.

indicates that the *Kenyapithecus* humerus is more gracile than that of extant great apes, but less gracile than gibbons (McCrossin, 1994a). In terms of these proportions, it is most similar to arboreal anthropoids such as pitheciines, atelines, and colobines, and well as to *Pliopithecus, Dendropithecus,* and *Dryopithecus. Proconsul* and *Sivapithecus* humeri are more robust.

The head of the *Kenyapithecus* proximal humerus is directed posteriorly as in most anthropoids, but differs from extant apes and spider monkeys, which have a more medially directed humeral head (Figs. 4 and 5). The only two other proximal humeri known for Miocene hominoids, *Dendropithecus macinnesi* or *Proconsul africanus* (Gebo *et al.,* 1988) and *Nyanzapithecus pickfordi* (McCrossin, 1992), are also posteriorly directed like *Kenyapithecus.* Claims of a hominoidlike posteromedial orientation of the humeral head of *P. africanus* juvenile KNM-RU 2036 (Napier and Davis, 1959; Andrews, 1985; Ward, 1993) are highly speculative because the proximal end is not preserved on this specimen and the humerus shaft is distorted. The same is equally true of *Sivapithecus,* since no proximal humeri are known for the genus.

Like proximal humeri of *Dendropithecus macinnesi* or *P. africanus* and *Nyanzapithecus pickfordi,* that of *Kenyapithecus* lacks features shared by all living

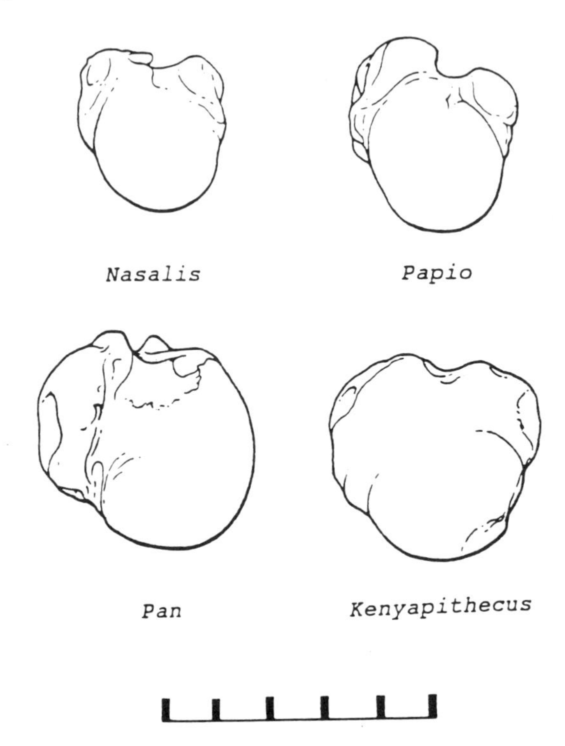

Fig. 5. Humerus of *Nasalis larvatus* (upper left), *Papio hamadryas* (upper right), *Pan paniscus* (lower left), and *Kenyapithecus africanus* (lower right) in proximal view. Each scale segment = 1 cm.

hominoids, that are related to agile climbing and facultative arm-swinging (McCrossin, 1994a). Rather than being large and globose, the articular surface of the *Kenyapithecus* humeral head is quite flat proximally and its proximodistal height is low, as in terrestrial cercopithecoids. The greater tuberosity of the *Kenyapithecus* humerus is large, anterolaterally placed, and extends proximally slightly farther than the level reached by the articular surface of the head (Figs. 4 and 5), as in terrestrial cercopithecoids such as *Erythrocebus, Papio,* and *Theropithecus* (McCrossin, 1994a). In contrast, extant hominoids and arboreal cercopithecoids have humeral heads that extend above the greater tuberosity. The lesser tuberosity of the *Kenyapithecus* humerus is large and anteromedially positioned, a condition shared with Old World monkeys and nonateline platyrrhines, but not seen in modern hominoids, for which the lesser tuberosity is very small and anteriorly positioned (Fig. 5). The broad and shallow bicipital groove shared by *Kenyapithecus,* cercopithecoids, and nonateline New World anthropoids also differs from that of hominoids (except *Pongo*) and atelines, which have very narrow intertubercular sulci.

In cross section, the humerus shaft immediately distal to the head is characterized by a rounded deltopectoral crest anterolaterally, separated from a more clearly defined crest for *m. teres major* medially by the broad and shallow intertubercular sulcus. Like cercopithecoids and most ceboids, the humerus shaft of *Kenyapithecus* is anteriorly flexed approximately one-third of the way down its length, a result of strong development of the deltopectoral crest (Clark and Leakey, 1951). Humerus shafts of *Aegyptopithecus zeuxis* (Fleagle and Simons, 1982), *Proconsul africanus* (Napier and Davis, 1959), *Griphopithecus darwini* (Zapfe, 1960), *Sivapithecus sivalensis* (Pilbeam *et al.*, 1990), and *Australopithecus afarensis* (Lovejoy *et al.*, 1982) also possess the moderate degree of anterior flexion seen in the humerus shaft of *Kenyapithecus*. Napier and Davis (1959) have suggested that among primates flexion of the humerus shaft is associated with quadrupedal modes of locomotion. Essentially straight humerus shafts, in contrast, are seen in the slender loris, atelines, *Pliopithecus* (Zapfe, 1960), and extant hominoids. *Dryopithecus fontani* from Saint Gaudens is said to possess a more weakly developed deltopectoral crest and less anterior flexion of the shaft than that of *Kenyapithecus* (Clark and Leakey, 1951; Pilbeam and Simons, 1971), but because of the juvenile age of the specimen and postmortem crushing of its proximal shaft, it is not known whether this perceived shape is actually characteristic of the species.

The distal humerus from Fort Ternan exhibits an unusual combination of features, some of which are shared derived similarities with living hominoids (Fig. 6). The trochlea resembles modern hominoids in being broader

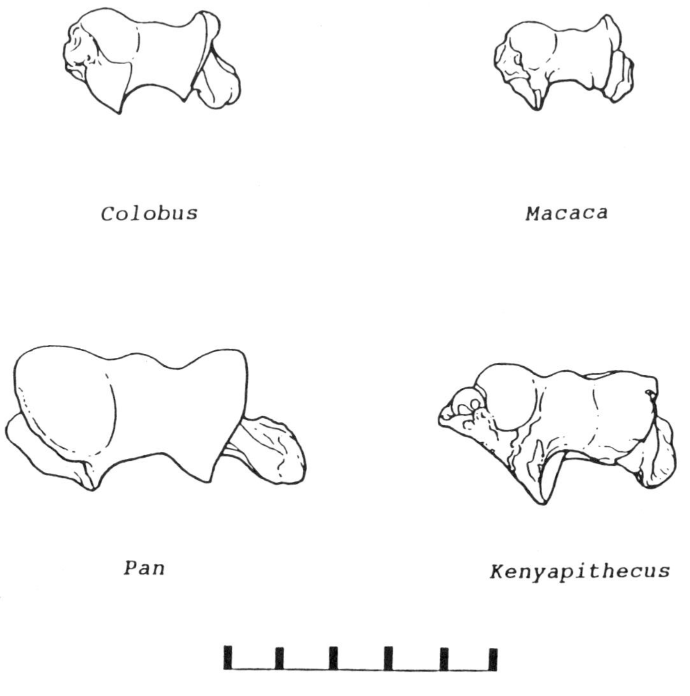

Fig. 6. Humerus of *Colobus guereza* (upper left), *Macaca nigra* (upper right), *Pan paniscus* (lower left), and *Kenyapithecus wickeri* (KNM-FT 2751, lower right) in distal view. Each scale segment = 1 cm.

than the capitulum and in having a well-developed keel laterally (Morbeck, 1983). Among cercopithecoids, in contrast, the trochlea is narrower than the capitulum and possesses a strong keel medially (Napier and Davis, 1959). A deep groove sets the trochlea apart from the globular capitulum of KNM-FT 2751. Thus, a hominoidlike *zona conoidea* for reception of the radial head is present on the *Kenyapithecus* distal humerus. In terms of its depth and breadth, the *zona conoidea* of *K. wickeri* is most comparable to that of *Hylobates,* being exiguously shallower and broader than that of *Pan.* The *zona conoidea* enhances stability of pronation and supination movements of the forearm by providing a secure articulation for the rim of the radial head.

The most distinctive feature of the *Kenyapithecus* distal humerus is the strong posterior inclination of the medial epicondyle (Fig. 6). Among hominoids and platyrrhines, the entepicondyle is large and medially directed, reflecting the premium placed on digital grasping (Fleagle and Simons, 1982). In cercopithecoids, especially the semiterrestrial and terrestrial species, in contrast, the medial epicondyle is abbreviated and posteromedially oriented (Birchette, 1982). This shortening and posterior reflection of the medial epicondyle in ground-dwelling cercopithecines has been related to a reduction in the mass of the carpal and digital flexors which take their origin from the entepicondyle, and to an increase in the moment arm of *m. pronator teres* around the axis of pronation (Birchette, 1982). The angle of posterior reflection of the medial epicondyle seen in *K. wickeri* (54°) is most comparable to values of *Erythrocebus patas* (mean = 51°, Fleagle and Simons, 1982). Medial epicondyles are more laterally directed in other Oligo-Miocene noncercopithecoid catarrhines, including *Aegyptopithecus zeuxis* (23–27°), *Pliopithecus vindobonensis* (13°), *Simiolus enjiessi* (26°), *Dendropithecus mancinnesi* (28°), and *Dryopithecus brancoi* (35°) (Fleagle and Simons, 1982; Rose *et al.*, 1992).

The olecranon of *Kenyapithecus,* as demonstrated by a small proximal fragment of the left ulna recovered from Maboko in 1992, is relatively long and retroflected as in terrestrial and semiterrestrial cercopithecines (arboreal Old World Monkeys have long but straight olecranons) and unlike the extremely reduced olecranon of modern hominoids. *Kenyapithecus* resembles *Proconsul nyanzae* in this feature, but the olecranon is incompletely known for other Miocene large-bodied hominoids. In addition, the presence of a greater amount of articular surface on the lateral (than medial) side of the anconeal process in *Kenyapithecus* is shared with terrestrial cercopithecoids, but this process tends to be symmetrical in arboreal cercopithecoids (Birchette, 1982).

Less is known about the hand than about the arm bones of *Kenyapithecus.* A pisiform of *Kenyapithecus* collected from Maboko Island in 1994 retains a distinct facet for the styloid process of the ulna, as in *Proconsul* and the Old World monkeys, but unlike modern hominoids. The one metacarpal known for *K. africanus* from Maboko, that of the third digit, has a strong dorsal transverse ridge adjacent to the distal end. Similar transverse dorsal ridges found on third metacarpals of *Pan* and *Gorilla* (but not *Hylobates* or *Pongo*) were said to be a derived feature unique to knuckle-walkers, and therefore a potential indicator of this locomotor pattern in fossil taxa (Tuttle, 1967). However, transverse dorsal ridges are present in large terrestrial and semiterrestrial papionins, especially *Mandrillus*. Hence, the dorsal ridge may be indicative of digitigrade hand postures in the broad sense, including both the dorsal digitigrady of knuckle-walking apes and the palmar digitigrady of terrestrially adapted cercopithecoids. During these forms of locomotion, the dorsal ridge prevents hyperextension at the metacarpophalangeal joint. The distal end of the *Kenyapithecus* third metacarpal is heart shaped, the dorsal aspect being narrower than the palmar side, as in nonhominoids and nonatelines, and in contrast to the equilateral distal ends of extant hominoids.

A complete pollicial proximal phalanx of *Kenyapithecus*, in recent collections from Maboko, resembles *Pan* rather than *Pongo* in having a deeply excavated and relatively large facet for *m. adductor pollicis*. Two intermediate phalanges from Maboko are relatively short and stout with little dorsoventral curvature of the shaft and well-marked flanges on the ventral surface for insertion of the digital flexors, similar to phalanges of semiterrestrial and terrestrial catarrhines such as *Cercopithecus, Mandrillus,* and *Pan* (McCrossin, 1994a). They differ from longer and more gracile intermediate phalanges of *Pliopithecus, Proconsul,* and generally arboreal extant primates. Three complete terminal phalanges from Maboko Island have distal ends that are moderately narrow and pointed with a prominent tubercle on the ventral side, more similar to those of *Pliopithecus* and extant cercopithecoids than to the broader ends and flatter ventral tubercles of extant hominoids.

The femur of *Kenyapithecus* is known from four specimens: a right shaft lacking the proximal and distal ends (BMNH M 16332/BMNH M 16333), a left proximal end lacking most of the greater trochanter (BMNH M 16331), a fragmentary left shaft (BMNH M 16330), and a left distal end preserving the lateral condyle and most of the patellar groove (recovered in 1992). All four specimens are very similar in size and morphology, and may represent a single individual. Our reconstruction confirms earlier impressions of the bone's long and slender proportions (Fig. 4; Clark and Leakey, 1951; Ruff *et al.*, 1989). Its estimated length, approximately 277 mm, is greater than that of the humerus, indicating a humerofemoral length index of 95 for *Kenyapithecus* (*contra* Simons and Pilbeam, 1972). Comparable mean values for humerofemoral indices are observed in *Macaca, Papio, Theropithecus,* and *Pan.* An appreciably longer humerus (and/or much shorter femur) is seen in *Oreopithecus* (Schultz, 1960), *Hylobates, Gorilla,* and *Pongo.*

The *Kenyapithecus* femoral head is relatively small, being most similar in this respect to cercopithecines (McCrossin, 1994a). The neck is long relative to the superoinferior diameters of the head and neck and is angled high relative to the shaft, like ceboids and hominoids, but unlike the relatively shorter femoral necks and lower femoral neck–shaft angles seen in cercopithecoids (Clark and Leakey, 1951). As in ceboids and hominoids, and unlike cercopithecoids, the femoral head projects farther proximally than the greater trochanter and does not exhibit extension of articular surface onto the posterior aspect of the neck. On the distal end of the femur, the patellar groove is quite broad and shallow and the lateral condyle is thicker (anteroposteriorly) relative to its breadth than in hominids, but narrower than in baboons (McCrossin, 1994a). Several of these features are seen on the femora of *Proconsul nyanzae,* although the femur of Kenyapithecus is more gracile and has a smaller head than that of the early Miocene hominoid (Ruff *et al.,* 1989). The *Kenyapithecus* femur is also similar to that of *Dryopithecus* from Eppelsheim in its overall proportions, the relatively small head, high neck angle, posterior neck tubercle, moderate development of the gluteal tuberosity, and shape of the midshaft cross section (Clark and Leakey, 1951).

A well-preserved patella of *Kenyapithecus* was discovered at Maboko in 1992 (McCrossin, 1994a). It resembles New World anthropoids and hominoids, with mediolaterally broad proportions and a short nonarticular extension distally. Cercopithecoids have narrower patellae. However, much of the difference between cercopithecoid and other anthropoid patellae is related to the greater length of their nonarticular distal extensions, a feature related to their greater emphasis on extension at the knee joint. The *Kenyapithecus* patella is thin anteroposteriorly, like platyrrhines and hominoids and unlike the thick patellae of cercopithecoids. A broad and thin patella with a short nonarticular distal extension is also seen in *Pliothecus* (Zapfe, 1960) and *Oreopithecus* (Schultz, 1960), and probably characterized the last common ancestor of extant catarrhines.

A right distal tibia referable to *Kenyapithecus,* preserving the shaft and distal end, was collected from Maboko Island by Owen and MacInnes in the 1930s, and is part of the collections at the British Museum of Natural History (McCrossin, 1994a). In virtually all aspects, its anatomy is similar to that described for *Proconsul nyanzae* (KNM-RU 1939). Tibiae of both species differ from extant catarrhines in having a shaft that tapers evenly to the distal end, rather than broadening abruptly. The articular surface for the astragalar trochlea is thick (anteroposteriorly) and broad (mediolaterally), as in cercopithecoids, but unlike the anteroposteriorly compressed and mediolaterally broad articulation of atelines and hominoids. Other aspects of the *Kenyapithecus* distal tibia, including presence of a medial keel separating lateral and medial sulci for the astragalar trochlea, are also shared by the Miocene apes and cercopithecoids, but are not seen in modern hominoids or in some platyrrhines. A distal fibula of *K. africanus* found at Maboko in 1994 resembles that of hylobatids in having an astragalar facet that faces mediodistally. Among extant great apes, this facet faces farther distally, whereas in cercopithecoids it faces more medially.

An astragalus of *Kenyapithecus* collected at Maboko Island in 1992 is quite comparable to that of other Miocene large-bodied hominoids, including *Proconsul, Afropithecus, Dryopithecus,* and *Sivapithecus.* One interesting aspect of the bone is the proximally concave and distally convex topography of the articular facet for the anteromedial facet of the calcaneum (sustentaculum tali). A similarly shaped facet is seen in *Afropithecus* (KNM-WK 18120) and occasionally in

Pan paniscus. In contrast, a uniformly convex contour is seen in *Colobus, Hylobates*, and *Pan troglodytes*.

The proximal ends of a left first metatarsal, right second metatarsal, left fourth metatarsal, and right fifth metatarsal of *Kenyapithecus* have been recovered from Maboko since 1992. The first metatarsal is quite robust (measured by comparing the dorsoplantar diameter of its proximal end to the maximum mediolateral breadth of the astragalus), as in gibbons and unlike much more gracile first metatarsals of *Proconsul nyanzae* (based on KNM-RU 1743 and 5872) and most extant catarrhines (McCrossin, 1994a). In plantar view, the medial part of the entocuneiform facet of the *Kenyapithecus* hallucial metatarsal is quite flat, as in *Papio hamadryas*, and unlike the proximally recurved medial extension seen in *Colobus guereza* and *Pan troglodytes*. The absence of articular surface wrapping medially around the entocuneiform indicates that the hallux of *Kenyapithecus* was habitually adducted as in baboons. The Maboko ape appears to lack the distinct prehallux facet seen in many platyrrhines, *Aegyptopithecus, Proconsul*, and *Hylobates*. In addition, the peroneal tubercle is large, indicating powerful adduction of the hallux and eversion of the foot. The mesocuneiform facet on the second metatarsal of *Kenyapithecus* is narrow like that of cercopithecoids and unlike the broad facet seen in hominoids. Otherwise, metatarsals of *Kenyapithecus* are broadly similar to those known for *Proconsul*.

Taxonomic Diversity and Phylogenetic Position

Features previously used to place *K. africanus* in a different genus and sometimes family than *K. wickeri* relied heavily on comparison of female maxilla KNM-FT 46 and male maxilla BMNH M. 16649, as well as on small fossil samples available prior to 1988 which sometimes included specimens misattributed to the genus (Pickford, 1985). When the two type maxillae are examined in the context of documented variation within extant species, perceived differences in the orientation of the upper canine socket, height of the root of the zygomatic process above molar alveoli, palate depth, maxillary sinus development, and P^4 breadth all fall within the range of modern hominoid species. Suggested differences in the orientation and lateral flare of the origin of the zygomatic (Pickford, 1985) need to be verified with more complete material. Proposed differences in the number of labial P^3 roots in the two specimens (Pickford, 1985) seem to have been based on a misreading of Leakey (1967) and are false. In addition, as new fossils from Maboko expand knowledge of variation in the lingual relief of the upper central incisor, transverse crest acuity of the P^4, cingulum development on upper and lower molars, and structure of the P_3 for *K. africanus*, Fort Ternan and Maboko dentitions are found to greatly overlap in size and in morphology.

Based on the available craniodental evidence, we conclude that variation in the crania, dentition, and distal end of the humerus shaft of *K. wickeri* and *K. africanus* do not exceed levels of variation observed within species of living hominoids. It is tempting to view the large-bodied hominoids from Fort Ternan and Maboko Island as representing a single species. However, since knowledge of variation within the sample from Fort Ternan is very limited, we refrain from formally synonymizing *K. wickeri* and *K. africanus* until further remains of the type species are found. It should be emphasized that *K. wickeri* and *K. africanus* share at least one derived characteristic, a strongly proclined symphyseal axis and procumbent lower incisors, to the exclusion of other Miocene (including *Proconsul, Griphopithecus, Sivapithecus,* and *Ouranopithecus*) and Recent hominoids. Although comparable elements are not preserved for the two species, both have terrestrial adaptations of the postcrania that are not seen in other Miocene hominoids.

Relative to early Miocene apes, including *Afropithecus* and *Proconsul, Kenyapithecus* exhibits many features of the mandible and dentition that are derived toward the extant hominoid condition. *Kenyapithecus* was the first hominoid in the fossil record with a mandible exhibiting a simian shelf resulting from its larger inferior than superior transverse torus, a posteriorly directed genioglossal pit, and a relatively short and wide corpus (McCrossin and Benefit, 1993a). This configuration is shared with *Sivapithecus* and *Ouranopithecus,* as well as with extant apes (Ward and Brown, 1986; Bonis and Melentis, 1980). Relative to *Afropithecus, Proconsul,* and other early Miocene apes, *Kenyapithecus* exhibits extreme reduction to complete absence of upper molar lingual cingula, lacks buccal cingula on deciduous premolars, and retains lower molar buccal cingula on only two third molars. Cingula are more frequent on the deciduous premolars and permanent molars of *Griphopithecus* and *Dryopithecus,* but are largely absent in *Sivapithecus, Ouranopithecus,* and extant apes. The trend toward increasing the thickness of molar enamel is documented as beginning with *Kenyapithecus* (Martin, 1986). Compared with *Proconsul* and the so-called small-bodied apes of the early Miocene, the upper premolars of *Kenyapithecus* are large relative to the size of the upper first molar, a feature shared with extant great apes and other large-bodied middle Miocene hominoids. The dp_3s of *Kenyapithecus* are derived relative to early Miocene ape dp_3s and closely resemble *Pan* and *Gorilla* in having a much shorter and less steeply inclined preprotocistid, less extension of labial enamel onto the mesial root, a distinct metaconid that is positioned close to the protoconid, and greater development of the hypoconid and entoconid (McCrossin and Benefit, 1993a). This morphology is shared with *Dryopithecus laietanus* and the supposed hominid *Ardipithecus ramidus,* but *Sivapithecus, Pongo,* and indisputable hominids have more molariform dp_3s. We consider these traits as derived relative to early Miocene apes, but primitive for both the middle Miocene ape clade and for the last common ancestor of modern hominoids.

From the perspective of the postcrania, however, *Kenyapithecus* and all middle Miocene apes for which postcrania are known (*Dryopithecus* and *Sivapithecus*), excluding *Oreopithecus,* are no more derived toward the extant hominoid condition than is the early Miocene ape *Proconsul*. The development of the lateral trochlear keel and well-defined zona conoidea are perhaps the most important features linking *Proconsul, Kenyapithecus, Dryopithecus,* and *Sivapithecus* to Hominoidea (Napier and Davis, 1959). The distal humeri of *Aegyptopithecus, Pliopithecus,* and *Dendropithecus* lack these derived features and are more similar to platyrrhines (Fleagle, 1983).

All early to late middle Miocene hominoids for which the proximal humerus is known, with the possible exception of *Oreopithecus,* lack the medially directed humeral head, deep and narrow bicipital groove, and small and anteriorly positioned lesser tuberosity, that are uniquely shared by extant apes and hominids within the Catarrhini. Therefore, they lack perhaps the most definitive characteristic of modern hominoids, the rotary shoulder joint. Although proximal humeri of other middle Miocene hominoids are unknown, comparison of the proximal end of *Dryopithecus* and *Sivapithecus* humerus shafts with that of *Kenyapithecus* indicates that the same was true of these Eurasian hominoids (McCrossin, 1994a). In addition, from combined samples of *Kenyapithecus, Dryopithecus,* and *Sivapithecus,* it is apparent that these apes also lacked the reduced olecranon, intra-articular meniscus between the ulnar styloid process and the pisiform, semispherical proximal capitate, well-developed hamulus, symmetrical and weakly grooved astragalar trochlea, and weakly concave facet for the medial malleolus on the astragalus, which are shared by all modern apes and hominids (Pilbeam *et al.*, 1990; Conroy and Rose, 1983; Morbeck, 1983; Spoor *et al.*, 1992). Consequently, the middle Miocene radiation of large-bodied apes, which was highly successful in terms of its diversity and geographic distribution, can be considered as broadly ancestral to, but not true members of, the clade to which extant hominoids and fossil hominids belonged.

Adaptive Reasons for the Appearance of Derived Features

Derived features of the maxilla, mandible, and dentition of *Kenyapithecus* appear to be functionally related. Among extant anthropoids, only the bearded saki (*Chiropotes*), the uakari (*Cacajao*), and to a lesser extent the orangutan (*Pongo*) consistently show the same suite of craniodental features that are characteristic of *Kenyapithecus* (Table I). In spite of the fact that the pitheciines have thin-enameled molars, the diets of *Chiropotes* and *Cacajao* are composed almost entirely of hard fruits and dehiscent seeds and nuts (van Roosmalen *et al.*, 1988). Rather than crush the hard fruits between their molars, *Chiropotes* and *Cacajao* use their robust upper canines, enlarged premolars, and robust and strongly procumbent lower incisors to break open the tough outer coats of nuts and fruits with hard seeds (van Roosmalen *et al.*, 1988). The robust construction of the *Kenyapithecus* and *Cebupithecia* mandibular symphyses, especially the presence of a massive inferior transverse torus, is part of a func-

Table I. Dentognathic Similarities of *Kenyapithecus*, *Chiropotes*, and *Cacajao*, Indicative of Independently Acquired Adaptations for Sclerocarp Foraging and Seed Predation[a]

A. Root of maxillary zygomatic process positioned anteriorly, at M^1
B. Canine jugum prominent and postcanine fossa strongly developed
C. I^{1-2} strongly heteromorphic
D. C^1 robust, tusklike, and externally rotated
E. Upper premolar occlusal area enlarged (relative to occlusal area of M^1)
F. Mandible with strongly proclined symphyseal axis, well-developed inferior transverse torus, and robust corpus
G. I_{1-2} mesiodistally narrow, labiolingually thick, high crowned, and strongly procumbent
H. Molar cusps low, rounded, and covered with crenulated enamel

[a]From McCrossin (1994a).

tional complex related to specialized use of the lower incisors. It seems that the inferior transverse torus evolved to resist anteroinferiorly directed bending moments during incisal biting.

It is suggested that the appearance of many dentognathic features that typify the middle Miocene apes and the last common ancestor of modern apes and hominids are plausibly correlated with increased exploitation of hard fruits and nuts, cracked open by a specialized anterior dentition (McCrossin and Benefit, 1993a,b, 1994). The robust mandibular corpus and strong lateral buttressing first seen in *Kenyapithecus* among Miocene hominoids is consistent with a seed predation or nut-cracking adaptation, contributing to the withstanding of strong occlusal loading during the comminution of hard objects between thick-enameled posterior dentition (Kay, 1981; Hylander, 1984). Aspects of this nut-cracking adaptation of the anterior dentition are seen in the early Miocene taxon *Afropithecus*, which shares upper incisor heteromorphy, robust and externally rotated canines, and large upper premolars with *Kenyapithecus* and later great apes. It seems that apes initially accomplished a shift from a soft to hard fruit diet by changing their anterior dentition, followed by later adaptations of the mandible.

The postcrania of *Kenyapithecus* are similar to those of other Miocene apes (except *Oreopithecus*) in lacking modern hominoidlike suspensory adaptations of the forelimb. However, postcranial elements of *Kenyapithecus* differ from those known for other Miocene apes in exhibiting features adapted for a semiterrestrial digitigrade mode of locomotion (Table II). *Kenyapithecus* shares with semiterrestrial and terrestrial cercopithecoids, a flattened rather than strongly domed humeral head, presumably related to increased stability of the glenohumeral joint, particularly during the stance phase of pronograde quadrupedalism (Birchette, 1982). The large, anteriorly placed, and elevated height of the *Kenyapithecus* greater tuberosity closely resembles that of semiterrestrial cercopithecoids such as *Presbytis entellus* and *Macaca nemestrina*, and may be adapted to enhance forceful protraction of the arm by providing

Table II. Semiterrestrial Adaptations of the Postcranial Skeleton of *Kenyapithecus*

Element	Condition	Functional implications
Humerus	A. Greater tubercle is positioned anteriorly and projects slightly farther proximal than articular surface of the head	Elongation of lever arm of *m. supraspinatus* for rapid protraction of the arm; restriction of rotational movements at glenohumeral joint
	B. Articular surface of the head is proximodistally short and flattened proximally	Stabilization of stance postures during pronograde quadrupedal progression
	C. Crest extending distally from lesser tubercle is proximally restricted	Abbreviation of lever arm of *m. teres major*, reduction of medial rotation of arm, indicating adducted arm postures
	D. Medial epicondyle is short and posteriorly reflected	Reduction of mass of forearm and digital flexor musculature; enhancement of action of forearm pronator musculature
	E. Olecranon fossa is deep, with strongly developed and keeled lateral border	Stabilization of humeroulnar joint in fully extended and pronated forearm postures
Ulna	F. Olecranon is relatively long and moderately retroflected	Enhanced thrust of forearm extension produced by *m. triceps brachii* when elbow is in postures approaching maximum extension
Third metacarpal	G. Dorsal transverse ridge above distal end	Prevents hyperextension of the metacarpal–phalangeal joint during digitigrade locomotion
Femur	H. Relatively small articular suface of head	Restriction of range of thigh excursion during abduction at hip
First metatarsal	I. Medial portion of entocuneiform facet almost planar, does not curve proximally	Restriction of range of abduction at hallucial tarsometatarsal joint, indicating habitually adducted hallux
Intermediate phalanx	J. Short and stout, with slight shaft curvature	Reduction of capabilities for digital grasping, improved resistance of compressive forces in digitigrade hand postures

elongation of the moment arm of *m. supraspinatus* (Birchette, 1982) or for stabilizing the shoulder during the stance phase of quadrupedalism (Larson and Stern, 1992). The shortening and strong posterior inclination of the medial epicondyle on the *Kenyapithecus* distal humerus appears to be related to a reduction in the mass of carpal and digital flexors which take their origin from the entepicondyle as seen in semiterrestrial and terrestrial cercopithecines (Birchette, 1982). The extremely broad lateral epicondyle of KNM-FT 2751 most closely resembles distal humeri of *Papio, Mandrillus,* and *Theropithecus,* and may reflect a greater mass of carpal and digital extensors in *Kenyapithecus.* A humeral–femoral length index of 95 is derived from our reconstruction of the *K. africanus* limb bones. This value indicates that fore-

and hindlimbs were used fairly equally during locomotion, as occurs in the terrestrial catarrhines *Macaca, Papio, Theropithecus,* and *Pan,* which happen to have the same humeral–femoral length index as *Kenyapithecus.* The short, stout, and straight intermediate phalanges of *Kenyapithecus* are typical of semiterrestrial and terrestrial monkeys and apes (Tuttle, 1967). *Kenyapithecus* shares a strong transverse dorsal ridge on the proximal end of the third metacarpal with large terrestrial hominoids (*Pan* and *Gorilla*) and semiterrestrial papionins (*Papio* and *Mandrillus*). Such a ridge prevents hyperextension of the knuckle joint during various forms of digitigrade terrestrial locomotion, including knuckle-walking. The hallux of *Kenyapithecus* appears to have been adducted, as indicated by the flat, rather than curved entocuneiform facet on the first metatarsal as in baboons, and unlike the proximally recurved and medially extended facets of primates with strongly abducted big toes of arboreal monkeys and all modern apes.

We conclude that a macaquelike (rather than chimpanzeelike) version of digitigrade terrestrial quadrupedalism characterized *Kenyapithecus.* As for macaques, the middle Miocene ape may have employed palmigrade hand postures when in the trees, but were digitigrade when walking or running on the ground. This new interpretation contradicts previous claims that the postcrania "from Maboko Island . . . indicate little change from the generalized arboreal quadrupedalism present in the early Miocene hominoids like *Proconsul* (Andrews, 1992, p. 643), and that no morphological indicators of terrestrial locomotion are known for Miocene hominoids (Rose, 1993).

In addition to many postcranial features *Kenyapithecus* shares with terrestrial cercopithecoids, it differs from these species in the relative size of the humeral head, gracility of the humeral shaft, the breadth of the trochlea, the development of the lateral trochlear keel, the depth of the zona conoidea, and the angle of the femoral neck of *Kenyapithecus.* These features may relate to scansorial activities, including forelimb-dominated movements such as hoisting, stable pronation and supination of the forearm throughout the entire range of elbow flexion and extension, as well as hindlimb participation in vertical climbing. Thus, it seems likely that *Kenyapithecus* retained the quintessentially primate ability to climb trees for sleeping, to flee predators, and to gain access to fruits.

Conclusions: Implications for the Evolutionary History of Modern Apes

Recent analyses suggest that the large-bodied hominoids of the middle–late Miocene are members of the great ape and human clade (Ward and Pilbeam, 1983; Kelley and Pilbeam, 1986; Martin, 1986; Brown and Ward, 1988). Almost all of the features used to ally *Kenyapithecus, Dryopithecus,* and *Sivapithecus* with living great apes involve the jaws and teeth. According to

Andrews (1985, p. Table 3), *Kenyapithecus, Dryopithecus,* and *Sivapithecus* share four derived features with the last common ancestor of great apes and humans: (1) robust canines, (2) robust and not bilaterally compressed P_3s, (3) wider tooth rows, and (4) a deep symphysis with a larger inferior than superior torus. In addition, *Kenyapithecus* and *Sivapithecus* are viewed as sharing derived thick-enameled molars with the reconstructed ancestor to great apes and humans (Andrews, 1985). As is well known, *Sivapithecus* shares several resemblances of the interorbital and subnasal regions with *Pongo* (Ward and Pilbeam, 1983; Ward and Brown, 1986).

Because of the functional correlation between many of these characteristics (upper incisor heteromorphy, upper canine robusticity and external rotation, maxillary premolar enlargement, mandibular incisor procumbency, and development of a strong inferior transverse torus) and adaptations for predation on fruits with hard seeds and nuts (Table I), the possibility arises that craniodental similarities between *Kenyapithecus* and the last common ancestor of extant great apes and humans are convergently acquired adaptations for sclerocarp foraging and seed predation, rather than uniquely shared derived features of the great ape and human clade. Alternatively, adaptations for consumption of hard fruits and nuts may have characterized the last common ancestor of *Kenyapithecus* and modern hominoids (McCrossin and Benefit, 1993a,b, 1994). If so, these features would be primitive retentions, while the adaptations for a diet of ripe fruit and young leaves seen in living hylobatids are derived. In this case, resemblances between *Kenyapithecus, Dryopithecus, Sivapithecus,* and extant great apes would reflect a phenomenon whereby the jaws and teeth of the last common ancestor of living apes more closely resemble those of great apes than those of gibbons. Support for this latter interpretation comes from a recent reexamination of the ancestral craniofacial morphotype of catarrhines (Benefit and McCrossin, 1991, 1993b). Reference to a newly reconstructed ancestral morphotype of Old World higher primates based on inclusion of archaic catarrhines (Benefit and McCrossin, 1991, 1993b) indicates that contrary to prevailing wisdom (e.g., Delson and Andrews, 1975), hylobatids may possess several character states that are derived with respect to the last common ancestor of living hominoids, including an abbreviated premaxilla, homomorphic upper incisors, slender and monomorphic canines, a highly sectorial P_3, a small inferior transverse torus, and thin molar enamel.

The general absence of modern hominoid postcranial features in the Miocene large-bodied hominoids may be taken as evidence of mosaic evolution, with great ape craniodental features appearing prior to postcranial advancements. This situation may indicate that postcranial features related to suspensory locomotion evolved independently and in parallel in four different hominoid clades: (1) *Oreopithecus,* (2) *Hylobates,* (3) *Pongo,* and (4) African apes and humans (Pilbeam *et al.,* 1990). Inclusion of *Kenyapithecus, Dryopithecus,* and *Sivapithecus* (collectively) in the great ape and human clade involves parallel acquisition of several postcranial features, including: (1) a medially directed head of the humerus, (2) a deep and narrow bicipital groove, (3) a small and anteriorly positioned lesser tuberosity, (4) a reduced olecranon, (5) a semispherical proximal capitate, (6) a well-developed hamulus, (7) a symmetrical and weakly grooved astragalar trochlea, and (8) a weakly concave facet for the medial malleolus on the astragalus.

Based primarily on the presence of an allegedly gorillalike subnasal pattern in the Baragoi palate referred to *K. africanus* (Ishida, 1986) as well as the strong inferior transverse torus, robust canines, enlarged premolars, and thick molar enamel thought to ally the material from Fort Ternan and Maboko Island with great apes and humans (Andrews, 1985; Martin, 1986), it is tempting to view *Kenyapithecus* as a member of the clade that gave rise to gorilla, chimpanzees, and humans (Brown and Ward, 1988). If true, then a pattern of semiterrestrial locomotion, such as knuckle-walking, might have characterized early members of this clade prior to the emergence of modern hominoidlike configurations of the shoulder, elbow, and wrist that have traditionally been linked with brachiation (Simons and Pilbeam, 1972). This notion may derive support from the fact that certain extant hominoid postcranial characteristics, such as a deep olecranon fossa with extension of the trochlear surface onto the lateral wall and an anteroproximally facing trochlear notch of the ulna, are shared with terrestrial and semiterrestrial Old World monkeys but not with the suspensory atelines (*Ateles, Lagothrix,* and *Brachyteles*). Hence, these features may relate to a premium being placed on full extension at the elbow during terrestrial quadrupedalism as a preadaptation to the extended forelimb postures employed during suspensory locomotion.

We suggest an alternative hypothesis regarding the relationships of *Kenyapithecus, Dryopithecus,* and *Sivapithecus.* Setting aside the possibility of parallel acquisition of modern hominoidlike morphologies by Oreopithecidae (McCrossin, 1992), it is most likely that the specializations of the limb and axial skeleton seen in extant hominoids evolved once. If, in fact, the shared postcranial features of gibbons and great apes are the result of common inheritance, then the absence of these features in the large-bodied middle Miocene apes indicates that they diverged prior to the last common ancestor of living hominoids. The last common ancestor of all modern hominoids, including the gibbon, would have evolved more recently than 14 Ma. Our consideration of presently available craniodental and postcranial evidence leads us to strongly favor the latter interpretation.

The appearance of a pitheciine-like dietary specialization for seed predation and a semiterrestrial pattern of locomotion in *Kenyapithecus* foreshadows similar adaptations reconstructed for the last common ancestor of great apes and humans generally, and of *Gorilla, Pan,* and *Homo* specifically. Although in our view it is not a member of the African ape and human clade, *Kenyapithecus* may be the most appropriate model for the pre-australopithecine phase in human evolution because it lacks the uniquely derived features of the gorilla, chimpanzees, and modern humans. The presence of these hominidlike tendencies in *Kenyapithecus* should help to clarify the emergence of hominids from the Miocene hominoid radiation.

ACKNOWLEDGMENTS

We thank the Office of the President of the Republic of Kenya and the National Museums of Kenya for permission to excavate on Maboko Island. We are also grateful for the help and encouragement provided by Richard Leakey, Meave Leakey, F. Clark Howell, and John Fleagle. We are indebted to David Begun, Carol Ward, and Michael Rose for inviting us to contribute to this volume. We also thank our Maboko Island field crew, including Blasto Onyango, the late Paul Odera Abong, Kate Blue, Stephen Gitau, Brad Watkins, and Tony Hynes, as well as the curatorial staff at the National Museums of Kenya, especially Mary Muungu, Emma Mbua, Frederick Kyallo, and Alfreda Ibui. We gratefully acknowledge generous support from the National Science Foundation, L. S. B. Leakey Foundation, Fulbright Collaborative Research Program, Wenner–Gren Foundation for Anthropological Research, Rotary International Foundation, Office of Research Development and Administration of Southern Illinois University, the Boise Fund of Oxford University and the R. H. Lowie Fund of the University of California at Berkeley. We thank Tom Gatlin for the illustrations in Figures 3–6.

References

Andrews, P. 1985. Family group systematics and evolution among catarrhine primates. In: E. Delson (ed.), *Ancestors: The Hard Evidence*, pp. 14–22. Liss, New York.

Andrews, P. 1992. Evolution and environment in the Hominoidea. *Nature* **360:**641–646.

Andrews, P., and Walker, A. 1976. The primate and other fauna from Fort Ternan, Kenya. In: G. Ll. Isaac and E. R. McCown (eds.), *Human Origins: Louis Leakey and the East African Evidence*, pp. 279–304. Benjamin, Menlo Park, CA.

Begun, D. R. 1987. *A Review of the Genus Dryopithecus*. Ph.D. thesis, University of Pennsylvania.

Begun, D. R. 1992. Miocene fossil hominoids and the chimp–human clade. *Science* **257:**1929–1933.

Benefit, B. R., and McCrossin, M. L. 1991. Ancestral facial morphology of Old World higher primates. *Proc. Natl. Acad. Sci. USA* **88:**5267–5271.

Benefit, B. R., and McCrossin, M. L. 1993a. New *Kenyapithecus* postcrania and other primate fossils from Maboko Island, Kenya. *Am. J. Phys. Anthropol. Suppl.* **16:**55–56 (Abstr.).

Benefit, B. R., and McCrossin, M. L. 1993b. Facial anatomy of *Victoriapithecus* and its relevance to the ancestral cranial morphology of Old World monkeys and apes. *Am. J. Phys. Anthropol.* **92:**329–370.

Benefit, B. R., and McCrossin, M. L. 1994. Comparative study of the dentition of *Kenyapithecus africanus* and *K. wickeri*. *Am. J. Phys. Anthropol. Suppl.* **18:**55 (Abstr.).

Birchette, M. G. 1982. *The Postcranial Skeleton of Paracolobus chemeroni*. Ph.D. thesis, Harvard University.

Bonis, L. de, and Melentis, J. 1980. Nouvelles remarques sur l'anatomie d'un Primate hominoide du Miocene: *Ouranopithecus macedoniensis*. *C. R. Acad. Sci.* **290:**755–758.

Brown, B., and Ward, S. 1988. Basicranial and facial topography in *Pongo* and *Sivapithecus*. In: J. H. Schwartz (ed.), *Orang-utan Biology*, pp. 247–260. Oxford University Press, London.

Clark, W. E. L., and Leakey, L. S. B. 1951. The Miocene Hominoidea of East Africa. *Br. Mus. Nat. Hist. Fossil Mamm. Afr.* **1:**1–117.

Conroy, G. C., and Rose, M. D. 1983. The evolution of the primate foot from the earliest primates to the Miocene hominoids. *Foot & Ankle* **3:**342–364.

Delson, E., and Andrews, P. 1975. Evolution and interrelationships of the catarrhine primates. In: W. P. Luckett and F. S. Szalay (eds.), *Phylogeny of the Primates*, pp. 405–446. Plenum Press, New York.

Fleagle, J. G. 1983. Locomotor adaptations of Oligocene and Miocene hominoids and their phyletic implications. In: R. L. Ciochon and R. S. Corruccini (eds.), *New Interpretations of Ape and Human Ancestry*, pp. 301–324. Plenum Press, New York.

Fleagle, J. G., and Simons, E. L. 1982. The humerus of *Aegyptopithecus zeuxis:* A primitive anthropoid. *Am. J. Phys. Anthropol.* **59:**175–193.

Gebo, D. L., Beard, K. C., Teaford, M. F., Walker, A., Larson, S. G., Jungers, W. L., and Fleagle, J. G. 1988. A hominoid proximal humerus from the early Miocene of Rusinga Island, Kenya. *J. Hum. Evol.* **17:**393–401.

Hylander, W. 1984. Stress and strain in the mandibular symphysis of primates: A test of competing hypotheses. *Am. J. Phys. Anthropol.* **64:**1–46.

Ishida, H. 1986. Investigation in northern Kenya and new hominoid fossils. *Kagaku* **56:**220–226.

Kay, R. F. 1981. The nut-crackers: A new theory of the adaptation of the Ramapithecinae. *Am. J. Phys. Anthropol.* **55:**141–151.

Kay, R. F., and Simons, E. L. 1983. A reassessment of the relationship between later Miocene and subsequent Hominoidea. In: R. L. Ciochon and R. S. Corruccini (eds.), *New Interpretations of Ape and Human Ancestry*, pp. 577–624. Plenum Press, New York.

Kelley, J. J., and Pilbeam, D. R. 1986. The dryopithecines: Taxonomy, anatomy and phylogeny of Miocene large hominoids. In: D. R. Swindler and J. Erwin (eds.), *Comparative Primate Biology, Vol. 1: Systematics, Evolution and Anatomy*, pp. 361–441. Liss, New York.

Kinzey, W. G. 1992. Dietary and dental adaptations in the Pitheciinae. *Am. J. Phys. Anthropol.* **88:**499–514.

Larson, S. G., and Stern, J. T. 1992. Further evidence for the role of supraspinatus in quadrupedal monkeys. *Am. J. Phys. Anthropol.* **87:**359–363.

Leakey, L. S. B. 1962. A new lower Pliocene fossil primate from Kenya. *Ann. Mag. Nat. Hist.* **13:**689–696.

Leakey, L. S. B. 1967. An early Miocene member of Hominidae. *Nature* **213:**155–163.

Leakey, L. S. B. 1968. Lower dentition of *Kenyapithecus africanus*. *Nature* **217:**827–830.

Leakey, R. E. F., and Leakey, M. G. 1986. A new Miocene hominoid from Kenya. *Nature* **324:**143–146.

Lovejoy, C. O., Johanson, D. C., and Coppens, Y. 1982. Hominid upper limb bones recovered from the Hadar Formation: 1974–1977 collections. *Am. J. Phys. Anthropol.* **57:**637–650.

McCrossin, M. L. 1992. An oreopithecid proximal humerus from the middle Miocene of Maboko Island, Kenya. *Int. J. Primatol.* **13:**659–677.

McCrossin, M. L. 1994a. *The Phylogenetic Relationships, Adaptations, and Ecology of Kenyapithecus*. Ph.D. dissertation, University of California at Berkeley.

McCrossin, M. L. 1994b. Semi-terrestrial adaptations of *Kenyapithecus*. *Am. J. Phys. Anthropol. Suppl.* **18:**142–143 (Abstr.).

McCrossin, M. L., and Benefit, B. R. 1993a. Recently recovered *Kenyapithecus* mandible and its implications for great ape and human origins. *Proc. Natl. Acad. Sci. USA* **90:**1962–1966.

McCrossin, M. L., and Benefit, B. R. 1993b. Clues to the relationships and adaptations of *Kenyapithecus africanus* from its mandibular and incisor morphology. *Am. J. Phys. Anthropol. Suppl.* **16:**143 (Abstr.).

McCrossin, M. L., and Benefit, B. R. 1994. Maboko Island and the evolutionary history of Old

World monkeys and apes. In: R. S. Corruccini and R. L. Ciochon (eds.), *Integrative Paths to the Past: Paleoanthropological Advances in Honor of F. Clark Howell*, pp. 95–122. Prentice–Hall, Englewood Cliffs, NJ.

Martin, L. B. 1986. Relationships among extant great apes and humans. In: B. Wood, L. Martin, and P. Andrews (eds.), *Major Topics in Primate and Human Evolution*, pp. 161–187. Cambridge University Press, London.

Morbeck, M. E. 1983. Miocene hominoid discoveries from Rudabanya: Implications from the postcranial skeleton. In: R. L. Ciochon and R. S. Corruccini (eds.), *New Interpretations of Ape and Human Ancestry*, pp. 369–404. Plenum Press, New York.

Napier, J. R., and Davis, P. R. 1959. The fore-limb skeleton and associated remains of *Proconsul africanus*. *Br. Mus. Nat. Hist. Fossil Mamm. Afr.* **16:**1–69.

Pickford, M. 1982. New higher primate fossils from the middle Miocene deposits at Majiwa and Kaloma, western Kenya. *Am. J. Phys. Anthropol.* **58:**1–19.

Pickford, M. 1985. A new look at *Kenyapithecus* based on recent discoveries in western Kenya. *J. Hum. Evol.* **14:**113–143.

Pilbeam, D. R., and Simons, E. L. 1971. Humerus of *Dryopithecus* from Saint Gaudens, France. *Nature* **229:**408–409.

Pilbeam, D. R., Rose, M. D., Barry, J. C., and Shah, S. M. I. 1990. New *Sivapithecus* humeri from Pakistan and the relationship of *Sivapithecus* and *Pongo*. *Nature* **348:**237–239.

Rose, M. D. 1993. Locomotor anatomy of Miocene hominoids. In: D. Gebo (ed.), *Postcranial Adaptation in Nonhuman Primates*, pp. 252–272. Northern Illinois University Press, De Kalb.

Rose, M. D., Leakey, M. G., Leakey, R. E., and Walker, A. C. 1992. Postcranial specimens of *Simiolus enjiessi* and other primitive catarrhines from the early Miocene of Lake Turkana, Kenya. *J. Hum. Evol.* **22:**171–237.

Ruff, C. B., Walker, A., and Teaford, M. F. 1989. Body mass, sexual dimorphism and femoral proportions of *Proconsul* from Rusinga and Mfangano Islands, Kenya. *J. Hum. Evol.* **18:**515–536.

Schultz, A. H. 1960. Einige Beobachtungen und Masse am Skelett von *Oreopithecus* im vergleich mit anderen catarrhinen Primaten. *Z. Morphol. Anthropol.* **50:** 136–149.

Simons, E. L., and Pilbeam, D. R. 1972. Hominoid paleoprimatology. In: R. H. Tuttle (ed.), *The Functional and Evolutionary Biology of Primates*, pp. 36–64. Aldine, Chicago.

Simons, E. L., and Pilbeam, D. R. 1978. *Ramapithecus* (Hominidae, Hominoidea). In: V. J. Maglio and H. B. S. Cooke (eds.), *Evolution of African Mammals*, pp. 147–153. Harvard University Press, Cambridge, MA.

Spoor, C. F., Sondaar, P. Y., and Hussain, S. T. 1992. A hominoid hamate and first metacarpal from the Late Miocene Nagri Formation of Pakistan. *J. Hum. Evol.* **21:**413–424.

Tuttle, R. H. 1967. Knuckle-walking and the evolution of hominoid hands. *Am. J. Phys. Anthropol.* **26:**171–206.

van Roosmalen, M. G. M., Mittermeier, R. A., and Fleagle, J. G. 1988. Diet of the bearded saki (*Chiropotes satanas chiropotes*): A neotropical seed predator. *Am. J. Primatol.* **14:**11–35.

Ward, C. V. 1993. Torso morphology and locomotion in *Proconsul nyanzae*. *Am. J. Phys. Anthropol.* **92:**291–328.

Ward, S. C., and Brown, B. 1986. The facial skeleton of *Sivapithecus indicus*. In: D. R. Swindler and J. Erwin (eds.), *Comparative Primate Biology, Vol. 1: Systematics, Evolution and Anatomy*, pp. 413–452. Liss, New York.

Ward, S. C., and Pilbeam, D. R. 1983. Maxillofacial morphology of Miocene hominoids from Africa and Indo-Pakistan. In: R. L. Ciochon and R. S. Corruccini (eds.), *New Interpretations of Ape and Human Ancestry*, pp. 211–238. Plenum Press, New York.

White, T. D., Suwa, G., and Asfaw, B. 1994. *Australopithecus ramidus*, a new species of early hominid from Aramis, Ethiopia. *Nature* **371:**306–312.

Zapfe, H. 1960. Die Primatenfunde aus der miozanen Spaltenfullung von Neudorf an der March (Devinska Nova Ves), Tschechoslowakei. Mit anhang: Der Primatenfund aus dem Miozan von Klein Hadersdorf in Niederosterreich. *Schweiz. Palaeontol. Abh.* **78:**1–293.

The Taxonomy and Phylogenetic Relationships of *Sivapithecus* Revisited

13

STEVE WARD

Introduction

Sivapithecus was resident in South Asia for over 5 million years. Its first appearance (ca. 12.7 Ma) and its last occurrence (ca. 6.8 Ma) in Siwalik sediments are likely to have coincided with global climatic and tectonic/eustatic events. Based on all available material, mostly limited to teeth, it appears that *Sivapithecus* experienced little significant anatomical change throughout this long period of time, nor is there evidence that the genus was extensively speciose. We do know that it was not a very common taxon in the Siwaliks. Assuming limited taphonomic bias in collecting, the genus probably never represented more than 1% of the overall mammalian community. For these reasons, the taxonomic status of *Sivapithecus* is still very much an open question, and we still have a rather limited understanding of the functional biology of its masticatory and locomotor systems. However, important new specimens attributable to *Sivapithecus* have been recovered over the last several years from Pakistan, and these new fossils, in conjunction with intensive pro-

STEVE WARD • Department of Anatomy, Northeastern Ohio Universities College of Medicine, Rootstown, Ohio 44272.

Function, Phylogeny, and Fossils: Miocene Hominoid Evolution and Adaptations, edited by Begun *et al.* Plenum Press, New York, 1997.

grams of magnetostratigraphic and sedimentological correlation have done much to narrow the focus on discussion. Nevertheless, neither the fossils nor the contextual data have brought consensus to the question of the relation ships of *Sivapithecus* to other Miocene large hominoid taxa outside of South Asia, or to the living great apes and humans. It is my intention here to evaluate the placement of *Sivapithecus* within the Hominoidea using all available fossil material, including some new specimens not yet formally described.

Background

Until recently, it was *au courant* when reviewing the history of modern paleoanthropology, to hold that Elwyn Simons "resurrected" a hypothesis proposed by Lewis in 1932 that some smaller Miocene hominoid species from the Siwaliks were plausible hominid ancestors. What subsequently came to be known as the *Ramapithecus* debate had a profound and lasting effect on paleoanthropology. Many of the contributors to this volume were drawn to the field by this debate, and most others were directly involved. While the outcome of the debate can be said by some to be unresolved, the way morphology is used in phylogeny reconstruction, the application of engineering principles to skeletal biology, the extension of molecular genetic approaches in primate phylogenetic reconstruction, and how fieldwork is planned and executed were all profoundly affected by this discussion. Between 1960 and the mid-1970s the debate surged, driven in part by tensions in the area of taxonomic methodology, and by intense interest in hominid discoveries in East Africa. Then, in 1979 a pivotal event in the *Ramapithecus* debate ensued: the recovery by David Pilbeam's group of a partial face that seemed to definitely resolve the issue. As is now well known, GSP 15000 appeared to share a set of patently derived cranial/facial features with modern orangutans. While the *Pongo*-like features were a surprise, Pilbeam and others were already by this time coming to doubt the hominid status of *Ramapithecus*, by then referred to the appropriate nomen *Sivapithecus*. Since 1979, GSP 15000, as well as other hominoid fossils from the Siwaliks have been subjected to intense scrutiny. Many new avenues of inquiry emerged in paleoanthropology as a result. Much of the impetus behind the analysis of primate enamel ultrastructure was originally the result of interest in the Siwalik fossils (Gantt *et al.*, 1977; Martin, 1985). Detailed comparative anatomical assessments of the *Sivapithecus* maxilla, especially the subnasal region (Andrews and Cronin, 1982; Ward and Kimbel, 1983; Ward and Pilbeam, 1983), the postcranial skeleton (Rose, 1983), mandibular anatomy (Brown, 1989), and life-history parameters (Kelley *et al.*, 1995) have done much to document the evolutionary relationships of *Sivapithecus*. Throughout this period, all workers directly involved with the original material adhered to a general consensus that *Sivapithecus* represented a sister taxon to modern orangs, or otherwise embraced the robusticity of a *Sivapithecus–Pongo* clade. While some of us were less committed to this view than others, we felt that the characters present in the facial skeleton offered a compelling case for a *Sivapithecus–Pongo* clade, while at the same time recognizing that *Sivapithecus* possessed many features that were clearly not shared with modern orangs at all. As we noted on several occasions, *Sivapithecus* was not a Miocene orangutan.

Ironically, exactly 10 years after the recovery of GSP 15000, two new fossils were found in Pakistan that have fomented a new period of revisionism in the systematics of *Sivapithecus*. Two partial humeri, one from the very productive Sethi Nagri locality (Y311), and the other from the nearby Chinji-level locality Y76 near Bhilomar, were collected during the 1989 field season. While both humeri lacked their heads, their distal joint surfaces were sufficiently complete to attribute them to *Sivapithecus*. The distal humeral morphology of *Sivapithecus* is quite apelike, with a globular capitulum, and spool-like trochlea. It came as a surprise when the proximal half of the new humeri were similar in overall form to large cercopithecoids. From the deltopectoral crest to the anatomical neck, they clearly lacked any of the derived humeral morphology shared by all living great apes and humans.

The implications of these observations (Pilbeam *et al.*, 1990) cannot be underestimated. Correctly identifying the phylogenetic relationships of *Sivapithecus* is not a trivial exercise. As has been pointed out by a number of investigators, should *Sivapithecus* be unambiguously identified as a sister taxon of the living orangutan, several important consequences must follow. First, orangutan ancestry would be anchored to middle Chinji levels in Pakistan, particularly at locality 750 (Kappelman *et al.*, 1991). In theory, it would therefore be possible to calibrate hominoid phylogenetic "clocks." Based on restriction mapping, DNA hybridization, and other approaches, an arbitrary 12.5 Ma "start" for a *Pongo* clade would permit coarse estimates of Gorilla–chimpanzee–human divergence times, although more precise issues, such as the possible occurrence of di- and trichotomies, would not be illuminated. Also deriving from the long estrangement of the South Asian large hominoid radiation from presumed African ancestors is the progress of homoplastic morphologies, particularly in the structure of the proximal humerus and scapulothoracic regions. Did in fact the loss of medially torqued and posteriorly retroflexed proximal humeri, and their replacement by columnar shafts and globular heads with low tuberosities occur as a unitary event, or independently two, three, or even more times? And how do these new specimens influence the interpretation of craniodental evidence? Pilbeam (1996) has recently noted with a certain amount of justification that paleontologists are often "craniophilic" in their approach to phylogeny reconstruction. There is a tendency to rely on skulls and teeth as the more highly valent pool of evidence. Should in fact the anatomy of the face by considered immune from homoplastic change? Clearly not. Finally, operating on the assumption of strict monophyly in post-22-Ma large hominoids, the issue of African taxa plausibly ancestral to a *Sivapithecus–Pongo* clade remains curiously and vexingly open. In order to make progress in resolving these questions, existing fossil collections from all parts of the Old World Miocene should be reexamined, and heretofore seldom-used character complexes must be carefully assessed, particularly for their homoplastic potential. While *Sivapithecus* has been as exhaustively studied as any Miocene hominoid, there are elements of the face, permanent and deciduous dentition, and postcranium that are currently under review by myself and a number of my colleagues. The results of these inquiries show promise in elucidating on anatomical grounds, the relationships of *Sivapithecus* and the living orangutan.

Sivapithecus in Context: Stratigraphic Range, Siwalik Vegetation History, and Community Structure

The earliest known occurrence of *Sivapithecus* in the Siwaliks is at Harvard-GSP locality 750 situated in middle Chinji sediments approximately 10 km west of Chinji Village. Two maxillary molars and a molar fragment were recovered from this locality and are clearly attributable to *Sivapithecus* based on criteria used to assign all other large hominoid isolated teeth from the Siwaliks to this genus. Using magnetostratigraphic techniques, Kappelman *et al.* (1991) placed this locality in chron 5A.r-1 with an original age estimate of 12.46–12.49 my. Recent revisions in the Geomagnetic Reversal Time Scale (GRTS) have increased this estimate to a lower date of approximately 12.75 my. Despite concerted and intensive collecting in middle to lower Chinji exposures, during the 1980s we have been unable to document the occurrence of *Sivapithecus* in Pakistan earlier than this date. It is becoming increasingly evident then that locality 750 marks the period very close to the time *Sivapithecus* appeared in South Asia.

Unlike it earliest appearance, the last occurrence of *Sivapithecus* in the Siwaliks is somewhat more ambiguous. There are several upper Dhok Pathan specimens that occur late in the Siwalik late Miocene near Khaur Village in Pakistan (von Koenigswald, 1983) but the youngest material is probably from Haritalyangar in the Indian Siwaliks. The revised GRTS suggests a latest date of between 8.0 and 8.5 my. It has been suggested that there are problems with the Haritalyangar correlations (Kelley and Pilbeam, 1986), and there is no doubt that a last occurrence datum for *Sivapithecus* in South Asia has been the subject of some confusion. The Haritalyangar sediments have not benefited from the more rigorous sampling program that has progressed continuously in the Pakistani Siwaliks for almost two decades and until such a program is executed a last occurrence of about 7.0 Ma is probably the most reasonable estimate. The last occurrence of large hominoid *sensu stricto* appears to be *Gigantopithecus bilaspurensis,* at ca. 6.3 Ma. However, there is a general consensus that this form is not attributable to *Sivapithecus.*

Habitats occupied by *Sivapithecus* are difficult to reconstruct. For a number of reasons, plant fossils are rare in the mid to late Miocene Siwaliks sequence. There are, however, some indirect approaches that reveal some broad details of habitat composition. The vegetation history of the Pakistani Siwaliks throughout the stratigraphic range of *Sivapithecus* has been the subject of intensive scrutiny over the last several years, especially as it relates to patterns of global climatic change, regional tectonic history, and the history of regional faunal change. The evidence marshaled to bring into focus the overall floral communities in which *Sivapithecus*, as well as the bulk of the Siwalik fauna lived, derives from several sources: dental anatomy, dental wear, and enamel microwear of browsing and grazing taxa, and more recently, analysis of stable carbon isotopes in paleosols and in the apatite of mammalian dental enamel.

During the mid to late 1970s, when "*Ramapithecus*" was widely viewed as an early hominid, broadly deterministic models of human origins current at that time mandated that "hominids" were creatures of dry open savanna woodlands. Thus, it was natural for the short-snouted, obtusely cusped, and thick-enameled *Ramapithecus* to be associated with hypothetical grasslands in the middle to late Siwalik Miocene. Kay's 1982 paper on "ramapithecine" masticatory biology was instrumental in showing that the dentognathic anatomy of *Ramapithecus* need not necessarily be associated with open-country small-object feeding, but rather could be equally well suited to the exploitation of a variety of food resources, including nuts. Following the dental emancipation of *Ramapithecus* (now almost universally referred to as *Sivapithecus*), Kappelman (1988) building on the work of Badgley and Behrensmeyer (1980) and their colleagues showed that the predominant habitat of *Sivapithecus* for most of its 5-my tenure in the Siwaliks was largely forest and woodland. This conclusion was based on several lines of evidence, all of it indirect: taphonomic reconstruction of the localities themselves, assessment of the sedimentary history of the major formational units, and careful documentation of the composition of the macro-and microfaunal elements in the recovered assemblages. Detailed work on postcranial anatomy of the bovids (Kappelman, 1988) and dental microwear of the giraffids (Morgan and Solunias, 1990) began to suggest the presence of patchy local habitats, but with forests of various types dominating the overall picture. The role of grasses in habitat structure became uncertain. At this point it was becoming clear from several lines of evidence that faunal change in the Siwaliks was related to broader patterns of climatic change in the circum Indian Ocean basin, and that fluctuating ratios of carbon isotopes in Siwalik paleosols might yield evidence concerning the composition of floral communities. One hypothesis that emerged in the mid to late 1980s posited the development of marked seasonality in the Siwalik region including the evolution of monsoonal systems. The rationale here was that ongoing orogeny following the Indian plate's collision with the Asian continental crust lifted the Tibetan Plateau to its current height by 10–7 Ma triggering monsoonal atmospheric cycles. In an important series of papers, Cerling *et al.* (1989, 1993) and Quade *et al.* (1989) identified a strong isotopic signal in Siwalik paleosols suggesting a major shift in plant communities from predominantly C_3 to C_4 photosynthetic pathways.

This shift occurred at about 7.0 Ma, and is consistent with a change from warmer, more humid forests and woodlands, to drier, more open grasslands. Changes in stable oxygen isotopes also supported this model. It was the consensus of geochemists working both with continental as well as with oceanic benthic sediments that the development of the Asian monsoon near the end of the Miocene best accounted for these findings. Further, the composition and patterns of turnover in terrestrial mammalian taxa seemed to generally support this model. For example, Leiberman (1993) has shown that by middle Chinji times, many bovids have distinct hypoplasias in their dental cementum suggestive of seasonal changes in deposition of this tissue. The development of seasonality in South Asia can be shown to be linked with certain faunal events. However, two recent reports suggest that our understanding of Miocene Siwalik habitats that *Sivapithecus* occupied for the last half of the Miocene is not as clear-cut as previously thought.

With respect to a definitive shift from predominantly C_3 to C_4 plant communities, Morgan *et al.* (1994) have provided evidence suggesting that Siwalik plant communities were more diverse through time than earlier isotopic analyses had suggested. Morgan and her colleagues analyzed carbon isotopes in the enamel apatite sampled from seven large mammalian families found in the middle to late Miocene in both Kenya and Pakistan. While they did find an isotopic signal in Pakistan supporting dominance of C_4 grasses after 7.0 Ma, they found a C_4 signal at 9.4 Ma as well. This finding strongly suggests a pattern of local environmental "patchiness" or heterogeneity than indicated by Quade's and Cerling's work. Further, the data from Kenya show a far less distinct transition from C_3 to C_4 habitats, suggesting that broad generalizations concerning Indian Ocean basin climate dynamics are fraught with uncertainty. Finally, in a recent analysis of Himalayan fracture zones, Coleman and Hodges (1995) have found that uplift of the Tibetan Plateau was already close to its definitive altitude by about 14 Ma, much earlier than usually accepted. This would mean that if regional climatic patterns, particularly the evolution of seasonality and the monsoon, were in part driven by Himalayan orogeny, then we should be seeing evidence of this in the Siwalik isotopic record and by extension, in faunal community structure. We are left at present to conclude that the habitat of *Sivapithecus* in the mid to later Miocene did indeed change through time, with forests giving way to ever-increasing incursions of C_4 grasses some time after 10.0 Ma. It is unlikely, however, that one simple model of climatic and correlative floral change will suffice to clearly define Siwalik paleoenvironments.

Sivapithecus: The Data Base

Given the attention devoted to *Sivapithecus* since the turn of the century, it is remarkable how little we really know of its anatomy. Figure 1 shows sche-

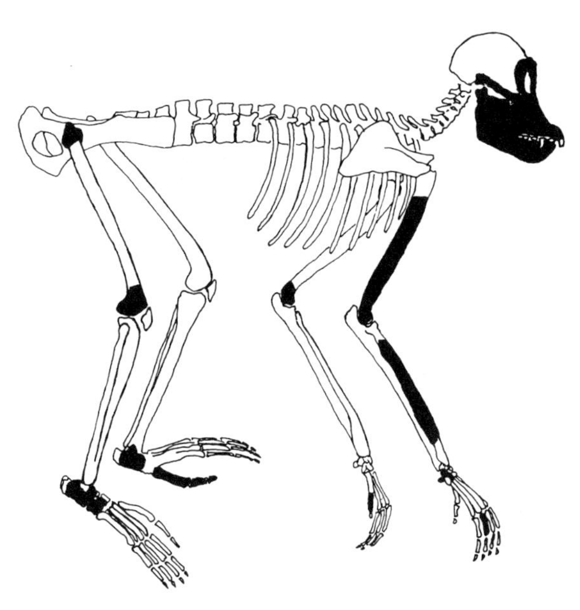

Fig. 1. Schematic representation of the skeletal anatomy of *Sivapithecus*. Known parts are shown in black.

matically that we have detailed knowledge of the adult (and to a lesser extent, the juvenile) mandible, all of the permanent teeth, and about half of the deciduous teeth. We have a good understanding of the maxilla, including size and anatomical variation, but the middle and upper parts of the face are only known from one individual, namely, GSP 15000. We know nothing of the cranial base or vault, nothing at all about the vertebral column, ribs, scapula, and pelvis. In the forelimb, we now have a reasonably good understanding of the humeral shaft and distal humerus, but importantly, not the proximal end. Little is known of the forearm. We have about 70% of one juvenile radius, no ulna, and about half of the carpal bones. There are several metacarpal and manual phalangeal specimens in varying states of preservation. With respect to the forelimb, then, we lack very important functional and phylogenetic information. In the hand, we do not know if there was an os centrale, which *Pongo* retains and other modern large hominoids have fused with the scaphoid. We also know nothing about the ulnar styloid–pisotriquetral complex, or the proximal and distal radioulnar joints. There are no relevant fossils to allow us to assess possible patterns of thumb reduction, and it is not possible to identify brachial/antebrachial proportions. The situation is little better in the lower extremity. There are some fragmentary proximal femurs, but they are insufficiently preserved to be very useful in functional reconstructions. A very well preserved distal femur was recovered on the Potwar Plateau in 1992 (Pilbeam, personal communication; MacClatchy, 1995) and should prove to be very helpful in establishing some kinematic parameters of the knee joint.

There are bits of femur shaft, but no tibia or fibula. Identification of an intermembral index for any of the species of *Sivapithecus* is thereby forestalled. We are on slightly better terms with the foot. Most of the tarsal elements are known from single individuals, although there is unfortunately no endocuneiform. There are several metatarsal fragments and pedal phalanges. There is also a partial hallux, which is a very useful specimen. Compared with what is currently known about the skeletal anatomy of *Proconsul, Afropithecus,* and what is coming to light about *Kenyapithecus,* this is clearly a limited sample on which to base meaningful comparisons and functional assessments. However, there is sufficient anatomy preserved to permit evaluation of several important joint complexes, dental anatomy and proportions as well as elements of facial form.

Facial and Gnathic Anatomy

The range of shared and presumably derived facial features supporting a *Sivapithecus–Pongo* clade has been extensively discussed and reviewed since the initial description of GSP 15000. The premaxillary alveolar clivus, subnasal/palatal topography, interorbital constriction, frontal invasion by the maxillary sinus, supraorbital rim/temporal line complex, and contact of the postglenoid process to the tympanic tube are all characters that are shared with modern orangs. But what of the rest of the face and mandible? It has also been emphasized elsewhere that the mandibular anatomy of *Sivapithecus* is markedly dissimilar in all essential components (symphysis, corpus, and ramus) to the orangutan mandible (Brown, 1989). The only apparent similarity in the two taxa is the tall and vertical ramus, which correlates strongly with posterior facial height. These relationships probably reflect important topographic and functional relationships in the facial region.

Unlike *Pongo, Sivapithecus* shares with earlier Miocene apes a very long midface segment. The distance between the estimated position of rhinion and glabella in GSP 15000 is remarkably long. Even with the loss of the rostral end of the nasal bones, it is quite clear that the Potwar face had very long nasal bones. This is a profoundly different pattern of facial architecture than is present in most modern apes and all hominids. An illustrative way of comparing large hominoid facial topography is to examine the relationships of the nasal capsule to the floors of the orbits. (Fig. 2) In orangs, rhinion is located at a variable level below the orbital margins. In some female individuals, it lies higher, as in humans. There are some male orangs in which the length of the midface segment defined by a cord between rhinion and a tangent defined by the lowest point of the inferior orbital margins is reminiscent of *Sivapithecus.* This is particularly true of some Sumatran specimens. However, the overall contribution of nasal bones to total facial height in all orangs is considerably less than that for the one *Sivapithecus* specimen where it is possible to evaluate

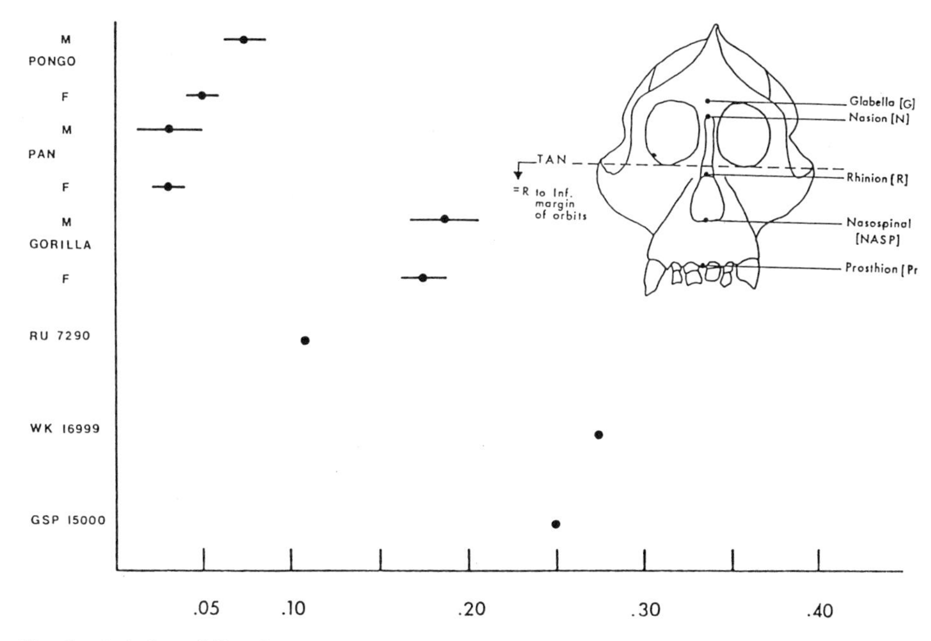

Fig. 2. Relative midface length in large hominoids expressed as a ratio of rhinion–infraorbital tangent length to nasospinale–nasion length. See text for discussion.

the anatomy. The topography of the Mt. Sinap face (MTA 2125) most often attributed to *S. meteai* appears to be very similar to GSP 15000. However, evidence is emerging suggesting that this taxon may in fact represent a different genus (Begun and Gülec, 1995; Begun, personal communication).

Facial height in chimpanzees and gorillas reveals somewhat different arrangements (Fig. 3). Chimpanzees have short nasal bones, both in terms of their absolute length and as a contribution to facial height. Moreover, there is little sexual dimorphism manifest in adult chimpanzees. Gorillas, on the other hand, have long nasal bones, especially the males. When the distance between rhinion and the inferior transorbital line is expressed as a ratio of nasal bone length, male gorillas are more similar to Miocene apes. At present it is unclear to what extent body size plays a role in these relationships, but Begun and Gülec (1995) normalized nasal bone lengths by maxillary second molar breadth for several Miocene hominoid taxa and found a relatively consistent relationship. Not surprisingly, gorillas are dimorphic in this character, although less so than orangs.

The long midface of *Sivapithecus* could represent a primitive feature in hominoids, although in the earliest material such as the *P. heseloni* face from Rusinga (KNM-RU 7290), the nasal bones appear relatively short. This could be a function of body size, a primitive feature, shared with small-bodied apes, or a correlate of unknown functional/adaptive biology. Later in time both the *Moroto polatl* and the *Afropithecus* cranium from Kalodirr have facial propor-

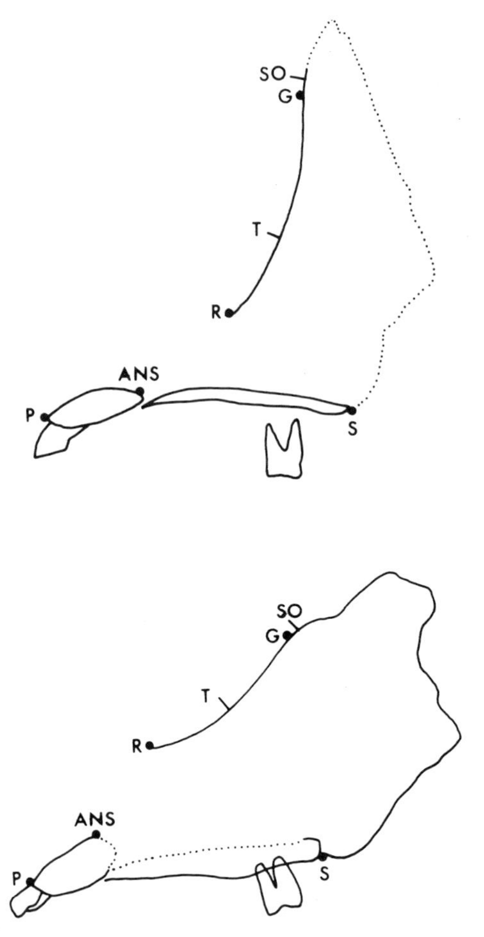

Fig. 3. Face component proportions in *Sivapithecus* (GSP 15000) and *Afropithecus* KNM-WK 16999. For abbreviations designating landmarks, see Fig. 2. Note the separation of rhinion (R) and the infraorbital transverse plane (T).

tions very similar to those of *Sivapithecus,* although overall topographic anatomy is very different. Unfortunately, the form of the nasal bones in *Kenyapithecus* is either unknown, or remains undescribed. Later in the Miocene, both *Ouranopithecus* and *Lufengpithecus* appear to possess more "modern"-appearing midface relationships. The recent revision of Siwalik chronostratigraphy, now placing some *Sivapithecus* specimens earlier in time, may help account for the observed distribution of morphology.

Another element of facial architecture in which *Sivapithecus* and orangs differ involves the disposition of the maxillary sinuses. As has been noted earlier by several workers, orangs have a highly pneumatized face, with the maxillary sinuses encroaching on the canine roots in adults. In *Sivapithecus,* the maxillary sinuses are much more restricted, and from all appearances less aggressive in their invasion of the maxillary body. The maxillary sinus does approximate the canine alveolus anteriorly, and invades the zygomatic process superolaterally but in general antral inflation of the maxilla is not as pronounced as is the case in mature orangs.

The mandible of *Sivapithecus* is remarkable for diversity in size and shape between species and through time, while retaining an overall gestalt characteristic for the genus. While many of these issues have been dealt with elsewhere (Brown, 1989, this volume), it is appropriate here to reiterate earlier observations that of all known cranial parts, the mandible is least similar to *Pongo* than any other part of the skull. The Siwalik mandibles have impressions for the anterior digastric muscles, have a robust but compact inferior transverse torus on the lingual surface of the symphysis, and with few exceptions have relatively broad corpora when compared to height. The most notable similarity with respect to orangs is the possession of tall and vertical mandibular rami. This arrangement is probably associated with the pronounced airorhynchy and deep posterior face that is evidenced in GSP 15000.

One aspect of *Sivapithecus* jaw morphology that has been all but overlooked is the structure of the juvenile mandible. Two specimens are known: GSP 11536, a left corpus and symphysis attributed by Kelley (1988) to *S. parvada* from locality Y311 at Sethi Nagri, and GSP 12709, a corpus with partial rami probably best attributed to *S. indicus* from locality Y260 in the general vicinity of Khaur. The Sethi Nagri specimen is now dated at approximately 10 my, while Y260 located at the "U" sand level is currently dated at 8.8 my. Both specimens retain unerupted teeth, and reasonably good buccal and lingual corpus contours and are of a comparable developmental age based on tooth eruption status. They are notable for their robusticity (Fig. 4). The symphyses show evidence of both inferior and superior tori. The few other Miocene juvenile mandibles that have been extensively described are from Kenya. In terms of size and developmental status, SO 542 from Songhor is fairly similar to the Siwalik specimens. Probably attributable to *Proconsul major*, KNM SO 502 has a bulbous inflated corpus and an internal symphyseal surface with only one prominent torus. Thus, as is the case with *Sivapithecus*, the characteristic symphyseal profile of larger *Proconsul* individuals appears to be established at an early developmental age. The recently described juvenile *Kenyapithecus* mandible from Maboko (McCrossin and Benefit, 1993) shows evidence of two symphyseal tori and a long, sublingual planum. While similar to the two Siwalik juvenile mandibles in overall corpus robusticity, the Maboko juvenile has a much more obliquely inclined symphysis and a longer sublingual planum. In both *Sivapithecus* juveniles, the sublingual plane is shorter as a proportion of lingual symphyseal length, and appears to be thicker as well. Finally, the symphyses of GSP 11536 and GSP 12709 are more vertically disposed than is the case for all known earlier juvenile large hominoid mandibles, with angulation to the alveolar plane of the corpus being 47 and 51°, respectively. These values are well within ranges reported by Brown (1989)

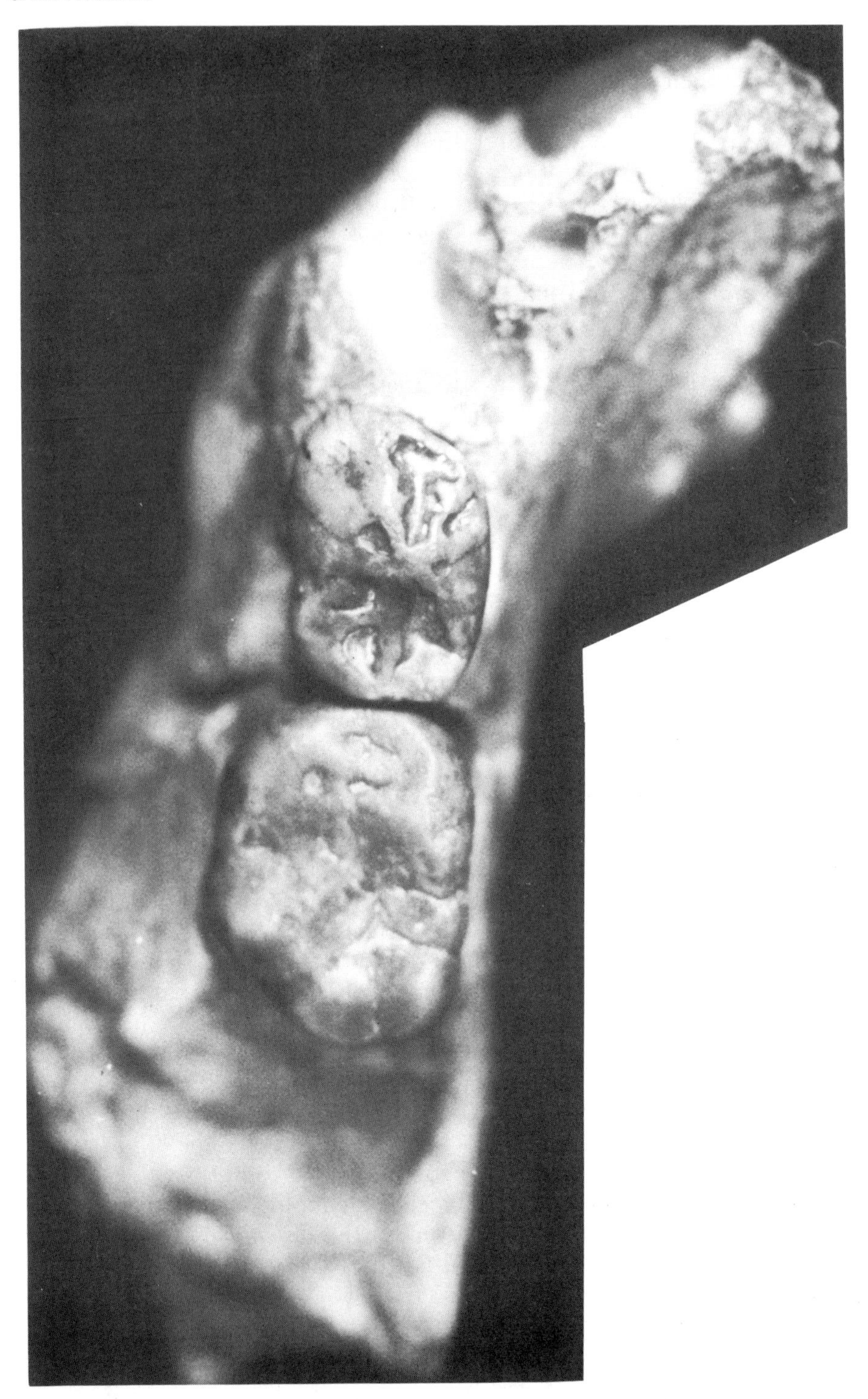

Fig. 4. Occlusal view of GSP 11536, showing the deciduous third and fourth premolars.

for *Sivapithecus* adults, suggesting that this relationship does not change appreciably during postnatal growth and replacement of the deciduous teeth by their permanent successors.

The mandibular rami in the Siwalik juvenile mandibles also provide some clues concerning facial form and growth during the dental eruption cycle. The rami are hafted onto the corpus at a relatively forward position, foreshadowing what we know to be the typical adult condition. The geologically younger of the two Siwalik specimens, GSP 12709, has a right ramus that is intact anteriorly, preserving most of the coronoid process. As is the case in GSP 15000, this ramus is tall and vertically oriented, and not dissimilar to the ramal relationships in *Pongo* at this stage of development. While still a matter of speculation, it is possible that the corpus–ramus topography preserved in the mandible of GSP 12709 is a manifestation of a facial growth trajectory that results in the pronounced airorhynchy shared by *Sivapithecus* and *Pongo*.

The dentition of *Sivapithecus* has been the focus of interest for over 100 years. Most of this attention has been directed to the permanent dentition, with the small but interesting deciduous tooth sample never having benefited from detailed analysis. The sample consists of a dP3 and a dP4 from GSP 11536, a right mandibular canine and right dP3 from GSP 12709, and a probable dM2 (GSP 17919). While limited in scope, the elements that are preserved are quite useful. The canine is small, conical, and remarkable for the amount of cingulum formation present. There is swelling all around the base of the crown, with an excavated area on the labial surface, a pronounced ridge lingually, and a heel distally. There are few deciduous Miocene canines with which to compare the Siwalik specimen, but it appears that the *Sivapithecus* deciduous canine is more robust than those of *Proconsul*, particularly the juvenile from Meswa Bridge (KNM ME-2).

The deciduous third premolars from the two Pakistani juvenile mandibles are similar in overall morphology, but differ in size. They both have expanded trigonids, with distinct metaconids. The metaconids are connected to the taller, more massive protoconids by a short protolophid. Two long crests, the preprotocristid and premetacristid, extend from their respective cusps mesially to the mesial marginal ridge, delimiting large mesial foveae. There is very limited extension of the protoconid onto the mesial root. The talonids are unremarkable, bearing a small hypoconid and vestigial swelling corresponding to entoconids. There are certain similarities between the Siwalik dP3's and the occlusal morphology of the *Kenyapithecus* dP3 described by McCrossin and Benefit (1993). The Maboko specimen, and *Griphopithecus* from Turkey also have distinct metaconids, with a consequent oval to rectangular crown shape, and show limited extension of the protoconid onto the mesial root. As pointed out by McCrossin and Benefit, the *Sivapithecus* dp3's differ from the middle Miocene hominoid, and are similar to *Pongo* in the expansion of the mesial fovea, and length of the mesial crests (Fig. 5).

The dP4 of *S. parvada* (GSP 11536) is remarkable for its similarity to modern orangs. The crown is completely molariform, with the talonid being

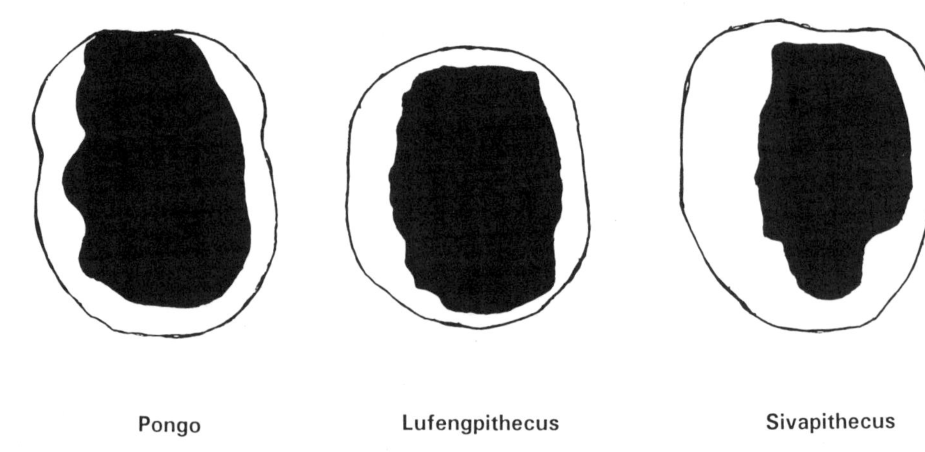

Fig. 5. Molar occlusal fovea area for mandibular second molars. See text for discussion.

only minimally wider than the trigonid. The trigonid represents about one-third of total crown surface area. The arcuate shape of the paracristid, which connects the protoconid and metaconid along the mesial margin of the crown, is essentially identical to the condition described by Swarts (1988) for *Pongo*. The hypoconid is massive, the cristid obliqua is strongly developed, and the well-formed hypoconulid is in a distal position. The entoconid is also strongly expressed and an accessory tubercle is situated between it and the metaconid, foreshadowing similar accessory cusp formation in adult *S. parvada*. The Siwalik dP4 differs from *Proconsul* in the expansion of its trigonid region and central location of the hypoconulid. It differs from *Kenyapithecus* in having an expanded talonid basin, and more massive primary cusps.

As noted previously, the permanent dentition of *Sivapithecus* has been the subject of extensive descriptive treatment for over 100 years. This includes general descriptions of crown morphology, metrics, enamel thickness, and subocclusal morphology. There are some elements of anterior tooth morphology and size relationships that are reminiscent of *Pongo*, while for the most part, the postcanine teeth show more anatomical resemblance to *Australopithecus* (particularly *A. afarensis*) and *Ouranopithecus* than anything else. The exceptions are of some interest, however, and will be considered below.

The maxillary incisors of GSP 15000 have been noted to be quite heteromorphic, and this has been viewed as a robust character linking *Sivapithecus* and *Pongo*. A new specimen from Sethi Nagri (GSP 46460) reported by Kelley *et al.* (1995) ostensibly supports this contention. The specimen was recovered *in situ* and consists of unerupted maxillary central and lateral incisors. Given the size of the teeth, it is clearly attributable to the largest *Sivapithecus* species, *S. parvada* Kelley. The two incisors are strongly heteromorphic. However, as Kelley *et al.* (1995) have shown, there are fossil and living hominoids that evince a greater degree of size disparity in maxillary incisor proportions.

Thus, while there are morphological features present in GSP 46460 that are plausibly shared with *Pongo,* significant size disparity between I1 and I2 could well prove to be a primitive feature in middle to late Miocene hominoids.

The lingual morphology of the two incisors comprising GSP 46460 is also instructive, and perhaps more useful in considerations of *Sivapithecus* systematics. The arrangement of the basal tubercle on the central incisor, as well as the form and distribution of the lingual enamel crenulations are more similar to *Pongo pygmaeus* than they are to any other large hominoid. This was somewhat of a surprising finding, as most other *Sivapithecus* maxillary incisor morphology differs from that of modern orangs in several important respects. Whether the pattern present in *S. parvada* is a function of the very large size of this species, or is indicative of a shared–derived relationship with modern orangs remains to be determined.

Molar morphology does not provide any real evidence supporting a possible *Sivapithecus–Pongo* clade. But while the occlusal anatomy of *Sivapithecus* permanent molars does not appear to bear any similarity at all to *Pongo,* the molars of *Lufengpithecus* do. Orangs and *Lufengpithecus* have flattened molar crowns, peripheralized cusp apices, and seem to share a florid pattern of enamel wrinkling. Analysis of these and other characters (Ward *et al.*, 1991) confirmed significant differences in the molar anatomy of *Sivapithecus* and *Pongo,* but likewise raised questions about the apparent similarities between orangs and the Lufeng teeth. The most notable distinction between *Sivapithecus* and the other two forms involves the relative size of the molar occlusal foveae. When calculated as a ratio of total crown area, the Siwalik molars have restricted foveae, amounting to about 65% of crown area. In orangs and *Lufengpithecus,* the occlusal foveae are relatively large, representing over 80% of total crown area. The combination of muted cusp contours and occlusal crenulations appears to establish a strong phenetic relationship in the Lufeng teeth and that of subfossil and modern orangs. However, careful study reveals that both the primary and secondary fissues in Lufeng molars are significantly shallower than those of *Pongo,* in which the crenulations extend to a depth close to the dentinoenamel junction. The fissures in *Sivapithecus* can be quite prominent in unworn teeth, but they are coarser overall, and are even shallower than those of *Lufengpithecus.* Finally, based on a limited sample of 14 maxillary and mandibular *Lufengpithecus* molars, three-dimensional morphometrics derived from high-resolution laser scanning shows evidence of intracrown differences in topography and cusp relationships. While still the subject of ongoing investigation, a hypothesis positing parallelism/convergence in molar occlusal design accounting for orang and *Lufengpithecus* similarities in molar occlusal design could well be supported.

Overall, evidence provided by the dentition and mandible provides scant support for a *Sivapithecus–Pongo* clade. The same can be said for midface proportions and pneumatization. However, some aspects of this evidence, particularly deciduous premolar morphology, and maxillary incisor heteromorphy can be considered supportive of a sister taxon relationship although as previously noted, heteromorphic maxillary incisors can also be viewed as a large hominoid synapomorphy. The question now becomes, how robust are these characters, and more importantly, to what extent are they reflections of parallel/convergent trends in hominoid evolution?

Postcranials

As noted previously, the record of *Sivapithecus* postcranial anatomy is quite limited. The most significant assessments of this material have been published by Rose (1983, 1986, 1988, 1989). Other important contributions have been made by Langdon (1986) and Spoor *et al.* (1991). The recovery of the new humeri (Pilbeam *et al.*, 1990) and preliminary assessment of the distal femur (MacClatchy, 1995) have increased our ability to reconstruct basis elements of locomotor function. More detailed analyses of these specimens are under way, and a major review of all postcranial elements attributable to *Sivapithecus* is in progress (Madar, 1994 and personal communication). However, some general details are now coming into focus.

First, there is presently no evidence that *Sivapithecus* was significantly orthograde in its positional behavior. While additional forelimb and vertebral specimens would be helpful here, the forelimb, carpal, and hand anatomy presents a complex array of features that makes comparisons with any one extant taxon almost meaningless. Certain aspects of proximal and distal humeral morphology, to the extent they are known, mirror earlier Miocene taxa in having more primitive anthropoid features proximally, with more modern anatomy in the elbow region. Functionally, there is evidence of stability in all of the joints in the regions of the elbow, as suggested by the depth of the zona conoidea and trochlea, but there is also some evidence of an increased range of pronation and supination through a wide range of flexion and extension of the elbow (Rose, 1989; Madar, 1994). Madar's innovative analysis of the geometric properties of GSP 30754 has provided new and interesting insights into the functional milieu of this humerus. She showed that the maximum mediolateral (ML) bending strength of the Sethi Nagri humerus is greater than the anteroposterior (AP) bending strength. This pattern also occurs in gorillas, macaques, and hylobatids; orangs, chimpanzees, and humans have greater humeral bending strength in the AP plane. Madar suggests that the modern apelike elbow region of *Sivapithecus,* in association with geometric properties (overall torsional and bending strength greater than for living hominoids) could hint at a "novel behavioral adaptation on primitive proximal humeral morphology" (Madar, 1994). It should be noted, however, that the Sethi Nagri humerus (GSP 30754) is very robust, and this probably influences the results of sectional/geometric analyses of the shaft. A more gracile humerus from locality Y76 (GSP 30730) is unfortunately too crushed for a reliable geometric assessment Moyà-Solà and Köhler (1996) claim to have reconstructed a cast of this specimen, and have suggested on the basis of this reconstruction two distinctive locomotor patterns for *Sivapithecus:* climbing and terrestrial preferences for the larger species (*S parvada*), and a higher frequency of arboreal/branch walking for the smaller species, *S. sivalensis.* At present, the notion that two disparate locomotor behavioral patterns characterized *Sivapithecus* is not widely accepted, largely because of lack of supporting evidence from other elements of the postcranium. However, other primates, such as lemurs, can have very different locomotor repertoires and substrate preferences while sharing a common anatomical ground plan (Ward and Sussman, 1979). Thus, the recovery of additional postcranial material from all *Sivapithecus* species is essential.

In the hand, the capitate and hamate are reported by Rose (1994) to be consistent with extensive mobility in the midcarpal joint complex, although certain components of hamate morphology suggest limited extension. The fingers and toes both suggest that powerful grasping was an important function of the cheridia. The new distal femur is still under analysis, but its condylar morphology is consistent with a broad range of flexion/extension, with the possibility of notable tibial rotation in the knee joint complex. Finally, several new phalanges are currently under analysis (Kelley, personal communication) and it can be expected that more information will soon be available concerning the functional anatomy of the hand and foot.

The known postcranial anatomy of *Sivapithecus* supports an overall skeletal reconstruction of a large primate with a deep and narrow chest, robust forelimbs, and cheridia with mobile carpal and tarsal regions, and grasping pollical and hallucial elements. Given the rather large range of body size within and between *Sivapithecus* species, it will continue to be difficult to clearly identify a categorical locomotor reconstruction. Moreover, it is distinctly possible that modern analogues do not exist as such. Continued analysis of currently known specimens using bioengineering approaches should prove very useful in pursuing these questions, but until evidence of complete joints such as the shoulder, hip, and proximal carpal is recovered, more questions than answers will remain.

Sivapithecus: Taxonomic and Phylogenetic Problems

Like most other genera comprising the Siwalik faunal assemblage, *Sivapithecus* was in all likelihood an immigrant taxon (Barry *et al.*, 1985). If this assumption is true, it is unknown from whence *Sivapithecus* or its antecedents may have come. The best current guess posits incursion of *Sivapithecus* into South Asia from the west, shortly after 13 Ma. Alternatively, some of the paleoecological evidence suggests a pattern of east-to-west vegetation change that would favor migration of forest forms, such as *Sivapithecus,* from East Asia toward the developing Himalayan forefront. Other scenarios are of course possible, but the essential fact remains that *Sivapithecus* has no unambiguous ancestors. Nor does it appear to share many features with any other Miocene hominoid other than *Ankarapithecus*. The latter form is currently undergoing revision, involving reassessment of earlier material from Sinap (heretofore referred to the nomen *Sivapithecus meteai*). At present it is not certain whether *Ankarapithecus* will emerge as a sister taxon of *Sivapithecus* or alternatively prove to be linked to later taxa in Greece and Europe.

The only known African taxon that could conceivably be related to *Sivapithecus* is *Kenyapithecus*. McCrossin and Benefit (1994) have made an interesting case for eliminating *Kenyapithecus* from the ancestry of modern great apes and humans. However, that argument hinges on the assumption that the two species of this middle Miocene genus, *K. africanus* and *K. wickeri*, are in fact both species of the same genus. Some workers have expressed doubt that this is the case (Harrison, 1992). *K. wickeri*, in the few facial parts preserved from Fort Ternan, does bear some similarity to smaller *Sivapithecus* material, and indeed, this was an important factor in attributing the relevant Fort Ternan fossils to *Ramapithecus*. Nevertheless, *Sivapithecus*, like *Ouranopithecus*, and *Lufengpithecus* for the present remains an isolated taxon, with no plausible antecedents earlier in the Miocene.

Apart from uncertain ancestry, the occurrence of local evolutionary change in *Sivapithecus* during its 5.5-my tenure in the Siwaliks is poorly understood. This dilemma of course directly impacts species-level taxonomic assignments. In addition, Kelley (1986, 1988) has identified the potential effects of time-averaging in the Siwalik sequence, and how stochastic problems can amplify potential errors in evaluating metric variation patterns, which in turn complicates alpha taxonomy. We have at least one advantage in confronting these problems, and that is the ever-improving calibration of the entire Siwalik sequence. Based on the presently accepted Siwalik time scale, the following picture emerges. There are three reasonably well-documented species of *Sivapithecus* in the Siwaliks. Size is the primary criterion of species assignment. The earliest species is *S. sivalensis*. It is followed by the larger taxon *S. indicus*. There is size overlap between these two taxa. Both forms were apparently present until shortly after 8.0 Ma, when *Sivapithecus* disappears from the Siwalik record. Another species, *S. parvada* Kelley, 1989, is distinguished by its very large size relative to the previously mentioned species, Curiously, *S. parvada* is known from only one locality (Sethi Nagri) and is dated at about 10 Ma. Interspersed throughout the Indo-Pakistani Siwaliks there are scattered specimens that may complicate this basic three-species model. The most intriguing of these is *Sivasimia chinjiensis* Chopra 1983. This is a partial upper molar from Chinji-equivalent levels in the Ramnagar area. The tooth is quite crenulated and has an occlusal fovea that appears much less restricted that those of *Sivapithecus*. In addition, three Siwalik specimens that span much of the stratigraphic range of *Sivapithecus* in Indo-Pakistan were assembled to the hypodigm *S. simonsi* by Kay in 1982. The type, GSI D-298, was recovered from Kundal Nala in late Chinji sediments, while the Domeli mandible (BMNH 15423) from lower Nagri levels and a maxilla from Haritalyangar (GSI D-185) occur later in time. Once again, the primary criterion for species assignment was tooth size, as well as certain components of mandibular corpus size and shape.

No matter how many, or how few species of *Sivapithecus* occupied the Siwaliks through time, a troubling point must be raised. There is an unspoken, but pervasive acceptance of the fact that all of these species were very similar, especially craniodentally, and changed little through time. In reality, we have made broad assumptions about species change in *Sivapithecus*, generally assuming that all species had faces like GSP 15000, and that all of the teeth, mandibles, and postcranials are scaled variants of each other. In reality, we cannot say for certain whether or not the earliest *Sivapithecus* was in fact very similar to the last. My examination of the Chinji maxilla (GSP 16075) caused me to conclude that from what was preserved, the premaxillary/subnasal pattern was the same arrangement present in GSP 15000 (Raza *et al.*, 1983). I see no reason to change that conclusion based on a recent reexamination of the specimen. Nevertheless, while it appears that there was very little evolutionary change in *Sivapithecus* between 12.7 and 6.8 Ma, this assumption is based on less than an ideal sample.

Conclusions

The problems that preclude an unambiguous taxonomic placement of *Sivapithecus* within Hominoidea underscore some fundamental challenges confronting evolutionary systematics. As has been noted by Pilbeam (1996) and Andrews and Pilbeam (1996), it is certainly clear that characters used to assess the affinities of fossil taxa tend to be poorly understood from a functional and/or developmental viewpoint. There is no doubt that the molecular genetics of most of our traits are little understood, and this is obviously a major problem in reconstructing robust phylogenetic hypotheses. Since the discovery of GSP 15000, and until the recovery of the new humeri from Pakistan, the affinities of *Sivapithecus* seemed to be as well established as one could hope for in a fossil primate. A *Sivapithecus–Pongo* clade was concordant with some presumptive divergence chronologies, was geographically salubrious, and served as a "phylogenetic comfort zone" for many workers, as it linked a Miocene genus with a living great ape. For the reasons enumerated above and more recently by Pilbeam (1996), a *Sivapithecus–Pongo* clade is now less assured. Pilbeam (1996) has suggested that almost all Miocene taxa with the exception of *Oreopithecus* and the possible exception of *Ouranopithecus* represent an "arachaic" radiation of large hominoids that are phylogenetically unrelated to modern great apes and humans. There may be evidence to support this notion in recent attempts to calibrate evolutionary rates in selected genes. Other workers fret about the number of reversals that must be invoked to rectify certain phylogenetic trees, or are concerned about the unknown but ever-lurking specter of homoplasy in the evolutionary history of hominoids. Recent work on homeotic genes has suggested that Dollo's law of irreversibility may be struck down, and perhaps the role of homoplasy in complicating phylogeny reconstructions will also be better understood. After a careful reassessment of all of the available evidence relevant to the taxonomic status of *Sivapithecus*, I take the view that a *Sivapithecus–Pongo* clade remains the strongest phylogenetic hypothesis. It satisfactorily accounts for an impressive array of shared features in the face, face-cranial hafting, and deciduous dental anatomy. It accepts a distinctive pattern of postcranial anatomy that from a functional standpoint permits an opportunistic use of multiple habitat and substrate types. And finally it places a large hominoid in South Asia in the late Miocene at a time when significant faunal change occurred, perhaps presaging emigration from the Siwalik region of forest-dwelling taxa. This view can be justifiably criticized as craniophilic. But a *Sivapithecus–Pongo* clade must be refuted by stronger evidence than any that has heretofore been brought forward.

Note Added in Proof

Alpagut *et al.* (1996) have recently reported the recovery of a new partial hominoid skull from the Sinap Formation. The new specimen (AS95-500) is relatively undistorted in the parts preserved, unlike the earlier face from Sinap (MTA 2125). Observations on the new skull by Alpagut *et al.* support the view that *Ankarapithecus* is not a member of a putative *Sivapithecus–Pongo* clade, but is best viewed as an early member of the great ape-human clade.

Acknowledgments

Many colleagues have been instrumental in contributing to my work on *Sivapithecus* over the past 15 years, sharing their ideas, their data, and the joys as well as the pitfalls of fieldwork. For their generosity and inspiration I thank them all, but especially John Barry, Barbara Brown, Will Downs, John Kappelman, Jay Kelley, David Pilbeam, and Mahmood Raza. Michele Morgan shared unpublished data and helpful comments. Finally, to David Begun, Mike Rose, and Carol Ward, the editors of this volume, I extend thanks for their Job-esque patience.

References

Alpagut, B., Andrews, P., Fortelius, M., Kappelman, J., Temizsoy, I., Celebi, H., and Lindsay, W. 1996. A new specimen of *Ankarapithecus meteai* from the Sinap Formation of central Anatolia. *Nature* **382:**349–351.

Andrews, P. 1971. *Ramapithecus wickeri* mandible from Fort Ternan, Kenya, *Nature* **230:**192–194.

Andrews, P. J., and Cronin, J. 1982. The relationships of *Sivapithecus* and *Ramapithecus* and the evolution of the orang-utan. *Nature* **297:**541–546.

Andrews, P. J., and Pilbeam, D. 1996. The nature of the evidence. *Nature* **379:**123–124.

Badgley, C., and Behrensmeyer, A. K. 1980. Paleoecology of Middle Siwalik sediments and faunas, northern Pakistan. *Palaeogeogr. Palaeoclimatol. Palaeoecol.* **30:**133–155.

Barry, J. C., Johnson, N. M., Raza, S. M., and Jacobs, L. L. 1985. Mammalian faunal change in the Neogene of southern Asia and its relation to global climate and tectonic events. *Geology* **13:**367–640.

Begun, D. R., and Gülec, E. 1995. Restoration and reinterpretation of the fossil specimen attributed to *Sivapithecus meteai* from Kaymeak (Hassorien) central Anatolia, Turkey. *Am. J. Phys. Anthropol. Suppl.* **20:**63.

Brown, B. 1989. *The Mandibles of Sivapithecus*. Ph.D. dissertation, Kent State University.

Cerling, T. E., Quade, J., Wang, Y., and Bowman, J. R. 1989. Carbon isotope in soils and paleosols as ecology and paleoecology indicators. *Nature* **341:**138–139.

Cerling, T. E., Quade, J., and Wang, Y. 1993. Expansion of C4 ecosystems as an indicator of global ecological change in the Miocene. *Nature* **361:**344–345.

Chopra, S. R. K. 1983. Significance of recent hominoid discoveries from the Siwalik Hills of India. In: R. L. Ciochon and R. S. Corruccini (eds.), *New Interpretations of Ape and Human Ancestry*, pp. 415–417. Plenum Press, New York.

Coleman, M., and Hodges, K. 1995. Evidence for Tibetan Plateau uplift before 14 myr from a new minimum age for east–west extension. *Nature* **374:**49–51.

Gantt, D. G., Pilbeam, D., and Stewart, G. 1977. Hominoid enamel prism patterns. *Science* **198:**1155–1157.

Harrison, T. 1992. A reassessment of the taxonomic and phylogenetic affinities of the fossil catarrhines from Fort Ternan, Kenya. *Primates* **33:**501–522.

Kappelman, J. 1988. Morphology and locomotor adaptations of the bovid femur in relation to habitat. *J. Morphol.* **198:**119–130.

Kappelman, J., Kelley, J., Pilbeam, D., Sheikh, K. A., Ward, S., Anwar, M., Barry, J. C., Brown, B., Hake, P., Johnson, N. M., Raza, S. M., and Shah, S. M. I. 1991. The earliest occurrence of *Sivapithecus* from the middle Miocene Chinji Formation of Pakistan. *J. Hum. Evol.* **21:**61–73.

Kay, R. F. 1981. The nut-crackers—A theory of the adaptations of the Ramapithecidae. *Am. J. Phys. Anthropol.* **55:**141–151.

Kay, R. F. 1982. *Sivapithecus simonsi,* a new species of Miocene hominoid, with comments on the phylogenetic status of Ramapithecinae. *Int. J. Primatol.* **3:**113–173.

Kelley, J. 1986. *Paleobiology of Miocene Hominids*. Ph.D. thesis, Yale University.

Kelley, J. 1988. A new large species of *Sivapithecus* from the Siwaliks of Pakistan. *J. Hum. Evol.* **17:**305–324.

Kelley, J., and Pilbeam, D. 1986. The dryopithecines: Taxonomy, comparative anatomy, and phylogeny of Miocene large hominoids. In: D. R. Swindler and J. Irwin (eds.), *Comparative Primate Biology, Vol. 1: Systematics, Evolution, and Anatomy*, pp. 361–411. Liss, New York.

Kelley, J., Anwar, M., McCollum, M. A., and Ward, S. C. 1995. The anterior dentition of *Sivapithecus parvada,* with comments on the phylogenetic significance of incisor heteromorphy in Hominoidea. *J. Hum. Evol.* **28:**503–517.

Langdon, J. H. 1986. Functional morphology of the Miocene hominoid foot. *Contrib. Primatol.* **22:**1–255.

Leiberman, D. 1993. Life history variables preserved in dental cementum microstructure. *Science* **261:**1162–1164.

McCrossin, M. L., and Benefit, B. R. 1984. Maboko Island and the evolutionary history of the Old World monkeys and apes. In: R. S. Corruccini and R. L. Ciochon (eds.), *Integrative Paths to the Past: Paleoanthropological Advances in Honor of F. Clark Howell,* pp. 95–121. Prentice–Hall, Englewood Cliffs, N.J.

McCrossin, M. L., and Solunias, N. 1993. Recently recovered *Kenyapithecus* mandible and its implications for great ape and human origins. *Proc. Natl. Acad. Sci. USA* **90:**1962–1966.

MacClatchy, L. M. 1995. Postcranial adaptations in Miocene hominoids. *J. Vertebr. Paleontol.* **IV**(*Suppl.* 3) 41A.

Madar, S. 1994. Humeral shaft morphology of *Sivapithecus. Am. J. Phys. Anthropol. Suppl.* **20:**140.

Martin, L. 1985. Significance of enamel thickness in the hominoid evolution. *Nature* **314:**260–263.

Morgan, M. E., and Solunias, N. 1990. Reconstruction of Miocene paleoenvironments in the Siwaliks from tooth microwear analysis of herbivores. *J. Vertebr. Paleontol.* **IV**(*Suppl.* 3)**:**36A.

Morgan, M. E., Kingston, J. D., and Marino, B. D. 1994. Carbon isotopic evidence for the emergence of C4 plants in the Neogene from Pakistan and Kenya. *Nature* **367:**162–165.

Moyà-Solà, S., and Köhler, M. 1996. A *Dryopithecus* skeleton and the origins of great-ape locomotion. *Nature* **379:**156–159.

Pilbeam, D. 1996. Genetic and morphological records of the Hominoidea and hominid origins: A Synthesis. *Mol. Phylogenet. Evol.* **5:**155–168.

Pilbeam, D., Rose, M. D., Barry, J. C., and Shah, S. M. I. 1990. New *Sivapithecus* humeri from Pakistan and the relationship of *Sivapithecus* and *Pongo. Nature* **348:**237–239.

Quade, J., Cerling, T. E., and Bowman, J. R. 1989. Development of Asian monsoon revealed by marked ecological shifts during the last Miocene in northern Pakistan. *Nature* **342:**163–165.

Raza, S. M., Barry, J. C., Pilbeam, D., Rose, M. D., Shah, S. M. I., and Ward, S. C. 1983. New hominoid primates from the middle Miocene Chinji Formation, Potwar Plateau, Pakistan. *Nature* **406:**52–54.

Rose, M. D. 1983. Miocene hominoid postcranial morphology: Monkey-like, ape-like, neither, or both? In: R. L. Ciochon and R. S. Corruccini (eds.), *New Interpretations of Ape and Human Ancestry,* pp. 415–417. Plenum Press, New York.

Rose, M. D. 1986. Further hominoid postcranial specimens from the Late Miocene Nagri Formation of Pakistan. *J. Hum. Evol.* **15:**333–367.

Rose, M. D. 1988. Another look at the anthropoid elbow. *J. Hum. Evol.* **17:**193–224.

Rose, M. D. 1989. New postcranial specimens of catarrhines from the Middle Miocene Chinji Formation, Pakistan. *J. Hum. Evol.* **18:**131–162.

Rose, M. D. 1994. Quadrupedalism in Miocene hominoids. *J. Hum. Evol.* **26:**387–411.

Spoor, C. F., Sondaar, P. Y., and Hussain, S. T. 1991. A hominoid hamate and first metacarpal from the Late Miocene Nagri Formation of Pakistan. *J. Hum. Evol.* **21:**413–424.

Swarts, J. D. 1988. Deciduous dentition: Implications for hominoid phylogeny. In: J. H. Schwartz (ed.), *Orangutan Biology,* pp. 263–270. Oxford University Press, London.

von Koenigswald, G. H. R. 1983. The significance of hitherto undescribed Miocene hominoids from the Siwaliks of Pakistan in the Senkenberg Museum, Frankfurt. In: R. L. Ciochon and R. S. Corruccini (eds.), *New Interpretations of Ape and Human Ancestry,* pp. 517–526. Plenum Press, New York.

Ward, S. C., and Kimbel, W. H. 1983. Subnasal alveolar morphology and the systematic position of *Sivapithecus. Am. J. Phys. Anthropol.* **61:**157–171.

Ward, S. C., and Pilbeam, D. R. 1983. Maxillofacial morphology of Miocene hominoids from Africa and Indo-Pakistan. In: R. L. Ciochon and R. S. Corruccini (eds.), *New Interpretations of Ape and Human Ancestry,* pp. 211–238. Plenum Press, New York.

Ward, S. C., and Sussman, R. H. 1979. Correlates between locomotor anatomy and behavior in two sympatric species of *Lemur. Am. J. Phys. Anthropol.* **50:**575–590.

Ward, S. C., Beecher, R., and Kelley, J. 1991. Postcanine occlusal anatomy of *Sivapithecus, Lufengpithecus,* and *Pongo. Am. J. Phys. Anthropol. Suppl.* **12:**181 (abstr.).

Phyletic Affinities and Functional Convergence in *Dryopithecus* and Other Miocene and Living Hominids

14

DAVID R. BEGUN and LÁSZLÓ KORDOS

Introduction

Dryopithecus provides a good case history illustrating the disagreement and confusion over Miocene hominoid systematics described in the introduction to this volume. *Dryopithecus* is currently known from well-preserved maxilla and other portions of the cranium, from larger numbers of isolated teeth, and from well-preserved postcrania, mostly from Hungary (Rudabánya) and Spain (Can Llobateres and Can Ponsic). It is for the most part fossils from Spain and Hungary that permit a more detailed analysis of the functional anatomy and phylogenetic affinities of *Dryopithecus*. In this chapter we will briefly review the anatomy of *Dryopithecus* in comparison with other hominoids. The functional and behavioral implications of this anatomy will be

DAVID R. BEGUN • Department of Anthropology, University of Toronto, Toronto, Ontario M5S 3G3, Canada. LÁSZLÓ KORDOS • The Hungarian Geological Museum, H-1143 Budapest, Hungary.

Function, Phylogeny, and Fossils: Miocene Hominoid Evolution and Adaptations, edited by Begun *et al.* Plenum Press, New York, 1997.

briefly discussed and a phylogenetic alternative chosen on the basis of functional and phylogenetic criteria (see introduction to the volume and below). The implications for the evolution of hominid dietary and positional behavior will be discussed.

Methods

The approach taken here is to combine cladistic systematics with functional anatomy by using the latter, more complex hypotheses (see introduction to the volume), to resolve ambiguities in the patterns detected by cladistic methods. In this study character polarities are assigned by character analysis involving only the outgroup criterion. Characters are identified on the basis of availability, previously demonstrated usefulness, consistency within operational taxonomic units (the taxa analyzed), and decoupling (two or more characters cannot *always* occur together). The characters chosen were present in the largest number of fossil taxa, with all characters codable for *Dryopithecus*. The outgroup was defined based on previous work (Begun, 1992a, 1994) as *Proconsul,* with *Hylobates* and the Cercopithecoidea as additional outgroups to resolve uncertainties or in the event that data were not available from fossils. Character states were collected on original specimens of fossil and extant anthropoids, on casts of a few additional specimens of *Sivapithecus,* and on casts and photographs of *Australopithecus*. Character states are unordered, i.e., characters with more than two states can change in any order. This is to avoid any *a priori* limitation on the direction of evolutionary transformations.

The patterns of relationship consistent with the fewest overall number of character state transitions (maximum parsimony) were calculated using Hennig86 (Farris, 1988). A consistency index (CI) was calculated as the ratio of the number of character states over the total number of character state transitions, with the preferred hypothesis maximizing this index. In addition to the maximum parsimony hypotheses, hypotheses with up to 5% lower CI were also sought from the data, using the branch switching function of Hennig86. This is consistent with the statistical standard of rejecting samples as being derived from different populations if they are less than 5% different in some statistical parameter. Although the assumptions behind the statistical null hypothesis (e.g., normal distribution) do not apply to the consistency index calculation, both are probability statements. Since we know evolution is not parsimonious, we should expect a margin of error with maximum parsimony, and 5% was borrowed from statistics. It may prove necessary in the future to modify this figure (see below).

Homoplasies are identified based on the preferred hypothesis as those character state positions that are inconsistent with the preferred parsimony cladogram. Homoplasies are not identified *a priori. A posteriori,* homoplasies revealed by the parsimony criterion are examined in order to attempt to explain their most likely distributions. Clusters of homoplasies at particular nodes are examined for functional consistency, that is, to determine the extent to which they as a group correspond to a particular function or potential source of selection. Phylogenetic hypotheses within 5% of the maximum CI value are compared to test for functional consistency. Less parsimonious cladograms within this range are considered equally likely as most parsimonious cladograms if functional criteria suggest that the parsimony criterion alone is potentially misleading. This can occur when disproportionate numbers of characters come from single anatomical regions, or when characters are observed never to covary. With multiple maximally parsimonious hypotheses, functional criteria may be used to decide on a single preferred hypothesis. Character states and their polarities in each taxon are listed in Table I. A selection of the cladograms generated by Hennig86 are listed in Figs. 1–3.

European Miocene Hominid Morphology

The European Miocene hominids *Oreopithecus, Dryopithecus*, and *Ouranopithecus* share a large number of characters with *Sivapithecus, Lufengpithecus*, and living hominids not found in any other middle Miocene hominoid or in any early Miocene form (Andrews, 1985, 1992; Andrews and Martin, 1987; Begun, 1992a,b; Delson, 1985). Among other hominoids, *Kenyapithecus sensu stricto (Kenyapithecus wickeri)* shares a few characters with the more derived taxa. These include elongated premolars and molars, labiolingually expanded upper central incisors, reduced P3 heteromorphy, reduced incidence of molar cingula, reduced male upper canines at the crown cervix, and increased height of the zygomatic root of the maxilla (Andrews and Martin, 1987; Begun, 1987, 1992a). *Sivapithecus, Dryopithecus, Ouranopithecus*, and *Lufengpithecus* share additional features with hominids, including some degree of elongation of the premaxilla (relative to molar size), accompanied by overlapping of the premaxilla and palatine process of the maxilla, a further increase in the height of the maxillary zygomatic processes (except *Ouranopithecus*, see below), a broad nasal aperture base, further increase in incisor labiolingual robusticity, raised P_4 talonids, and M_1 closer in size to M_2. *Sivapithecus, Oreopithecus*, and *Dryopithecus* share a number of derived characters of the forelimb with living great apes (postcrania are few and only briefly described for *Ouranopithecus* and *Lufengpithecus*). The distal end of the humerus is mediolaterally broad, anteroposteriorly flattened, with distinct radial and coranoid fossa. The medial and lateral olecranon pillars are closer in size than in earlier forms, and the trochlea are broad and more symmetrical, with a more strongly projecting lateral trochlear keel, a well-defined zona conoidea, and a large, spherical capitulum. The proximal ulna (not known in *Sivapithecus*) has a strongly keeled trochlear notch, a broad lateral trochlear articular surface, a large, laterally oriented radial facet, and a reduced olecranon process (known only in *Oreopithecus*). The phalanges of *Sivapithecus, Oreopithecus, Ouranopithe-*

Table I. Character States Used in This Analysis[a,b]

EAEMH	*Dryopithecus*	Griph.	*Siva.*	*Pon.*	*Our.*	*Gor.*	*Pan*	*Aust.*
labiolingually thin incisors	thick incisors	1	1	1	1	1	1	1
broader incisors	narrow	0	0	0	0	0	0	0
peg-shaped I^2	0	0	0	0	0	0	1	1
I^2 with cingulum	lacks cingulum	0	1	1	1	1	1	1
upper *I* heteromorphy	0	0	1[1]	1	0	0	2[2]	2
vertically implanted canines	0	0	1[3]	1	0	0	0	0
larger canines/postcanine	reduced canines	1	1	1	2[4]	1	1	2
robust canines	compressed canines	1	1	1	1	1	1	1
tall male canine crown	0	0	0	0	1[5]	0	0	1
narrow canine cingula	thick, rounded cingula	1	1	1	1	1	1	1
P^3 cusp heteromorphy	reduced heteromorphy	1	1	1	1	1	1	1
tall, narrow P^3 paracones	low, rounded	1	1	1	1	1	1	1
narrow P_3	broad P_3	1	1	1	1	1	1	1
no P_3 mesiolingual beak	P_3 with beak	1	1	1	1	1	1	1
no P_3 metaconid	P_3 metaconid	1	1	1	1	1	1	1
short P_4	Longer P_4	1	1	1	1	1	1	1
low P_4 talonids	high P_4 talonids	1	1	1	1	1	1	1
large M^{2-3} metacones	reduced	0	1	1	1	1	1	1
small M^1	large M^1	0	1	1	1	1	1	1
large M^3	0	0	0	1[6]	0	0	1	0
steep molar sides	0	0	0	0	0	0	1[7]	1
tall molar cusps	0	1[8]	1	1	1	0	0	1
high dentine penetrance	0	1[9]	1	0	1	0	0	1
broader molars	elongated molars	1	1	1	1	1	2[10]	2
proportioned anterior/posterior dentition	0	0	0	0	1[11]	0	0	1

stronger cingula	reduced cingula	2[12]	1	1	1	1	1	1
thin palatine process	thick	1	1	1	1	1	1	1
round greater palatine foramen	0	0	1[13]	1	0	0	0	0
short nasoalveolar clivus	clivus > 150% M^1	?	2[14]	2	1	3[15]	4[16]	4
flat alveolar premaxilla	biconvex	?	0	0	1	1	1	1
short nasal premaxilla	long	?	1	1	1	1	2[17]	2
incisive fenestration	incisive canal present	?	1	1	1	1	1	1
no canal	incisive canal broad	?	2[18]	2	1	1	3[19]	3
incisive fossa opposite *C*	incisive fossa distal to *C*	0	1	1	1	1	2[20]	2
small maxillary sinus	larger	1	1	2[21]	?	2	2	2
inferior maxillary sinus floor	superior	0	1	1	1	1	1	1
smooth subnasal fossa	stepped	?	2[22]	2	1	1	1	1
narrow, notched nasal aperture base	broad, flat	?	0	0	1	1	1	1
deep canine fossa	shallow	0	0	0	1	1	1	1
shallow maxilla	deep	0	1	1	1	1	1	1
low zygomatic root	high	0	1	1	0	1	1	1
thin zygomatic root	thick	1	1	1	1	1	1	1
curved lateral malar surface	0	?	1[23]	1	0	0	0	0
narrow at frontozygomatic	thicker	?	0	1	1	1	0	0
broad interorbital	0	?	1[24]	1	0	0	0	0
broad orbits	more elongated	?	2[25]	2	1	0	1	1
broad nasal bones at nasion	narrower	?	1	1	1	1	1	1
indistinct glabella	inflated	?	0	0	1	1	1	1
no supraorbital torus	torus present	?	0	0	1	2[26]	2	2
no supraciliary ridges	ridges present	?	1	1	1	1	1	1
shallow or absent supratoral sulcus	0	?	0	0	0	1[27]	1	1
frontal sinus above nasion	above and below	?	0	0	?	1	1	1

(*continued*)

Table I. *(Continued)*

EAEMH	*Dryopithecus*	Griph.	*Siva.*	*Pon.*	*Our.*	*Gor.*	*Pan*	*Aust.*
large frontal sinuses	small	?	2[28]	2	?	3[29]	3	3
vertical frontal squama	0	?	0	0	0	1[30]	1	0
convex or flat facial profile	0	?	1[31]	1	0	0	0	0
narrow temporal fossa	broader	?	0	0	?	1	1	1
inion well above glabella	lower	?	?	1	?	1	2[32]	2
stronger external occipital protuberance	reduced	?	?	1	?	1	1	1
superiorly oriented nuchal plane	posteriorly	?	?	1	?	2[33]	2	2
smaller brain	larger	?	?	1	?	1	1	1
shorter neurocranium	elongated	?	?	0	?	1	1	1
shallow glenoid fossa	deep	?	0	0	?	1	1	1
larger articular tubercle	small	?	0	0	?	1	1	1
low entoglenoid process	prominent	?	0	0	?	1	1	1
narrow entoglenoid process	broad	?	0	0	?	1	1	1
unfused articular and tympanic temporal	?fused	?	0	0	?	1	1	1
retroflexed humeral shaft	anteroflexed	0	0	1	?	1	1	1
deep humeral shaft	broad	0	1	1	?	1	1	1
prominent deltopectoral plane	reduced	0	0	1	?	1	1	1
small lateral trochlear keel	large	0	1	1	?	1	1	1
shallow zona conoidea	deep	0	1	1	?	1	1	1
posterior medial epicondyle	medial	0	1	1	?	1	1	1
narrow trochlear notch	broad	0	1[34]	1	?	1	1	1
no ulnar trochlear notch keel	present	0	?	1	?	1	1	1
small, anterior radial notch	large, lateral	0	?	1	?	1	1	1

deep ulnar shaft	shallow	0	?	1	?	1	1	1
olecranon process strong	reduced	0	?	1	?	1	1	1
shallow lunate scaphoid facet	deep	?	0	0	?	1	1	1
straighter[35] phalanges	curved	0	1	1	?	1	1	1

[a]Characters were excluded if not known in at least two fossil ingroups. Character states of EAEMH (East African early Miocene hominoids) were coded as 0. *Dryopithecus* character states were coded as 1 unless otherwise indicated. Thus, 0 refers to character states shared with EAEMH and 1 to character states shared with *Dryopithecus*, where the state in *Dryopithecus* is not 0. Derived character states not found in *Dryopithecus* are marked by footnotes where they appear first in the table. Griph. = *Kenyapithecus* samples from Maboko, Majiwa, Kaloma, and Fort Ternan, and samples from Paşalar and Çandir; *Siva.* = *Sivapithecus* (India and Pakistan); *Pon.* = *Pongo*; *Our.* = *Ouranopithecus*; *Gor.* = *Gorilla*; *Aust.* = *Australopithecus afarensis* or *A. africanus* where unknown in *A. afarensis*.

[b]Derived character states not found in *Dryopithecus* (see text for more details):

1. increased upper incisor heteromorphy
2. greatly reduced heteromorphy
3. laterally flared upper canines
4. very reduced canines
5. female morph canine crown
6. reduced M^3
7. flared molar sides
8. low, rounded cusps
9. low dentine penetrance
10. shorter molars lacking cingulum
11. enlarged postcanine teeth
12. intermediate cingulum development, especially on lower molars
13. more elongated greater palatine foramina
14. clivus much longer
15. clivus length slightly longer than state 1
16. clivus length intermediate between 2 and 3
17. intermediate nasal premaxilla length
18. canal extremely reduced in caliber
19. canal of intermediate caliber between states 1 and 2
20. incisive fossa distal to P^3
21. maxillary sinus larger still
22. smooth, occluded subnasal floor
23. flat, anteriorly facing malar surface
24. narrow interorbital space
25. very elongated orbits
26. strongly developed torus
27. broad supratoral sulcus
28. sinuses absent
29. sinuses very large in the interorbital space
30. more horizontal frontal squama
31. concave facial profile
32. inion very low on neurocranium
33. more inferiorly oriented nuchal plane
34. based on the morphology of the humeral trochlea
35. phalanges with well-developed secondary shaft characters (curvature, well-marked fibrous flexor sheath ridges, ventral keels or grooves)

cus, and *Dryopithecus* are robust with powerfully developed secondary shaft characters (e.g., flexor sheath ridges, strong longitudinal shaft curvature). Recently discovered postcranial material of *Dryopithecus* from Spain (Moyà-Solà and Köhler, 1996) indicates a very modern but generic hominid morphology similar to *Oreopithecus* (see Harrison and Rook, this volume), with a broad thorax, elongated forelimbs, short hindlimbs, large hands, large, spherical femoral heads, and a number of other characters related to arboreal mobility. This specimen also confirms the conclusion that the more primitive femur from Eppelsheim often attributed to *Dryopithecus* cannot belong to this taxon, and is most likely a large pliopithecid (see Begun, 1989, 1992b, for a discussion of this issue).

These represent a large number of shared derived characters of the hominid clade including these fossil forms. However, many additional traits suggest several mutually exclusive hypotheses of relationship within the hominids. *Sivapithecus* and *Pongo* share with chimps and fossil humans very large upper central incisors, and still more elongated premaxilla with longer, reduced caliber incisive canals. *Sivapithecus* but not *Pongo* also shares narrow lateral orbital pillars with chimps and humans, and thick enamel, low dentine penetrance and low, rounded molar cusps with *Kenyapithecus, Ouranopithecus,* and hominini (*Australopithecus* and *Homo*). *Pongo* but not *Sivapithecus* shares with all living hominids enlarged maxillary sinuses not found in any fossil form. *Sivapithecus* also lacks the rounded humeral shaft cross section, shaft anteroflexion, more medially oriented (or less posteriorly deflected) medial epicondyle, and possible proximal shaft torsion of *Dryopithecus* and all living hominids including *Pongo.*

The torsion of the humeral head of *Dryopithecus* is debatable (e.g., Begun, 1992b; Rose, this volume). Rose (this volume) notes that on HGP 3, the shaft of *Dryopithecus fontani* from St. Gaudens, the morphology of the base of the lesser tubercle, the buttress for the posterior pole of the head, and the shallowness of the bicipital groove all suggest a more posteriorly oriented humeral head, as in nonhominoids. The two features of the anterior half of the humerus (bicipital groove and lesser tubercle base) are problematic because this area is damaged on the specimen. The lips of the bicipital groove are abraded, and the groove itself is somewhat crushed. It was probably deeper than preserved on the specimen, and probably even deeper than in *Pongo,* which combines a shallow groove with head torsion (though less torsion than in African apes and humans; Evans and Krahl, 1945). The bicipital groove does face more anteriorly than laterally (i.e., it is displaced medially), a position associated with medial head torsion in living hominoids (Aiello and Dean, 1990; Begun, 1992b; Pilbeam and Simons, 1971). As noted above, the new *Dryopithecus* postcrania from Spain suggest a broad thorax for *Dryopithecus,* which is associated with a medial torsion of the humeral shaft (Rose, this volume). This suggests that the St. Gaudens specimen was more modern as well.

Sivapithecus and *Pongo* lack patent incisive canals, broad incisive fossae, stepped subnasal fossae, shallow canine fossae, projecting glabella, and large frontal sinus between glabella and nasion, all found in *Ouranopithecus* (excluding the frontal sinus, unknown for *Ouranopithecus*), *Dryopithecus*, and hominines. *Dryopithecus* also shares with hominines an elongated neurocranium, broad infratemporal fossa, deep glenoid fossa, small articular tubercle, and a broad, prominent entoglenoid process. These last few characters are not known for *Ouranopithecus*.

Gorilla shares with *Pan*, *Homo*, and *Australopithecus* more projecting glabella, more strongly developed supraorbital tori and supratoral sulci, larger frontal sinuses, a fused os centrale and other characters of the hand (Lewis, 1989), and a number of characters of the foot related to terrestrial plantigrady (Gebo, 1992). *Gorilla* shares with *Homo*, *Australopithecus*, and possibly *Griphopithecus* (Begun, 1992b) a reduced brachial index. *Gorilla* also shares with *Pan* a number of characters related to knuckle-walking, horizontal frontal squama, and broad supratoral sulci. *Gorilla* shares with *Dryopithecus* and other Miocene hominids peg-shaped upper lateral incisors, long P_4–M_1, a relatively large M_3, a deep maxillary alveolar process, long nasal portion of the premaxilla, and a higher position of inion. *Gorilla* shares with *Dryopithecus* narrower upper central incisors, a relatively large incisive fossa, shorter premaxilla and incisive canal, the latter also of large caliber, tall molar cusps, and steep-sided molars. *Dryopithecus* shares with African apes thinner enamel with higher dentine penetrance, and with humans, more vertical frontal squama and less pronounced supratoral sulci.

Ouranopithecus shares many of the same characters with hominines found in *Dryopithecus* (see Table I). In addition, however, *Ouranopithecus* lacks the high zygomatic root of *Sivapithecus*, *Dryopithecus*, and most hominines, while it shares molar megadontia (relative to other teeth) and strongly reduced canines with *Australopithecus*, and hyperthick enamel and low zygomatic roots with robust autralopithecines. Table I summarizes the distributions of characters in these taxa.

Character Analysis

Different combinations of the characters noted above suggest different phylogenetic interpretations. Even after accounting for the similarities among forms that are caused by the retention of primitive features (many of the character states shared by *Gorilla* and *Dryopithecus* fall into this category), many inconsistencies persist. Regardless of the phylogeny ultimately chosen, there will remain a large number of characters that cannot be accounted for by a single transition, and that must therefore have arisen more than once (homoplasies). In the data presented above, most of the uncertainty concerns the position of *Sivapithecus/Pongo*, *Dryopithecus*, *Ouranopithecus*, and *Gorilla*.

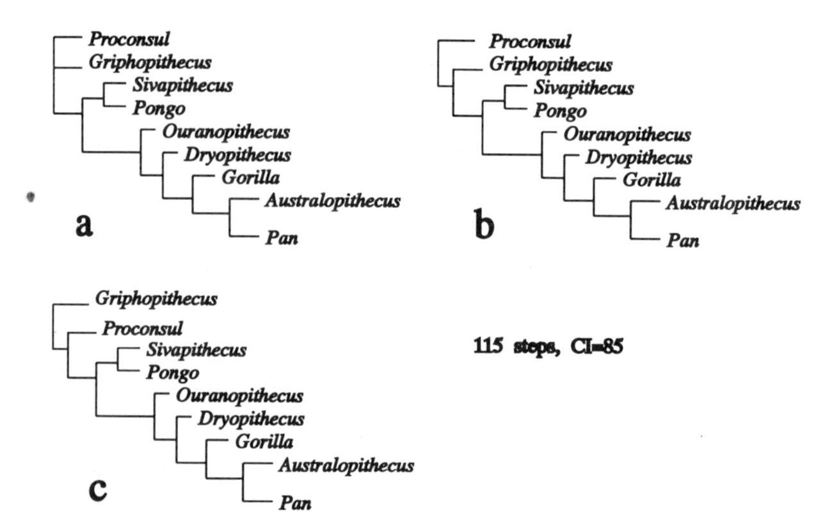

Fig. 1. Maximum parsimony cladogram. a, unresolved at root; b, *Proconsul* as the outgroup; c, *Griphopithecus* (representing the griphopiths, which also includes *Kenyapithecus*) as the outgroup. See text for discussion.

If the characters shared between *Sivapithecus/Pongo* and hominines are synapomorphies, then the similarities between *Dryopithecus* and hominines must be homoplasies. Similarly, if the characters shared among *Dryopithecus/Ouranopithecus* or *Dryopithecus* alone and hominines are synapomorphies, then those shared between *Sivapithecus/Pongo* and hominines must be homoplasies. As noted above, given that there is no clear-cut criterion for establishing *a priori* which characters will be homoplasies and which will be homologies, a cladogram was generated to identify the maximally parsimonious pattern of branching and to identify homoplasies, the smallest possible number of characters inconsistent with this phylogenetic hypothesis. The results are presented in Fig. 1.

In this example, there are two hypotheses of relationship with the same CI. In one (Fig. 1b), *Proconsul* is the sister clade to all other hominoids, and in the other (Fig. 1c), the sister clade is the griphopiths. However, two reasons suggest that Fig. 1b represents the more likely of these two hypotheses. First, much data are missing from the griphopith column, obscuring its relationship to other forms. Second, and more importantly, the griphopith group is almost certainly paraphyletic (Begun, 1992b), since it minimally includes the taxa *Griphopithecus* and *Kenyapithecus*, and there is some evidence to place each of these in different clades (Alpagut *et al.*, 1990; Andrews and Martin, 1987; Begun, 1992b; McCrossin and Benefit, 1993).

The parsimony criterion is unappealing to many paleobiologists because of its dependence on numbers of characters rather than biological or evolutionary significance of characters. At this stage of the analysis, functional anatomical analysis provides an opportunity to introduce biologically meaningful criteria into the decision-making process for choosing among competing cladograms, going beyond the pure logic of Occam's razor. Taking Fig. 1b, the most parsimonious cladogram, as a point of departure, other hypotheses of relationship can be considered. As noted above, other cladograms with a CI 5% or less lower than the maximum parsimony cladogram could potentially be considered valid competing hypotheses. For these data, too many additional cladograms that fit the 5% criterion (CI $\geq$ 81, or $\leq$ 121 steps) can be found to feasibly examine in detail here, so a revised 2% criterion (equivalent to 98% confidence interval) was used. But, even at 5%, none of the additional hypotheses break up the *Pan–Australopithecus* clade (a *Pan–Gorilla* clade requires at least 124 steps and an *Australopithecus–Ouranopithecus* clade 123, both over the 5% criterion).

One clade in Figs. 1 and 2 often cited in previous work on Miocene hominoid systematics, the *Sivapithecus–Pongo* clade, can be disrupted with about a 2.5% change in consistency. The most parsimonious cladogram with *Sivapithecus* as the outgroup to the other homonids including *Pongo* is 118 steps (CI = 83, Fig. 3e). The most parsimonious cladogram with either *Sivapithecus, Pongo,* or both as the outgroup to the African apes and humans is 130 (CI = 75).

Clades with one to three more steps above the most parsimonious cladogram mostly vary in their placement of *Dryopithecus, Ouranopithecus,* and *Gorilla* relative to one another. Of the two resolved cladograms of 116 steps (CI = 84, Fig. 2), one places *Ouranopithecus* rather than *Dryopithecus* as the outgroup to the African ape and human clade (Fig. 2a), and the other places *Ouranopithecus* and *Dryopithecus* as sister taxa in a clade that is the sister clade to African apes and humans (Fig. 4). Four more cladograms were found with 118 steps (CI = 83, Fig. 3), all varying only in the taxic constitution of the sister clade to chimps and humans. One places *Ouranopithecus* alone as the

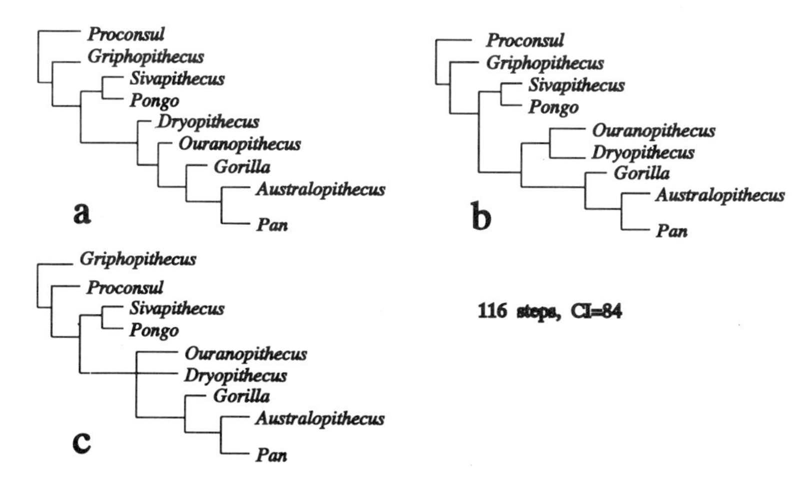

Fig. 2. Next most parsimonious cladograms. See text for discussion.

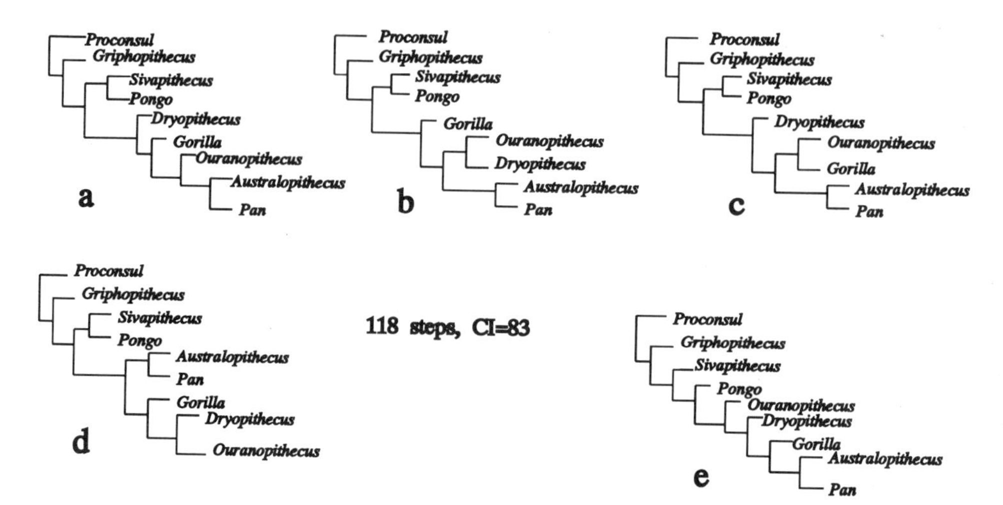

Fig. 3. Five next most parsimonious cladograms after those in Fig. 2. See text for discussion.

sister clade (Fig. 3a), another the combined *Ouranopithecus–Dryopithecus* clade (Fig. 3b), a third the combined *Gorilla–Ouranopithecus–Dryopithecus* clade (with *Gorilla* the sister to *Dryopithecus–Ouranopithecus,* fig. 3d), and a fourth the combined *Ouranopithecus–Gorilla* clade (Fig. 3c).

Seven more clades were found with 119 steps, and six more with 120 steps. They all place *Dryopithecus* between *Gorilla* and the chimp–human clade, or place *Ouranopithecus* as the sister clade to all other hominids.

Functional Anatomy and the Development of Hominine Characters in Dryopithecus and Ouranopithecus

Dryopithecus and *Ouranopithecus* differ primarily from the *Sivapithecus/Pongo* clade in sharing hominine characters of the periorbital region and neurocranium. In nearly all cases, however, these characters are less well developed in the fossil forms than in living hominines or *Australopithecus.* It might be tempting on the basis of these differences to conclude that these characters, including frontal sinuses between nasion and glabella (not known in *Ouranopithecus*), projecting glabella, supraorbital tori, supratoral sulci, and elongated premaxilla, developed independently in *Dryopithecus/Ouranopithecus* and hominines. In addition, many of these characters are said to be better developed in *Ouranopithecus* than in *Dryopithecus,* suggesting to some that *Ouranopithecus* is more closely related to hominines, in particular to *Gorilla,* than is *Dryopithecus* (Dean and Delson, 1992; see also Fig. 3c). In fact, *Dryopithecus* and *Ouranopithecus* are quite similar in these areas where both are known, and the differences can probably be attributed to size (Begun, 1995).

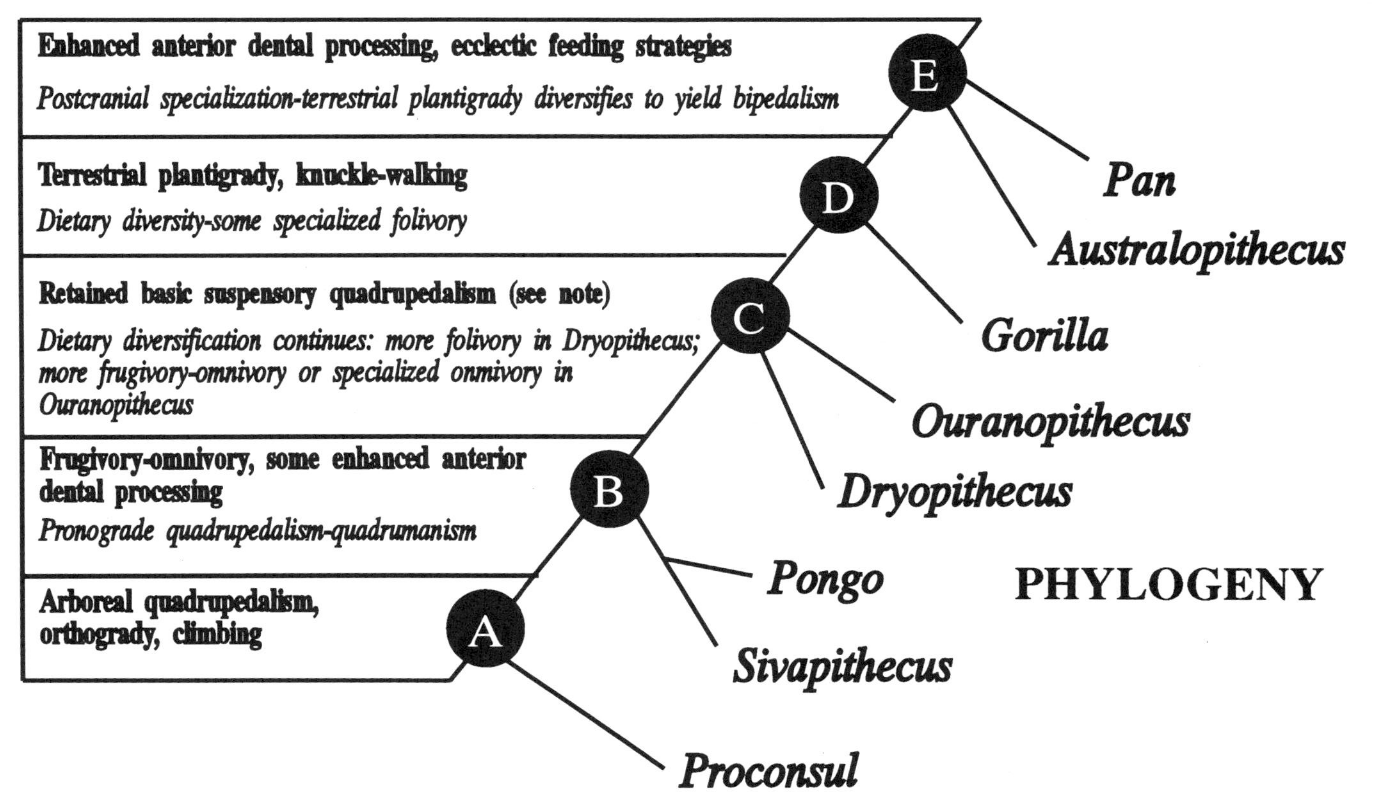

Fig. 4. Combined functional and phylogenetic results. The cladogram from Fig. 2b with functional anatomical characters flagged at their appropriate nodes. Bold represents synapomorphies at each node. Italic represents autapomorphies. Note: Suspensory quadrupedalism is also associated with *Oreopithecus* and thus probably evolved before node C. Uncertainties about details of the anatomy of *Oreopithecus* resulting from damage to the specimens and limited access to original material preclude a definitive placement of the genus in this analysis.

Dryopithecus males are morphologically closer to *Ouranopithecus* males (only the male periorbital region is known) than are *Dryopithecus* females. The differences between *Dryopithecus* and *Ouranopithecus* are mirrored by similar differences in degree of development of the same structures in gorillas relative to chimps. There is no detectable difference in premaxillary length or premaxillary–palatine process overlap between the two fossil forms (*contra* Andrews, 1992), mostly because of the poor state of preservation of this area in *Ouranopithecus*. Thus, the only differences between the two fossil taxa occur in areas in which distinctions also occur in living forms, where the structures are more strongly developed in the larger taxon. This analogy to living forms that are also of differing sizes suggests that the differences between *Dryopithecus* and *Ouranopithecus*, which are known to differ considerably in size, are also specifically related to size, a functional (allometric) observation, and not to degree of relationship, a phylogenetic criterion, with living forms.

The fact that the development of such characters as the supraorbital torus or glabella projection is less in *Ouranopithecus* than in similarly sized gorillas (females) and less in *Dryopithecus* than in similarly sized chimps (females as well) indicates a phylogenetically meaningful transformation. The two more closely related forms (chimps and gorillas, see below) share a more similar morphology despite differences in size because they are more closely related. The fossil taxa share similar morphologies despite size differences because they are more primitive. This suggests that there is some functional integrative phenomenon that causes the coevolution of increases in several different periorbital characters. The premaxilla, which is at the small end of the range of variation of the hominids in the fossil forms, may also be included in this phenomenon. The function of increased development of periorbital structures is unclear, but it may be related to increases in facial length that accompanies an increase in premaxillary length, or it may be related to the functions of anterior dental processing that may be related to increases in incisor size and premaxillary length. The morphology of orangs and *Sivapithecus* demonstrates that robust periorbital morphology need not necessarily accompany increases in facial length, however, suggesting that another phenomenon, such as klinorhynchy, may be at work (Shea, 1988). The consistently smaller or less developed anatomy of these structures in the fossil forms suggests a related functional/evolutionary hypothesis. These characters are poorly developed because they represent the initial phases of the appearance of these traits in hominids. As the hominines evolve, these characters become more strongly developed. The phylogenetic data suggest that this degree of development of these characters is primitive because they are also found with clearly primitive characters for living hominines and australopithecines in *Dryopithecus* and *Ouranopithecus*. The combining of functional and phylogenetic criteria toward the understanding of the evolutionary history of character complexes is to be preferred over classical approaches, in which characters that are present but poorly developed are assumed to have preceded the same characters that are more strongly developed, because of the effects of some unstated emergent functionality. That a clear-cut unidirectional transformation sequence from small to large tori, for example, is inadequate, is exemplified by the diversity in this feature in *Dryopithecus, Ouranopithecus, Pan, Gorilla, Australopithecus,* and *Homo.*

Beyond Parsimony: Choosing among Hypotheses Based on the Functional Anatomy of Ouranopithecus Jaws and Teeth

One aspect of the most parsimonious hypothesis, that *Ouranopithecus* is more primitive than *Dryopithecus* (Fig. 1), rests on the interpretation of a single character, the position of the zygomatic root of the maxilla. As noted above, the zygomatic root is positioned high, above the maxillary alveolar process, in most hominids, whereas it is low in *Ouranopithecus, Oreopithecus,* early to middle Miocene hominoids, hylobatids, and most other primates. The root of the zygomatic process is also low in some specimens of robust *Australopithecus* (KNM-ER 406, SK 48, KNM-WT 17000). If this trait in *Ouranopithecus* is homologous to the primitive condition, as indicated by its presence in all outgroups, then *Ouranopithecus* would lack a derived feature of the *Dryopithecus*–hominine clade and should be considered more primitive. This requires that the high zygomatic root of *Sivapithecus/Pongo* and the low root of *Australopithecus* each be homoplasies. If the trait is a homoplasy in *Ouranopithecus,* and merely converges on the primitive condition, then several hypotheses of relationship between *Dryopithecus* and *Ouranopithecus* are possible. In this latter scenario, the presence of a high root in *Sivapithecus* is a synapomorphy secondarily lost in *Ouranopithecus,* while the low root in *Australopithecus* is again interpreted as homoplastic. Both options require the same number of steps to account for the distribution of zygomatic root height character states.

The functional anatomy of this region of the face suggests a plausible resolution. A low zygomatic root position has at least two distinct implications for jaw biomechanics. In nonhominids, the effect is to lower the attachment site of the superficial masseter muscle to a position closer to the molar tooth row. This is the case because the root originates low on the alveolar process and projects mostly laterally and posteriorly, but not superiorly, to the zygomatic arch, where the muscle attaches. Thus, the origin of the masseter muscle is not displaced superiorly to the extent seen in hominids. In australopithecines with a low zygomatic root the orientation is more superolateral. The root arises low on the alveolar process but is quite long, so that the origin of the masseter (near the zygomatico-maxillary suture) is still displaced superiorly. In *Ouranopithecus* it is not clear which of these patterns is represented because the root of the zygomatic is broken well inferior to the suture.

A lower zygomatic root associated with an approximation of the origin and insertion of the masseter (as in nonhominids) allows a relatively small space for the muscle. Since the masseter is multipennate, a shorter distance means fewer muscle fibers, because these are oriented at an angle to the overall line of action of the muscle. If the insertion of the masseter is displaced from the origin, as in forms with relatively deep mandibles, the effect of a low origin of the masseter may be mediated. This may to some extent have been the case in larger *Proconsul (P. nyanzae, P. major)* and possibly *Oreopithecus.*

Lowering the zygomatic root, when not associated with a lower origin of the masseter, as in some robust australopithecines, may in contrast be a reaction to bending moments and shear, indicative of very high bite forces, and thus powerful masseter muscles. The gradual superolateral slope of the inferior edge of the zygomatic process from a low position on the maxilla may be necessary to reinforce the alveolar process as a response to these stresses. This is particularly true when associated with a superiorly displaced masseter origin on a high and laterally flared zygomatic arch, as in robust australopithecines, because the larger space for the masseter implies more power, and the lateral displacement suggests higher bending and shear moments on the alveolar process.

Additional evidence from the anatomy of *Ouranopithecus* suggests that maximizing occlusal bite force and mandibular translation across the maxilla, was a specific adaptive strategy of this taxon. Molar surface areas are huge in *Ouranopithecus,* and the molar cusps are broad and rounded. The third molar is consistently the largest tooth in *Ouranopithecus,* as in most australopithecines, whereas in most other hominids the M_3 is reduced. These characters are all plausibly related to high bite forces and rotary movements of the mandible (Kay, 1981). The deep and robust mandibles of *Ouranopithecus* suggest a response to longitudinal and transverse bending and to longitudinal axial torsion, again implying powerful mastication (Hylander, 1979, 1988). The enamel of *Ouranopithecus* is hyperthick (Andrews and Martin, 1991; Kay and Ungar, this volume), suggesting an abrasive diet, often associated with the need for powerful mastication. The idea that all of these characters are plausibly related to powerful mastication is reinforced by the observation that all of the same features are also present in robust australopithecines. These hominids are widely believed to have specialized in the development of a huge masticatory apparatus capable of generating tremendous forces throughout the jaws and teeth (Hylander, 1988; Rak, 1983; Robinson, 1956; Tobias, 1967; Wolpoff, 1973). Australopithecines also share with *Ouranopithecus* a further decrease in male canine size and in premolar sectoriality. These too may be related to the masticatory strategy of both taxa.

In contrast, the low zygomatic roots of nonhominids (see above) are not associated with a suite of characters apparently related to powerful mastication, and may in fact be functionally related to lower masticatory power. Thus, the functional anatomical evidence suggests that the low zygomatic root of *Ouranopithecus* is part of a complex of features related to powerful mastication, and is best interpreted as a parallelism with robust australopithecines and a convergence with more primitive hominoids, which have low zygomatic roots for different reasons. Thus, a combination of functional and phylogenetic criteria suggests that the cladograms in Fig. 2a or b may be preferable to the maximum parsimony cladogram (Fig. 1), which differs by only one step. The additional step involves molar cusp height, which requires three steps with *Dryopithecus* as the outgroup to the African apes and humans, and four with *Ouranopithecus* as the outgroup. Independent reduction of molar cusp height in *Ouranopithecus* probably accompanies increased enamel thickness and is functionally related to the masticatory strategy described above.

Homoplasies in *Ouranopithecus* masticatory anatomy allow for the placement of this taxon as a sister clade to African apes and humans, but this alone cannot resolve relations between *Ouranopithecus* and *Dryopithecus*. Between the cladograms in Fig. 2a and b, the main difference is the composition of the outgroup to the African apes and humans. The number of steps for each is the same, and there are no differences in numbers of steps for each character. It is interesting that although both fossil taxa are reasonably well known, they cannot be resolved from one another. This is probably related to the fact that the cranial anatomy of *Ouranopithecus* is not as well known and the postcrania are very poorly represented. More complete representation of the lower face, neurocranium, basicranium, and especially postcranium in *Ouranopithecus* will probably resolve this polychotomy, but for now that must await future discoveries (Fig. 2c).

The remaining cladograms in Fig. 3 all have at least one thing in common. They all break up the African ape–human clade with two additional steps, which requires that characters shared exclusively among these forms must have evolved independently in *Gorilla*. Many of these are included in the data analyzed here, but one group of characters, those related to knuckle-walking, is not, because these traits are not well represented in the fossil record. However, circumstantial evidence suggests that knuckle-walking was not an important aspect of the positional repertoire of any known Miocene hominoid, including *Ouranopithecus* and *Dryopithecus*. *Dryopithecus* possesses proximal and intermediate phalanges that are relatively long, slender, and strongly curved. The phalanges of *Ouranopithecus* are basically similar though larger in most dimensions. This is unlike the shorter, stouter, more robust, less curved, phalangeal shafts of African ape phalanges. The metacarpals of *Dryopithecus* also lack indications of knuckle-walking such as dorsal ridges, though these are not always found on the metacarpal heads of known knuckle-walkers (Inouye, 1992). The tarsal bones of *Dryopithecus* lack any indications of terrestrial plantigrady, and in fact are more suggestive of specialized arboreal functions (Morbeck, 1983, and personal observations). The hamate of *Dryopithecus* lacks characters of the metacarpal articular surface and dorsal surface found in many chimp and gorilla specimens related to stability at the carpometacarpal joint in extended postures. The hand bones of other Miocene hominoids also completely lack any indications of knuckle-walking (Beard *et al.*, 1986; Morbeck, 1975, 1983; Napier and Davis, 1959; Rose, 1983, 1988, 1992; Sarmiento, 1987). This suggests that knuckle-walking evolved after the divergence of Miocene hominoids but before the separation of chimps and gorillas, and is thus a synapomorphy of a clade that includes the African apes. It is likely that this clade would also have included humans, which share several characters of the hands with African apes (e.g., fused os centrale, shorter, stouter metacarpals and phalanges, larger carpometacarpal joints oriented more normal to the long axis of the digits). The presence of these additional synapomorphies of the African ape and human clade would make it even more difficult to accommodate hypotheses of relationship among hominids that place *Dryopithecus* or *Ouranopithecus* closer to chimps and humans than *Gorilla*.

Functional Anatomy and the Homoplasies of Sivapithecus/Pongo

The phylogenetic hypotheses represented by Fig. 2a and b still contain a substantial number of homoplasies, as indicated by the CI values. Further analysis is required to explain the distribution of these characters. The largest number of homoplasies in a single clade occurs in *Sivapithecus/Pongo*. The Asian great apes together are similar to chimps and australopithecines in the substantial increase in premaxillary length, the reduction of incisive canal caliber, the increase in incisive canal length, more posterior position of the incisive fossa, and mesiodistally expanded upper central incisors. Incisive canal length and incisive fossa position are strongly correlated to premaxillary length in hominids, though this need not necessarily be the case, since the canal is formed by both a lengthening of the premaxilla and its overlapping of the palatine process of the maxilla. Incisive canal caliber is also linked to the other characters, though less so, given the diversity of this characteristic in hominids of similar and differing premaxillary lengths (i.e., some evidence of decoupling). Whether these characters are considered as individual traits or as one complex character, they represent a striking homoplasy with chimps and australopithecines. That they are homoplasies is established by the fact that a much larger number of similarities exist among *Dryopithecus, Ouranopithecus, Gorilla, Pan,* and *Australopithecus* (see above). In addition, the premaxilla of Asian great apes is different in morphological details. For example, it is usually even longer and more horizontal in orientation, and nearly fused to the palatine process of the maxilla such that the subnasal floor is completely smooth (Ward and Pilbeam, 1983). In other hominids the premaxilla is separated from the palatine process midsagittally by a well-defined step and a large incisive fossa. Also, the upper incisors differ between both groups of hominids in that the upper centrals are dramatically expanded in the Asian forms, while both upper incisors are expanded in chimps and humans.

The presence of this set of homoplasies in *Sivapithecus/Pongo* can be explained by a hypothesis of similarity in function. Both the Asian great apes and the chimp/human clade have large, mesiodistally expanded upper central incisors. The chimp/human clade has functionally expanded the incisor region even further by increasing the mesiodistal length of the upper lateral incisors, which are distinctively spatulate in this group. The relative size of the incisors is correlated to diet, as has been shown repeatedly by a number of workers (Hylander, 1975; Kay and Covert, 1984; Kay and Hylander, 1978). Relatively large incisors are correlated to frugivorous diets in primates. Frugivorous or omnivorous primates tend to have larger incisors because they are used to process food items. The enhanced anterior dental processing of food that is associated with frugivory is also suggested by the more frugivore-like morphology of the molars of *Sivapithecus–Pongo*, *Pan*, and *Australopithecus* and by the overall similarities in microwear noted among all of these taxa (Teaford and Walker, 1984) as well. These taxa have molars with low, rounded cusps and relatively shallow basins (more strongly developed in the thickly enameled fossil forms), in contrast to *Dryopithecus* and *Gorilla*, which have narrower incisors and molars with taller, more peripheralized cusps surrounding broad, deep occlusal basins. It is plausible to suggest that the increase in the premaxilla in Asian great apes and the chimp/human clade is related to the increase in the size of the incisors, which are of course housed within alveoli of the premaxilla. The parallel acquisition of an enlarged incisor region in large-bodied hominoids may have led to a parallel increase in the size of the premaxilla in response to higher levels of stress. Reduction of incisive canal caliber independently in both groups may represent a response to high levels of shear, bending, and torsion of the premaxilla that might be expected from the powerful biting of objects of various sizes and hardness between the incisors.

An alternative hypothesis, that *Gorilla* premaxillae are secondarily shortened, that the *Sivapithecus–Pongo* and *Pan–Australopithecus* similarities are synapomorphies, and that *Dryopithecus* and *Ouranopithecus* retain the primitive condition of the premaxilla, is also functionally internally consistent, but requires an additional set of homoplasies from anatomical regions outside the premaxilla and nasal fossa. *Ouranopithecus* and *Dryopithecus* would have to have developed similarities to African apes and humans in the periorbital region (see above) independently, or *Sivapithecus* would have to have reverted to the primitive condition. *Dryopithecus* would also have to have developed similarities with African apes and humans in neurocranial and temporal bone morphology (see Table I and Begun, 1992a, 1994; Kordos and Begun, 1996) independently as well. These characters together are not so clearly integrated functionally as are the suite of proposed homologies of the premaxilla and nasal fossa in *Sivapithecus–Pongo*.

Another interesting set of homoplasies occurs in the postcranium of *Sivapithecus*. The distal humerus of *Sivapithecus* shares with all hominids a broad, deep trochlea, well-formed lateral trochlear ridge (less well formed in *Pongo* and variable in hominini), deep zona conoidea, medially directed medial epicondyle, broad, shallow olecranon fossa, equal-sized medial and lateral olecranon pillars, and a flattened distal shaft (Begun, 1992a, 1994; Pilbeam *et al.*, 1990; Rose, 1988). However, *Sivapithecus* also has a retroflexed humeral shaft, a flattened deltoid plane, mediolaterally compressed proximal shaft, less proximal humeral shaft torsion, and mediolateral bowing of the shaft (Pilbeam *et al.*, 1990), all of which are found on the humeral shafts of early and middle Miocene forms such as *Proconsul, Kenyapithecus,* and *Griphopithecus* (=*Austriacopithecus*), as well as large baboons and other cercopithecoid monkeys. On face value, these characters would seem to represent primitive features retained in *Sivapithecus* and missing from *Dryopithecus.* This would require substantial homoplasy in forelimb anatomy between *Pongo* and the African ape and human clade. However, it is also possible that the shared morphology of the *Pongo, Dryopithecus,* and African ape and human forelimb is primitive for all hominids including the *Sivapithecus–Pongo* clade, and that "primitiveness" of the *Sivapithecus* humerus is autapomorphic (not homologous to the primitive condition seen in *Proconsul*). A third possibility is that some aspects of the *Sivapithecus* humerus are primitive (e.g., low proximal shaft torsion) and the alternative character state (more torsion) homoplastic in *Pongo* and the African ape/human clade, while others (e.g., massive, flat deltopectoral plane, pronounced shaft retroflexion, medial bowing) are autapomorphic for *Sivapithecus.* The detailed differences among these competing hypotheses are beyond the scope of this chapter (Rose, this volume; S. Ward, this volume). Whichever patter of homoplasy is ultimately considered more likely, it is clear that substantial amounts of parallel evolution characterize the orang lineage.

Early Hominid Phylogeny and the Evolution of Hominid Functional Complexes

The evidence of systematics provides an independent framework with which to reconstruct the evolving anatomy of the hominid lineage (Fig. 4). In this chapter we have focused on the evidence of the gnathic complex and the forelimb, but other evidence is available from the foot. In this final section, we will restrict comments to the jaw and forelimb.

The earliest evidence of the modern hominoid forelimb appears with *Proconsul,* which has a slightly developed lateral trochlear keel and zona conoidea on the distal end of the humerus. In other features of the distal articular morphology and the shaft, early Miocene forms are primitive, retaining traits commonly found in a diversity of primate above-branch generalized arboreal quadrupeds. *Kenyapithecus* from Fort Ternan and specimens attributed to *Kenyapithecus* from Maboko basically conform to this pattern. Phylogenetically, the earliest indications of modern hominoidlike suspensory morphology of the humerus appear with the temporally relatively late *Oreopithecus.* The morphology of the humerus suggests a wider array of limb positions, with shoulder abduction especially well developed compared with earlier forms. Ranges of flexion–extension at the elbow are increased, and the humeroulnar articulation is stabilized in the entire range of this motion by a complex arrangement of keels. The humeroradial articulation is displaced laterally and posteriorly, enhancing pronation–supination. Limb positions were more routinely flexed and abducted at the shoulder and extended at the elbow, and movement probably occurred primarily below branches. The subsequent radiation of hominids for which humeral morphology is known (*Sivapithecus, Dryopithecus,* living hominids) retains these basic suspensory characteristics, despite several apparently independent shifts to terrestriality (?*Sivapithecus,* hominines).

Gnathic anatomy appears to be more diverse and specialized during hominid evolution. *Oreopithecus* is highly specialized in gnathic anatomy, consistent with strongly folivorous feeding preferences (Harrison, 1987), though very similar to other hominids in postcranial anatomy. In the few postcranial remains known from *Dryopithecus,* the basic pattern remains, though substantial changes occur in gnathic anatomy. *Dryopithecus* shares with all hominids (except *Oreopithecus*) changes in premaxillary, incisor, and canine morphology, canine–postcanine proportions and premolar–molar proportions indicative of an important shift in feeding strategies. The functional implications of incisor size increase along with studies of fossil and living hominoid microwear suggest that this change involved some form of specialized frugivory including a significant anterior dental processing component not seen in more primitive catarrhine frugivores (*Aegyptopithecus, Proconsul*). In the early phases of this dietary diversification, *Dryopithecus* focused on relatively soft, nonabrasive fruits, probably with a significant percentage of leaves in the diet, as suggested by its relatively thin enamel, high dentine penetrance, pointed molar and premolar cusps, and narrow, tall-crowned, and labiolingually thick incisors. Fruits and other arboreal food sources may have been extracted from tough protective coverings with the tall, robust incisors and relatively daggerlike canines, but there is no evidence of hard object feeding or abrasives. This is consistent with the view that *Dryopithecus* was strongly arboreal and probably did not rely very heavily on food from terrestrial sources. *Ouranopithecus,* in contrast, seems to have taken a different direction, with increases in masticatory robusticity, postcanine tooth size, and hyperthick enamel indicative of high levels of abrasion and very tough and/or hard food items. It would be interesting to know to what extent *Ouranopithecus* was terrestrial. Its diet and paleoecology both suggest higher levels of terrestriality, as does its body size, which was quite large. *Lufengpithecus* is very similar to *Dryopithecus* in incisor, premaxillary, canine, and postcanine morphology and in characters of the frontal related to the attachments of the temporalis muscle. It was also similar to *Dryopithecus* in having a preference for forested habitats, as indicated by the presence of both taxa in lignites at three localities, and by additional evidence of forested conditions at other localities (Andrews *et al.,* this volume). It probably shared details of dietary adaptations with *Dryopithecus,* with some local differences, accounting for the unique characters of the *Lufengpithecus* face and molar occlusal morphology

(Schwartz, this volume). *Sivapithecus* appears to have become more specialized in positional behavior, while retaining primitive characters of the dentition that first appear in *Kenyapithecus*. A major change does occur in the anterior palate and dentition of *Sivapithecus* indicative of some specialized anterior dental processing, but microwear suggests few differences from chimps (Teaford and Walker, 1984), and molar morphology (Kay, 1981, 1985) suggests a generalized pattern suitable to the feeding strategies of many different primates (*Cebus, Kenyapithecus, Australopithecus, Homo*).

The appearance at different times in hominid evolution of various anatomical features found in living hominids, as noted here concerning the forelimb and lower face, is an example of the wider phenomenon of mosaic evolution (Dobzhansky *et al.*, 1977). The phylogenetic data presented here provide evidence of mosaic evolution in the timing and rate of development of modern characters. In addition to the apparent differences in the timing and rate of evolutionary developments between the forelimb and cranium in hominids, as noted above, other evidence of mosaicism exists within the cranium alone.

One example is the evolution of the premaxilla and subnasal region. A modern hominid premaxilla appears first in the *Sivapithecus*/*Pongo* lineage (Ward and Pilbeam, 1983), at about 12 Ma (Kappelman *et al.*, 1991). The premaxilla and its subnasal relationship to the maxilla is dramatically different in this lineage from that in any other primate, and appears to have developed quite rapidly from the more primitive morphology characteristic of *Proconsul* and *Hylobates*. Following its sudden appearance, the Asian premaxillary morphology has changed relatively little and can in this sense be considered conservative. In contrast, the modern form of the African ape and human (australopithecine) premaxilla is not known until the appearance of *Australopithecus afarensis* and *A. anamensis*, at 4 to 4.2 Ma (Johanson *et al.*, 1982; Leakey *et al.*, 1995). In addition, the morphological transformation of the African ape and human premaxilla and subnasal region appears to have been more gradual, with a number of recognizable intermediate steps. The phylogenetic data suggest that the African morphology also evolved from a primitive form such as *Proconsul*, but through several intermediate steps including the form seen in *Dryopithecus, Ouranopithecus*, and possibly *Lufengpithecus*, to the more elongated form in *Gorilla*, and finally to the most derived morphology of *Pan* and *Australopithecus*. The evolution of the African premaxillary morphology was thus more complex, or less conservative, and occurred over a longer period. It is interesting in this context to consider the most recent, and perhaps most dramatic modification to this region. This is the replacement in *Homo*, very early in ontogeny, of the premaxilla with the maxilla. That such a transformation was possible may be related to the fact that the region has been a site of frequent changes during its evolution, i.e., that an emergent or epigenetic factor has contributed to the development of the human premaxilla and subnasal region. It should be emphasized that this conclusion is not based on an *a priori* assessment of directionality in character state transformations, as the character states analyzed here were treated as unordered (see above).

Other aspects of the upper jaw have been very conservative during hominid evolution. The maxilla proper has been less dynamic, with only a few changes since the appearance of its modern hominid form with *Dryopithecus* and *Sivapithecus* (see above). The same is true for the teeth. Major changes in upper lateral incisor morphology in chimps and humans and in canine morphology in *Ouranopithecus, Ardipithecus* (White *et al.*, 1994), and *Australopithecus* are exceptions in an otherwise conservative hominid dentition. Though slow and subtle, changes in some teeth have produced distinctive morphologies that are particularly suitable for taxonomic differentiation at several levels. Begun and Kordos (1993) have used minor changes in M_3 morphology in *Dryopithecus* to diagnose species, and Kordos (1990) has shown that M_3 morphology can also diagnose hominoid lineages at the supragenetic level. As Atchley *et al.* (1992) have recently noted with regard to the evolution of the mammalian mandible, the hominid cranium is also a complex structure composed of numerous functional units, both integrated but also somewhat independent of one another. A phylogenetic hypothesis can reveal patterns of mosaic evolution that may ultimately become more understandable with increasing knowledge of the functional anatomy and developmental biology of the cranium.

Conclusions

European Miocene hominids add considerably to our understanding of early hominid behavioral and phyletic diversity, and provide evidence relevant to hypotheses of relations among great apes and humans. Regardless of the preferred hypothesis of relationship, a large number of homoplasies must be part of the evolutionary history of the hominids. In contrast to strict or transformed cladistic approach, the methods used here rely on both cladistic methodology for character analysis (determination of polarity, with homology as the operational null hypothesis), and functional morphology to explain homoplasies and choose rationally among competing equally or nearly equally parsimonious hypotheses. Based on these methods, a phylogenetic hypothesis one step less parsimonious than the maximum parsimony cladogram is considered the most likely, the single additional step being explicable on functional grounds. Homoplasies were relatively widespread, however, affecting both craniodental and postcranial characters. Of the 17 homoplasies identified in the maximum parsimony hypothesis, 15 were craniodental (of 85 character states, 18%) and 2 (of 13 character states, 15%) were postcranial. The most homoplastic characters were molar cusp morphology and dentine penetrance. For these data postcranial anatomy is no more vulnerable to parallelism or convergence than craniodental data. If anything, it is slightly less so, though for cranial and postcranial data with more complete representation across all taxa the differences may disappear.

The data analyzed here suggest that *Dryopithecus* and *Ouranopithecus* together represent the unresolved sister clade to the African apes and humans, with *Sivapithecus* as the immediate outgroup. The homoplasies revealed by this hypothesis are explicable as sets of parallelisms in dietary strategy and positional behavior in various lineages. The observation of parallelism is consistent with other evidence of diet, gnathic morphology, microwear, forelimb functional anatomy, and behavior in living and fossil forms. The phylogenetic hypothesis also reveals the sequence of fossil forms during hominid evolution, the functional anatomy of which provides evidence for the sequence of changes in positional behavior and diet that have characterized this lineage. Hominids diversify following a shift toward more suspensory positional behaviors as suggested by the forelimb and pedal evidence. Dietary specializations are widespread and parallelism in gnathic and dental morphology occurs several times, while postcranial anatomy remains more constant in the Miocene, with fewer examples of homoplasy, despite significant diversity in positional behavior among living forms.

Phylogenetic analysis provides a framework for interpreting functional anatomical or behavioral events in the evolutionary history of the hominids. The functional anatomy of various hominid taxa provides a framework for interpreting or explaining character state distributions that cannot be explained by systematics, and for choosing among phylogenetic hypotheses.

References

Aiello, L. C., and Dean, C. 1990. *An Introduction to Human Evolutionary Anatomy.* Academic Press, London.

Alpagut, B., Andrews, P., and Martin, L. 1990. New Miocene hominoid specimens from the middle Miocene site at Paşalar. *J. Hum. Evol.* **19:**397–422.

Andrews, P. 1985. Family group systematics and evolution among catarrhine primates. In: E. Delson (ed.), *Ancestors: The Hard Evidence,* pp. 14–22. Liss, New York.

Andrews, P. 1992. Evolution and environment in the Hominoidea. *Nature* **360:**641–646.

Andrews, P., and Martin, L. B. 1987. Cladistic relationships of extant and fossil hominoids. *J. Hum. Evol.* **16:**101–118.

Andrews, P., and Martin, L. 1991. Hominoid dietary evolution. *Philos. Trans. R. Soc. London Ser. B* **334:**199–209.

Atchley, W. R., Cowley, D. E., Vogel, C., and McLellan, T. 1992. Evolutionary divergence, shape change, and genetic correlation in the structure of the rodent mandible. *Syst. Biol.* **41:**196–221.

Beard, K. C., Teaford, M. F., and Walker, A. 1986. New wrist bones of *Proconsul africanus* and *P. nyanzae* from Rusinga Island, Kenya. *Folia Primatol.* **47:**97–118.

Begun, D. R. 1987. *A Review of the Genus Dryopithecus.* Ph.D. dissertation, University of Pennsylvania.

Begun, D. R. 1989. A large pliopithecine molar from Germany and some notes on the Pliopithecinae. *Folia Primatol.* **52:**156–166.

Begun, D. R. 1992a. Miocene fossil hominids and the chimp–human clade. *Science* **257:**1929–1933.

Begun, D. R. 1992b. Phyletic diversity and locomotion in primitive European hominids. *Am. J. Phys. Anthropol.* **87:**311–340.

Begun, D. R. 1994. Relations among the great apes and humans: New interpretations based on the fossil great ape *Dryopithecus. Yearb. Phys. Anthropol.* **37:**11–63.

Begun, D. R. 1995. Late Miocene European orang-utans, gorillas, humans, or none of the above? *J. Hum. Evol.* **29:**169–180.

Begun, D. R., and Kordos, L. 1993. Revision of *Dryopithecus brancoi* Schlosser, 1901, based on the fossil hominoid material from Rudabánya. *J. Hum. Evol.* **25:**271–285.

Dean, D., and Delson, E. 1992. Second gorilla or third chimp? *Nature* **359:**676–677.

Delson, E. 1985. Catarrhine evolution. In: E. Delson (ed.), *Ancestors: The Hard Evidence,* pp. 9–13. Liss, New York.

Dobzhansky, T., Ayala, F. J., Stebbins, G. L., and Valentine, J. W. 1977. *Evolution.* W. H. Freeman & Company, San Francisco.

Evans, F. G., and Krahl, V. E. 1945. The torsion of the humerus: A phylogenetic study from fish to man. *Am. J. Anat.* **76:**303–337.

Farris, J. S. 1988. *Hennig 86 Reference, Version 1.5* [software manual].

Gebo, D. L. 1992. Plantigrady and foot adaptation in African apes: Implications for hominid origins. *Am. J. Phys. Anthropol.* **89:**29–58.

Harrison, T. 1987. The phylogenetic relationships of the early catarrhine primates: A review of the current evidence. *J. Hum. Evol.* **16:**41–80.

Hylander, W. L. 1975. Incisor size and diet in anthropoids with special reference to the Cercopithecidae. *Science* **189:**1095–1098.

Hylander, W. L. 1979. Mandibular function in *Galago crassicaudatus* and *Macaca fascicularis:* An *in vivo* approach to stress analysis of the mandible. *J. Morphol.* **159:**253–296.

Hylander, W. L. 1988. Implications of *in vivo* experiments for interpreting the functional significance of "robust" australopithecine jaws. In: F. E. Grine (ed.), *Evolutionary History of the "robust" Australopithecines,* pp. 55–83. Aldine de Gruyter, New York.

Inouye, S. E. 1992. Ontogeny and allometry of African ape manual rays. *J. Hum. Evol.* **26:** 459–485.

Johanson, D. C., Taieb, M., and Coppens, Y. 1982. Pliocene hominids from the Hadar Formation, Ethiopia (1973–1977): Stratigraphic, chronologic, and paleoenvironmental contexts, with notes on hominid morphology and systematics. *Am. J. Phys. Anthropol.* **57:**373–402.

Kappelman, J., Kelley, J., Pilbeam, D., Sheikh, K. A., Ward, S., Anwar, M., Barry, J. C., Brown, B., Hake, P., Johnson, N. M., Raza, S. M., and Shah, S. M. I. 1991. The earliest occurrence of *Sivapithecus* from the middle Miocene Chinji Formation of Pakistan. *J. Hum. Evol.* **21:**61–73.

Kay, R. F. 1981. The nut-crackers: A theory of the adaptations of the Ramapithecinae. *Am. J. Phys. Anthropol.* **55:**141–151.

Kay, R. F. 1985. Dental evidence for the diet of *Australopithecus. Annu. Rev. Anthropol.* **14:**315–341.

Kay, R. F., and Covert, H. H. 1984. Anatomy and behavior of extinct primates. In D. J. Chivers, B. A. Wood, and A. Bilsborough (eds.), *Food Acquisition and Processing in Primates,* pp. 467–508. Cambridge University Press, London.

Kay, R. F., and Hylander, W. L. 1978. The dental structure of mammalian folivores with special reference to Primates and Phalangeroidea. In G. G. Montgomery (ed.), *The Ecology of Arboreal Folivores,* pp. 173–192. Smithsonian Institution Press, Washington, DC.

Kordos, L. 1990. Analysis of tooth morphotypes of Neogene hominoids. *Anthropol. Hung.* **21:**11–24.

Kordos, L., and Begun, D. R. 1996. A new reconstruction of RUD 77, a partial cranium of *Dryopithecus brancoi* from Rudábanya, Hungary (submitted for publication).

Leakey, M. G., Feibel, C. S., McDougall, I., and Walker, A. 1995. New four-million-year-old hominid species from Kanapoi and Allia Bay, Kenya. *Nature* **376:**565–571.

Lewis, O. J. 1989. *Functional Morphology of the Evolving Hand and Foot.* Clarendon Press, Oxford.

McCrossin, M. L., and Benefit, B. R. 1993. Recently recovered *Kenyapithecus* mandible and its implications for great ape and human origins. *Proc. Natl. Acad. Sci. USA* **90:**1962–1966.

Morbeck, M. E. 1975. *Dryopithecus africanus* forelimb. *J. Hum. Evol.* **4:**39–46.

Morbeck, M. E. 1983. Miocene hominoid discoveries from Rudabánya: Implications from the postcranial skeleton. In: R. L. Ciochon and R. S. Corruccini (eds.), *New Interpretations of Ape and Human Ancestry,* pp. 369–404. Plenum Press, New York.

Moyà-Solà, S., and Köhler, M. 1996. A *Dryopithecus* skeleton and the origins of great ape locomotion. *Nature* **379:** 156–159.

Napier, J. R., and Davis, P. R. 1959. The forelimb skeleton and associated remains of *Proconsul africanus. Br. Mus. Nat. Hist. Fossil Mamm. Afr.* **16:**1–69.

Pilbeam, D. R., Rose, M. D., Barry, J. C., and Shah, S. M. I. 1990. New *Sivapithecus* humeri from Pakistan and the relationship of *Sivapithecus* and *Pongo. Nature* **348:**237–239.

Pilbeam, D. R., and Simons, E. L. 1971. Humerus of *Dryopithecus* from Saint Gaudens, France. *Nature* **229:**406–407.

Rak, Y. 1983. *The Australopithecine Face.* Academic Press, New York.

Robinson, J. T. 1956. The dentition of the Australopithecinae. *Mem. Tvl. Mus.* **9:**1–179.

Rose, M. D. 1983. Miocene hominoid postcranial morphology: Monkey-like, ape-like, neither, or both? In: R. L. Ciochon and R. S. Corruccini (eds.), *New Interpretations of Ape and Human Ancestry,* pp. 405–417. Plenum Press, New York.

Rose, M. D. 1988. Another look at the anthropoid elbow. *J. Hum. Evol.* **17:**193–224.

Rose, M. D. 1992. Kinematics of the trapezium–1st metacarpal joint in extant anthropoids and Miocene hominoids. *J. Hum. Evol.* **22:**255–266.

Sarmiento, E. 1987. The phyletic position of *Oreopithecus* and its significance in the origin of the Hominoidea. *Am. Mus. Novit.* **2881:**1–44.

Shea, B. T. 1988. Phylogeny and skull form in the hominoid primates. In J. H. Schwartz (ed.), *Orang-utan Biology,* pp. 233–245. Oxford University Press, London.

Teaford, M. F., and Walker, A. C. 1984. Quantitative differences in dental microwear between primate species with different diets and a comment on the presumed diet of *Sivapithecus. Am. J. Phys. Anthropol.* **64:**191–200.

Tobias, P. V. 1967. *The Cranium and Maxillary Dentition of Australopithecus (Zinjanthropus) boisei, Vol. 2: Olduvai Gorge.* Cambridge University Press, London.

Ward, S. C., and Pilbeam, D. R. 1983. Maxillofacial morphology of Miocene hominoids from Africa and Indo-Pakistan. In: R. L. Ciochon and R. S. Corruccini (eds.), *New Interpretations of Ape and Human Ancestry,* pp. 211–238. Plenum Press, New York.

White, T., Suwa, G., and Asfaw, B. 1994. *Australopithecus ramidus:* A new species of early hominid from Aramis, Ethiopia. *Nature* **371:**306–312.

Wolpoff, M. H. 1973. Posterior tooth size, body size, and diet in South African gracile australopithecines. *Am. J. Phys. Anthropol.* **39:**375–394.

The Phylogenetic and Functional Implications of *Ouranopithecus macedoniensis*

15

LOUIS DE BONIS and GEORGE KOUFOS

The anatomy of fossil mammals is the key to understanding their place among beings, living or fossil. Anatomical characters allow one to both recognize phyletic positions and provide hypotheses on the modes, processes, and even causality of evolution. Most anatomical changes must correspond more or less to a change in adaptation. The new features are on the one hand derived, in a phyletic sense, and on the other hand they are adaptations to a new way of life. These new features help to characterize and to distinguish the younger taxa that emerge from a lineage. For a different way of life, organs must have different functions and, generally, a different shape or a different size.

A paleontological work must begin with a careful study of characters to identify the plesiomorphic, or primitive characters for the lineage under study, from the apomorphic, derived or evolved characters. This results from comparative studies with fossil and extant taxa thought to be related to the fossils being studied. Only the apomorphic characters can point out special relationships between different taxa because they are inherited features from common ancestors, even if these ancestors are yet unknown.

In a certain extent, these derived characters correspond to a change in

---

LOUIS DE BONIS • Laboratoire de Gébiologie, Biochronologie, Paléontologie Humaine, 86022 Poitiers Cedex, France. GEORGE KOUFOS • Laboratory of Geology, University of Thessaloniki, 540 06 Thessaloniki, Greece.

Function, Phylogeny, and Fossils: Miocene Hominoid Evolution and Adaptations, edited by Begun *et al.* Plenum Press, New York, 1997.

the structure of organisms and a change in the functions that link a living being with its environment. Changes in functions and behavior can correspond to changes in the environment. Some changes can reflect local variations in climate or vegetation but some others must correspond to global changes and/or large-scale faunal turnover.

The late Miocene (Vallesian) hominoid *Ouranopithecus macedoniensis* was found in Macedonia (northern Greece) and is known from several specimens including a partial skull, maxillae, mandibles, and isolated teeth. Many features of these fossils indicate that they belong to a peculiar genus different from other Miocene or recent hominoids. A careful study and comparison of these remains can help to place *O. macedoniensis* into the evolutionary tree of the hominoids. Here we first consider the main characters of this genus and those that may be apomorphic relative to other Miocene primates. Then we look for links between characters and environment and try to demonstrate a global faunal change during the late Miocene that could explain the changes of anatomy and function in the hominoid superfamily.

Geological Setting and Dating of O. macedoniensis

Remains of *O. macedoniensis* come from two localities of the Axios Valley (Macedonia, Greece), Ravin de la Pluie and Xirochori, which belong to Nea Messimbria formation (Bonis *et al.*, 1988). That formation, which consists of hard red clays, silts, sandstones, and gravels, has been dated as late Vallesian mammal age (MN 10 = Late Miocene) from the mammalian fauna it contains (Bonis *et al.*, 1992). It is overlain by the Vathylakkos formation and the Dytiko formation, which have been dated as early Turolian (MN 11) and late Turolian (MN 13), respectively. A late Vallesian age of about 9 million years old is estimated on the basis of local stratigraphy, biostratigraphic comparisons, and paleomagnetic data (Kondopoulo *et al.*, 1992). *O. macedoniensis* is also found at the "Nikiti 1" locality, in the Chalkidiki peninsula of Macedonia. This locality is situated in the Nikiti Formation, which consists of gravels, pebbles, and sands capped by redbeds. The complete fauna from the locality also indicates a late Vallesian to early Turolian age (MN 10–MN 11) (Koufos *et al.*, 1991; Koufos, 1993).

Main Characters of O. macedoniensis

Dental Characters

Sexual Dimorphism

O. macedoniensis is a large-sized hominoid but the sample allows for the recognition of two different sets of individuals. They are separated by the general body size, estimated from jaws and teeth, as well as by the relative size of the canines and the shape and proportions of the P_3s. One group consists of relatively small individuals with relatively short canines and more rounded P_3s; the other one is composed of larger individuals with relatively larger canines and slightly more elongated P_3s. Individuals in the first group are certainly female, and those in the second are males, of a single species. On both male and female P_3s the protocristid (or metalophid) is slightly obliquely placed relative to the labial cristid (less on females than males). These teeth are not sectorial and the honing facet, for the upper canine, is always lacking.

Differences in body size and the relative size of the canines between males and females vary from one recent hominoid species to another. Differences can be very substantial, as in *Gorilla,* or minimal, as in hylobatids. The variation of body size between males and females of *O. macedoniensis* does not exceed the values found in a recent large hominoid species as *Gorilla gorilla.* Sexual dimorphism is very common among hominoid primates. It exists in recent species as well as in the fossil hominoids such as *Proconsul.* We can suggest that substantial sexual dimorphism is primitive for this superfamily. But on the other hand, however great the sexual dimorphism may be in *Ouranopithecus,* the male and female canines are reduced compared with the canines of *Proconsul* or with those of recent large hominoids except humans (Bonis and Koufos, 1993).

Other Morphological Dental Characters

The upper premolars are not very heteromorphic. Both have a flattened and rounded paracone and a small height difference between paracone and metacone. The molars are large with a very small vestigial cingulum, and there are some accessory cusps on the unworn upper and lower M3.

Thickness of Enamel

In most recent hominoid species, the cheek teeth are thinly enameled. These teeth are adapted to a diet composed primarily of relatively soft foods like fruits, leaves, or buds that are found easily throughout the year in tropical or equatorial evergreen forests. *Pongo pygmaeus,* the orangutan, has slightly thickened enamel despite the fact that it is now a forest dweller. On the opposite extreme, humans have thickly enameled cheek teeth. These teeth are not adapted to the cooked food diet of recent humans, but are a remnant atavistic character inherited from our Mio-Pliocene ancestors. Thick enamel and robust jaws seem to be adaptive characters for primates with a diet partly composed from hard foods like seeds, nuts, roots, or tubers. It is linked with a more open environment than tropical evergreen forests, with a more seasonal climate including an annual dry period (Simons, 1964; Jolly, 1970; Kay, 1981) as was suggested for the late genus "*Ramapithecus.*"

The phylogenetic implications of thick enamel are more puzzling. It could be considered as a plesiomorphic character for recent hominoids except hylobatids (Pongidae and Hominidae). In this case, the thin enamel of African apes, *Pan* and *Gorilla,* is a derived character (Martin, 1983, 1985; Andrews and Martin, 1987). The common (Miocene?) ancestors of humans, *Gorilla, Pan,* and *Pongo* would have had thick enamel. Humans would have retained fairly thick enamel while *Pongo* would have become thinner, and both *Gorilla* and *Pan* thinner still.

On the other hand, it is possible to imagine another scenario for evolution of hominoid tooth enamel. The Middle Miocene, between 17 and 13 Ma, was a period of drastic climatic change in both southern and northern hemispheres, with some botanic turnover. This is for instance the time when forests decreased and Graminae spread in North America, linked probably with a drier climate, which corresponded to a major change in the evolution of horses. The family Equidae was composed of browser species during the Eocene, Oligocene, and Early Miocene. During the middle Miocene the subfamily Equinae (grazing horses) arose, which, with hypsodont cheek teeth, is adapted to these new environments. In Eurasia, the family Bovidae adapted to similar environments at the same time, also entering Africa. It would be reasonable to suggest that some primate lineages also adapted to more open environments. Primate are not grazing animals but they can feed in an open environment by enamel thickening. This could have been the case for *Afropithecus,* with thickened enamel and a long "baboonlike" snout, as well as *Kenyapithecus* with its two species *K. wickeri* and *K. africanus,* if these species actually belong to the same genus. It would also be the case for the lineages leading to *Sivapithecus, Griphopithecus,* or *Ouranopithecus,* whose ancestors came from Africa. Thick enamel, following this hypothesis, would be a result of parallel adaptive trends among several primate lineages adapted to more or less similar environments. In fact, *Kenyapithecus* gives a fairly good idea of what could have been the ancestral morphotype of these genera and of recent hominines. If the beginning of enamel thickening occurred in several lineages, some of them becoming extinct and some others being present in the extant faunas, we have to seek other shared derived characters to distinguish each of them. Until now we failed to do that in part because hominoid fossils are generally incomplete and in part because the separate lineages of Miocene hominoids are increasingly similar when they are observed shortly after their split from their common stem.

Jaw Biomechanics

Reconstructing jaw biomechanics from fragmentary fossil specimens is a very difficult task, but some indications of function are preserved in the jaw and tooth morphology of *Ouranopithecus.* Muscle attachment sites can often be defined by the traces they leave on fossilized bone. On the mandible of *Ouranopithecus* a crest running along the gonial region (*tuberositas massetericus*) indicates a strong attachment of the *m. masseter superficialis (lamina prima),* which was thus quite powerful. On the ramus the limit between the *masseter intermedius* and the *masseter profundus* is not present. Although present on some fossil human mandibles, the robust gonial crest of *Ouranopithecus* looks more apelike than humanlike, but the general chewing pattern, especially the temporomasseter complex, does not differ much among higher primates (Gaspard, 1972). Differences in dental patterns of occlusion are more significant. Occlusal facets on the dentition provide good evidence of the pattern of mastication. Monkey and ape occlusal patterns are guided or constrained by the length of the canines and by the honing complex between the upper canine and the P_3, which hinders lateral jaw movements before the end of occlusion. The honing facet, a large, vertical facet on the anterobuccal face of the P_3, is so named because it hones or sharpens the distal edge of the upper canine, and is typical of most catarrhines. The pattern in *Ouranopithecus* is quite different. The P_3 of *Ouranopithecus* is principally worn from the tip down, producing a horizontal facet more mesially and an obliquely oriented posterior facet. These facets indicate a significant amount of anteroposterior and lateral jaw movement, a different pattern and therefore probably a different diet from that of apes. This P_3 wear pattern in *Ouranopithecus* is made possible by the dramatic reduction in canine size compared with *Proconsul* (Bonis and Koufos, 1993), or recent apes. On the other hand, the canine–premolar honing complex in *Ouranopithecus* is quite similar to that of the australopithecine *Australopithecus afarensis*. A number of *A. afarensis* P_3 specimens are single cusped, as is typical of apes including *Ouranopithecus*, but they always lack the honing facet seen in all apes except *Ouranopithecus*. The canines of *A. afarensis* are more reduced in size than are those of *Ouranopithecus*, but larger than in other hominines. Several *A. afarensis* specimens from Hadar are very similar to the specimens of *Ouranopithecus* in the shape of the wear facets as well. It is possible that this similarity correspond to a similarity in jaw movement and probably as well to similar diets in both forms.

Shape of the Mandibular Condyle

The shape and size of the mandibular condyle are directly linked with masticatory movements. A very small condyle, relative to the size of the mandible, indicates relatively weak masticatory muscles and a low masticatory activity as in *Orycteropus*, for instance, whose feeding activity is principally based on its long, sticky, and wormlike tongue, with which it eats termites or ants. Ungulates have generally powerful masseter and pterygoid muscles with an anteriorly directed muscle force vector, and large anteroposterior and lateral jaw movements so as to browse or graze their vegetal food. They have large, slightly convex condyles with a relatively long anteroposterior radius of curvature. The glenoid fossa is also slightly concave and sometimes it is even absolutely flat. In contrast, carnivores have a posteriorly directed muscle force vector with powerful masseter and temporal muscles, and large vertical movements of the mandible, to cut the meat or to crush the bones. They have a cylindrical mandibular condyle with a short radius of curvature, which works into a very deep and anteroposteriorly narrow glenoid fossa.

Primates are less diverse in the shape of their mandibular condyles and temporomandibular joints (TMJ) but there are some differences among the extant hominoids. Living apes have a large slightly convex mandibular condyle with a large radius of curvature. The glenoid fossa is large and flat without a clear anterior process. *Homo* has a different TMJ with a more developed anterior process and a deeper and concave fossa in front of the postglenoid process. This structure corresponds to a relatively narrow mandibular condyle when compared with that of apes, which have larger, less convex (larger radius of curvature) condyles. Australopithecines could be considered as intermediate, for these characters, between apes and humans (James, 1960; Picq, 1990). *Australopithecus afarensis* is more apelike than more recent hominines in TMJ (Picq, 1990) and condylar morphology. The index of the anteroposterior diameter of the mandibular condyle divided by its mediolateral breadth provides some indication of the shape of the condyle, although the radius of curvature is not expressed by these measurements. For recent humans the index can vary from 32 to 53, with a mean of 43.6 (Fig. 1). Fossils of the genus *Homo* display the same range of variation, as do the few measured specimens of *Australopithecus*. On the other hand, if the same index is computed for *Gorilla*, the values can vary from 45 to 62 with a mean of 51.5 (Fig. 1). These values indicate a mandibular condyle that is more robust with a larger articular surface. The same is true for *Pan*, with indices running from 47 to 64 and a mean of 53.6 (Fig. 1). We did not compute the same index for *Pongo* because of the important cranial and dental differences that separate *Ouranopithecus* from the pongids. The value for the single *Ouranopithecus* mandible specimen with a preserved mandibular condyle is 43. This value is outside of the range of variation of all but *Homo* (Fig. 1). Student's *t* test confirms the statistically significant difference between the value for *Ouranopithecus* and African apes, and also shows there is no statistically significant difference between *Ouranopithecus* and *Homo* (Table I). The *Ouranopithecus* mandibular condyle does not belong to a population like that of *Gorilla* or *Pan* but it is

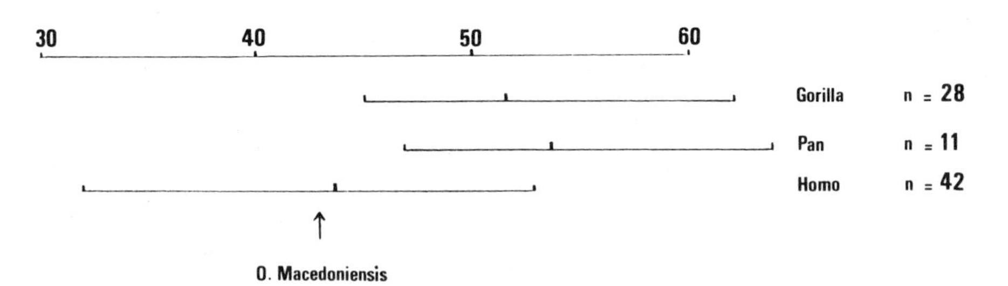

Fig. 1. Variation (means and ranges) in the mandibular condyle index (anteroposterior/lateral diameters) of *Gorilla, Pan,* and *Homo*. Note the value of the same index for *Ouranopithecus macedoniensis* (arrow).

Table I. Student's *t* Test of the Null Hypothesis that the Mandibular Condyle Index in *Ouranopithecus macedoniensis* (43.00) Is Likely to Be Found within Samples of Modern Hominoids at $p = 0.001$

Taxon	Mean	*S*	*t*	Null hypothesis
Gorilla	51.57	4.47	10.2	Rejected
Pan	53.66	5.18	6.19	Rejected
Homo	43.64	5.10	0.82	Not rejected

compatible with that of the hominines (*Australopithecus* and *Homo*). The shape of the mandibular condyle would probably give some indications as to the kind of chewing and the type of diet, and, in this case, the functional mechanics of the *Ouranopithecus* jaws would have been more or less similar to that of the more recent hominines.

Phyletic Position of Ouranopithecus

The phyletic position of *Ouranopithecus* can be established from several characters that we consider as shared derived characters with Plio-Pleistocene hominines (see Bonis and Koufos, 1994). In summary, these characters are: rounded cusps of the upper premolars; nearly symmetrical P_3; nearly homomorphic upper premolars; reduced canines; rounded basal section of the crowns of the canines; no honing facet on the P_3; short, broad P_3 crowns; very thick enamel of the jugal teeth; and vertical profile of the upper face or the small glabella depression on the continuous supraorbital torus. This would indicate a splitting between hominines and African apes earlier than supposed by some molecular studies on recent primates. But it is obvious that a "molecular clock" must be calibrated from fossil evidence. The most recent estimates for the dating of the split between the two groups does not take into account recent discoveries of presumably anthropoid primates and even catarrhines in the middle and late Eocene of North Africa (Bonis *et al.*, 1988; Godinot and Mahboubi, 1992) as well as the presence of primates which could belong to the stem group of anthropoids in the Paleocene of Morocco (Sigé *et al.*, 1990).

Paleoenvironment of O. macedoniensis

It is always difficult to reconstruct the environment from a fossil fauna. Sometimes, different studies of the same locality can give different results, as is the case for the African middle Miocene site of Fort Ternan (Shipman *et al.*,

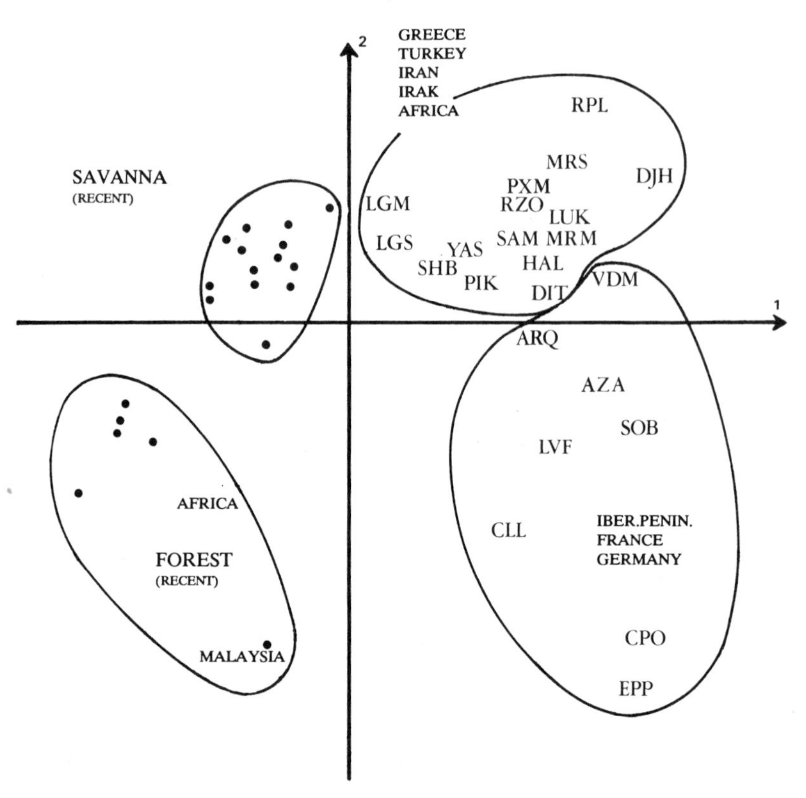

Fig. 2. Plotting on the plane of the 1 and 2 axis of a factorial (multivariate) correspondence analysis of different recent and fossil mammalian localities. Variates are the number of species for the different mammalian taxonomic categories.

Solid circles = recent faunas.

ARQ = Arquillo (L. Mioc.), Span; AZA = Azambujeira (L. Mioc.), Spain; CLL = Can Llobateres (L. Mioc.), Spain; CPO = Can Ponsic (L. Mioc.), Spain; DIT = Dytiko (L. Mioc.), Greece; DJH = Jebel Hamrin (L. Mioc.), Iraq; EPP = Eppelsheim (L. Mioc.), Germany; HAL = Halmyropotamos (L. Mioc.), Greece; LGM = Langebaanweg-QSM (E. Plioc.), South Africa; LGS = Langebaanweg-PPM (E. Plioc.), Africa; LUK = Lukeino (L. Mioc.), Kenya; LVF = Los Valles di Fuentiduena (L. Mioc.), Spain; MRM = Maraghe-middle (L. Mioc.), Iran; MRS = Maraghe-upper (L. Mioc.), Iran; PIK = Pikermi (L. Mioc.), Greece; PXM = Prochoma (L. Mioc.), Greece; RPL = Ravin de la Pluie (L. Mioc.), Greece; RZO = Ravin des Zouaves 5 (L. Mioc.), Greece; SAM = Samos (L. Mioc.), Greece; SHB = Sahabi (L. Mioc.), Libya; SOB = Soblay (L. Mioc.), France; VDM = Venta del Moro (L. Mioc.), Spain; YAS = Yassören (L. Mioc.), Turkey. Note the difference between recent and fossil faunas on the first axis and the separation between recent savanna and forest faunas on the second axis. The Late Miocene western European mammalian faunas have the same values as the recent forest faunas on the second axis, except the Turolian faunas of Arquillo and Venta del Moro, while the Greek, Turkish, Iraqian, Iranian, or African fossil faunas plot near the savanna faunas vis-à-vis the second axis. Note also the extreme position of the Ravin de la Pluie among the savannalike faunas.

1981; Shipman, 1986; Pickford, 1987; Cerling *et al.*, 1991, 1992; Retallack, 1992a). But the faunal composition itself can give very useful indications for localities younger than the middle Miocene, because the taxonomic categories of the fauna (from family to order) are almost the same as those of the recent faunas. A factor analysis applied to the taxonomic composition of the recent faunas separates very well forested faunas from savanna faunas (Bonis *et al.*, 1993). If late Miocene faunas are introduced into the analysis, we can show a similar separation among different localities (Fig. 2). Some fossil faunas follow the recent forest faunas on the graph and we can conclude that they have a similar composition and they lived in the same environment. Other fossil faunas, including the locality of Ravin de la Pluie, plot near the recent savanna faunas in Fig. 2, and represent savannalike environments.

Ravin de la Pluie differs from most of the other Vallesian localities from western Europe whose faunas contain numerous forest-dweller mammals. The Vallesian layers of northern Greece correspond to a change in the climate of the eastern Mediterranean, with an arrival of several species adapted to a more open environment. The primate *Ouranopithecus* bears characters that fit fairly well with the new landscape. The robustness of its jaws and the enamel thickness of the cheek teeth allowed it to adopt a diet different from that of its forest-dweller ancestors. The changing of the environment caused a functional change in the jaw mechanics and in the diet and perhaps was a first step toward the Plio-Pleistocene Homininae.

References

Andrews, P., and Martin, L. 1987. Cladistic relationships of extant and fossil hominoids. *J. Hum. Evol.* **16:**101–118.

Bonis, L. de, and Koufos, G. 1993. The face and the mandible of *Ouranopithecus macedoniensis:* Description of new specimens and comparisons. *J. Hum. Evol.* **24:**469–491.

Bonis, L. de, and Koufos, G. 1994. Phyletic relationships and taxonomic assessment of *Ouranopithecus macedoniensis* (Primates, Mammalia).*Current Primatology* **1:**3 *Ecology and Evolution:* 295–301.

Bonis, L. de, Jaeger, J. J., Coiffait, B., and Coiffait, P. E. 1988. Découverte du plus ancien primate catarrhinien connu dans l'Eocène supérier d'Afrique du Nord. *C. R. Acad. Sci.* **306:**929–934.

Bonis, L. de, Bouvrain, G., and Koufos, G. 1988. Late Miocene mammal localities of the lower Axios Valley (Macedonia, Greece) and their stratigraphic significance. *Mod. Geol.* **13:**141–147.

Bonis, L. de, Bouvrain, G., Geraads, D., and Koufos, G. 1992. Diversity and paleoecology of Greek late Miocene mammalian faunas. *Palaeogeogr. Palaeoclimatol. Palaeoecol.* **91:**99–121.

Bonis, L. de, Bouvrain, G., Geraads, D., and Koufos, G. 1993. Multivariate study of late Cenozoic mammalian faunal compositions and paleoecology. *Paleontol. Evol.* **24–25:**87–95.

Cerling, T. E., Quade, J., Ambrose, S. H., and Sikes, N. E. 1991. Fossil soils, grasses and carbon isotopes from Fort Ternan, Kenya: Grassland or woodland? *J. Hum. Evol.* **21:**295–306.

Cerling, T. E., Kappelman, J., Quade, J., Ambrose, S. H., Sikes, N. E., and Andrews, P. 1992. Reply to comment on the paleoenvironment of *Kenyapithecus* at Fort Ternan. *J. Hum. Evol.* **23:**371–377.

Gaspard, M. 1972. *Les Muscles Masticateurs du Singe à l'Homme.* Maloine, Paris.

Godinot, M., and Mahboubi, M. 1992. Earliest known simian primates found in Algeria. *Nature* **357:**324–326.

James, W. W. 1960. *The Jaws and Teeth of Primates.* Pitman Medical Publishing, London.

Jolly, C. 1970. The seed-eaters: A new model of hominid differentiation based on baboon analogy. *Man* **5:**5–26.

Kay, R. F. 1981. The nut-crackers: A new theory of the adaptation of the Ramapithecinae. *Am. J. Phys. Anthropol.* **55:**141–151.

Kondopoulo, D., Sen, S., Koufos, G., and Bonis de, L. 1992. Magneto and biostratigraphy of the late Miocene mammalian locality of Prochoma (Macedonia, Greece). *Paleont. Evol.* (Barcelona) **24–25:**135–139.

Koufos, G. D., Syrides, G., Koliadimou, K. F., and Kostopoulos, D. 1991. Un nouveau gisement de Vertébrés avec hominoïde dans le Miocène supéreur de Macédoine (Grèce). *C. R. Acad. Sci.* Paris, ser. II, **313:**691–696.

Koufos, G. D. 1993. Mandible of *Ouranopithecus macedoniensis* (Hominidae, Primates) from a new late Miocene locality of Macedonia (Greece). *Am. J. Phys. Anthrop.* **91:**225–234.

Martin, L. 1983. *The relationships of the Later Miocene Hominoidea.* Ph.D. dissertation, University of London.

Martin, L. 1985. Significance of enamel thickness in hominoid evolution. *Nature* **314:**260–263.

Pickford, M. 1987. Fort Ternan (Kenya) paleoecology. *J. Hum. Evol.* **16:**305–309.

Picq, P. 1990. *L'articulation temporo-mandibulaire des Hominidés.* Cahiers de Paléoanthropologie, CNRS, Paris.

Retallack, G. J. 1992a. Middle Miocene fossil plants from Fort Ternan (Kenya) and evolution of African grasslands. *Paleobiology* **18(4):**383–400.

Retallack, G. J. 1992b. Comment on the paleoenvironment of *Kenyapithecus* at Fort Ternan. *J. Hum. Evol.* **23:**363–369.

Shipman, P. 1986. Paleoecology of Fort Ternan reconsidered. *J. Hum. Evol.* **15:**193–204.

Shipman, P., Walker, A., Van Couvering, J. A., Hooker, P. J., and Miller, J. A. 1981. The Fort Ternan hominoid site, Kenya: Geology, age, taphonomy and paleoecology. *J. Hum. Evol.* **10:**49–72.

Sigé, B., Jaeger, J. J., Sudre, J., and Vianey-Liaud, M. 1990. *Altiatlasius koutchii* n. g., n. sp. primate omomyidé du Paléocène supérieur du Maroc et les origines des Euprimates. *Palaeontographica* **214:**31–56.

Simons, E. 1964. On the mandible of *Ramapithecus. Proc. Natl. Acad. Sci. USA* **51:**528–535.

Enigmatic Anthropoid or Misunderstood Ape? The Phylogenetic Status of *Oreopithecus bambolii* Reconsidered

16

TERRY HARRISON and LORENZO ROOK

Introduction

The phylogenetic status of *Oreopithecus bambolii* from the late Miocene of Italy has been a source of much debate since the species was first described in 1872. This observation in itself is hardly surprising, since most fossil primates known since the end of the last century have acquired a complicated history of ideas on their taxonomic and phylogenetic placement. What is so unusual about *Oreopithecus*, however, is that this debate has continued to the present,

This chapter is dedicated to the memory of Johannes Hürzeler (1908–1995) whose profoundly important contribution to the study of *Oreopithecus* and the Baccinello faunas has influenced both of us to follow the same path. The "keeper of the abominable coalman" may no longer be with us, but his remarkable discoveries will undoubtedly continue to inspire and excite the imagination of future generations of vertebrate paleontologists.

TERRY HARRISON • Department of Anthropology, New York University, New York, New York 10003. LORENZO ROOK • Dipartimento di Scienze della Terra, Università di Firenze, Florence, Italy.

Function, Phylogeny, and Fossils: Miocene Hominoid Evolution and Adaptations, edited by Begun *et al.* Plenum Press, New York, 1997.

and there are no indications from the current literature that its phylogenetic status is close to being resolved (e.g., Delson, 1988; Harrison, 1991; Andrews, 1992; Begun, 1994). The problem is especially perplexing because *Oreopithecus* is one of the best-known fossil primates. It is easy to comprehend how researchers might have difficulties establishing the relationships of fossil taxa based on one or two isolated teeth or just a few jaw fragments, but *Oreopithecus* is known from an almost complete subadult skeleton, several partial skeletons, and dozens of relatively complete mandibles and crania. We find ourselves, therefore, in the uncomfortable position of not being able to rely on the excuse favored by most paleontologists in this situation, that the solution to the problem lies in finding more and better material. In the case of *Oreopithecus* we have all the material we need; the shortcomings are not in the available evidence, but in the way that we view it. So why is it that several generations of primate paleontologists have failed to agree on the evolutionary status of *Oreopithecus*? A review of the literature clearly shows that part of the problem is as much sociological as it is scientific, involving a complex interplay of different philosophies, politics, and personalities that are difficult to tease apart from the purely empirical evidence. The consequence of these and other contributing factors is that *Oreopithecus* is perceived to be an "enigmatic anthropoid" (Delson, 1987), one that does not readily conform to our expectations of extinct hominioids based on other lines of evidence. However, is it really that *Oreopithecus* represents a piece of the puzzle that does not fit, or is it simply because the limitations that we impose on our expectations of hominoid evolution are too narrow, and that *Oreopithecus* is being made to fit the wrong puzzle altogether? We suspect that it is the latter that represents the crux of the *Oreopithecus* problem. Of the various factors that have served to confound recent attempts to resolve the phylogenetic relations of *Oreopithecus*, three can be identified that we believe have had a particularly profound impact.

First, much of the recent discussion on the phylogenetic status of *Oreopithecus* has centered around whether or not the genus should be recognized as a cercopithecoid (Szalay and Delson, 1979; Delson, 1979, 1987; Rosenberger and Delson, 1985; Harrison, 1986a, 1987a, 1991; Sarmiento, 1987). However, the postcranial anatomy establishes beyond a doubt the hominoid* affinities of *Oreopithecus*, and few workers remain unconvinced by the overwhelming weight of this evidence (Stern and Jungers, 1985; Susman, 1985; Harrison, 1986a, 1987a, 1991; Sarmiento, 1987, 1988; Rose, 1988, 1993; Fleagle, 1988; Senut, 1989; Martin, 1990). Although the debate on the super-

*The following taxonomic terminology is used throughout the text to refer to major *extant* hominoid groups: Hominoidea (hominoids) = Hylobatidae (hylobatids) + Hominidae (hominids); Hominidae (hominids) = Ponginae (pongines) + Homininae (hominines): Ponginae (pongines) = *Pongo* (orangutan); Homininae (hominines) = African apes + humans. Hominini (hominins) = humans. Further details on the taxonomy of hominoids are given in Table VI.

familial relationships of *Oreopithecus* has ultimately proved to be a useful academic exercise for establishing the hominoid affinities of *Oreopithecus* (after all, few other noncercopithecoid catarrhines whose hominoid status might justifiably be questioned, have been subject to the same level of critical scrutiny), it has had the unfortunate consequence of diverting attention away from the more pertinent question of how *Oreopithecus* is related to other fossil and extant hominoids. The earlier debates concerning whether or not *Oreopithecus* was a hominid *sensu lato* had a similar effect, in that critics focused simply on refutation of its hominid status, rather than on the more constructive enterprise of presenting viable alternatives (Hürzeler, 1954, 1958, 1960, 1962, 1968; von Koenigswald, 1955; Remane, 1955, 1965; Kürth, 1956, Straus, 1957, 1958, 1963; Butler and Mills, 1959; Preuschoft, 1960; Heberer, 1961; Trevor, 1961; Simpson, 1963).

Second, even though the consensus view is that *Oreopithecus* is a hominoid, it is a very specialized one, and this has been perceived by most workers as a serious impediment to establishing its precise relationships among the hominoids. The dentition and certain features of the cranium of *Oreopithecus* are highly derived; evidently part of an adaptive complex associated with an ecological niche that has no close analogue among living primates. Autapomorphies are obviously of limited utility in establishing phylogenetic relationships, but the problem in the case of *Oreopithecus* is compounded by the fact that the unique specializations all occur in the dentition and face. These are the regions that primate paleontologists rely on to establish the alpha-taxonomy and phylogenetic relationships of fossil taxa. The outcome of the debate over the phylogenetic status of *Oreopithecus* might have been entirely different if *Oreopithecus* had had a cranium and teeth identical to the modern orangutan, but conversely possessed bizarre specializations of the postcranium. In this case, there probably would have been little difficulty in recognizing it as a peculiar pongine. Without recourse to appropriate dental or cranial comparisons, primate paleontologists have been hesitant to position *Oreopithecus* in relation to other fossil and extant hominoids. For example, Andrews (1992), in a recent review of hominoid phylogeny, reached the following conclusion about *Oreopithecus:*

> All agree on the highly derived nature of the teeth, and it is the postcranial adaptations that are of particular interest, for *Oreopithecus* shares with living hominoids a number of apparent postcranial synapomorphies. Some of these characters are also present in *Dryopithecus*, and they have been used to support a relationship of this genus with the living great apes, but their presence in *Oreopithecus*, which is so highly specialized cranially, must cast some doubt on this interpretation. (p. 645)

Basically, Andrews is arguing that the craniodental specializations of *Oreopithecus* call into question the likelihood that the postcranial features shared by *Oreopithecus* and extant hominids are, in fact, truly valid synapomorphies. In the light of the preceding discussion, the logical contradictions of this statement are self-evident.

As a consequence of its uniqueness, *Oreopithecus* has generally been included in a distinct family, the Oreopithecidae, within the Hominoidea (Schwalbe, 1915; Butler and Mills, 1959; Hürzeler, 1962; Simpson, 1963; Straus, 1963; Leakey, 1963; Hoffstetter, 1982; Harrison, 1986b, 1991). This is a neat taxonomic finesse, one that is often employed by paleontologists to deal with divergent taxa, but it sidesteps the critical issue of having to determine the sister-group relationships of *Oreopithecus*. After all, specialization alone should not negate *Oreopithecus* as a sister taxon to a more conservative extant or extinct group of hominoids. As Harrison (1991) has pointed out previously, *Oreopithecus* is an oddity among a clade of primate oddities. If degree of specialization is invoked as a reason for failing to resolve the phylogenetic relationships of *Oreopithecus,* then why is it that we have been successful in this regard in dealing with other specialized hominoid lineages, such as the hominins, pongines, and hylobatids? Clearly, other factors are contributing to the problem.

One other difficulty arises from the fact that primte paleontologists doggedly persist in being fixated on the search for ancestral–descendant relationships, despite the wider usage in recent years of cladistic concepts and methods. *Oreopithecus,* being a specialized and endemic form, is clearly not directly ancestral to any later hominoids, and to some researchers this means that *Oreopithecus* has little relevance for providing insight into hominoid evolution in general. However, the same argument could equally be applied to *Pongo* or *Gorilla*. In the broader perspective of a cladistic phylogenetic analysis, where sister-group relationships are preeminent, even specialized forms, once their precise relationships have been established, can be extremely important in helping to redefine morphotypes and to clarify relationships between neighboring taxa. Since the Hominoidea includes few living species, all of which exhibit high frequencies of autapomorphic features, the introduction of fossil taxa, especially those as completely known as *Oreopithecus,* has the potential to greatly enhance our capabilities to reconstruct more accurate morphotypes (Harrison, 1991, 1993).

Third, and perhaps most importantly, it has been difficult to resolve the relationships of *Oreopithecus* because it is much better known than any other later Miocene hominoid. This might seem like a contradiction. Even if we take into account the well-known paleontological paradox that in practice the more one knows the less one understands, it is still reasonable to assume that with better-known taxa it should be easier to establish their relationships than it is for those that are poorly known. However, this relies on the availability of suitable comparative material. Until recently, *Oreopithecus* was the only fossil hominoid from the later part of the Miocene that was known from relatively complete cranial and postcranial material. Comparisons between *Oreopithecus* and other later Miocene hominoids were primarily restricted to aspects of the teeth and jaws. As discussed above, since these are the most specialized anatomical regions of *Oreopithecus,* it is hardly surprising that these comparisons provided few clues to help clarify the relationships of *Oreopithecus*. The most extensive comparisons possible have been with extant hominoids and with primitive catarrhines. These have established that *Oreopithecus* is quite unlike any of the living hominoids in its cranium and dentition, but yet, at the same time, it is unique in being the only fossil primate that shares the entire suite of specialized postcranial features that characterizes the extant hominoid clade. Reference to the partial skeletons of *Proconsul* from the early Miocene of East Africa, and *Pliopithecus* from the middle Miocene of Europe, have only served to confuse the situation further. Both of the latter taxa are craniodentally much more "hominoidlike" than *Oreopithecus,* while their postcranials are more conservative, being much closer to the primitive catarrhine morphotype than are those of *Oreopithecus.* Since *Proconsul* was widely accepted to be an ancestral great ape, at least until the late 1970s, the co-occurrence of a hominoidlike postcranium in *Oreopithecus,* with distinctly unhominoid-like teeth, was, therefore, a difficult combination of features to reconcile. There were two possible options available. Either *Oreopithecus* was an advanced hominoid, with a postcranial skeleton more derived in the direction of extant hominoids than any other known fossil catarrhine, but with particular craniodental specializations, or alternatively it was a bizarre early offshoot that acquired its hominoidlike specializations of the postcranium independently of later hominoids. The latter alternative was the one favored by most leading primate paleontologists of the day, who in turn assumed that the postcranial features shared by modern hominoids had been developed independently a number of times in different lineages, and that these were not necessarily indicative of close phyletic relationship (Clark and Thomas, 1951; Napier and Davis, 1959; Simons, 1962, 1967; Pilbeam, 1969). This was a viewpoint that prevailed until well into the 1970s, and one that lingers even today. It stems from the ill-conceived notion that craniodental features are better indicators of relationship than are postcranial features, because the latter, which provide the basis for reconstructing locomotor behavior, are much more prone to functional convergence. However, as Harrison (1982, 1986b, 1987a,b, 1991) and many other researchers have argued, the postcranial features and character complexes shared by extant hominoids are so detailed and so pervasive that they are extremely unlikely to be the product of convergent evolution. Unfortunately for *Oreopithecus,* the historical bias toward using craniodental features for establishing relationships has tended to obfuscate the phylogenetic significance of the postcranial features.

As noted above, one of the main problems hampering attempts to clarify the status of *Oreopithecus* is that, until recently, detailed comparisons of the cranial and postcranial features were primarily limited to primitive catarrhines, such as pliopithecids and proconsulids, and to extant hominoids. Lacking suitable comparative material of other later Miocene hominoids from Eurasia and Africa has made it difficult to establish just what to expect conservative hominids to look like. New discoveries of relatively complete crania of several hominoid taxa from Europe and Asia (i.e., *Dryopithecus, Graecopithecus,* and *Lufengpithecus*), and the recent important discovery of a partial postcranial skeleton of *Dryopithecus* from Spain, have dramatically improved our understanding of the anatomy and morphological diversity among later Miocene Eurasian hominoids (Wu, 1984; Kordos, 1987; Begun *et al.*, 1990; Bonis *et al.*, 1990; Bonis and Koufos, 1993; Moyà-Solà and Köhler, 1993, 1995, 1996; Begun, 1994). In this context, the phylogenetic status of *Oreopithecus* becomes more readily apparent. Preliminary comparisons indicate that cranially and postcranially, *Oreopithecus* conforms quite closely to the general pattern seen in other European later Miocene hominoids, and it appears to be especially similar to *Dryopithecus. Oreopithecus,* the enigmatic anthropoid, turns out to be not so enigmatic after all. It is simply another late Miocene European hominoid with its own unique specializations. It is ironic, but perhaps not entirely unexpected, that the key to eventually resolving the relationships of *Oreopithecus* rested not with *Oreopithecus* itself, but with the discovery of additional information about contemporary hominoids. In retrospect, we can now recognize that the missing piece of the puzzle that made *Oreopithecus* so enigmatic in the first place, was not the fact that it was too specialized, but that we did not know as much as we thought we knew about other Miocene hominoids. In the final analysis, even though recent discoveries of fossil hominoids from Eurasia have contributed to a better understanding of the phylogenetic relationships of *Oreopithecus,* it is *Oreopithecus,* with its newly established status as the best-known Eurasian hominid, that takes on special significance for providing important new insights into hominoid evolution.

The main aim of this chapter is to present a summary of the morphological evidence available for assessing the phylogenetic status of *Oreopithecus.* Obviously, a thorough review of the anatomy of *Oreopithecus,* and detailed comparisons with other fossil hominoids, are beyond the scope of the present study. Our intention is to focus primarily on the major characteristics that we consider (or have been considered by others in the past) as important features for establishing the relationship of *Oreopithecus* to extant hominoids. This is a crucial initial step in reassessing the status of *Oreopithecus,* and it provides a preliminary foundation from which to direct more detailed comparative studies and phylogenetic analyses. In the following sections of the text, we begin by presenting a brief overview of the chronology, paleoecology, and biogeography of *Oreopithecus,* to provide essential contextual information for interpreting its evolutionary history, and then finally we examine the anatomical basis for assessing its phylogenetic relationships.

Chronology, Paleoecology, and Biogeography

In order to better appreciate why *Oreopithecus bambolii* is such a distinctive hominoid, it is essential to understand the biogeography and paleoecology of the fossil sites at which *Oreopithecus* has been recovered. This provides the necessary contextual backdrop, so that the specializations seen in *Oreopithecus* can be interpreted within their appropriate adaptive and environmental milieu.

The distinctive late Miocene vertebrate fauna associated with *Oreopithecus* has been recovered from a small cluster of localities in the Maremma region of southern Tuscany since the latter part of the last century. The fauna exhibits a high level of endemism and an unusual community structure, which suggests an insular environment. Until recently, this fauna was known exclusively from the Maremma sites. However, a small collection of undescribed fossil vertebrates from a lignite mine at Serrazzano (Pomarance, Pisa), recently "rediscovered" in a drawer at the Museum of the University of Florence, provides evidence of the occurrence of an *Oreopithecus*-bearing fauna in a sedimentary basin to the north of Maremma. In addition, *Oreopithecus* and an associated "maremmian" fauna was recently recovered from late Miocene sediments at Fiume Santo in northern Sardinia (Cordy and Ginesu, 1994), thus indicating a western geographic extension of the fauna into Sardinia, and possibly also into Corsica. Together these localities comprise a distinct Tusco-Sardinian paleobioprovince (Hürzeler and Engesser, 1976; Azzaroli *et al.*, 1987). The paleobioprovince was disrupted during the Messinian by the intense tectonic activity associated with the final orogenic phases of the formation of the Apennine chain, and southern Tuscany was eventually incorporated into peninsular Italy (Torre *et al.*, 1995).

The mixed zoogeographical affinities of the Tusco-Sardinian faunal province testify to the fact that the region, probably made up of a small group of islands on the northern fringe of the Tethys, was periodically connected to continental Europe and possibly to Africa. It can be deduced from the available geological evidence that the fauna migrated across a mobile and folded belt that extended from western Liguria southwards across the western part of the present Tyrrhenian sea to join northern Tunisia (Boccaletti *et al.*, 1987, 1990). This connection allowed faunal migrations from Europe and possibly Africa into the Tusco-Sardinian bioprovince by a combination of island sweepstakes and by direct movements across ephemeral land bridges (Sondaar, 1977, 1986; Harrison and Harrison, 1989). It is also possible that African mammals arrived via Kabylie and the Sardo-Corsican massif before the formation of the Tunisian–Sardinian strait at the Astaracian–Vallesian boundary (Esu and Kotsakis, 1983). The recent discovery of a mandible of *Stegotretrabelodon syrticus* in pre-Messinian deposits near Vibo Valentia (Calabria, southern Italy) provides further support for the existence during the Tortonian of a land corridor from North Africa across the Sicily–Calabria arch (Torre *et al.*, 1995). It seems likely, however, that this connection between North Africa and the Tusco-Sardinian region was never fully emergent, since there appears to have been no direct interchange of mammals between continental Europe and Africa via this particular route, only intermittent migrations from both directions into the Tusco-Sardinian region.

Engesser (1989), based on studies of the fauna from the site of Baccinello, has hypothesized that at least two dispersal events must have taken place prior

0to the occurrence of the oldest faunal assemblage (correlated with MN 11, early Turolian). The first event involved glirids and ochotonids from Europe, and probably occurred sometime during the middle–late Miocene, since these taxa already show high levels of endemism in the earliest fossil horizon. The second dispersal event, this time from Africa, is inferred from the appearance of alcelaphine and neotragine bovids. Previously, it was believed that *Oreopithecus* was also part of this influx of African mammals into the region (Harrison, 1985, 1986b, 1987a; Harrison and Harrison, 1989; Engesser, 1989). Indeed, such an inference was supported by the occurrence of fossil catarrhines at early and middle Miocene sites in East Africa with distinctive dental adaptations very similar to those of *Oreopithecus* (Harrison, 1985, 1986b). The best known of these is *Nyanzapithecus,* which has been considered to be a primitive oreopithecid. Although this still remains an option, new evidence suggests that an alternative phylogenetic scheme is now more probable. One of the initial problems in linking *Nyanzapithecus* with *Oreopithecus* was that it implied that a specialized and advanced lineage of hominids had already diverged in Africa from the basal hominoid stock by 19 Ma, which is inconsistent with the timing of divergence events based on evidence from other fossil catarrhine groups. Moreover, the recovery of more complete cranial material from Europe in recent years has demonstrated that *Oreopithecus* is remarkably similar in its detailed morphology to contemporary European hominids, especially to *Dryopithecus* (see further discussion of this topic below). As a result, *Oreopithecus* is more likely to be a descendant of a Miocene European species, like the majority of mammals in the Tusco-Sardinian fauna, than the terminal branch of a dentally specialized hominoid lineage that had diverged in Africa by the early Miocene. We now prefer to consider *Nyanzapithecus* as a derived representative of the Proconsulidae that developed a series of dental specializations in parallel with *Oreopithecus.*

Apart from a number of regional studies (e.g., Giannini *et al.,* 1972; Martini and Sagri, 1993), the geological setting of late Miocene localities in southern Tuscany has not been studied in any detail; only the stratigraphy of the Baccinello–Cinigiano basin has been well documented (De Terra, 1956; Lorenz, 1968; Benvenuti *et al.,* 1995). Since the late 1950s, paleontological research in the area was undertaken by Dr. J. Hürzeler (Natural History Museum, Basel), who obtained a large collection of fossil mammals from the lignite mine at Baccinello prior to its closure in 1959 (Hürzeler, 1958; Hürzeler and Engesser, 1976). Geological mapping and paleontological prospecting at surface outcrops in the area has been subsequently carried out by researchers from the Natural History Museum in Basel and the University of Florence.

The stratigraphic succession of the Baccinello–Cinigiano basin, with a series of sediments about 400 m thick, has been studied most recently by Benvenuti *et al.* (1995). Different faunal assemblages of fossil vertebrates (V0, V1, V2, and V3) and fossil mollusks (F1 and F2) have been identified in the sedimentary succession (Gillet *et al.,* 1965; Hürzeler, 1975; Hürzeler and Engesser, 1976). The three youngest vertebrate assemblages (V0 to V2) include about 20 species of mammals, almost all of which are unique to the Tusco-Sardinian sites. The high level of endemism of the faunas, in conjunction with their low taxonomic diversity, the predominance of specialized bovids, the tendency for development of hypsodonty and large body size in some of the rodents, and the absence of nonlutrine carnivores are all indicative of an insular environment (Azzaroli *et al.*, 1987; Harrison and Harrison, 1989; Rook *et al.*, 1996). The youngest fossil vertebrate faunal assemblage, V3, is entirely different from the previous ones, and is composed of species common to contemporary latest Miocene European localities (Hürzeler and Engesser, 1976; Engesser, 1989; Rook *et al.*, 1991; Rook and Torre, 1995). The high level of endemism of the V1 and V2 assemblages prevents direct biochronologic correlations with other European sites. However, the V0 fauna includes *Huerzelerymys vireti*, which permits a reliable correlation with European sites assigned to MN 11 (early Turolian), while the V3 fauna is most comparable to that from European sites correlated with MN 13 (late Turolian). This serves to constrain the age of V1 and V2 assemblages to middle or late Turolian. Potassium-argon dates on samples from an intercalated tuff give an estimated age of 8.4 Ma for the V2 fauna (Hürzeler and Engesser, 1976; Hürzeler, 1987; Harrison and Harrison, 1989; Engesser, 1989; Rook *et al.*, 1996).

The V1 fauna from Baccinello occurs in a lignite layer. This assemblage is considered equivalent to the *Oreopithecus*-bearing faunas from coal mines at Casteani, Montemassi, and Ribolla in southern Tuscany (Hürzeler and Engesser, 1976; Azzaroli *et al.*, 1987). The remains of *Oreopithecus bambolii* are extremely abundant in V1, and this species represents one of the commonest mammals at the site. Until recently, the occurrence of *Oreopithecus* in the V2 faunal assemblage at Baccinello was uncertain, in contradistinction to its relative abundance in broadly equivalent levels at Monte Bamboli. The only known specimen from Baccinello was a cranial fragment collected by Hürzeler in 1956 (Hürzeler and Engesser, 1976). However, the recent recovery of some teeth and a mandible of *O. bambolii* from horizons above V1 has confirmed the occurrence of this species in younger strata at Baccinello (Rook, 1993; Harrison and Rook, 1994; Rook *et al.*, 1996).

Evidence for a primarily aquatic setting and a humid forested environment is provided by the extensive lignite accumulations, the common occurrence of skeletal remains in anatomical connection, the abundance of fossil crocodiles, chelonians, and freshwater mollusks, and the occurrence of otters. Palynological analyses on lignite samples associated with the V1 assemblage indicate that the vegetation was representative of a mixed lowland mesophytic forest (Harrison and Harrison, 1989). The area was evidently poorly drained, and the forested areas were interspersed with numerous freshwater pools and shallow lakes (Teichmüller, 1962; Harrison and Harrison, 1989; Benvenuti *et al.*, 1995). The inferred ecological requirements of the fauna, however, are not entirely consistent with this paleoenvironmental reconstruction. For example, the degree of hypsodonty of the bovids and of some of the rodents suggests a preference for drier and more open habitats. Similarly, increased size and hypsodonty in several bovid and rodent lineages during the time interval between V1 and V2 assemblages (Hürzeler, 1983; Engesser, 1983, 1989) indicate that the paleoecology at Baccinello was becoming increasingly drier. This is in agreement with evidence from palynology (Benvenuti *et al.*, 1995) and the general scarcity of turtles and crocodiles in the V2 faunal assemblage. Interestingly, there is also a corresponding decline in the abundance of *Oreopithecus* in V2, which might imply a relatively narrow ecological preference by this taxon for swampy, forested habitats. However, the lignites at Monte Bamboli, which are broadly equivalent in age to V2, contain numerous remains of *Oreopithecus* and aquatic vertebrates, and this suggests that the drier conditions at Baccinello were part of a local phenomenon, rather than representative of wider regional changes in the vegetation, climate, or landscape.

Phylogenetic Relationships

As indicated in the introduction, recent debates about the phylogenetic relationships of *O. bambolii* have mainly focused on whether or not *Oreopithecus* is a cercopithecoid or a hominoid. With such uncertainty surrounding even the superfamilial status of *Oreopithecus,* it is hardly surprising that little or no attention has been paid to assessing the possible relationships of *Oreopithecus* within the Hominoidea. Nevertheless, the available evidence is adequate to resolve this problem, and we present here a brief review of the main morphological features that establish that *Oreopithecus* is not only a hominoid, but also a hominid. However, we start by reviewing the unique specializations of *Oreopithecus* because it is these features that have been cited as the most serious impediment to resolving the phylogenetic relationships of *Oreopithecus.*

Unique Specializations of Oreopithecus

In many respects the cranium of *Oreopithecus* is consistent with the inferred primitive catarrhine morphotype (Harrison, 1987a,b, 1991). The face is relatively short and broad, the palate is long and quite narrow, the cheek tooth rows converge slightly anteriorly, the orbits are subcircular and are situated far anteriorly on the face, the interorbital region is relatively wide, the nasal aperture is quite narrow with its maximum breadth located at midheight, there is only a slight overlap between the inferior margin of the orbit and the superior rim of the nasal aperture in the horizontal plane, the anterior root of the zygomatic arch is located close to the alveolar margin of the cheek teeth, and the neurocranium is low and globular (Fig. 1).

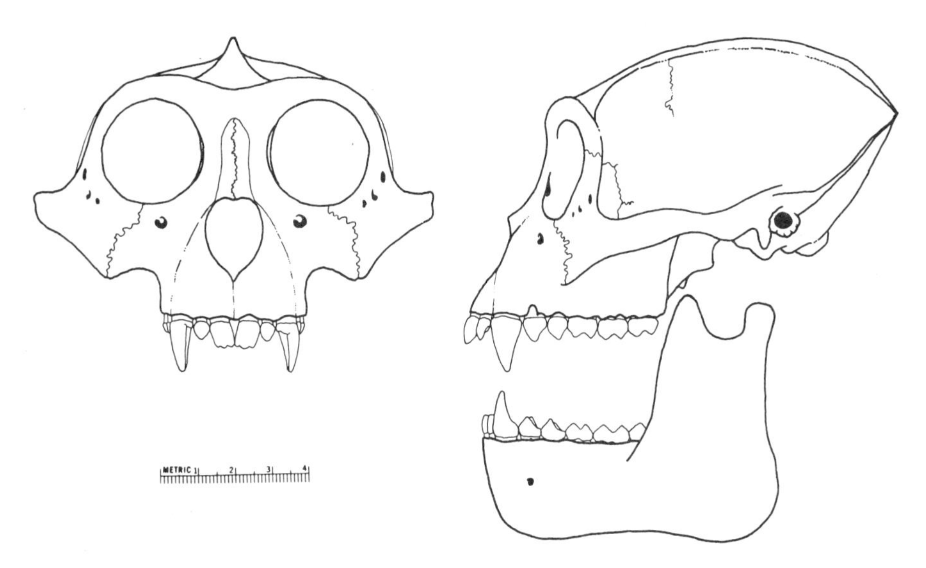

Fig. 1. Composite reconstruction of the skull of *Oreopithecus bambolii* (male individual). Left, anterior view; right, lateral view. Bar = 4 cm.

However, superimposed on these primitive features is a suite of cranial specializations unique to *Oreopithecus*, and these are presumably functionally related to adaptations for powerful chewing and folivory (Szalay and Delson, 1979; Harrison, 1987a, 1991). Undoubtedly, the most distinctive characteristics of *Oreopithecus* are found in the dentition, and, like the cranium, it provides further evidence that *Oreopithecus* occupied a specialized dietary niche. The numerous distinctive craniodental features of *Oreopithecus* have been described in some detail previously (Hürzeler, 1949, 1951, 1958, 1968; Butler and Mills, 1959; Straus, 1963; Szalay and Berzi, 1973; Szalay and Delson, 1979; Harrison, 1986b, 1987a), so the following discussion provides only a brief summary of the main specializations.

Dentition

The lower molars and dP_4 are specialized in the following features: (1) the crowns are relatively long and narrow; (2) buccolingual waisting occurs midway along the length of the crown; (3) the cusps are voluminous with relatively high relief; (4) the occlusal crests are short but well developed; (5) a distinct paraconid is present, at least on dP_4 and M_1; (6) a small conule, the protoconulid, is usually present midway along the length of the preprotocristid, at its junction with the mesiobuccal crest of the metaconid; (7) the oblique crest linking the metaconid and protoconulid divides the trigon basin into two small foveae; (8) the mesial fovea is narrow and restricted in size; (9) the protoconid–metaconid and hypoconid–entoconid are generally arranged in transverse pairs; (10) a well-developed mesoconid (= centroconid) is present; (11) the hypometacristid and hypoprotocristid are directed distally at an oblique angle to join the mesoconid; (12) the cristid obliqua is strongly obliquely oriented, and terminates at the mesoconid; (13) the buccal and lingual notches are well developed; (14) the buccal cingulum is very reduced or absent; (15) the hypoconulid is positioned close to the midline of the tooth, and is relatively reduced in size on dP_4–M_2; (16) small subsidiary tubercles occur frequently on the distal margin of the crown on either side of the hypoconulid; (17) the talonid is very elongated in M_3; and (18) there is a steep increase in size from M_1 to M_3.

Other specializations of the lower dentition include: (1) the lower incisors have three well-developed mammelons; (2) P_3 tends to be bicuspid, with a well-developed metaconid in most specimens; (3) P_4 is longer than broad, with two subequal cusps separated by a deep V-shaped notch; and (4) dP_3 is an elongate molariform tooth.

The upper molars and dP^4 are derived in the following respects: (1) the crown is longer than broad, narrows distally, and is slightly waisted midway along its length; (2) the crown is relatively high, and the cusps are voluminous, with high relief; (3) the paraconule is well developed and situated close to the midline of the tooth; (4) there is a distinct parastyle on the mesial margin of the crown; (5) the paraconule–parastyle complex is borne on a mesially extended cingular shelf; (6) the mesial fovea is restricted to a small, pitlike depression; (7) the trigon basin is very restricted because of the narrowness of the crown and the voluminous nature of the cusps; (8) there is a deep V-shaped buccal notch; (9) there is a well-developed metaconule on the crista obliqua; (10) the prephypocone crista passes mesiobuccally from the hypocone to terminate at or close to the metaconule; (11) a transverse crest links the hypocone and metacone to define a small triangular pit just distal to the crista obliqua; (12) there is often a distinct midline conule on the distal cingular shelf that presumably represents the homologue of the metastyle; (13) the lingual cingulum is prominent mesially, but is narrow and discontinuous lingually; and (14) the upper molars increase in size from M^1 to M^3.

Other specializations of the upper dentition include: (1) the I^1 is very robust, with a great buccolingual thickness at the base of the crown, and little mesiodistal expansion toward the apex of the crown; (2) the unworn apex of I^1 is thick, and it bears a number of small mammelons, of which the central one is particularly prominent; (3) the lingual cingulum on I^1 is extremely well developed, very elevated, and it bears a prominent cusplike tubercle that reaches almost to the apex of the crown; (4) wear on the upper and lower central incisors suggests that overbite was the common condition; (5) marked differential wear on the anterior teeth and cheek teeth suggests that preparation of food items using the incisors and canines was an especially significant behavior; (6) the upper premolars tend to be ovoid in occlusal outline, with a protocone that is only slightly lower than the paracone, and a distinct lingual cingulum on both P^3 and P^4; (7) dC^1 has a distinct tubercle on the lingual cingulum, and is double-rooted; and (8) dP^3 is an elongate, molariform tooth.

Cranium and Mandible

As might be anticipated for a primate with such distinctive teeth, the skull also demonstrates a unique suite of features. Some of the key autapomorphies distinguishing *Oreopithecus* from other catarrhines can be summarized as follows: (1) the face is relatively very short, particularly for such a large-bodied catarrhine; (2) the upper incisors are only slightly procumbent, with the roots implanted almost vertically in the relatively abbreviated premaxilla (most similar to humans among extant anthropoids); (3) the maxillary sinus is extremely restricted (unlike extant hominoids, but similar to Old World monkeys); (4) the nasal process of the maxilla forms a short, flaring winglike support for the nasal bones (similar to that in hylobatids, and superficially to that in humans); (5) the anterior root of the zygomatic arch is placed far forward on the face, vertically above P^4/M^1 (as in extant primates adapted for powerful chewing); (6) the zygomatic arch is slightly upwardly curved, dorsoventrally deep, and strongly laterally flaring (similar to extant primates adapted for folivory or powerful chewing); (7) the glenoid fossa is relatively short and deep, bordered anteriorly by a prominent articular eminence (similar, to some extent, to the specialized pattern in *Homo*); (8) the neurocranium is relatively small, and the estimated cranial capacity indicates a degree of encephalization that falls at the lowest end of the range for modern catarrhines (Harrison, 1989); (9) the temporal lines in both males and females are well defined, and they converge anteriorly, just behind the supraorbital torus, to produce a small frontal trigone and a pronounced sagittal crest (cresting in females is very rare in extant catarrhines, but does occasionally occur in *Theropithecus* and *Gorilla*); (10) the lambdoid crest is extremely well developed; (11) the lower incisor roots are implanted more or less vertically in the mandible; (12) the external surface of the mandibular symphysis is steeply inclined to almost vertical (as in *Symphalangus*); (13) the mandibular corpus is relatively robust below the cheek teeth; (14) the mental foramen, located midway down the corpus below P_3, is higher and more anteriorly placed than in any other catarrhine (with the exception of *Homo*); (15) the ramus has an expanded gonial area, and its posterior margin ascends almost vertically (as in primates with specialized folivorous diets); (16) the anterior margin of the ramus has a distinct overlap with M_3 in lateral view; (17) the articular condyle of the mandible is broad, mediolaterally strongly convex, with no extension posteriorly for articular contact with the postglenoid process (this is a unique combination of features among catarrhines, but is most similar to the derived condition in *Homo* and *Symphalangus*) (Harrison, 1987a, 1989; Rook *et al.*, 1996) (see Fig. 1).

Postcranium

Oreopithecus appears to have few unique postcranial specializations that distinguish it from the primitive hominoid or hominid morphotypes. Several features of the elbow, pelvis, and foot might possibly be construed as autapomorphies, but these are relatively minor details, and it is remarkable just how closely *Oreopithecus* conforms to the inferred primitive hominoid morphotype from which the more specialized postcranial patterns of the extant hominids have been derived (Harrison, 1987a, 1991; Sarmiento, 1987).

Implications

This review of the peculiarities of the teeth and skull highlights the morphological uniqueness of *Oreopithecus*. Despite the attention that has been directed in the literature to the importance of these features for understanding the phylogenetic relationships of *Oreopithecus*, they are really much more helpful for reconstructing its paleobiology. The craniodental evidence suggests that *Oreopithecus* had a unique suite of adaptations associated with a specialized dietary behavior. The individual specializations of the cranium and mandible listed above can be grouped into three main derived structural–functional complexes: (1) development of larger and more anteriorly placed masticatory musculature; (2) shortening and increased robusticity of the lower face and mandible; and (3) relative reduction in the size of the neurocranium. These structural–functional complexes are, in turn, components of a larger functional–behavioral complex, to which the dental specializations can be added, that can be associated with a particular dietary niche seemingly involving powerful chewing and the processing of leaves and other fibrous plant materials. There are no close analogues among extant primates for reconstructing the dietary behavior of *Oreopithecus*, and other mammals with similar dental specializations, such as suids, tayassuids, and macropodid marsupials, may provide more useful models. It is evident from the nature of its dental and cranial specializations, however, that *Oreopithecus* is not the end product of some failed attempt by an obscure hominoid stranded on a small group of islands in the northern Tethys to emulate the adaptive success of its phyletic cousins, the cercopithecids. The direct antecedents of Miocene cercopithecids were, after all, "dental apes," craniodentally very similar to the basal stock from which the *Oreopithecus* lineage diverged, that successfully evolved cercopithecid dental specializations. We also know from other mammal groups, such as the macropodid and phalangerid marsupials, and the subfossil Malagasy indrioids, that convergence on the cercopithecid dental pattern can be achieved to quite remarkable degrees, even when originating from very different ancestral patterns. We can be sure from this that *Oreopithecus* is not a monkeylike ape that failed to make the grade. It is something entirely different; a unique and fascinating animal that provides new insights into the past adaptive diversity of catarrhine primates.

Aspects of the local ecology, or perhaps the increased availability of certain niches as a result of the impoverished insular faunal community, allowed *Oreopithecus* access to resources that required unusual specializations for a primate to exploit. Even though we have a reasonably good understanding of the general ecology and vegetation that existed in the Baccinello region during the late Miocene (Harrison and Harrison, 1989), it is impossible to ascertain the specific types of plant foods that *Oreopithecus* might have exploited. Nevertheless, a number of pieces of evidence might allow us to speculate about some general aspects of the dietary behavior of *Oreopithecus*: (1) *Oreopithecus* is an exceedingly common element of the fauna at sites such as Baccinello, and we can deduce from this that *Oreopithecus* probably occurred in quite high densities; (2) the vegetation, as reconstructed from palynological data, indicates a warm temperate woodland that would have been subject to seasonal fluctuations in temperature and precipitation, even after accounting for the possible ameliorating effects on the climate related to the close proximity of the Tethys. These two pieces of evidence suggest that *Oreopithecus* was primarily exploiting a locally abundant food resource that was available throughout the year, such as the leaves (and perhaps seasonally also the cones and seeds) of evergreen trees and shrubs. This interpretation would be consistent with the craniodental specializations and relative brain size, as well as with evidence from dental microwear and molar shearing-crest development (Dirks, personal communication; Ungar and Kay, 1995). Another possibility is that *Oreopithecus* was exploiting aquatic or wetland plants, such as water lilies, reeds, sedges, cattail, pondweeds, horsetails, and stoneworts, all of which are abundantly represented in the pollen spectrum from Baccinello (Harrison and Harrison, 1989). However, in this case, we might have anticipated a more derived postcranium, with greater specialization, perhaps, for terrestrial locomotion or hindlimb suspension that would have enabled *Oreopithecus* to forage close to the water's edge.

As mentioned above, the craniodental specializations are of limited utility for establishing the phylogenetic relationships of *Oreopithecus*. In fact, because the specializations occur mainly in the cheek teeth and lower face, the main source of characters used to establish the phylogenetic and taxonomic relationships of fossil primates, it has had a confounding effect on interpretations of the relationships of *Oreopithecus*. However, because the specializations are so distinctive they can be helpful in recognizing taxa that are closely related to *Oreopithecus*. Harrison (1986b), for example, has previously made a case to link *Oreopithecus* with *Nyanzapithecus* (and possibly also *Rangwapithecus*) from the early and middle Miocene of East Africa on the basis of synapomorphies of the cheek teeth. The upper premolars of *Nyanzapithecus* resemble those of *Oreopithecus* in being relatively narrow and ovoid in shape, the two cusps are voluminous, crowded together, and similar in height, and a lingual cingulum is developed on both P^3 and P^4. The upper molars are very similar in their detailed structure, and they share the following specializations: the crown is longer than broad, becoming narrower distally; the four main cusps are inflated and restrict the size of the occlusal basins; the mesial fovea is small and its surrounding structures reorganized; the prehypocone crista passes directly from the hypocone to the crista obliqua; and the crowns increase in size from M^1 to M^3. P_4 is similar in that the crown is long and narrow, and the two cusps are elevated, subequal in height, transversely aligned, and separated by a deep V-shaped sulcus. The lower molars resemble *Oreopithecus* in being relatively long and narrow, and in having inflated cusps, restricted occlusal basins, and a poorly developed buccal cingulum.

On the basis of these resemblances, Harrison (1986b, p. 279) concluded that "[T]he degree of similarity of the molars and premolars of *Nyanzapithecus* and *Oreopithecus* is so marked, and the specializations they share so distinctive, that there can be little doubt that the two taxa are closely phyletically related." This led Harrison (1986b) to include *Nyanzapithecus* together with *Oreopithecus* in a distinct hominoid family, the Oreopithecidae. However, while this scheme still remains an option, we are now more inclined to accept that the dental similarities shared by *Nyanzapithecus* and *Oreopithecus* are homoplasies, rather than synapomorphies. This is mainly related to the fact that new discoveries of *Dryopithecus* from Rudabánya (Hungary) and Can Llobateres (Spain) have established for the first time just how similar this taxon is in its cranial morphology to *Oreopithecus*. We acknowledge the possibility that many of the shared features of the cranium of *Dryopithecus* and *Oreopithecus* could well be retained primitive hominid characteristics. However, given the degree of similarity, it now seems likely that *Oreopithecus* is more closely related to *Dryopithecus* than it is to *Nyanzapithecus*. The recently recovered partial skeleton of *Dryopithecus* from Spain provides further support for a close phyletic relationship with *Oreopithecus* (Moyà-Solà and Köhler, 1996). In fact, *Dryopithecus*, or a close European relative of *Dryopithecus*, could well be a candidate for an ancestral taxon from which *Oreopithecus bambolii* was derived. Until detailed comparisons with other late Miocene Eurasian hominids can be undertaken, we provisionally accept the proposal that *Oreopithecus* represents a highly specialized and insular member of the Dryopithecinae. In addition to the phylogenetic and taxonomic implications, this finding also has important consequences for understanding the biogeographic relationships of *Oreopithecus*. It signifies that the direct ancestor of *Oreopithecus* was an immigrant into the Tusco-Sardinian bioprovince from continental Europe rather than from Africa as was originally supposed.

As noted by Harrison (1991), *Oreopithecus* retains a phylogenetically conservative and functionally generalized postcranium conforming closely to the ancestral hominid morphotype. *Oreopithecus* appears to be less derived in most features of its postcranial skeleton than extant hominids, and as such, it provides a close approximation to the morphotype from which the more specialized postcranial patterns of the extant taxa were derived (Harrison, 1986b, 1991; Sarmiento, 1987).

Derived Features Shared by Oreopithecus *and Extant Hominoids*

A number of authors have previously shown that *Oreopithecus* shares an extensive series of important synapomorphies of the postcranium with the extant hominoids (Hürzeler, 1958, 1968; Schultz, 1960; Knussmann, 1967;

Stern and Jungers, 1985; Susman, 1985; Harrison, 1986a,b, 1987a, 1991; Sarmiento, 1987, 1988; Rose, 1988, 1993). These synapomorphies are so numerous and so detailed that there can be little doubt that they indicate a close phylogenetic relationship. Craniodental features tend to be less helpful for establishing relationships, in this case, mainly because of the large number of unique traits, and because of the paucity of characters that serve to distinguish the inferred primitive hominoid and catarrhine morphotypes.

Cranium and Dentition

One of the main reasons why stem catarrhines, such as *Pliopithecus, Dendropithecus, Turkanapithecus,* and *Proconsul,* have often been regarded in the past as hominoids, is that their teeth and skulls bear a closer phenetic similarity to extant hominoids than they do to the cercopithecids. However, much of this similarity reflects the retention of primitive catarrhine characteristics (Harrison, 1982, 1987b). In fact, there are relatively few craniodental specializations that can be considered to distinguish the ancestral catarrhine morphotype from the ancestral hominoid morphotype. Nevertheless, Harrison (1987b) provided a short list of derived features shared by extant hominoids that are not found among stem catarrhines. The most important of these are: (1) a relatively broad palate anteriorly, associated with mesiodistally broadened upper incisors; (2) upper molars slightly broader than long, with low rounded cusps, and a reduced lingual cingulum; (3) P_3 with a moderately short honing face, and somewhat reduced sectorial function; (4) lower molars relatively broad, with reduced buccal cingulum; and (5) M_3 reduced in size relative to M_2. One important point to stress from the outset, however, is that a number of these features are functionally associated with adaptations for a frugivorous diet, and it is uncertain to what extent they represent a derived character complex shared primitively by all extant hominoids, or one developed independently several times as a consequence of functional convergence. Even so, the modified P_3 and the reduction of the molar cingulum are probably independent characters that serve to link extant hominoids. In most of these respects *Oreopithecus* is far too derived to be able to be compared meaningfully with the condition in hominoids. In fact, *Oreopithecus* retains several features that are apparently closer to the primitive catarrhine condition than that seen in modern hominoids (i.e., a palate that narrows slightly anteriorly, relatively small incisors, long and narrow lower molars, and an M_3 that is much larger than M_2), but, as discussed above, these are all likely to be part of a suite of derived features that are unique to *Oreopithecus.* Similarly, extant hominoids have upper molars that are relatively narrower than those of primitive catarrhines, and *Oreopithecus* is further specialized in having upper molars that are distinctly longer than broad.

Of all of the hominoid synapomorphies, the structure of P_3 is probably the most useful, since the canine–anterior premolar complex of *Oreopithecus* appears to be less derived than the rest of its dentition. Cercopithecoids and hominoids have developed divergent P_3 morphologies with respect to the primitive catarrhine morphotype (Harrison, 1987b). The latter is inferred to include a relative high-crowned P_3 with a moderate degree of enamel extension onto the mesial root of the tooth, as in *Proconsul, Dendropithecus,* and other early Miocene catarrhines from East Africa. Cercopithecids, on the other hand, are more derived in having a relatively elongated and low-crowned P_3, and a pronounced extension of enamel onto the mesial root. This produces a very long honing face for occlusion with the upper canine, especially in males. Hominoids, by contrast, particularly hominids, have relatively much shorter honing faces on P_3, and they also exhibit a more restricted range of sexual dimorphism than in cercopithecids (Harrison, 1987b; Robinson, 1996). In this respect, the P_3 in *Oreopithecus,* with its relatively short mesial honing face, is similar to the condition in extant hominoids, and is closest to the derived hominid pattern (Table I).

Another feature that may prove to be significant in *Oreopithecus* is the relative degree of pneumatization of the auditory region. Catarrhines are more derived than other primates in having a relatively less inflated bulla, even when body size is taken into consideration. Even so, Old World monkeys still retain a distinctly domed bony mass medial to the auditory process, while in hominoids this region has a low topography, and tends to be relatively flat. The morphology in *Oreopithecus* is comparable to the derived condition seen in extant hominoids.

Table I. Relative Length of the Mesial Honing Face on P_3 in Fossil and Extant Catarrhines[a]

	Sex	*N*	Mean	Range	S.D.
Hominids[b]	F	27	134.0	93.2–167.7	17.69
	M	30	144.5	102.3–182.9	19.16
Hylobatids[c]	F	20	151.1	119.0–175.0	13.50
	M	35	150.8	123.8–194.9	16.07
Colobines[d]	F	25	181.8	150.0–220.5	21.77
	M	31	221.6	155.8–304.7	37.59
Cercopithecines[e]	F	30	215.9	156.9–307.8	36.26
	M	34	287.8	217.1–383.3	43.81
Oreopithecus	M + F	8	132.3	100.0–156.3	16.12
Proconsul[f]	M + F	15	156.4	128.8–182.4	16.69
Dendropithecus[g]	M + F	12	149.5	136.0–169.2	10.40
Propliopithecus[g]	M + F	5	165.1	146.4–191.9	18.43

[a]Index = length of mesial honing face × 100/buccolingual breadth of crown.
[b]Includes *Pongo pygmaeus* (n = 13); *Pan troglodytes* = (n = 19); *Gorilla gorilla* (n = 25).
[c]Includes *Hylobates* spp. (n = 42); *Symphalangus syndactylus* (n = 13).
[d]Includes *Colobus guereza* (n = 18); *Presbytis* spp. (n = 12); *Nasalis larvatus* (n = 14); *Trachypithecus cristatus* (n = 12).
[e]Includes *Papio anubis* (n = 16); *Macaca* spp. (n = 35); *Cercopithecus aethiops* (n = 13).
[f]Data from Andrews (1978).
[g]Data from Harrison (1982).

Furthermore, Grine *et al.* (1985) have demonstrated that the ultrastructural features of the enamel on the molars of *Oreopithecus* (i.e., intermediate thick enamel produced by rapid amelogenesis, and a predominance of prism development pattern 3) also support affinities with the extant hominoids.

Postcranium

Oreopithecus shares numerous specialized features of the postcranium with extant hominoids, and these provide the main basis for recognizing a close phylogenetic relationship (Harrison, 1991). The list of derived characters that *Oreopithecus* shares with the inferred primitive hominoid morphotype is extensive, but these can be grouped into a more manageable series of structural–functional complexes (Table II). From these we may deduce that the last common ancestor of hominids was a large-bodied arboreal primate, adapted for vertical climbing on large-diameter supports, for bridging, and for forelimb suspension on smaller caliber supports (Harrison, 1987a, 1991). Since the entire suite of postcranial adaptations in *Oreopithecus* and hominoids appears to be associated with an integrated behavioral complex, it is certainly possible that these shared features could have been developed independently

Table II. Features of the Postcranial Skeleton That *Oreopithecus* Shares with the Primitive Hominoid Morphotye Organized into Structural–Functional Complexes[a]

1. Strongly differentiated usage of the forelimb and hindlimb
   - Forelimbs considerably longer than the hindlimbs
   - Humeral head large in relation to the size of the femoral head
2. An increased potential for raising the forelimb above the head
   - Mediolaterally broad and dorsoventrally shallow thorax
   - Low angle between the neck and corpus of the superior ribs
   - Broad and deeply concave glenoid fossa of scapula
   - Acromion process long and widely flaring, with a relatively long neck
   - Coracoid process of scapula robust and elongated
   - Low gleno-axillary angle
   - Head of humerus large, globular, and elevated well above the tuberosities
   - Slight to moderate humeral torsion
   - Deltoid insertion located relatively distally on the shaft of the humerus
3. An increased potential for full extension and powerful flexion of the forelimb at the elbow
   - Elongated and robust coracoid process on the scapula
   - Deep, narrow, and well-defined bicipital groove on the proximal humerus
   - Radial neck relatively long
   - Deep olecranon fossa on the distal humerus, with a high incidence of fenestration
   - Strongly waisted trochlea with well-developed median trochlear keel on the distal humerus
   - Very reduced olecranon process on the proximal ulna
   - Sigmoid notch of proximal ulna broad and posteriorly tilted, with a well-developed median keel

(continued)

Table II. (*Continued*)

4. A greater potential for circumduction at the shoulder joint and pronation–supination at the elbow and wrist joints
    - Broad and deeply concave glenoid fossa of scapula
    - Large and globular humeral head, projecting well above the tuberosities
    - Expansive and globular capitulum of distal humerus
    - Head of radius circular in proximal view, with symmetrical capitular depression
    - Broad radial notch on proximal ulna
5. An increased range of abduction–adduction and greater potential for powerful flexion at the wrist
    - Lunate and scaphoid with extensive articular surface for radius
    - Hamate with extremely well-developed unciform process
6. An increased potential for powerful manual grasping of large-diameter supports
    - Relatively short, mobile, and fully opposable pollex
    - Long and curved metacarpals and manual phalanges
    - Strongly developed palmar crests for the attachment of the fibrous sheaths of the flexor tendons
7. Adoption of a more orthograde posture
    - Robust cervical vertebrae with short, stout, and bifid neural spines
    - Mediolaterally broad and dorsoventrally shallow thorax
    - Clavicle moderately long and robust
    - Increased diameter and reduced length of the thoracolumbar vertebrae
    - Lumbar region shortened to five elements, with reduction in size of the articular and accessory processes
    - Transverse processes of lumbar vertebrae originate from the base of the pedicles
    - Increased number of sacral vertebrae to six elements
    - Relatively broad ilium
    - No external tail
8. An increased potential for full extension of the hip and knee joints
    - Ilium mediolaterally broad and coronally aligned
    - Prominent anterior inferior iliac spine
    - Large and deeply excavated fovea capitis on femoral head
9. Greater ranges of rotation of the hip and knee joints and inversion–eversion at the ankle joint
    - Globular femoral head, clearly distinct from the neck
    - Long and steeply inclined femoral neck
    - Patella relatively broad, and patella groove on femur broad and shallow
    - Distal end of femur anteroposteriorly short and broad
    - Structure of subtalar joint of calcaneus and talus
    - Cuboid with well-developed beaklike process for articulation with the distal calcaneus
10. An increased potential for body weight to be supported by a single hindlimb during climbing
    - Neck of femur relatively long and steeply inclined
    - Slightly laterally tilted femoral shaft
    - Medial condyle of femur slightly larger than the lateral condyle
11. An increased ability of the foot to grasp and to provide powerful push-off from large-diameter vertical supports
    - Tarsus relatively abbreviated, mobile and compact, with strong scars for interosseus ligaments
    - Prominent plantar tubercle on the heel of the calcaneus
    - Stout and fully abductable hallux
    - Moderately long and curved pedal phalanges

[a] Adapted from Harrison (1991).

as a consequence of convergent locomotor behaviors. However, the post-cranial similarities between *Oreopithecus* and extant hominoids are so wide-spread throughout the skeleton, and are so detailed, that the likelihood of this seems extraordinarily remote. It is true that some members of other primate lineages, such as the indrioids and the atelins, have developed similar locomotor adaptations, but their level of convergence is much less detailed, and in several complexes functional congruence has been achieved entirely through alternative structural pathways. The fact that only *Oreopithecus* and the living apes have developed this particular combination of postcranial specializations, and with such a remarkable degree of structural similarity in the individual characters, clearly indicates that *Oreopithecus* is phyletically closely related to the extant hominoids.

Implications

Since most of the previously recognized synapomorphies are confined to structural complexes of the postcranium, and since these are functionally associated with a particular mode of progression, it could be argued that such features are susceptible to high levels of homoplasy. On this basis, it has been suggested that *Oreopithecus* may have independently acquired a suite of post-cranial characters very similar to hominoids as a consequence of similar locomotor repertoires (Szalay and Delson, 1979; Szalay and Langdon, 1987; Delson, 1988). This is certainly a conceivable proposition. After all, the post-crania of atelins and certain subfossil Malagasy primates have converged to quite a remarkable degree on those of hominoids. If these distant relatives of hominoids have successfully developed similar structural–functional complexes in response to similar behaviors, then we might predict that *Oreopithecus*, a close phyletic relative, has the potential to develop convergences to an even greater degree. This proposal can be countered in two main ways. First, we reiterate the point made forcefully by Harrison (1987a, 1991) and by Sarmiento (1987) that the postcranial characteristics shared by *Oreopithecus* and the extant hominoids are so pervasive throughout the skeleton that it is almost impossible to consider that these could have been developed independently to such a remarkable degree of detail in *every* anatomical region. Second, hominoid synapomorphies are not restricted to features of the post-cranium in *Oreopithecus*. As discussed above, and in the next section, there are a number of characters of the dentition and cranium that represent important shared derived features linking *Oreopithecus* with extant hominoids. These include specializations of the P_3, incisive canal, and subarcuate fossa, features that are clearly unrelated to postcranial specialization and locomotor behavior, and provide independent confirmation of the hominoid affinities of *Oreopithecus*. The combined evidence from the craniodental and postcranial morphology is convincing enough to conclude that *Oreopithecus* is a hominoid, but this still leaves open the more difficult problem of establishing its relation-ships to other members of the superfamily.

Derived Features Shared by Oreopithecus and Extant Hominids

Harrison (1987a, 1991) and Andrews *et al.* (1996) have already made reference to several derived features that *Oreopithecus* shares with hominids, but the list of potential synapomorphies is much more extensive. Below, we present a summary of some of the more important features that indicate a close relationship between *Oreopithecus* and one or more of the extant hominids. Most of these features are postcranial specializations, but there are also several characters of the teeth and cranium.

Cranium and Dentition

Few derived features of the dentition are shared by *Oreopithecus* and extant hominids. The molars and incisors of *Oreopithecus* are obviously too specialized to yield any pertinent information, but the more generalized premolars do share several derived features with those of extant hominids. These include: (1) a relatively high incidence of a well-developed metaconid on P_3 (more than 80% in *Oreopithecus*); (2) the mesial honing face on P_3 is relatively very short (see Table I, and discussion above); (3) the protocone is only slightly less elevated than the paracone on the upper premolars (hylobatids and proconsulids are more primitive in retaining a greater differential between the two cusps; see Table III; and (4) the upper premolars are large in relation to the molars (hylobatids and proconsulids can be distinguished from extant hominids by having relatively much smaller premolars, and *Oreopithecus* is intermediate in this respect; see Table IV).

The cranium of *Oreopithecus* also has a number of derived features that it

Table III. Relative Height of Cusps on Upper Premolars[a]

		P^3			P^4	
Taxon	*N*	Mean	Range	*N*	Mean	Range
Hylobates moloch	10	60.8	45.3–72.7	10	83.9	61.9–100.0
Hylobates lar	24	62.3	46.8–80.0	24	82.5	58.1–109.1
Symphalangus syndactylus	12	67.0	48.3–89.1	12	89.7	74.4–100.0
Pongo pygmaeus	10	67.4	63.9–73.5	10	87.6	73.5–98.9
Gorilla gorilla	22	77.1	62.0–87.6	22	91.5	82.2–99.1
Pan troglodytes	18	69.1	56.4–86.2	18	86.7	72.1–94.0
Oreopithecus bambolii	6	86.0	78.6–95.0	10	97.0	84.8–114.0
Pronconsul spp.[b]	12	63.3	57.4–72.2	18	80.6	70.0–94.4
Dendropithecus macinnesi[c]	6	61.3	50.7–71.7	5	86.6	81.3–92.3
Propliopithecus spp.	2	64.9	63.6–66.2	4	84.7	81.0–87.5

[a]Index = height of protocone × 100/height of paracone.
[b]Data from Andrews (1978).
[c]Data from Harrison (1982, 1988).

Table IV. Relative Size of Upper Premolars[a]

Taxon	N	P3 Mean	P3 Range	N	P4 Mean	P4 Range
Hylobates moloch	10	56.6	47.0–67.3	10	53.4	46.6–64.7
Hylobates lar	26	59.4	47.3–72.5	26	58.1	48.4–71.2
Symphalangus syndactylus	12	53.2	46.2–65.5	12	54.4	43.2–66.5
Pongo pygmaeus	15	75.9	65.9–97.0	15	71.9	60.0–83.1
Gorilla gorilla	29	70.9	58.1–92.4	29	64.2	56.6–72.0
Pan troglodytes	20	70.0	53.6–89.7	20	62.9	56.1–79.0
Oreopithecus bambolii	10	62.7	53.9–70.7	10	58.9	54.2–66.5
Pronconsul spp.[b]	5	59.3	52.5–66.9	5	51.1	46.7–60.0
Dendropithecus macinnesi[c]	3	59.1	54.4–62.6	3	55.8	50.3–59.2
Propliopithecus spp.	4	48.0	41.8–53.3	4	40.7	37.8–44.4

[a]Index = length × breadth of upper premolar × 100/length × breadth of M^2.
[b]Data from Andrews (1978).
[c]Data from Harrison (1982, 1988).

shares with hominids. These include: (1) a relatively deep subnasal clivus (it is relatively deeper than in hylobatids and most cercopithecids, but it is still shallower than in all extant hominids); (2) the incisive canal is characterized by a pair of relatively large foramina, set well back from the alveolar margin of the upper central incisors, that opens anteriorly onto the palatal aspect of the premaxilla via distinct vestibular grooves (this pattern is closely similar to that seen in the extant African apes, especially *Gorilla,* and it presumably represents the primitive hominid condition); (3) lack of a distinct subarcuate fossa in the petrosal (hominids are unique among extant primates in the uniform absence of a subarcuate fossa); (4) the supraorbital torus is moderately thick and rounded, with a shallow supratoral sulcus and a slightly inflated glabellar region [the morphology is most similar to the African apes among extant primates, but differs in having less heavily inflated tori and a less distinct supratoral sulcus; however, it is closely comparable to the supraorbital morphology seen in *Dryopithecus*, and to a lesser extent in *Graecopithecus* (Kordos, 1987; Bonis *et al.*, 1990; Begun, 1992, 1994; Bonis and Koufos, 1993; Moyà-Solà and Köhler, 1995), and this is inferred to be a primitive hominid pattern]; (5) the presence of a small frontal sinus (as in *Dryopithecus,* this is considered to be a primitive hominid feature, structurally related to supraorbital torus development, and secondarily lost in pongines; the phylogenetic significance of this feature is rather complicated, however, as it seems to have been acquired independently in several lineages of extant anthropoids); and (6) the supraorbital foramen is entire, and it perforates the corpus of the supraorbital torus (a feature seen in high frequencies only in the African apes, humans, and *Cebus,* but is possibly structurally related to supraorbital torus development; see Msuya and Harrison, 1994). These characters, especially the structure of the incisive canal and the development of the subarcuate fossa, provide strong evidence that *Oreopithecus* is most closely related to the extant hominids.

Almost all of the craniodental features of *Oreopithecus* can be interpreted as either unique specializations or primitive hominid characters. There appear to be no features that can be confidently identified as shared derived characters linking *Oreopithecus* with any one particular group of extant hominids. Moyà-Solà and Köhler (1993) have recently suggested that *Dryopithecus*, which like *Oreopithecus* retains a basically conservative hominid facial pattern, has a derived zygomatic morphology that it shares uniquely with pongines. This includes a robust bone, with a rugose superior portion, and three zygomaticofacial foramina located high on its frontal process. Intriguingly, *Oreopithecus* has a very similar pattern, with a robust zygomatic, perforated by multiple foramina (modal number = 3) located well above the horizontal level of the inferior margin of the orbit. However, the significance of these features as potential synapomorphies of pongines is a contentious issue. While it is true that *Pongo* can be distinguished from the other extant hominoids by having a higher modal number of zygomaticofacial foramina, which tend to be located higher on the face, the purported polarity of the transformation sequence is not so evident when viewed in a wider comparative context. The inferiorly placed foramina in hylobatids, for instance, are almost certainly a product of their specialized orbital morphology, in which the lateral rims are slender and slightly protruding, and this condition is unlikely to be homologous with the zygomatic pattern typical of African apes and humans. By contrast, a similar condition to that seen in *Dryopithecus* and *Oreopithecus* is typically found in colobines, some cercopithecines, and in early fossil catarrhines, such as *Propliopithecus, Pliopithecus, Turkanapithecus*, and *Proconsul*. Moreover, multiple zygomaticofacial foramina are very common among extant primates, with three foramina being the modal number in *Pan, Symphalangus*, and many species of cercopithecids and platyrrhines (Msuya and Harrison, 1994). It seems more likely, therefore, that the presence of three prominent zygomaticofacial foramina located high on the frontal process of the zygomatic is a catarrhine symplesiomorphy, rather than a feature that is shared uniquely by certain pongines.

Postcranium

In addition to the cranial characters listed above, there is an extensive series of postcranial features that provides evidence that *Oreopithecus* is phyletically closely related to the hominids. This includes features of the axial skeleton, pelvis, hindlimb, and forelimb (see Table V). As noted above, the postcranium of *Oreopithecus* probably represents a close approximation to the primitive hominid morphotype. It certainly lacks any of the postcranial novelties associated with the pongines. However, there is a complex of shared derived features of the foot in *Oreopithecus* that is unique to the African apes

Table V. Synapomorphies of the Postcranium of Extant Hominids Also Shared by *Oreopithecus*

- Axial skeleton
  - Atlas vertebra
    - Posterior arch of neural canal robust, with well-developed dorsal tubercle
    - Narrow keyhole-shaped neural canal
  - Axis vertebra
    - Dens short and stout
    - Short, but well-developed neural spine
    - Foramen transversarium well separated from neural canal
  - Lumbar vertebrae
    - Accessory processes absent
  - Ribs
    - Neck short and relatively thick
- Forelimb
  - Ulna
    - Trochlear notch relatively broad
    - Coronoid process projects anteriorly much more than the olecranon beak, producing a backward-tilted trochlear notch
    - Trochlear notch with well-developed median keel
    - Proximal end of shaft with quadrilateral cross section
    - Scar for the insertion of brachialis deeply incised and located relatively far distally
  - Humerus
    - Trochlear forms a well-defined spool, with strong anteroposterior waisting
- Hindlimb
  - Pelvis
    - Relatively low, broad, and cranially flaring ilium
    - Ischial tuberosities absent
  - Femur
    - Shaft short and robust
  - Patella
    - Relatively broad
  - Talus
    - Shallow facet for medial malleolus of tibia
    - Dorsal articular surface relatively broad with a shallow trochlear and low rounded medial and lateral margins
    - Anterior subtalar facet is extremely short[a]
  - Calcaneum
    - Anterior segment very abbreviated[a]
  - Navicular
    - Proximodistally short and broad, with well-developed medial tuberosity[a]
  - Cuboid
    - Proximodistally relatively short[a]
    - Well-developed plantar tubercle on the calcaneal articular surface
    - Lacks a distinct facet for the sesamoid of the peroneus longus tendon
  - Cuneiforms
    - Proximodistally relatively very short[a]
    - Medial cuneiform with proximal end elevated with no distinct dorsal step[a]
  - Metatarsals
    - Relatively short and robust[a]

[a]Features shared exclusively with hominines (African apes and humans).

and humans among extant catarrhines (Table V). These mainly relate to a marked abbreviation of the distal segment of the tarsus and increased robusticity of the metatarsus (Riesenfeld, 1975; Szalay and Delson, 1979; Szalay and Langdon, 1987; Harrison, 1987a, 1991; Sarmiento, 1987). Previous authors have explained these apparent synapomorphies either as an independent acquisition that has converged on the African ape pattern (Szalay and Langdon, 1987), or as components of the ancestral hominid morphotype in which secondary reversion to a more primitive state has occurred in *Pongo* (Sarmiento, 1987; Harrison, 1991). However, since both of these assessments are strongly influenced by prior interpretations of the phylogenetic position of *Oreopithecus,* it should not be entirely ruled out that these features might represent important synapomorphies linking *Oreopithecus* with the extant hominines. Intriguingly, in light of the suggestion that *Oreopithecus* and *Dryopithecus* might be closely related, the pedal specializations that *Oreopithecus* shares with the African apes and humans could be used to provide additional independent support for the hominine affinities of *Dryopithecus,* a view that has been recently advocated by Begun (1992, 1994).

Implications

Shared derived features of the dentition, cranium, and postcranium provide good evidence to support the inference that *Oreopithecus* is closely related to the extant hominids. Even though *Oreopithecus* is the most complete fossil hominid known, like other late Miocene Eurasia apes, it has proved extremely difficult to establish its sister-group relationships within the Hominidae. The dentition and cranium represent a combination of autapomorphies and characteristics that are probably close to the primitive hominid morphotype. We can identify no derived craniodental characters that would convincingly link *Oreopithecus* with either the extant pongines or hominines.

As noted by previous authors (Harrison, 1987a, 1991; Sarmiento, 1987), the postcranium of *Oreopithecus* also appears to conform to a remarkable degree to the inferred primitive hominid morphotype. However, it does possess a combination of features, especially in the proportions and morphology of the foot, that are found only among the African apes and humans among extant catarrhines. This evidence can be interpreted in several different ways: (1) the characters are shared derived features of hominids that indicate a close relationship between *Oreopithecus* and hominines; (2) the characters are shared primitive features of hominids that have secondarily reverted back to a more primitive condition in *Pongo;* and (3) the characters have been independently acquired in *Oreopithecus* and extant hominines. The first of these three alternatives obviously represents the most parsimonious, because it avoids the need to invoke homoplasy. However, it is important to emphasize that the characters shared by *Oreopithecus* and extant hominines are part of a single functional complex, relating to the development of a short and highly mobile foot, and one that could be quite easily lost or gained independently in different hominoid lineages. There is no simple solution to this problem, but there is a definite direction in which to proceed. We can try to identify further derived characters from different structural–functional complexes that *Oreopithecus* might share with African apes and humans, thereby providing additional support for a close relationship between them, or alternatively we can recognize derived characters shared by all extant hominids that are not found in *Oreopithecus,* and these could be used to support the proposition that *Oreopithecus* is the sister group to all extant hominids.

There are a number of cranial characters that could be interpreted as shared derived features linking *Oreopithecus* with the hominines. For example, Begun (1992, 1994) has identified a number of features of *Dryopithecus,* but present also in *Oreopithecus,* as representing synapomorphies with the African apes and humans. These include a stepped subnasal fossa, a distinct incisive canal, a shallow canine fossa, a supraorbital torus with inflated glabella, and a small frontal sinus. However, as noted above, we prefer to identify these as primitive hominid features. Unfortunately, precise reconstructions of the primitive cranial morphotype for extant hominids have proved problematic, mainly because of the degree of specialization of the cranial morphology of pongines in relation to hominines. The "fuzzy" nature of the morphotype makes it difficult to discriminate between stem hominids and conservative hominines. This is at the root of current debates concerning the phyletic position of *Dryopithecus,* and it could also have important consequences for interpreting the status of *Oreopithecus.* Part of the solution to the problem clearly resides with the addition of fossil taxa to the analysis, as these potentially offer a wider range of morphologies from which to reconstruct possible transformation sequences and morphotypes. Nevertheless, there is an inevitable danger of succumbing to the pitfalls of circularity, whereby the placement of fossils determines the composition of the primitive hominid morphotype, and this in turn determines where to place the fossils.

A few features can be identified that might be of some significance in helping to establish *Oreopithecus* as the primitive sister group to extant hominids. First, unlike the extant great apes, *Oreopithecus* lacks a pronounced simian shelf. The internal surface of the mandible of *Oreopithecus* is buttressed by a low rounded superior transverse torus, situated about two-thirds down from the alveolar margin, and a bluntly rounded inferior transverse torus, almost equal in size to that of the superior torus. In the relatively weak expression of the transverse tori in general, and the dominance of the superior transverse torus over the inferior transverse torus, *Oreopithecus* closely resembles the pattern seen in hylobatids. Comparisons with other anthropoids suggest that this is the primitive catarrhine condition (Harrison, 1982). Since the primitive hominid morphotype can be inferred to have included a simian shelf, it can be argued that *Oreopithecus* is more primitive in this respect than all extant members of the clade. However, development of the symphyseal tori has almost certainly been influenced by the shortening and subsequent remodeling of the lower face in *Oreopithecus,* and this could have resulted in loss of the simian shelf, just as it was in the short-faced hominins.

Another feature in which *Oreopithecus* appears to be less derived than other hominids is in the structure of the distal articular surface of the metacarpals. In extant great apes the heads of the metacarpals are very broad dorsally, an adaptation that serves to provide greater stability of the metacarpophalangeal joint during full extension. In this respect, *Oreopithecus* is much more similar to the primitive catarrhine condition seen in nonhominoid primates, in which the distal articular surface narrows dorsally. However, there are two reasons why this feature might be of questionable utility as a synapomorphy of the extant hominids: (1) humans also retain a distal metacarpal pattern that is much closer to the primitive catarrhine condition than that seen in the great apes and (2) the specialized hominid pattern is also present in hylobatids, which, as discussed above, is best regarded as the sister taxon to the clade comprising *Oreopithecus* and extant hominids. Although Harrison (1982, 1987b) previously considered the quadrilateral and dorsally broadened metacarpal heads of hylobatids and great apes to be a shared primitive hominoid feature, its distribution among extant and fossil apes suggests that it was independently developed in different hominoid lineages.

In the absence of definitive shared derived characters to link *Oreopithecus* with either of the two extant hominid subfamilies, we tentatively consider it to be a stem hominid. Given the morphological similarities between *Oreopithecus* and *Dryopithecus,* we include these two conservative hominid taxa together in a separate subfamily, the Dryopithecinae. The taxonomy and phylogenetic scheme presented in Table VI and Fig. 2, respectively, reflect our current conception of the relationships of extant and fossil hominoids, and incorporate the revised interpretation of the status of *Oreopithecus* as discussed in this chapter.

Conclusions

In this chapter we have identified several main reasons why the phylogenetic status of *Oreopithecus,* often considered to be an "enigmatic anthropoid," has been particularly difficult to resolve. These include: (1) debates concerning whether or not *Oreopithecus* might be a cercopithecoid or a hominoid have diverted attention away from the more critical problem of determining its relationships within the Hominoidea; (2) in its cranium and dentition, the anatomical regions on which primate paleontologists rely heavily for alpha-taxonomic and phylogenetic assessments, *Oreopithecus* is highly specialized; and (3) until recently, *Oreopithecus* was the only Miocene hominoid known from relatively complete cranial and postcranial material, making it difficult to predict the general morphological pattern typical of conservative European late Miocene hominids. With more detailed information now avail-

Table VI. Classification of Hominoids Showing the Provisional Taxonomic Placement of *Oreopithecus*[a]

Superfamily Hominoidea
  Family Hylobatidae
      Hylobates
      Symphalangus
  Family Afropithecidae
      Afropithecus
      Heliopithecus
  Family Hominidae
    Subfamily Kenyapithecinae
      Kenyapithecus
      Griphopithecus
      ?Maboko hominid
    Subfamily Dryopithecinae
      Tribe Dryopithecini
        Dryopithecus
      Tribe Oreopithecini
        Oreopithecus
    Subfamily Ponginae
      Pongo
      Sivapithecus
      Gigantopithecus
      ?*Lufengpithecus*
    Subfamily Homininae
      Tribe Gorillini
        Gorilla
      Tribe Panini
        Pan
      Tribe Hominini
        Ardipithecus
        Australopithecus
        Paranthropus
        Homo
      Homininae *incertae sedis*
        Graecopithecus
    Hominidae *incertae sedis*
      Otavipithecus

[a]This taxonomic scheme generally follows that of Andrews (1985, 1992). The main differences are as follows: (1) the Proconsulidae is excluded from the Hominoidea, being recognized here as a group of stem catarrhines placed in their own superfamily, the Proconsuloidea (Harrison, 1987b, 1993); (2) Andrews's gradistically based stem hominid group, the Dryopithecinae, is subdivided into three monophyletic taxa, the Afropithecidae, Kenyapithecinae, and Dryopithecinae; (3) *Afropithecus*, and the apparently closely related *Heliopithecus*, are placed in a separate family, the Afropithecidae, which is tentatively retained in the Hominoidea, although it might eventually prove to be better placed in the Proconsuloidea; (4) *Oreopithecus* and *Dryopithecus* are placed together in the Dryopithecinae, but are separated at the tribal level to express their morphological distinctiveness; and (5) the Homininae is provisionally divided into three tribes until such time as a consensus is reached concerning the relationships between the main groups with the subfamily.

Fig. 2. Cladogram showing the inferred relationships between *Oreopithecus* and other extant and fossil catarrhines discussed in this chapter.

able on the anatomy of *Oreopithecus*, in conjunction with comparative anatomy from extant hominoids and recently recovered fossil hominid material from Europe, it has been possible to present here a reassessment of the phylogenetic affinities of *Oreopithecus*. The evidence clearly establishes that *Oreopithecus* is a hominoid, and that its closest affinities are with the extant hominids. Preliminary comparisons with fossil hominoids from Europe show that *Oreopithecus* is not such an enigmatic primate after all. In fact, it is comparable in many respects to the general morphology seen in other late Miocene hominids. The overall similarity to *Dryopithecus* is noteworthy, and this taxon may prove to be close to the ancestral form from which *Oreopithecus* was derived. Consequently, *Oreopithecus* is provisionally included along with *Dryopithecus* in the Dryopithecinae.

The main conclusions of this chapter can be summarized as follows:

1. *Oreopithecus* is associated with a distinctive Tusco-Sardinian fauna, whose high level of endemism and unusual community structure imply an insular environment. During the late Miocene the region probably consisted of a small group of islands in the northern Tethys, that was connected periodically to continental Europe and Africa by ephemeral land bridges. The Tusco-Sardinian fauna appears to have its strongest affinities with those from Europe, although the bovids probably originated in Africa.

2. Previously it has been suggested that *Oreopithecus* was also part of this influx of African mammals. This conclusion was primarily based on the occurrence at early and middle Miocene sites in East Africa of fossil catarrhines, such as *Nyanzapithecus*, with distinctive dental features like those of *Oreopithecus*. However, new evidence suggests that *Oreopithecus* is descended from a

more generalized hominid ancestor in Europe, and that *Nyanzapithecus* is probably a specialized proconsulid that developed similar dental adaptations in parallel to *Oreopithecus*.

3. Superimposed on the primitive hominid features of the skull and dentition in *Oreopithecus* is a suite of unique specializations functionally related to adaptations for powerful chewing and folivory, that suggests that *Oreopithecus* occupied a specialized dietary niche. There are no close analogues among modern primates, but other mammals with similar dental adaptations, such as suids, tayassuids, and macropodids, might provide useful models for reconstructing the dietary behavior of *Oreopithecus*.

4. Paleoecological reconstructions suggest that *Oreopithecus* probably had a relatively narrow ecological preference that included swampy, forested habitats. Seasonal fluctuations in the availability of plant resources, that can be inferred from the palynological evidence, probably necessitated that *Oreopithecus* was specialized in exploiting a locally abundant food resource that was available year-round, such as the leaves of evergreen trees and shrubs, or aquatic plants.

5. A combination of craniodental and postcranial features provides convincing evidence to support the hominid affinities of *Oreopithecus*. However, the problem of establishing whether or not it is a stem hominid or a taxon closely related to the extant pongines or hominines has proved more difficult to resolve. In the absence of definitive shared derived characters to link *Oreopithecus* with either of the two extant hominid subfamilies, we tentatively recognize it as a stem hominid.

6. New discoveries of *Dryopithecus* from late Miocene sites in Europe have revealed the marked similarity between this taxon and *Oreopithecus* in its cranial and postcranial morphology. This probably signifies a close phylogenetic relationship, but further comparisons are needed to document in detail the extent and nature of the similarity. Until such comparisons can be undertaken, we provisionally accept that *Oreopithecus* is closely related to *Dryopithecus*, and that it represents a highly specialized and insular member of the Dryopithecinae.

7. This chapter is intended to be an initial attempt to clarify the phylogenetic relationships of *Oreopithecus*, and hopefully it will provide a foundation from which to develop more sophisticated comparative studies and phylogenetic analyses. We believe that *Oreopithecus*, with its newly established status as the best-known Eurasian hominid, will now take on special significance as a pivotal taxon for future studies of hominid evolution.

Acknowledgments

We thank the editors for inviting us to prepare a contribution for this volume, and for their patience in waiting for the final product. We are extremely grateful to the late Johannes Hürzeler for allowing us to study the *Oreopithecus* material and the associated fauna in his care, and for the hospitality and encouragement that he extended to us during our many visits to Basel. We thank the following colleagues and graduate students for their help, advice, and comments during the course of this research: P. Andrews, A. Azzaroli, E. Baker, D. R. Begun, E. Delson, W. Dirks, B. Engesser, T. S. Harrison, C. J. Jolly, L. Kordos, M. Leakey, J. Manser, R. D. Martin, S. Moyà-Solà, C. Robinson, M. D. Rose, and D. Torre. We also acknowledge with thanks the following institutions for access to fossil and extant mammal collections in their care: Department of Mammalogy, American Museum of Natural History, New York; Departments of Palaeontology and Zoology, The Natural History Museum, London; Museo di Geologia e Paleontologia, University of Florence; Naturhistorisches Museum, Basel; National Museums of Kenya, Nairobi; and Institut Paleontològic, Sabadell. This work was supported by grants from the Leakey Foundation, the Boise Fund, and New York University to T.H., and Ministero dell'Universita e della Ricerca Scientifica e Tecnologica MURST to L.R. under the direction of D. Torre.

References

Andrews, P. 1978. A revision of the Miocene Hominoidea of East Africa. *Bull. Br. Mus. Nat. Hist. Geol.* **30:**85–224.

Andrews, P. 1985. Family group systematics and evolution among catarrhine primates. In: E. Delson (ed.), *Ancestors: The Hard Evidence,* pp. 14–22. Liss, New York.

Andrews, P. 1992. Evolution and environment in the Hominoidea. *Nature* **360:**641–646.

Andrews, P., Harrison, T., Delson, D., Bernor, R., and Martin L. in press. Distribution and biochronology of European and southwest Asian Miocene catarrhines. In: R. Bernor, V. Fahlbusch, and H. W. Mittmann (eds.), *Evolution, Chronology and Biogeographic History of Western Eurasian Later Neogene Mammal Faunas.* Columbia University Press, New York (in press).

Azzaroli, A., Boccaletti, M., Delson, E., Moratti, G., and Torre, D. 1987. Chronological and paleogeographical background to the study of *Oreopithecus bambolii. J. Hum. Evol.* **15:**533–540.

Begun, D. R. 1992. Miocene fossil hominids and the chimp–human clade. *Science* **257:**1929–1933.

Begun, D. R. 1994. Relations among the great apes and humans: New interpretations based on the fossil great ape *Dryopithecus. Yearb. Phys. Anthropol.* **37:**11–63.

Begun, D. R., Moyà-Solà, S., and Köhler, M. 1990. New Miocene hominoid specimens from Can Llobateres (Valles Penedes, Spain) and their geological and paleoecological context. *J. Hum. Evol.* **19:**255–268.

Benvenuti, M., Bertini, A., and Rook, L. 1995. Facies analysis, vertebrate paleontology and palynology in the Late Miocene Baccinello–Cinigiano basin (southern Tuscany). *Mem. Soc. Geol. Ital.* **48:**415–423.

Boccaletti, M., Cosentino, D., Deiana, G., Gelati, R., Lentini, F., Massari, F., Moratti, G., Pescatore, T., Porcu, A., Ricchetti, G., Ricci Lucchi, F., and Tortorici, L. 1987. Neogene dynamics of the peri-Tyrrhenian area in an ensialic context: Paleogeographic reconstructions. *Ann. Inst. Geol. Publ. Ungarn.* **70:**307–321.

Boccaletti, M., Ciaranfi, N., Cosentino, D., Deiana, G., Gelati, R., Lentini, F., Massari, F., Moratti, G., Pescatore, T., Ricci Lucchi, F., and Tortorici, L. 1990. Palinspastic restoration and paleogeographic reconstruction of the peri-Tyrrhenian area during the Neogene. *Palaeogeogr. Palaeoclimatol. Palaeoecol.* **77:**41–50.

Bonis, L. de, and Koufos, G. D. 1993. The face and the mandible of *Ouranopithecus macedoniensis:* Description of new specimens and comparisons. *J. Hum Evol.* **24:**469–491.

Bonis, L. de, Bouvrain, G., Geraads, D., and Koufos, G. 1990. New hominoid skull material from the late Miocene of Macedonia in northern Greece. *Nature* **345:**712–714.

Butler, P. M., and Mills, J. R. E. 1959. A contribution to the odontology of *Oreopithecus. Bull. Br. Mus. Nat. Hist. Geol.* **4:**1–26.

Clark, W. E. L., and Thomas, D. P. 1951. Associated jaws and limb bones of *Limnopithecus macinnesi. Br. Mus. Nat. Hist. Fossil Mamm. Afr.* **3:**1–27.

Cordy, J.-M., and Ginesu, S. 1994. Fiume Santo (Sassari, Sardaigne, Italie): un nouveau gisement à Oréopithèque (*Oreopithecidae, Primates, Mammalia*). *C. R. Acad. Sci. Ser. II* **318:**697–704.

Delson, E. 1979. *Oreopithecus* is a cercopithecoid after all. *Am J. Phys. Anthropol.* **50:**431–432.

Delson, E. 1987. An anthropoid enigma: Historical introduction to the study of *Oreopithecus bambolii. J. Hum. Evol.* **15:**523–531.

Delson, E. 1988. Oreopithecidae. In: I. Tattersall, E. Delson, and J. Van Couvering (eds.), *Encyclopedia of Human Evolution,* pp. 401–404. Garland, New York.

De Terra, H. 1956. New approaches to the problem of man's origin. *Science* **124:**1282–1285.

Engesser, B. 1983. Die jungtertiären Kleinsäuger des Gebietes der Maremma (Toskana, Italien). 1 Teil: Gliridae (Rodentia, Mammalia). *Ecologae Geol. Helv.* **76:**763–780.

Engesser, B. 1989. The Late Tertiary small mammals of the Maremma region (Tuscany, Italy). 2nd part: Muridae and Cricetidae (Rodentia, Mammalia). *Boll. Soc. Paleontol. Ital.* **29:**227–252.

Esu, D., and Kotsakis, T. 1985. Les vertébrés et les mollusques continéntaux du Tertiaire de la Sardeigne: paléobiogéographie et biostratigraphie. *Geol. Rom.* **22:**177–206.

Fleagle, J. G. 1988. *Primate Adaptation and Evolution.* Academic Press, New York.

Giannini, E., Lazzarotto, A., and Signorini, R. 1972. Lineamenti di geologia della Toscana meridionale. *Rend. Soc. Ital. Mineral. Petrol.* Fasc. Spec. **27:**33–168.

Gillet, S., Lorenz H. G., and Woltersdorf, F. 1965. Introduction à l'etude du Miocène supérieur de la région de Baccinello. *Bull. Serv. Carte Geol. Alsace Lorraine* **18:**31–42.

Grine, F. E., Krause, D. W., and Martin L. B. 1985. The ultrastructure of *Oreopithecus bambolii* tooth enamel: Systematic implications. *Am. J. Phys. Anthropol.* **66:**177–178.

Harrison, T. 1982. *Small-Bodied Apes from the Miocene of East Africa.* Ph.D. thesis, University of London.

Harrison, T. 1985. African oreopithecids and the origin of the family. *Am. J. Phys. Anthropol.* **66:**180.

Harrison, T. 1986a. The phylogenetic relationships of the Oreopithecidae. *Am. J. Phys. Anthropol.* **69:**212.

Harrison, T. 1986b. New fossil anthropoids from the middle Miocene of East Africa and their bearing on the origin of the Oreopithecidae. *Am. J. Phys. Anthropol.* **71:**265–284.

Harrison, T. 1987a. A reassessment of the phylogenetic relationships of *Oreopithecus bambolii* Gervais. *J. Hum. Evol.* **15:**541–583.

Harrison, T. 1987b. The phylogenetic relationships of the early catarrhine primates: A review of the current evidence. *J. Hum. Evol.* **16:**41–80.

Harrison, T. 1988. A taxonomic revision of the small catarrhine primates from the early Miocene of East Africa. *Folia Primatol.* **50:**59–108.

Harrison, T. 1989. New estimates of cranial capacity, body size and encephalization in *Oreopithecus bambolii. Am. J. Phys. Anthropol.* **78:**237.

Harrison, T. 1991. The implications of *Oreopithecus bambolii* for the origins of bipedalism. In: Y. Coppens and B. Senut (eds.), *Origine(s) de la Bipédie chez les Hominidés,* pp. 235–244. CNRS, Paris.

Harrison, T. 1993. Cladistic concepts and the species problem in hominoid evolution. In: W. H.

Kimbel and L. B. Martin (eds.), *Species, Species Concepts, and Primate Evolution,* pp. 345–371. Plenum Press, New York.

Harrison, T. S., and Harrison, T. 1989. Palynology of the late Miocene *Oreopithecus*-bearing lignite from Baccinello, Italy. *Palaeogeogr. Palaeoclimatol. Palaeoecol.* **76:**45–65.

Harrison, T., and Rook, L. 1994. The taxonomic, phylogenetic and biochronological implications of new *Oreopithecus* specimens from Baccinello V-2 (Maremma, Italy). *Am. J. Phys. Anthropol. Suppl.* **18:**102.

Heberer, G. 1961. *Die Abstammung des Menschen.* Athenaion, Constance.

Hoffstetter, R. 1982. Les Primates Simiiformes (= Anthropoidea) (compréhension, phylogénie, histoire biogéographique). *Ann. Paleontol.* **68:**241–290.

Hürzeler, J. 1949. Neubeschreibung von *Oreopithecus bambolii* Gervais. *Schweiz. Palaeontol. Abh.* **66:**1–20.

Hürzeler, J. 1951. Contribution à l'etude de la dentition de lait d'*Oreopithecus bambolii* Gervais, 1872. *Eclogae. Geol. Helv.* **44:**404–411.

Hürzeler, J. 1954. Zur systematischen Stellung von *Oreopithecus. Verh. Naturforsch. Ges. Basel* **65:**88–95.

Hürzeler, J. 1958. *Oreopithecus bambolii* Gervais: A preliminary report. *Verh. Naturforsch. Ges. Basel* **69:**1–47.

Hürzeler, J. 1960. The significance of *Oreopithecus* in the genealogy of man. *Triangle* **4:**164–174.

Hürzeler, J. 1962. Quelques réflexions sur l'histoire des anthropomorphes. *Colloq. Int. ÇNRS* **104:**441–450.

Hürzeler, J. 1968. Questions et réflexions sur l'histoire des anthropomorphes. *Ann. Paleontol.* **54:**1–41.

Hürzeler, J. 1975. L'age géologique et les rapports géographiques de la faune de mammifères du lignite de Grosetto (note préliminaire). *Colloq. Int. CNRS* **218:**873–876.

Hürzeler, J. 1983. Un alcélaphiné aberrant (Bovidé, Mammalia) des "lignites de Grosseto" en Toscane. *C. R. Acad. Sci. Paris* Sér II **269:**497–503.

Hürzeler, J. 1987. Lie Lutrinen (Carnivora, Mammalia) aus dem "Grosseto-Lignit" der Toscana. *Schweiz. Paläont. Abh.* **110:**25–48.

Hürzeler, J., and Engesser, B. 1976. Les faunes de mammifères néogènes du Bassin de Baccinello (Grosseto, Italie). *C. R. Acad. Sci. Sér. D* **283:**333–336.

Knussmann, R. 1967. Das proximale Ende der Ulna von *Oreopithecus bambolii* und seine Aussage über dessen systematische Stellung. *Z. Morphol. Anthropol.* **59:**57–76.

Kordos, L. 1987. Description and reconstruction of the skull of *Rudapithecus hungaricus* Kretzoi (Mammalia). *Ann. Hist. Nat. Mus. Natl. Hung.* **79:**77–88.

Kürth, G. 1956. *Oreopithecus bambolii* Gervais, ein Hominide von der Wende Miozän/Pliozän. *Naturwiss. Rundsch. Braunsgw.* **2:**57–61.

Leakey, L. S. B. 1963. East African fossil Hominoidea and the classification within the superfamily. In: S. L. Washburn (ed.), *Classification and Human Evolution,* pp. 32–49. Aldine, Chicago.

Lorenz, H. G. 1968. Stratigraphische und mikropaläontologische Untersuchungen des Braunkohlengebietes von Baccinello (Provinz Grosseto–Italien). *Riv. Ital. Paleontol. Stratigr.* **74:**147–270.

Martin, R. D. 1990. *Primate Origins and Evolution: A Phylogenetic Reconstruction.* Princeton University Press, Princeton, NJ.

Martini, I. P., and Sagri, M. 1993. Tectono-sedimentary characteristics of late Miocene–Quaternary extensional basins of northern Apennines, Italy. *Earth Sci. Rev.* **34:**197–223.

Moyà-Solà, S., and Köhler, M. 1993. Recent discoveries of *Dryopithecus* shed new light on evolution of great apes. *Nature* **365:**543–545.

Moyà-Solà, S., and Köhler, M. 1995. New partial cranium of *Dryopithecus* Lartet, 1863 (Hominoidea, Primates) from the upper Miocene of Can Llobateres, Barcelona, Spain. *J. Hum. Evol.* **29:**101–139.

Moyà-Solà, S., and Köhler, M. 1996. The first *Dryopithecus* skeleton: Origins of great ape locomotion. *Nature* **379:**156–159.

Msuya, C., and Harrison, T. 1994. The circumorbital foramina in primates. *Primates* **35:**231–240.
Napier, J. R., and Davis, P. R. 1959. The forelimb skeleton and associated remains of *Proconsul africanus*. *Br. Mus. Nat. Hist. Fossil Mamm. Afr.* **16:**1–69.
Pilbeam, D. R. 1969. Tertiary Pongidae of East Africa: Evolutionary relationships and taxonomy. *Yale Peabody Mus. Bull.* **31:**1–185.
Preuschoft, H. 1960. *Oreopithecus*—Mensch oder Menschenaffe? *Umschau* **17:**522–524.
Remane, A. 1955. Ist *Oreopithecus* ein Hominide? *Abh. Math. naturwiss. Kl. Akad. Wiss. Mainz* **12:**467–497.
Remane, A. 1965. Die Geschichte der Menschenaffen. In: G. Heberer (ed.), *Menschliche Abstammungslehre: Fortschritte der "Anthropogenie" 1863–1964,* pp.249–309. Gustav Fischer, Stuttgart.
Riesenfeld, A. 1975. Volumetric determination of metatarsal robusticity in a few living primates and in the foot of *Oreopithecus*. *Primates* **16:**9–15.
Robinson, C. 1996. The morphology of the anterior lower premolar in catarrhines and its phylogenetic implications. *Am. J. Phys. Anthropol. Supplement* **22:**220.
Rook, L. 1993. A new find of *Oreopithecus* (Mammalia, Primates) in the Baccinello basin (Grosseto, southern Tuscany). *Riv. Ital. Paleontol. Stratigr.* **99:**255–262.
Rook, L., and Torre, D. 1995. *Celadensia grossetana* nov. sp. (Cricetidae, Rodentia) from the late Turolian Baccinello–Cinigiano basin (Italy). *Geobios* **28:**379–382.
Rook, L., Ficcarelli, G., and Torre, D. 1991. Messinian carnivores from Italy. *Boll. Soc. Paleontol. Ital.* **30:**7–22.
Rook, L., Harrison, T., and Engesser, B. 1996. The taxonomic status and biochronological implications of new finds of *Oreopithecus* from Baccinello (Tuscany, Italy). *J. Hum. Evol.* **30:**3–27.
Rose, M. D. 1988. Another look at the anthropoid elbow. *J. Hum. Evol.* **17:**193–224.
Rose, M. D. 1993. Locomotor anatomy of Miocene hominoids. In: D. Gebo (ed.), *Postcranial Adaptation in Nonhuman Primates,* pp. 252–272. Northern Illinois University Press, DeKalb.
Rosenberger, A. L., and Delson, E. 1985. The dentition of *Oreopithecus bambolii:* Systematic and paleobiological implications. *Am. J. Phys. Anthropol.* **66:**222–223.
Sarmiento, E. E. 1987. The phylogenetic position of *Oreopithecus* and its significance in the origin of the Hominoidea. *Am. Mus. Novit.* **2881:**1–44.
Sarmiento, E. E. 1988. Anatomy of the hominoid wrist joint: Its evolutionary and functional implications. *Int. J. Primatol.* **9:**281–345.
Schultz, A. H. 1960. Einege Beobachtungen und Masse am Skelett von *Oreopithecus* im Vergleich mit anderen catarrhinen Primaten. *Z. Morphol. Anthropol.* **50:**136–149.
Schwalbe, G. 1915. Über den fossilen Affen *Oreopithecus Bambolii. Z. Morphol. Anthropol.* **19:**149–254.
Senut, B. 1989. *Le Coude Chez les Primates Hominoïdes: Anatomie, Fonction, Taxonomie et Évolution.* CNRS, Paris.
Simons, E. L. 1962. Fossil evidence relating to the early evolution of primate behavior. *Ann. N. Y. Acad. Sci.* **102:**282–294.
Simons, E. L. 1967. Fossil primates and the evolution of some primate locomotor systems. *Am. J. Phys. Anthropol.* **26:**241–254.
Simpson, G. G. 1963. The meaning of taxonomic statements. In: S. L. Washburn (ed.), *Classification and Human Evolution,* pp. 1–31. Aldine, Chicago.
Sondaar, P. Y. 1977. Insularity and its effects on mammal evolution. In: M. K. Hecht, P. C. Goody, and B. M. Hecht (eds.), *Major Patterns in Vertebrate Evolution,* pp. 671–707. Plenum Press, New York.
Sondaar, P. Y. 1986. The island sweepstakes. *Nat. Hist.* **95:**50–57.
Stern, J. T., and Jungers, W. L. 1985. Body size and proportions of the locomotor skeleton in *Oreopithecus bambolii. Am. J. Phys. Anthropol.* **66:**233.
Straus, W. L. 1957. *Oreopithecus bambolii. Science* **126:**345–346.
Straus, W. L. 1958. Is *Oreopithecus bambolii* a primitive hominid? *Anat. Rec.* **132:**511–512.
Straus, W. L. 1963. The classification of *Oreopithecus*. In: S. L. Washburn (ed.), *Classification and Human Evolution,* pp. 146–177. Aldine, Chicago.

Susman, R. L. 1985. Functional morphology of the *Oreopithecus* hand. *Am. J. Phys. Anthropol.* **66:**235.

Szalay, F. S., and Berzi, A. 1973. Cranial anatomy of *Oreopithecus. Science* **180:**183–185.

Szalay, F. S., and Delson, E. 1979. *Evolutionary History of the Primates.* Academic Press, New York.

Szalay, F. S., and Langdon, J. H. 1987. The foot of *Oreopithecus:* An evolutionary assessment. *J. Hum. Evol.* **15:**585–621.

Teichmüller, M. 1962. Die *Oreopithecus*-führende Kohle von Baccinello bei Grosetto (Toskana/ Italien). *Geol. Jahrb.* **80:**69–110.

Torre, D., Ficcarelli, G., Rook, L., Kotsakis, T., Masini, F., Mazza, P., and Sirotti, A. 1995. Preliminary observations on the paleobiogeography of the central Mediterranean during the late Miocene. In: *International Conference on the Biotic and Climatic Effects of the Messinian Event in the Circum-Mediterranean,* Benghazi, pp. 63–64.

Trevor, J. C. 1961. Is "the abominable coalman" in the human family? *New Sci.* **228:**816–818.

Ungar, P. S., and Kay, R. S. 1995. The dietary adaptations of European Miocene catarrhines. *Proc. Natl. Acad. Sci. USA* **92:**5479–5481.

von Koenigswald, G. H. R. 1955. Remarks on *Oreopithecus. Riv. Sci. Preist.* **10:**1–11.

Wu, R. 1984. The crania of *Ramapithecus* and *Sivapithecus* from Lufeng, China. *Cour. Forsch. Senckenberg* **69:**41–48.

Lufengpithecus and Hominoid Phylogeny

17

Problems in Delineating and Evaluating Phylogenetically Revelant Characters

JEFFREY H. SCHWARTZ

Introduction

In the 1970s, the first specimens of the large-bodied hominoid now referred to as *Lufengpithecus lufengensis* (Fig. 1) were discovered in the southern Chinese province of Yunnan, at the Shihuiba colliery site, which lies 9 km north of the town of Lufeng. The site, which is characterized by lignite deposits, is late Miocene (ca. 8 Ma) and thus approximately coeval with *Sivapithecus* sites in Turkey (Andrews and Tekkaya, 1980) and Indo-Pakistan (Pilbeam, 1982) and perhaps a few million years younger than the *Dryopithecus* sites in Hungary (Kordos, 1987; Kretzoi, 1975) and Spain (Moyà-Solà and Köhler, 1993).

Two virtually complete mandibles (both recovered in 1975), differing primarily in size, were referred to different taxa: a smaller *Ramapithecus lufengensis* and a larger *Sivapithecus yunnanensis* (Xu *et al.*, 1978; Xu and Lu, 1979). The latter differed from Indo-Pakistani *Sivapithecus* in having a wider and flatter digastric fossa and strongly wrinkled molar enamel. A partial,

JEFFREY H. SCHWARTZ • Department of Anthropology, University of Pittsburgh, Pittsburgh, Pennsylvania 15260.

Function, Phylogeny, and Fossils: Miocene Hominoid Evolution and Adaptations, edited by Begun *et al.* Plenum Press, New York, 1997.

Fig. 1. Partial skull of apparent male *Lufengpithecus lufengensis* (PA 644) (not to scale). The nasal bones are missing, thereby making the nasal aperture appear taller than it actually is. For other details, see text. (©Jeffrey H. Schwartz)

somewhat crushed skull (PA 644) discovered in 1978 was referred to *S. yunnanensis,* which Lu *et al.* (1981) argued had been ancestral to *Pongo,* a link between *Proconsul africanus* and *Pongo,* and also related to *Paranthropus boisei.* In 1981, a smaller more complete skull (PA 677) was referred to *R. lufengensis* (Wu *et al.,* 1983). Although the site of Lufeng has so far yielded hundreds of teeth and jaws, cranial and even postcranial fragments, the two partial crania and the two original mandibles remain the most informative specimens.

In 1983 Wu *et al.* suggested that the larger and smaller Lufeng hominoids represented, respectively, male and female individuals of the same taxon, the proper designation of which would be *Sivapithecus lufengensis;* the hypothesis of extreme sexual dimorphism has since been further developed morphologically and metrically by Kelley and Etler (1989; also Kelley and Xu, 1991). Wu *et al.* (1983, pp. 9–10) also argued that all *Sivapithecus* were related to *Pongo* because they shared the following features: upturned premaxillary region [apparently correlated with airorhynchy (e.g., see Brown and Ward, 1988)]; concave midfacial skeleton; marked temporal ridges emanating from the lateral portion of the supraorbital margin and paralleling the margin prior to angling posteriorly; orbits with supraorbital margins not confluent across glabella; narrow nasal aperture; deep canine fossae; laterally divergent upper canines and markedly curved upper central incisor roots with prominent jugae; moderately high-crowned cheek teeth; thick-enameled, occlusally crenulated molars lacking cingula.

But Wu *et al.* (1983) also delineated features that distinguished the Lufeng hominoid from other *Sivapithecus* as well as from *Pongo:* subrounded orbits; extremely wide interorbital region; concave interorbital and glabellar regions; concave to rounded nasoalveolar clivus; sub-"U"-shaped and slightly posteriorly divergent palate. Such distinctions later prompted Wu (1987, p. 271) to refer this hominoid to its own genus, *Lufengpithecus lufengensis,* to the diagnosis of which he added: face broad and short; hard palate broad, short, and shallow; orbits ovoid with somewhat angled outer corners and horizontal dimensions longest; posteriorly divergent dental arcade; molars with higher cusps and more occlusal wrinkling.

Although Wu did not address the potential phylogenetic relationships of this new genus, it would follow that if the craniodental features cited when *Lufengpithecus* was thought to be a species of *Sivapithecus* are both accurate and indicative of a close relationship with *Pongo,* then these features still are and do. The consequence of recognizing *Lufengpithecus* is that features hypothesized as being synapomorphic for *Sivapithecus* and *Pongo* cannot be derived at the level of the last common ancestor of these latter two taxa alone. Rather, these features would be synapomorphic at the level of a hypothetical ancestor shared by *Lufengpithecus, Sivapithecus,* and *Pongo.* And it is within this hypothesized clade that one must test the three alternative phylogenies: (1) *Sivapithecus* and *Pongo* are sister taxa; (2) *Lufengpithecus* and *Pongo* are sister taxa; (3) *Lufengpithecus* and *Sivapithecus* are sister taxa. As such, the issue then is not whether *Lufengpithecus* belongs to a *Sivapithecus–Pongo* clade, but whether, within this clade, *Sivapithecus* and *Pongo* are sister taxa. However, inasmuch as *Lufengpithecus* is not always included in phylogenetic analyses of hominoid relationships (e.g., Begun, 1992, 1994), or when it is, conflict remains about either its phylogenetic relationships or even if such can be determined (cf. Andrews, 1992; Kelley and Etler, 1988; Kordos, 1988), it seems necessary to address the basic phylogenetic question: To which hominoid(s) is *Lufengpithecus* most closely related?

Given the states of preservation of the specimens and the number of earlier contributions that refer to them (e.g., see review by Schwartz, 1990), little new description can be added here. However, the interpretation of the potential phylogenetic relationships of *Lufengpithecus* hinges on the determination of the polarity of the character states of its preserved morphologies. These features will be discussed comparatively among anthropoids by category, with the description of the feature in *Lufengpithecus* presented first. The comparative data derive in part from the literature, but primarily from study of more than 300 specimens of a diverse taxonomic array of nonhuman anthropoid primates, 500 modern human specimens, and either the originals or casts of relevant fossil material. With regard to extant taxa, ontogenetic series were studied whenever possible. For the sake of simplicity, representative taxa will be discussed throughout.

Comparative Morphology

1. Circumorbital region (including supraorbital torus and glabella): Lufengpithecus has low, mounded orbital rims and the orbits are separated by a broad but sunken glabellar region: it does not have a supraorbital sulcus.

Prior to discussing morphology in detail, it is important to make two points. First, neonates of a species typically do not display the supraorbital configuration or glabellar distension that might ontogenetically come to characterize the adults. Even in species with the most marked supraorbital distension [e.g., *Mandrillus* (Fig. 2), *Gorilla* (Fig. 5)], development of this region is not evident in extremely young individuals (e.g., prior to the eruption of M1). In taxa that ultimately attain less marked supraorbital distension (e.g., *Cercopithecus*), development is also often minimal in subadult individuals.

Second, although a depression or "dip" behind or above the orbits is created either when the supraorbital region is vertically distended or when the frontal is elevated to any degree above the level of the superior orbital margin, the resultant "sulci" are not equivalent. One could describe the slight curve in the transition from the superior orbital rims to the frontal in *Lufengpithecus* [as well as, e.g., *Turkanapithecus, Sivapithecus, Pongo, Homo neanderthalensis,* some *H. erectus* (e.g., Sangiran 17, the Solo skulls), *Paranthropus, Australopithecus, Cercopithecus, Presbytis, Colobus,* hylobatids, and most platyrrhines] as a "supraorbital sulcus" because it is a depression, albeit a slight one, situated above (even if slightly above) the highest level of the superior orbital rim. Thus, a supraorbital sulcus can be identified whether the superior orbital margins bear mounded rims, are distended anteriorly into a barlike torus (in which case it could be identified as a supratoral sulcus), or are almost featureless. On the other hand, in taxa [e.g., adult African apes, *Macaca, Mandrillus, Papio,* some *H. erectus* (e.g., Zhoukoudian), *H. ergaster*] in which the superior orbital margin is distended such that it rises above the floor of a depression behind it, the term "posttoral sulcus" is appropriate, especially because a vertically enlarged superior orbital rim can be classified as a torus. The anteroposterior length of a posttoral sulcus and its vertical depth are affected by the relative degrees of horizontal versus vertical distension of the supraorbital torus as well as elevation of the frontal. One can discuss the disposition of sulci relative to glabella similarly, i.e., in defining a "supraglabellar" or a "postglabellar" sulcus. Bearing these points in mind, we can turn to specifics of orbital rim development.

In contrast, for example, to *Gorilla,* which often develops a continuous, barlike supraorbital torus, *Cebus, Colobus, Cercopithecus,* and *Homo sapiens* acquire with growth little, if any, orbital distension. At most, the supraorbital

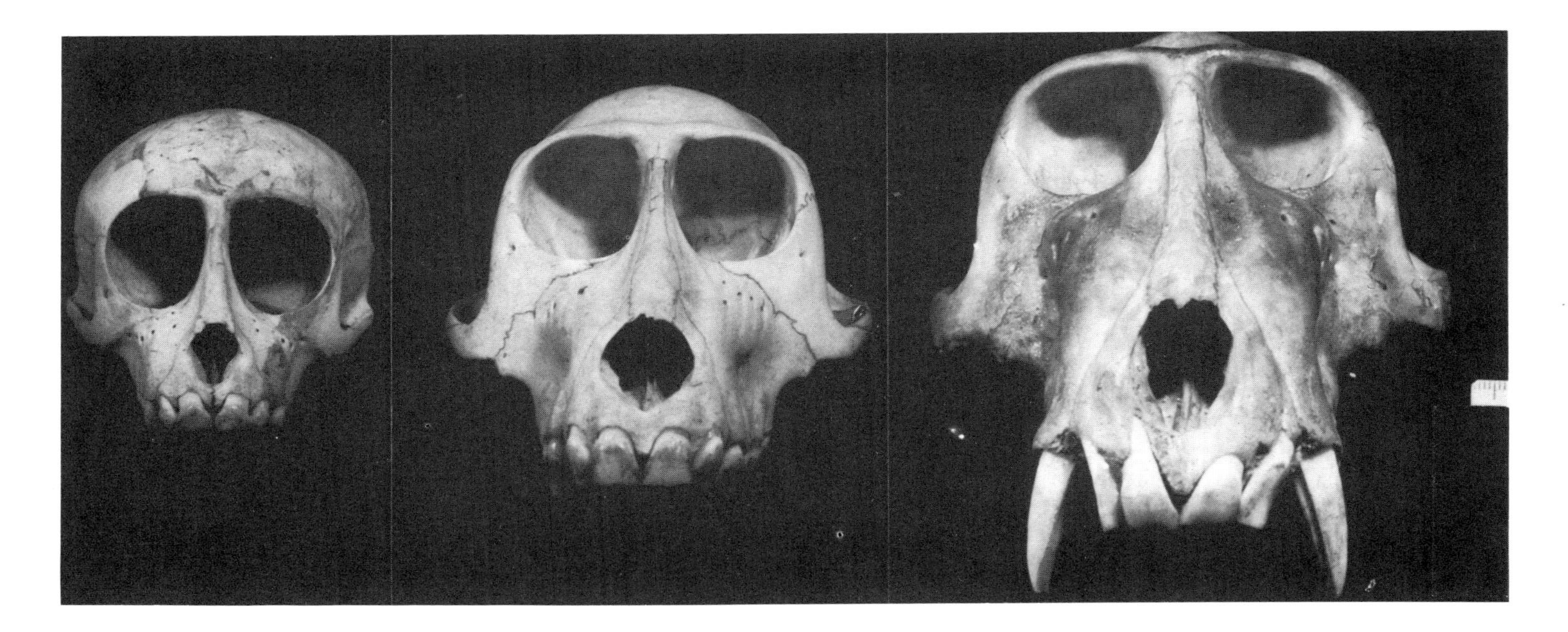

Fig. 2. *Mandrillus sphinx*. Infant (left; AMNH 99.1/1288): M1 erupted. Subadult (middle; AMNH 99.1/2776): M3s but not permanent canines erupting. Adult (right; AMNH 99.1/2049): M3s erupted. Scale in mm. (©Jeffrey H. Schwartz)

margin of adult *Colobus* and *Cercopithecus* is a thin, minimally anteriorly distended rim. Superficially, adult *Cebus* (Fig. 3) appear to display some circumorbital distension. However, this is an artifact induced by the temporal muscle, which leaves an elevated muscle scar around the perimeter of the orbit as a result of extreme midsagittal convergence; the midsagittally convergent temporal muscles create a sagittal crest that originates close to the region of glabella. As in African cercopithecines, the supraorbital–frontal region subtended between the temporal lines is in *Cebus* elevated or swollen to the height of the lines rather than forming the more commonplace supraorbital trigon, in which the temporal lines rise (even if slightly) above the level of the bone they subtend (e.g., as in *Turkanapithecus, Proconsul, Ankarapithecus, Dryopithecus, Victoriapithecus, Aegyptopithecus, Pongo, Gorilla, Paranthropus*). Otherwise in *Cebus* and African cercopithecines, the supraorbital configuration in midsagittal profile is convexly arcuate, with the curve flowing into the arc of the frontal bone. *Colobus, Cercopithecus,* and *Cebus* either do not, or only minimally, develop swelling in the region of glabella. *Proconsul* and *Cebus* may present similar midsagittal profiles (cf. Begun, 1994), but the similarity lies in both taxa developing a swollen glabellar region and not in the details of supra- or superolateral orbital morphology.

Among the Old World monkeys sampled, *Papio, Macaca,* and *Mandrillus,* for example, develop a barlike supraorbital torus that incorporates a distended or swollen glabellar region; in *Papio* glabella may be flexed superoinferiorly. Supraorbital development may occur earlier ontogenetically (at least relative to the eruption of the first permanent molar) in *Macaca* than in the other two taxa, but this needs to be investigated further. Supraorbital toral development, especially in *Papio* and *Mandrillus,* tends to straighten the superior orbital margin (such that the orbit looks like a "D" lying on its side with the straight side up) (Fig. 2). Thus, the supraorbital torus of these cercopithecids can be described as barlike in that not only is it horizontally oriented but it is also relatively uniformly thick from side to side. In these cercopithecids, the supraorbital torus projects noticeably anteriorly, but rises only slightly above the sulcus behind it. Because there is also little vertical elevation of the frontal, the posttoral sulcus is anteroposteriorly long but shallow.

Hylobatids are distinct in their development of thin, anteriorly protruding orbital rims. In juveniles, the medial portions of the rims do not project beyond the broad, flat glabellar region. Ontogenetically, however, the medial margins often become distended and thus in the adult project farther anteriorly than the somewhat domed glabellar region. Apparently in concert with the almost telescopic enlargement of the entire orbital rim, the hylobatid zygoma (which forms the lateral wall of the orbit) increases in size disproportionately and thus autapomorphically. Hylobatids do not have a supraorbital sulcus.

The African apes are distinguished among extant large-bodied hominoids by development of a relatively uniformly thick supraorbital torus that is

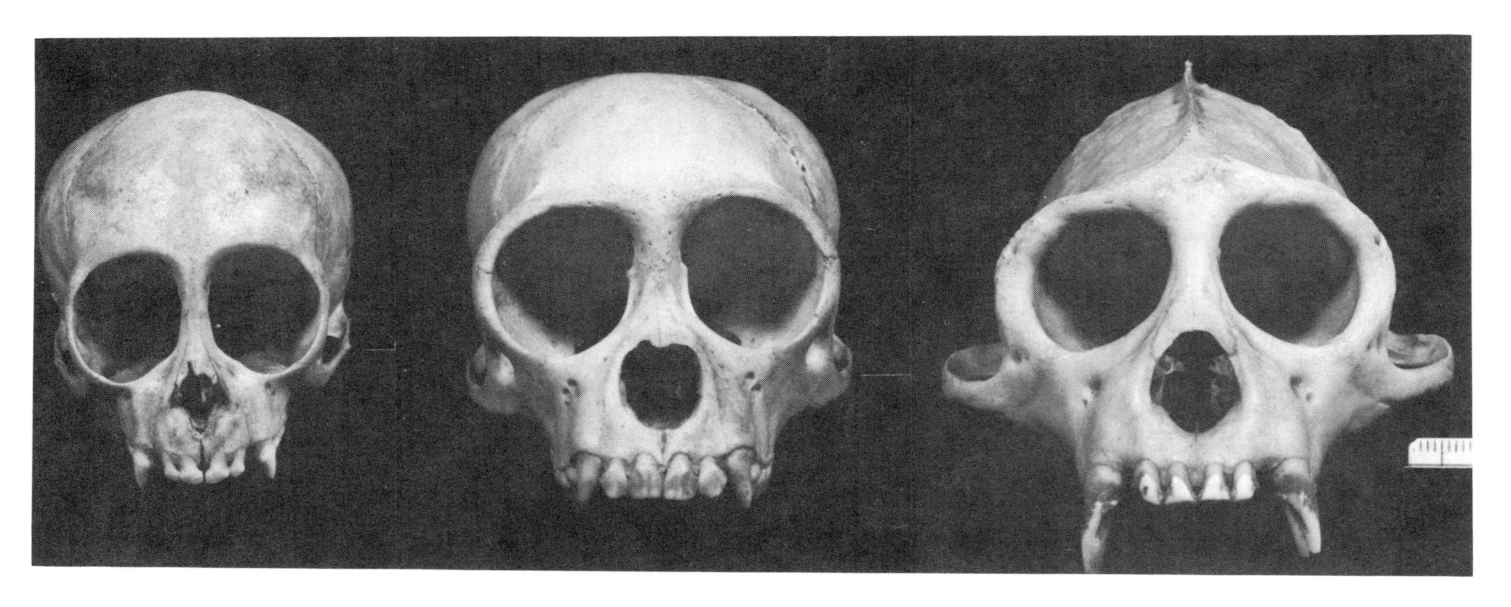

Fig. 3. *Cebus* sp. Infant (left; AMNH L. 85): dm3s erupted. Juvenile (middle; AMNH 99.1/2074): permanent canines half-erupted, M2s erupted, M3s not yet erupting. Adult (right; AMNH L. 41): M3s worn. Scale in mm. (©Jeffrey H. Schwartz)

continuous across glabella (Figs. 4 and 5). However, whereas in male *Gorilla* the supraorbital torus is often straight across from side to side and barlike (as described by Andrews, 1992; see Fig. 5), in *Pan* (Fig. 4) and female *Gorilla* it typically follows the rounded contour of the superior margin of the orbit. At glabella, the torus is typically straighter across in *Gorilla;* it often "dips" downward in *Pan.* Although the glabellar region broadens ontogenetically in both African apes, it tends to remain flatter in *Pan.* In both apes, the supraorbital torus projects anteriorly and is also variably distended vertically, albeit more noticeably in *Pan* than in *Gorilla.* Thus, the posttoral/postglabellar sulci of these hominoids can be variably shallow to moderately deep (Figs. 4 and 5). The depth of the sulcus is also affected by frontal elevation, which is most marked in *Pan* and female *Gorilla.*

Although it has been suggested that certain fossil taxa share similar supraorbital configurations with one or both African apes (e.g., Andrews, 1992; Begun, 1992, 1994; Dean and Delson, 1992), this is not the case in detail. For example, *Ouranopithecus* may have had a broad and swollen glabellar region, but it lacked a vertically distended or barlike torus and thus a posttoral sulcus (cf. Bonis and Koufos, 1993, p. 470). Descriptively, the superior orbital margins of *Ouranopithecus* are surmounted by moderately tall, uniformly thick but low moundlike rims that are confluent across a downwardly flexed glabellar region; there is also a sulcus above. The glabellar region of *Dryopithecus* may also have been broad, but it and the supraorbital regions were only slightly swollen and delineated above by merely faint sulci (cf. Begun 1992, 1994; Kordos, 1987; Moyà-Solà and Köhler, 1993). Frontal elevation and glabellar swelling are remarkably similar in *Dryopithecus* and *Proconsul* (cf. Begun, 1994; Walker *et al.,* 1983), although, in the former, the more medially arcuate temporal lines create a more restricted, shallow supraorbital trigon. Rather than being presumptively or incipiently toral, the superior orbital margins of *Dryopithecus* simply bear low moundlike rims (cf. Begun, 1994; Moyà-Solà and Köhler, 1993).

Orangutan orbital margins are apomorphic for extant anthropoids (including hylobatids) in that they bear low, mounded rims that are most prominent superiorly (Fig. 6). In further contrast to hylobatids, adults retain the juvenile configuration. Because orangutans also retain into adulthood the narrow interorbital region characteristic of the juvenile, the orbital rims (apomorphically) abut one another at the midline.

The orbital rims of *Sivapithecus, Ankarapithecus,* and *Lufengpithecus* are similar in detail to *Pongo* (e.g., Andrews and Cronin, 1982; Kelley and Pilbeam, 1986; Schwartz, 1990; Wu *et al.,* 1986). Thus, regardless of any hypothesis of primitiveness versus derivedness in catarrhine supraorbital toral development (e.g., no torus versus an anteriorly projecting versus a vertically distended bar), *Lufengpithecus Ankarapithecus, Sivapithecus* and *Pongo* appear to be synapomorphic in circumorbital configuration. The sunken glabellar region of *Lufengpithecus* is certainly autapomorphic among extant and known fossil anthropoids, although the primitiveness or derivedness of other config-

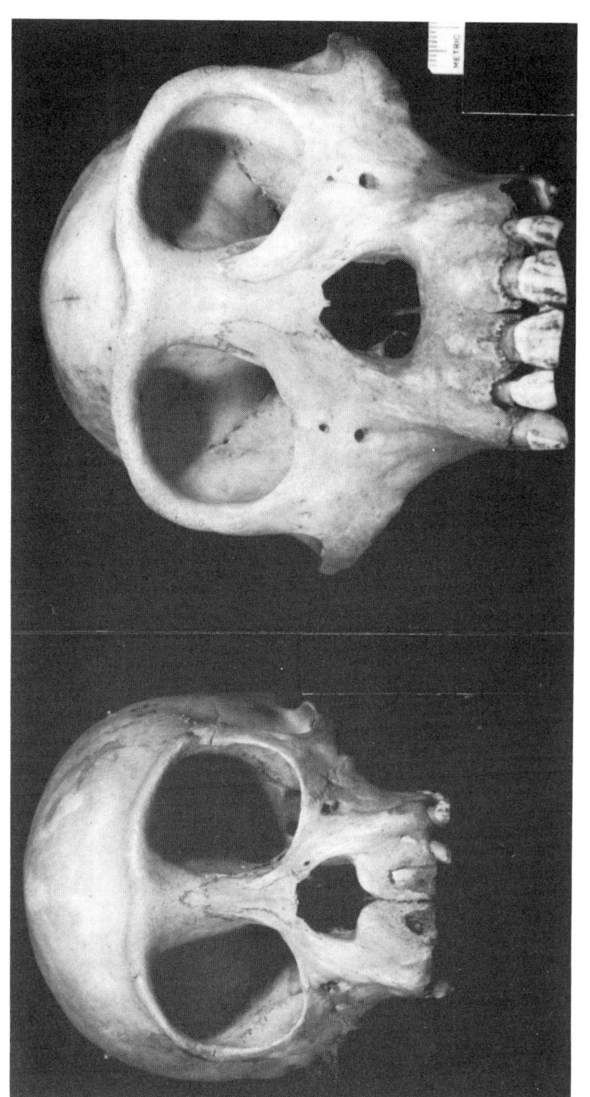

Fig. 4. *Pan troglodytes*. Infant (left; AMNH L. 19/ Ch. 7): dm2s erupted. Adult (right; AMNH L. 211/Ch. 14): M3s erupted. Scale in mm. (©Jeffrey H. Schwartz)

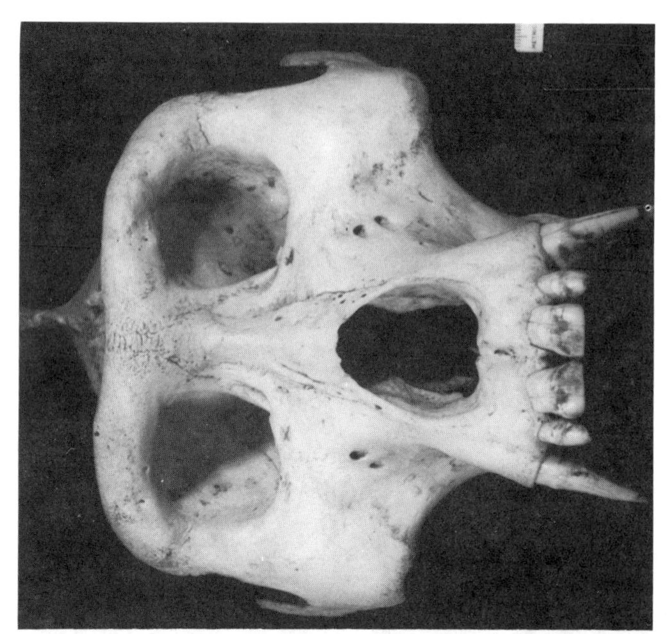

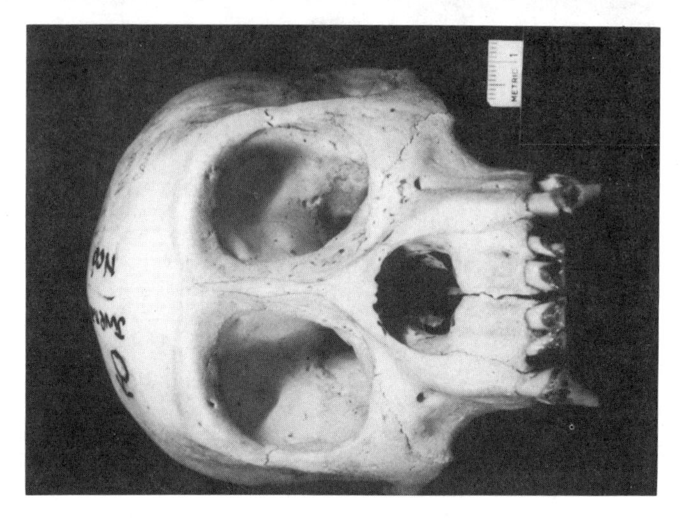

Fig. 5. *Gorilla gorilla*. Juvenile (left; AMNH 99.1/2251): dm2s and M1s erupting. Adult (right; AMNH 99.1/2044): M3s worn. (©Jeffrey H. Schwartz)

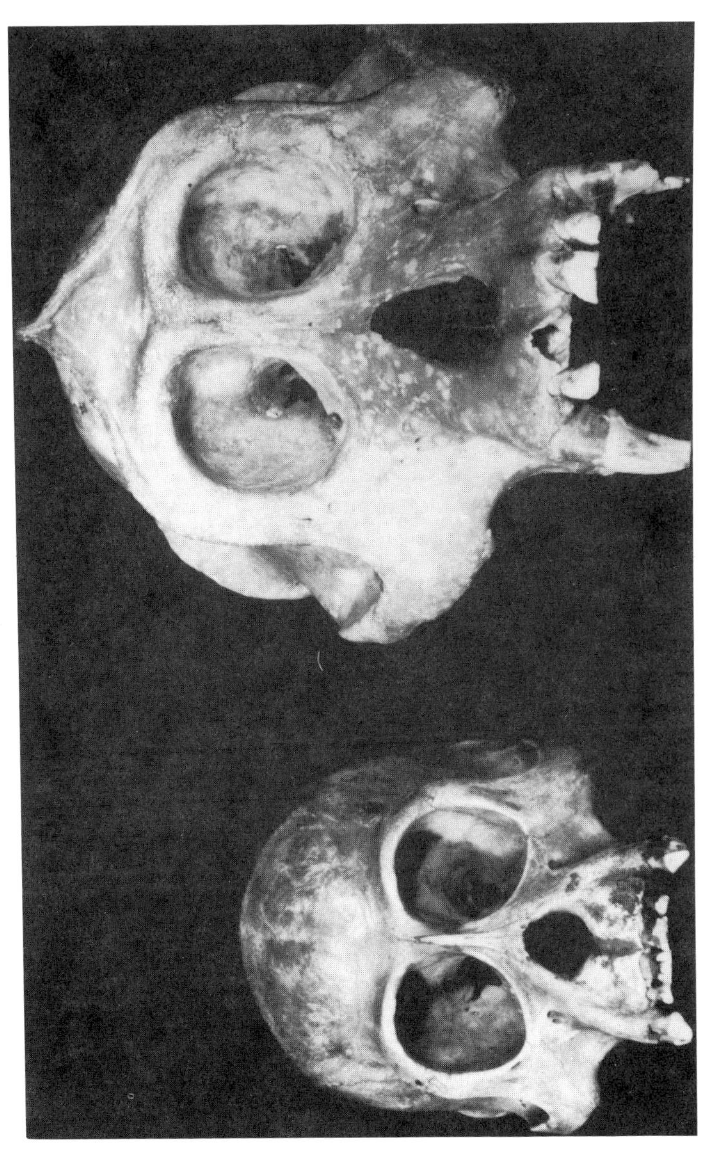

Fig. 6. *Pongo pygmaeus*. Infant (left; AMNH L. 28): M1s erupting. Adult (right; AMNH 140426): M3s worn. Not to scale (©Jeffrey H. Schwartz)

urations is unclear: for example, a rounded but somewhat low (as in *Proconsul, Dryopithecus,* various New and Old World monkeys, and apparently *Turkanapithecus* and *Afropithecus*) or depressed (as in *Ankarapithecus*) glabellar region, versus a more swollen one (as in other New and Old World monkeys). Seemingly synapomorphically, the glabellar regions of *Pongo* and *Sivapithecus* are obscured by the appression of right and left orbital rims.

2. *Thickening of lateral orbital margin: In Lufengpithecus the orbital rim is continuous with the frontal process of the zygoma; the zygoma broadens inferiorly alongside the orbit; an appearance of superolateral thickening is created by a distinct, raised, thick, and ridgelike temporal line that courses up, around, and over the lateral orbital margin.*

In general among catarrhines, and in contrast to prosimians and platyrrhines, the zygoma broadens inferiorly with differences between taxa being in the breadth of the zygoma inferiorly (e.g., compare the broad inferior zygoma of *Victoriapithecus* with the narrower inferior zygoma of *Cercopithecus*). Within the broader comparison among primates, the catarrhine configuration may represent another potential synapomorphy of this anthropoid clade. The development of lateral orbital margin thickening, however, is different in different taxa. It can be absolute (ontogenetic), expanded as a wall to the temporal muscle behind it, or built up via the surmounting of the rim by the temporal line superolaterally.

For example, in *Cebus* the lateral and superior portions of the orbital rim expand proportionately as a shelf anterior to the temporal muscle. The wider or broader adult lateral orbital margin maintains the characteristic shape of the juvenile, in which the outer edge follows the curvature of orbital shape. Thus, a superolateral "corner" is not thickened (there is no corner) and the base of the frontal process of the zygoma does not expand at the level of the inferior orbital rim (Fig. 3). In *Pongo,* on the other hand, the orbital margin superolaterally and especially superiorly thickens and the frontal process of the zygoma tends to broaden as it proceeds down the side of the orbit (Fig. 6). The amount of superolateral marginal thickening in *Pongo* is related to the degree to which the temporal lines are enlarged and expand up and around this part of the orbital rim (see also discussion of supraorbital costae in Clarke, 1977). For example, the temporal lines of some male *Pongo* extend as crestlike structures for much of the length of the supraorbital rims. *Dryopithecus, Ouranopithecus, Ankarapithecus, Sivapithecus,* and *Lufengpithecus* are similar to *Pongo* in lateral orbital morphology, including the superolateral course of the thickened temporal line along the supraorbital margin. *Lufengpithecus* is the most distinctive in temporal line prominence.

In *Victoriapithecus, Turkanapithecus, Afropithecus, Pan,* and *Gorilla,* as well as some New And Old World monkeys and hominids, the temporal lines may appear to thicken the superolateral margin of the orbit, but the position of these muscle scars remains posterior (even if slightly) to the supraorbital margin. In taxa with an anteriorly protruding supraorbital torus [and thus also a posttoral sulcus, e.g., *Papio, Mandrillus, Macaca, Pan,* and *Gorilla* (Figs. 2, 4, and 5)], the lateral margin of the torus (and thus the superolateral margin of the orbit) is accentuated from behind by the sulcus. In some *Gorilla,* the supraorbital torus is thickened medial to the downward turn of the zygomatic process of the frontal and/or just at the superolateral "corner" of the orbit.

Given the above, it appears that a hominid/African ape/*Ouranopithecus* clade within Anthropoidea cannot be delineated on the basis of thickening of the lateral orbital margin (cf. Andrews, 1992). In contrast, the possession by *Dryopithecus, Ouranopithecus, Ankarapithecus, Lufengpithecus, Sivapithecus,* and *Pongo* of a thickened temporal line that courses up and around the superolateral margin of the orbit (which also bears a low moundlike rim) emerges as potentially phylogenetically significant. *Ouranopithecus,* in which the thickness of the supraorbital mound continues around and down the side of the orbit, may be uniquely derived in this regard.

3. Orbital shape: The orbits of Lufengpithecus are horizontally ovoid/rectangular.

In most juvenile anthropoids the orbital margins are rounded or arced and are either vertically ovoid (taller than wide) or subcircular (height and width subequal). In many anthropoids, orbits become longer or wider ontogenetically, therefore changing from vertically ovoid to subcircular [e.g., *Cebus, Pan, Gorilla* (Figs. 3–5)], or from subcircular to longitudinally rectangular or elliptical (e.g., *Colobus, Papio*). In other anthropoids, however, the juvenile orbital shape is retained into the adult: subcircular (e.g., *Hylobates, Macaca*), vertically ovoid (*Pongo;* Fig. 6), or square or rectangular (humans). With regard to other shapes and changes, mandrills, for example, have natally longitudinally ovoid orbits that become rectangular in the adult.

The orbits of *Afropithecus* were taller than wide and bean-shaped (Leakey *et al.,* 1988b); *Sivapithecus,* tall and vertically ovoid (Pilbeam, 1982); *Ouranopithecus,* wide and horizontally rectangular (Bonis and Koufos, 1993); *Turkanapithecus,* square (Leakey *et al.,* 1988a); Rudabányan *Dryopithecus,* taller than wide and "D"-shaped (Kordos, 1987, 1991) and *Ankarapithecus,* about subequal in height and width (Alpagut *et al.,* 1996) and possibly "D"–shaped. The orbits of Can Llobateren *Dryopithecus* also look "D"-shaped but have been reconstructed as either taller than wide (Moyà-Solà and Köhler, 1993) or slightly wider than tall (Begun, 1994).

Although it seems reasonable to conclude in the absence of an ontogenetic series that adults with vertically ovoid or tall orbits or with horizontally rectangular or wide orbits possess differently derived configurations, it is difficult, in light of the above, to use orbital shape in the generation of any particular theory of relatedness. Thus, for example, the tall and vertically ovoid orbits of *Sivapithecus* and *Pongo* can be argued as synapomorphic for the two, but only after their relationship has been hypothesized on the basis of other, more clearly discernible potential synapomorphies. With regard to *Lufengpithecus,* the shape of the orbits are best interpreted as autapomorphic for it.

4. Interorbital distance: It is (extremely?) broad in Lufengpithecus.

With the exception of hylobatids and humans, extant anthropoid neonates typically have narrow interorbital regions. In some anthropoids (e.g., *Pongo,* cercopithecines, and some platyrrhines), the narrow neonatal configuration is retained into the adult (Figs. 2 and 6). In other anthropoids (e.g., the African apes, colobines, *Cebus,* and other platyrrhines), the interorbital region broadens ontogenetically (Fig. 3–5).

The broader comparison among extant anthropoids leads to the interpretations that: (1) humans and gibbons are derived (presumably independently so) in retaining the neonatally broad interorbital region into the adult; (2) the African apes and cercopithecines are primitively anthropoid in ontogenetically broadening the interorbital region; (3) *Pongo* and colobines are derived (presumably autapomorphically so) in retaining the neonatally narrow interorbital region into the adult (see also Delson, 1975, for Old World monkeys). Only if hylobatids are defined *a priori* as primitive relative to extant large-bodied hominoids (as is often the case, in morphological and molecular studies) could the retention into the adult of a neonatally broad interorbital region be conceived of as being primitive for the clade.

In the absence of relevant juvenile specimens, phylogenetic hypotheses concerning fossil taxa would be restricted to those in which adults possess one of two potentially apomorphic states: either a narrow interorbital region or one that is relatively broader than is typical of anthropoid primates. Thus, "narrow interorbital region" reemerges as a potential synapomorphy linking *Ankarapithecus, Sivapithecus,* and *Pongo* (cf. Alpagut *et al.,* 1996; Andrews and Cronin, 1982; Brown and Ward, 1988; Kelley and Pilbeam, 1986). The narrow interorbital region of *Victoriapithecus* must also be interpreted as apomorphic (at some hierarchical level within Anthropoidea) and not indicative of the primitive catarrhine state (cf. Benefit and McCrossin, 1991). On the other hand, such taxa as *Aegyptopithecus, Dryopithecus, Afropithecus, Turkanapithecus,* and *Ouranopithecus,* with their broad, but not excessively broad, interorbital regions remain primitive in this regard. Only if the interorbital region of *Lufengpithecus* is excessively broad (e.g., Wu *et al.,* 1983)—but it may not be (e.g., compared to *Pan;* Schwartz, 1990)—would it be (aut)apomorphic.

5. Nasal aperture: It is small and piriform in Lufengpithecus.

In most anthropoids (as well as prosimians) the nasal aperture of juveniles and adults is thin and tall in outline and is at least as narrow inferiorly as superiorly. Specific, common configurations are an upside-down triangle (e.g., *Macaca, Colobus*) or an ellipse [e.g., *Mandrillus, Cebus* (Figs. 2 and 3)]. *Proconsul* and *Victoriapithecus* retain this basic primitive configuration, but the nasal aperture is more diamond-shaped in outline.

In juvenile and adult hylobatids, the nasal aperture is primitively broader superiorly than inferiorly, but it is broader (i.e., less narrow) overall than is typical of nonhominoid anthropoids. From what is preserved (the inferior portion is missing), nasal aperture shape in *Turkanapithecus* appears similar to hylobatids.

In contrast to other extant anthropoids, extant large-bodied hominoids—from the juvenile to the adult—are distinctive in having a nasal aperture of different proportions: it is noticeably narrower superiorly than inferiorly and the wide inferior margin is relatively straight across. In *Pan, Gorilla,* and hominids, the nasal aperture is either roundly triangular or trapezoidal (Figs. 4 and 5), while in *Pongo* it is small and roundedly triangular [and often referred to as piriform (= pear-shaped)] (Fig. 6).

With regard to fossils, *Afropithecus* has been described as having a small, piriform nasal aperture (Leakey *et al.*, 1988b). Indeed, its nasal aperture is actually more pear-shaped compared with *Pongo* in that the superior portion is constricted relative to the base, which is slightly distended inferiorly at the midline; qualitatively, as in *Ankarapithecus*, the nasal aperture of *Afropithecus* appears relatively larger (especially taller relative to lower facial height) than the orangutan's. Within the confines of the comparisons here, *Afropithecus* might be synapomorphic with extant large-bodied hominoids in overall nasal aperture shape (i.e., narrower superiorly than inferiorly), but autapomorphic among anthropoids in having a truly pear-shaped nasal aperture. The shape and vacuity of the nasal aperture in *Ouranopithecus* and apparently Rudabányan *Dryopithecus* (only the floor of the nasal aperture is known) may be similar to *Pan* and *Gorilla*. In *Sivapithecus* and *Lufengpithecus*, the size and shape of the nasal aperture approximate those of *Pongo*. If we posit that a large, roundedly triangular or trapezoidal nasal aperture (typical of the majority of large-bodied hominoids) is primitive for the clade, then the small aperture of *Sivapithecus, Lufengpithecus,* and *Pongo* emerges as derived and potentially synapomorphic for them.

6. Lower facial triangle: The canine fossa is deep in Lufengpithecus and the markedly midsagittally angled canine "root" pillars delineate a triangular snout.

Depending on the length of the snout, and the height and robustness of the upper canine root, a diversity of anthropoids can be described as having a depression between the region of the canine root and the infraorbital region. The question is whether to identify this depression as a canine fossa. A consistent observation, however, is that in most anthropoids the canine roots (and/or the moderately stout pillars that continue up from the regions of the roots) do not veer or converge markedly toward the midline. Differences lie among the large-bodied hominoids.

Gorilla is similar to *Pan* in having relatively vertical canines (especially crowns) but root pillars that sometimes angle inward slightly (Fig. 4 and 5). These pillars do not course alongside the lateral margin of the nasal aperture and they do not set off the snout significantly from the rest of the lower facial skeleton. Juvenile *Pan* are similar to the adult, but in juvenile and the occasional adult *Gorilla* (e.g., see figure in Dean and Delson, 1992), a thickening may course up from the region of the upper canine root, arc inward, and wrap around the side of the broad nasal aperture. In these latter cases the snout thus appears semicircular in configuration and "pinched" off from the facial skeleton behind by a broad canine fossa (Fig. 5).

In *Pongo* (Fig. 6), the canine roots and relatively straight, long, and stout root pillars converge noticeably, paralleling the sides of the small triangular nasal aperture and further emphasizing the triangularly shaped snout, which is set off from the facial skeleton by a fairly deep but not particularly broad canine fossa. These features are retained from the juvenile into the adult and would seem to be apomorphic within Anthropoidea.

Ouranopithecus has a shallow canine fossa of moderate size. This hominoid, as well as *Turkanapithecus, Proconsul,* and apparently *Dryopithecus,* are also similar—but primitively so—to *Pan* and *Gorilla* as well as most anthropoids in having canine root pillars that neither converge markedly inward nor extend fully along the lateral margins of the nasal aperture (cf. Bonis and Koufos, 1993). A shallow to moderate fossa lies over the region of M^1 in *Turkanapithecus* and *Proconsul,* whereas *Dryopithecus* appears to have a true canine fossa. *Afropithecus* is interesting in that its canine roots/pillars are moderately convergent, but the pillars are neither markedly swollen nor do they extend fully along the sides of the nasal aperture. A restricted but relatively deep fossa lies over the region of the first molar, well back along the long snout. At present it is probably reasonable to conclude that *Afropithecus* is primitively anthropoid in root pillar stoutness, snout length, and (supramolar) fossa configuration, and presumably autapomorphic in canine root convergence. *Ankarapithecus* has slightly more medially convergent root pillars than *Afropithecus* that extend farther along the sides of the nasal aperture in the larger-canined MTA 2125 than in the smaller-canined AS 95-500 (cf. Alpagut *et al.,* 1996; Andrews and Tekkaya, 1980). Both specimens of *Ankarapithecus* have a pronounced canine fossa (ibid). *Sivapithecus* and *Lufengpithecus,* however, are presumably synapomorphic with *Pongo* in having markedly convergent canine roots/stout pillars that parallel the margins of a small nasal aperture. These three taxa also have moderately deep but constricted canine fossae.

7. Zygoma (maxillary/facial component) and maxilla (infraorbital component): The zygomatic/maxillary plane is vertically flat and deep and anteriorly facing in Lufengpithecus.

The typical anthropoid zygoma can be characterized thusly: (1) when viewed from above, it arcs posteriorly away from the lateral margin (and thus the plane) of the orbit and (2) when viewed from the side, the inferior margin of the facial root of the zygoma lies posterior to the inferior margin of the orbit and thus the zygoma is inferoposteriorly arcuate. This is noted clearly in New and Old World monkeys as well as in hylobatids. Although *Afropithecus* is often depicted with its orbits facing rather superiorly, orienting the facial skeleton to the plane of the cribriform plate (and thus the Frankfort horizontal) reveals that this hominoid had the primitive zygomatic configuration. *Turkanapithecus* retained both primitive states and *Proconsul* at least the first.

In general, the great ape zygoma faces anteriorly more than in most other anthropoids and that of *Pongo* is the most forwardly facing (Figs. 4 and 6). Viewed from the side, the inferior margin of the gorilla zygoma lies posterior to the plane of the inferior margin of the orbit. The inferior margin of the zygoma of *Pan* may be somewhat more anteriorly situated, but the bone retains the primitively inferoposteriorly arcing configuration. *Pongo* is unique among the great apes in that the inferior margin of its zygoma lies directly beneath or even slightly anterior to the inferior margin of the orbit. Thus, the anterior surface of its zygoma is flat and vertical and in the same plane as the orbit. Among anthropoids, the configuration of the orangutan zygoma would seem to be apomorphic.

Sivapithecus and *Lufengpithecus,* as well as *Ankarapithecus* (Alpagut *et al.,* 1996; Andrews and Tekkeya, 1980), (given what is preserved) *Ouranopithecus* (cf. Bonis and Koufos, 1993; Schwartz, 1990), and *Dryopithecus* (cf. Begun, 1994; Kordos, 1987; Moyà-Solà and Köhler, 1993) share, presumably synapomorphically with *Pongo,* a vertically flat and deep and anteriorly facing zygoma.

8. Zygomaticofacial foramina (number and position): Lufengpithecus had at least one large foramen situated superior to the inferior margin of the orbit.

Andrews and Cronin (1982) suggested that the possession by *Sivapithecus* and *Pongo* of multiple large zygomaticofacial foramina situated superior to the inferior margin of the orbit was synapomorphic for them. Schwartz (1990) pointed out that the male and female crania of *Lufengpithecus* as well as the female Rudabányan *Dryopithecus* could be described similarly, as later did Moyà-Solà and Köhler (1993) and Begun (1994) for Can Llobateren *Dryopithecus.*

Review here of the larger sample of anthropoids reveals, however, that the number of zygomaticofacial foramina is more variable from individual to individual than previously thought. The more consistent development of only one zygomaticofacial foramen appears to characterize only *Mandrillus* (Fig. 2) and *Colobus.* Thus, with regard to foramen number, neither can *Pan* and *Gorilla* be unequivocally distinguished from *Pongo* nor the latter united phylogenetically with *Sivapithecus* or any other taxon.

Foramen position is variable among New and Old World monkeys; i.e., in some individuals (of the same taxon) the foramen lies below, in others it is level with, and in yet other individuals it lies slightly above the inferior margin of the orbit (in these anthropoids, zygomaticofacial foramina should not be confused with another foramen that is often found in the zygomaticofrontal suture at the lateral margin of the orbital rim). In the African apes, however, the zygomaticofacial foramen (or foramina) more consistently lies below the inferior margin of the orbit, whereas in *Pongo* this foramen (or foramina) more consistently lies above the inferior orbital margin (Figs. 4 and 6). Among extant hominoids, *Pongo* is apomorphic in developing a large foramen (or foramina). In terms of foraminal position and size, *Sivapithecus, Lufengpithecus, Ankarapithecus,* and *Dryopithecus* (see above) emerge as potentially synapomorphic with *Pongo.* The region is insufficiently preserved in *Ouranopithecus* to be revealing.

9. Subnasal region: Lufengpithecus may have had a flat nasal floor and upwardly rotated premaxilla.

Brown and Ward (1988) and Kelley and Etler (1989) described the subnasal region of *Lufengpithecus* as similar to the African apes (stepped down nasoalveolar clivus; broad incisive fossa and canal in the floor of the nasal cavity), whereas Kordos (1988) concluded that it was synapomorphically like that of *Sivapithecus* and *Pongo* (extensive overlap of the maxilla by the posterior pole of the premaxilla; small incisive fossa and long, narrow incisive canal). Neither of these configurations is clearly evident in the specimens (Schwartz, 1990), but there are features of the upper incisors and the anterior margin of the premaxilla that contribute to a tentative partial reconstruction.

In the male Lufeng cranium, the upper central and lateral incisor crowns are oriented more orthally than procumbently, but the long roots of the central incisors are curved, paralleling the contour of the upper surface of the premaxilla. The preserved right lateral incisor and the alveoli of the other incisors in the female cranium present a similar picture. The curvature of the male's central incisor roots and the intact portions of the nasoalveolar region of the female indicate that the descent from this region into the floor of the nasal cavity was not steep. Although noted in the occasional chimpanzee (McCollum *et al.*, 1993), a relatively flat transition from the nasoalveolar clivus to the floor of the nasal cavity is typical of *Pongo* and apparently *Sivapithecus* (Ward and Kimbel, 1983). Recent cleaning of the *Ankarapithecus* specimen MTA 2125 (Begun and Gülec, 1995) revealed a mildly stepped nasoalveolar region (Alpagut *et al.*, 1996). If there is a correlation between having enlarged, curved upper incisor roots and a long, upwardly tilted premaxilla, as in *Ankarapithecus, Sivapithecus,* and *Pongo* (see Kelley and Pilbeam, 1986), then the same can be inferred for *Lufengpithecus.* Inasmuch as a markedly stepped down transition from the nasoalveolar clivus to the floor of the nasal cavity and an unrotated premaxilla [as in *Gorilla* (Ward and Kimbel, 1983) and apparently also *Ouranopithecus* (Bonis and Koufos, 1993; Bonis and Melentis, 1978, 1985)] appear to be primitive features for large-bodied hominoids (Martin, 1986; Schwartz, 1984), the configurations in *Ankarapithecus, Sivapithecus, Pongo,* and *Lufengpithecus* would be synapomorphic for them.

With regard to other aspects of the nasal region, the narrowness of the nasal aperture of *Lufengpithecus* mitigates against the presence of a broad or deep incisive fossa (= depression in the floor of the nasal cavity), as is seen in *Gorilla* (cf. Ward and Kimbel, 1983). But without evidence of the vomer it is difficult to place the "boundary" between the posterior pole of the premaxilla and the maxilla and, thus, to locate the two openings in the floor of the nasal cavity (one on either side of the vomer) that represent the apertures of the incisive canals. In *Pongo* and *Sivapithecus*, these apertures are situated farther back in the nasal cavity than in *Pan* (Ward and Kimbel, 1983). On the oral cavity side of the Lufeng palates, it is impossible to determine if there was a single incisive foramen (and, if so, if it was long and slitlike, as in *Pongo* and *Sivapithecus*) or some version of two incisive foramina (as in *Gorilla* and *Pan*) (Schwartz, 1983, 1988). Until these various apertures are identified, one cannot be certain about the length or vacuity or the incisive canal(s).

10. Paranasal sinuses: Lufengpithecus had superiorly expanded maxillary sinuses.

Brown and Ward (1988; see also Ward and Brown, 1986) suggested that *Lufengpithecus, Ouranopithecus,* and Rudabányan *Dryopithecus* (RUD-44) were similar to *Pongo* and *Sivapithecus* in having air cells in the interorbital region that derived from the maxillary rather than ethmoidal sinuses. Although Begun (1992) interpreted these air cells in RUD-44 as ethmoidal derivatives, and thus as "true" frontal sinuses (cf. Cave and Haines, 1940), Moyà-Solà and Köhler (1993) disagreed. For Can Llobateren *Dryopithecus* Begun (1994) described (p. 19) and illustrated (p. 21) a pair of large, symmetrical, unifocular air cells that are separated from one another by a midline bony septum that courses from nasion to a level below the supraorbital notch; at this point the air cells diverge and each courses over the medial portion of the supraorbital rim (the fully preserved left sinus tapers toward its end). Begun (p. 19) also suggested that, as judged from the right side, these sinuses communicated with the maxillary sinuses. And, indeed, these "frontal sinuses" are not dissimilar in their morphology and disposition to examples given by Schultz (1936, pp. 269–270) for superior maxillary sinus expansion in *Pongo,* but are dissimilar to the multifocular, asymmetrical air cells that constitute the frontal sinuses in the African apes and humans, in particular. In addition, Moyà-Solà and Köhler (1993) described a separate ethmoidal sinus in Can Llobateren *Dryopithecus.*

Since "superiorly expanded maxillary sinuses" has been proposed as a potential synapomorphy of *Pongo* and *Sivapithecus* (e.g., Ward and Kimbel, 1983), the issue of paranasal sinus development is not a trivial one: *Lufengpithecus* as well as *Ankarapithecus, Ouranopithecus,* and *Dryopithecus* could be united with *Pongo* and *Sivapithecus* by this character. Although Begun (1994) presents an interesting topographical argument for identifying a frontal sinus as an ethmoidal derivative (e.g., by its position relative to glabella), the criteria are not universally applicable to African apes or fossil hominids. On the basis of communicating drainage passages, however, humans and African apes develop ethmofrontal sinuses (Cave and Haines, 1940; Hershkovitz, 1977), apparently by primitive retention, as is further suggested by the presence of ethmofrontal air cells in *Proconsul africanus* (Walker and Teaford, 1989) and "*P. major*" [i.e., the upper jaw/lower facial specimen (UMP

62-11) from Moroto, Uganda (Pilbeam, 1969), which now seems to be referable to *Afropithecus* (e.g., Andrews, 1992), the most complete cranial specimen of which does have a sinus above the region of glabella (Leakey *et al.*, 1988b)]. If development of ethmofrontal sinuses (or at least ethmoidal air cells) is primitive for large-bodied hominoids and superiorly expanded maxillary sinuses apomorphic within this clade, the lack of a multifocular ethmoid in taxa with superiorly expanded maxillary sinuses could be interpreted not as a primitive retention, but as an apomorphic consequence of superior maxillary expansion.

11. The dentition.

Lufengpithecus shares with most large-bodied hominoids—but not *Dryopithecus, Proconsul,* or the extant African apes—the development of relatively thick molar enamel. In some features *Lufengpithecus* is most similar to *Pongo;* e.g., the upper central and all lower incisors are relatively high-crowned, molar occlusal foveae are less restricted (and cusp apices are situated at crown margins), molar crown relief is reduced, and enamel wrinkling dominates molar occlusal morphology (even in worn teeth) (Kelley and Pilbeam, 1986; Wu *et al.*, 1983).

Sivapithecus, Lufengpithecus, Ankarapithecus, and *Pongo* are similar in that, e.g., the upper incisors are strongly heteromorphic (in size and shape) with central incisors being somewhat spatulate, the upper canines are markedly angled outward (i.e., the root pillars converge midsagitally), and (with the apparent exception of *Ankarapithecus*) unworn molars are crenulated to some degree (Schwartz, 1990). *Dryopithecus* may have had relatively shorter and less laterally expanded upper central incisor crowns, but, as in *Sivapithecus, Lufengpithecus,* and *Pongo,* they bore strong lingual pillars (Begun, 1994). These four hominoids, as well as *Ankarapithecus* and *Ouranopithecus* lacked molar cingula.

Lufengpithecus appears to share dental apomorphies with at least *Sivapithecus* and *Pongo,* but finer phylogenetic resolution is not at present decipherable.

Conclusion

Clearly some of the problems attendant to unraveling the phylogenetic relationships of *Lufengpithecus* to other large-bodied hominoids—"large-bodied hominoid" itself being a phylogenetic hypothesis—are related to overwhelming autapomorphy, especially as reflected in its sunken superior interorbital, glabellar, and lower interorbital regions. No other known primates can be so described, including hylobatids. Also seemingly autapomorphic for *Lufengpithecus* is the development (in both male and female) of thickened, ridgelike, temporal lines, which, having coursed up and around the lateral orbital margins, are perfectly straight as they proceed posteriorly along the cranium. When viewed in the context of an overall morphological pattern rather than in terms of individual character states, *Lufengpithecus* might appear to be at the same time generally similar to all large-bodied hominoids (because of primitive retention) and yet similar to none specifically (because of autapomorphy) (cf. Kelley and Etler, 1989; Kelley and Pilbeam, 1986).

The autapomorphies and plesiomorphies aside, *Lufengpithecus* shares specific features with extant *Pongo* and Miocene *Sivapithecus:* nasal aperture size (small), convergence (marked) and extent (along the nasal margins) of upper canine root pillars, canine fossa depth (relatively deep), incisor crown height (high), molar enamel surface (crenulated), and some aspects of the nasal floor (flat) and premaxilla (upwardly rotated). If these features are apomorphic for *Pongo* and *Sivapithecus,* they are as well for *Lufengpithecus* and thus are potentially synapomorphic at the level of a hypothetical ancestor shared by these three taxa. Other cited synapomorphies of *Pongo* and *Sivapithecus* are found in *Lufengpithecus, Ankarapithecus,* and either *Ouranopithecus* or *Dryopithecus,* or both. These include orbital rimming (superior, low, mounds), zygomaticofacial foramen size (large) and position (above inferior orbital margin), orientation of zygoma (anteriorly directed and vertical), upper incisor size and shape relationship (heteromorphic), and upper central incisor crown morphology (broadly spatulate, with lingual pillar).

Although these five taxa do indeed appear to constitute a clade, the incompleteness of specimens of *Ouranopithecus* and *Dryopithecus* makes difficult resolving the sister relations of one or the other to the hypothesized *Pongo–Sivapithecus–Lufengpithecus–Ankarapithecus* clade. Tentatively, *Ouranopithecus* may be the sister taxon of this clade and *Dryopithecus* the sister taxon of all (Moyà-Solà and Köhler, 1993; Schwartz, 1990). Of course, more complete specimens of these fossil taxa as well as specimens of new taxa could very well demonstrate a different set of hierarchical relationships within the larger hypothesized *Pongo* clade. But the potential phylogenetic viability of "a *Pongo* clade" (see also Begun and Gülec, 1995; Moyà-Solà and Köhler, 1993) provides some fodder for a functional discussion of the craniofacial features of *Lufengpithecus* and other hominoids.

Consider the condition of airorhynchy, which has been described as variably present in *Pongo, Sivapithecus,* hylobatids, some Old and New World monkeys, and possibly *Aegyptopithecus* and *Pliopithecus* (e.g., see reviews in Brown and Ward, 1988; Schultz, 1968; Shea, 1988). Shea (1988), for example, has argued that the lack of a supraorbital torus and sulcus in these taxa is reflective of their airorhynchous condition and that the distribution among anthropoids of airorhynchy implies that this condition is primitive for hominoids. As such, Shea (1988) interprets klinorhynchy and the development of a supraorbital torus and sulcus as correlated and synapomorphic of hominids and African apes. Brown and Ward (1988), however, argue that airorhynchy and the associated craniofacial features of *Pongo* and *Sivapithecus* are derived among anthropoids and thus reflective of the relatedness of the former two taxa. Certainly, *Pongo* and *Sivapithecus* are very different craniofacially from virtually all other anthropoids (e.g., in frontal elevation, orbital rimming, interorbital distance, lower facial configuration, palatal thickening, upward premaxillary rotation and posterior distension). In addition, as illustrated here, one cannot generalize about supraorbital tori or sulci solely with regard to size or "presence" versus "absence." Thus, one should not conflate any amount of supraorbital thickening in conjunction with a shallow supraorbital sulcus (as, e.g., in *Turkanapithecus, Colobus, Paranthropus*) with a vertically distended supraorbital torus associated with a distinct posttoral sulcus (as, e.g., in *Mandrillus, Homo ergaster,* the African apes). This is relevant not only to claims of similarity in supraorbital torus and posttoral sulcus configuration in, for example, *Dryopithecus, Ouranopithecus,* African apes, and hominids (e.g., compare Andrews, 1992; Bonis and Koufos, 1993; Begun, 1992, 1994; Shea, 1988), but also to attempts to correlate supraorbital morphology with either klino- or airorhynchy (e.g., Ravosa and Shea, 1994; Shea, 1988). It may be that airorhynchous animals lack supraorbital tori and posttoral sulci, but so do many klinorhynchous animals, some of which lack any kind of supraorbital distension whatsoever [e.g., *Daubentonia,* marmosets, tamarins; see also Shea's (1988) illustration of *Cebus albifrons*]. In fact, very few klinorhynchous primates develop a true supraorbital torus and posttoral sulcus (*Mandrillus, Papio, Macaca, Pan, Gorilla, Homo ergaster,* some "*H. erectus*"), which suggests that such craniofacial hafting is not necessarily functionally correlated with the development of the torus and sulcus.

It appears that the only feature shared by all airorhynchous primates is dorsal deflection of the palate. Other features, such as upwardly oriented orbits or premaxillae, are not universal but taxon specific and may exist in the absence of an airorhynchous condition (e.g., the variably dorsally oriented orbits of *Nycticebus*). With regard to the features enumerated above that distinguish *Pongo* and *Sivapithecus* from other airorhynchous primates, it would be tempting to conclude that they were interrelated and functional correlates of the specific type of airorhynchy that these two hominoids develop. Palatal thickening and posterior premaxillary elongation may, however, be synapomorphies of large-bodied hominoids (Andrews, 1992; McCollum *et al.,* 1993; Schwartz, 1983) rather than a feature associated with airorhynchy, but upward rotation of the premaxilla could still be so correlated as might also be other features of *Pongo* and *Sivapithecus* (such as low-mounded rimmed and tall, ovoid orbits, narrow interorbital region, lack of ethmoidally derived frontal sinuses, superior expansion through the interorbital region of the maxillary sinus, anteriorly facing and vertical infraorbital/maxillary plane, outwardly rotated upper canine with strongly convergent root pillars, and deep canine fossae).

But just because we find these features together does not mean that they are all functionally or developmentally correlated. And this is where other taxa, such as *Lufengpithecus,* become even more relevant. Does the fact that *Lufengpithecus* had superiorly expanded maxillary sinuses or an anteriorly facing, deep, vertical infraorbital/maxillary plane mean that it, too, was airorhynchous? If upper canine root curvature is correlated with an upwardly rotated premaxilla, does this indicate that *Lufengpithecus* was airorhynchous? Clearly, the development of low-mounded rimmed orbits (as in *Lufengpithecus, Sivapithecus, Pongo,* as well as *Ankarapithecus, Ouranopithecus,* and *Dryopithecus*) is not correlated with the development of tall, ovoid orbits or the retention into the adult of a narrow interorbital region (*Sivapithecus, Pongo*). If *Lufengpithecus* was airorhynchous, then these latter features are not functional correlates of airorhynchy in (although they may be synapomorphies of) *Sivapithecus* and *Pongo.* Rather, *Pongo* and *Sivapithecus happen* to be airorhynchous hominoids which also *happen* to share the apomorphies of tall, ovoid orbits and paedomorphically narrow interorbital regions.

Even if we had more complete specimens of *Lufengpithecus,* and knew for certain that this hominoid had been airorhynchous, we would still have to try to sort out which features found together are functionally correlated and which just happen to be together, either because of a hierarchy of primitive retentions (including the apomorphies of the last common ancestor) or autapomorphy.

At present, the only clues to *Lufengpithecus* having been airorhynchous come from inferences about which features really are correlated with this configuration: a *Pongo*-like subnasal floor and curved upper incisor roots. If *Lufengpithecus* is the sister of (or at least a member of a clade that includes) *Pongo* and *Sivapithecus,* then, perhaps, (1) low-mounded rimmed orbits, (2) outwardly rotated upper canines with stout, medially convergent root pillars, (3) deep canine fossa, (4) anteriorly facing, flat and vertical infraorbital/maxillary plane, (5) superior maxillary sinus expansion, and (6) small, piriform nasal aperture are not just potential synapomorphies of these three hominoids, but also functional correlates of airorhynchy. However, characters 1, 2, and 3 describe not only these three taxa but also *Ouranopithecus* and *Ankarapithecus,* if not *Dryopithecus,* as well. If *Ouranopithecus* were klinorhynchous [but the claim was based on the assertion that this hominoid as well as the African apes and hominids possessed a barlike supraorbital torus and sulcus (Dean and Delson, 1992)], then features 4 and 5, although potentially synapomorphic for the hominoids that possess them, are only coincidentally (i.e., by way of primitive retention) associated with airorhynchy. However, the two preserved premaxillary regions of *Ouranopithecus* (Bonis and Melentis, 1978; Bonis and Koufos, 1993) appear to be upwardly rotated, suggesting that this feature might be a synapomorphy of *Ouranopithecus, Lufengpithecus, Sivapithecus,* and *Pongo* (to the exclusion of *Dryopithecus*), and, once more, begging the question of which morphology is functionally correlated with airorhynchy. Inasmuch as *Dryopithecus* appears not to have been airorhynchous, it would seem that little craniofacial morphology is specifically functionally correlated with dorsiflexion of the splanchnocranium.

Lufengpithecus is an important hominoid taxon. But its importance derives less from our ability to imbue it with scenarios of the functional significance of its morphology than from its effect on our reconstructions of the phylogenetic relationships of the large-bodied hominoids. And, at least with regard to craniofacial morphology, it appears that the particular theory of relationship informs the interpretation of functional morphology.

ACKNOWLEDGMENTS

I thank D. Begun, C. Ward, and M. Rose for the invitation to participate in their symposium and Wu Rukang (Institute of Vertebrate Paleontology and Paleoanthropology, Beijing) for allowing me to study the Lufeng specimens. Much of the comparative data were collected while a Kalbfleish fellow in the Department of Anthropology, American Museum of Natural History (AMNH). For access to specimens of anthropoids, I am grateful to I. Tattersall and J. Brauer (Department of Anthropology, AMNH), G. Musser and W. Fuchs (Department of Mammals, AMNH), and R. Thorington (Department of Mammals, National Museum of Natural History). I thank D. Begun and M. Rose for helpful criticisms of an earlier version of this manuscript. I would have liked to have been able to cite critical articles alluded to in an anonymous review of the manuscript, but B. Shea (whom I thank for discussion) and others have not (yet) written them. As always, P. Andrews and S. Ward provided stimulating discussion and debate.

References

Alapagut, B., Andrews, P., Fortelius, M., Kappelman, J., Temizsoy, I., Celebi, H., and Lindsay, W. 1996. A new specimen of *Ankarapithecus meteai* from the Sinap formation of central Anatolia. *Nature* **382:**349–351.

Andrews, P. 1992. Evolution and environment in the Hominoidea. *Nature* **369:**641–646.

Andrews, P., and Cronin, J. 1982. The relationships of *Sivapithecus* and *Ramapithecus* and the evolution of the orangutan. *Nature* **297:**541–546.

Andrews, P., and Tekkaya, I. 1980. A revision of the Turkish Miocene hominoid *Sivapithecus meteai*. *Paleontology* **23:**85–95.

Begun, D. R. 1992. Miocene fossil hominids and the chimp–human clade. *Science* **257:**1929–1933.

Begun, D. R. 1994. Relations among the great apes and humans: New interpretations based on the fossil great ape *Dryopithecus*. *Yearb. Phys. Anthropol.* **37:**11–63.

Begun, D. R. and Gülec, E. 1995. Restopration and reinterpretation of the facial specimen attributed to *Sivapithecus meteai* from Kayincak (Yassiören), central Antaolia, Turkey. *Amer. J. Phys. Anthropol.* (Suppl.) **20:**63–64.

Benefit, B. R., and McCrossin, M. L. 1991. Ancestral facial morphology of Old World higher primates. *Proc. Natl. Acad. Sci. USA* **88:**5267–5271.

Bonis, L. de, and Koufos, G. 1993. The face and the mandible of *Ouranopithecus macedoniensis:* Description of new specimens and comparisons. *J. Hum. Evol.* **24:**469–491.

Bonis, L. de, and Melentis, J. 1978. Les primates hominoïdes du Miòcene superior de Macedoine. *Ann. Paleontol. Vertebr.* **64:**185–202.

Bonis, L. de, and Melentis, J. 1985. La place du genre *Ouranopithecus* dans l'evolution des Hominides. *C. R. Acad. Sci.* **300:**429–432.

Brown, R., and Ward, S. 1988. Basicranial and facial topography in *Pongo* and *Sivapithecus,* In: J. H. Schwartz (ed.), *Orang-utan Biology,* pp. 247–260. Oxford University Press, London.

Cave, A. J. E., and Haines, R. W. 1940. The paranasal sinuses of the anthropoid apes. *J. Anat.* **74:**493–523.

Clarke, R. J. 1977. *The Cranium of the Swartkrans Hominid SK 847 and Its Relevance to Human Origins.* Ph.D. thesis, University of Witwatersrand, Johannesburg.

Dean, D., and Delson, E. 1992. Second gorilla or third chimp? *Nature* **359:**676–677.

Delson, E. 1975. Evolutionary history of the Cercopithecidae. In: F. S. Szalay (ed.), *Approaches to Primate Paleobiology,* pp. 167–217. Karger, Basel.

Hershkovitz, P. 1977. *Living New World Monkeys (Platyrrhini). Vol. 1* University of Chicago Press, Chicago.

Kelley, J., and Etler, D. 1989. Hominoid dental variability and species number at the late Miocene site of Lufeng, China. *Am. J. Primatol.* **18:**15–34.

Kelley, J., and Pilbeam, D. R. 1986. The dryopithecines: Taxonomy, comparative anatomy, and phylogeny of Miocene large hominoids. In: D. R. Swindler and J. Erwin (eds.), *Comparative Primate Biology,* Vol. I, pp. 361–411. Liss, New York.

Kelley, J., and Xu, Q. 1991. Extreme sexual dimorphism in a Miocene hominoid. *Nature* **352:**151–153.

Kordos, L. 1987. Description and reconstruction of the skull of *Rudapithecus hungaricus* Kretzoi (Mammalia). *Ann. Hist. Nat. Mus. Natl. Hung.* **79:**77–88.

Kordos, L. 1988. Comparison of early primate skulls from Rudabánya (Hungary) and Lufeng (China). *Anthrop. Hung.* **20:**9–22.

Kordos, L. 1991. Le *Rudapithecus hungaricus* de Rudabánya (Hongrie). *L'Anthropologie* **95:**343–362.

Kretzoi, M. 1975. New ramapithecines and *Pliopithecus* from the lower Pliocene of Rudabánya in north-eastern Hungary. *Nature* **257:**578–581.

Leakey, R. E. F., Leakey, M. G., and Walker, A. C. 1988a. Morphology of *Turkanapithecus kalakolensis* from Kenya. *Am. J. Phys. Anthropol.* **76:**277–288.

Leakey, R. E. F., Leakey, M. G., and Walker, A. C. 1988b. Morphology of *Afropithecus turkanensis* from Kenya. *Am. J. Phys. Anthropol.* **76:**289–307.

Lu, Z., Xu, Z., and Zheng, L. 1981. Preliminary research on the cranium of *Sivapithecus yunnanensis. Vertebr. Palasiat.* **19:**101–106.

McCollum, M. A., Grine, F. E., Ward, S. C., and Kimbel, W. H. 1993. Subnasal morphological variation in extant hominoids and fossil hominids. *J. Hum. Evol.* **24:**87–111.

Martin, L. 1986. Relationships among extant and extinct great apes and humans. In: B. A. Wood, L. Martin, and P. Andrews (eds.), *Major Topics in Primate and Human Evolution,* pp. 161–187. Cambridge University Press, London.

Moyà-Solà, S., and Köhler, M. 1993. Recent discoveries of *Dryopithecus* shed new light on evolution of great apes. *Nature* **365:**543–545.

Pilbeam, D. R. 1969. Tertiary Pongidae of East Africa: Evolutionary relationships and taxonomy. *Bull. Peabody Mus. Nat. Hist.* **31:**1–185.

Pilbeam, D. R. 1982. New hominoid skull material from the Miocene of Pakistan. *Nature* **295:**232–234.

Ravosa, M. J., and Shea, B. T. 1994. Pattern in craniofacial biology: Evidence from the Old World monkeys (Cercopithecidae). *Int. J. Primatol.* **15:**801–822.

Schultz, A. H. 1936. Characters common to higher primates and characters specific for man. *Quart. Rev. Biol.* **11:**259–83.

Schultz, A. H. 1968. The recent hominoid primates. In: S. L. Washburn and P. C. Jay (eds.), *Perspectives on Human Evolution,* Vol. 1, pp. 122–195. Holt, Rinehart & Winston, New York.

Schwartz, J. H. 1983. Palatine fenestrae, the orangutan, and hominoid evolution. *Primates* **24:**231–240.

Schwartz, J. H. 1984. Hominoid evolution: A review and a reassessment. *Curr. Anthropol.* **25:**655–672.

Schwartz, J. H. 1988. History, morphology, paleontology, and evolution. In: J. H. Schwartz (ed.), *Orang-utan Biology,* pp. 68–85. Oxford University Press, London.

Schwartz, J. H. 1990. *Lufengpithecus* and its potential relationship to an orang-utan clade. *J. Hum. Evol.* **19:**591–605.

Shea, B. 1988. Phylogeny and skull form in the hominoid primates. In: J. H. Schwartz (ed.), *Orang-utan Biology,* pp. 233–245. Oxford University Press, London.

Walker, A. C., and Teaford, M. 1989. The hunt for *Proconsul. Sci. Am.* **260:**76–82.

Walker, A., Falk, D., Smith, R., and Pickford, M. 1983. The skull of *Proconsul africanus:* Reconstruction and cranial capacity. *Nature* **305:**525–527.

Ward, S. C., and Brown, B. 1986. The facial skeleton of *Sivapithecus indicus.* In: D. R. Swindler and J. Erwin (eds.), *Comparative Primate Biology,* Vol. 1, pp. 413–452. Liss, New York.

Ward, S. C., and Kimbel, W. H. 1983. Subnasal alveolar morphology and the systematic position of *Sivapithecus. Am. J. Phys. Anthropol.* **61:**157–171.

Wu, R. 1987. A revision of the classification of the Lufeng great apes. *Acta Anthropol. Sin.* **6:**263–271.

Wu, R., Xu, Q., and Lu, Q. 1983. Morphological features of *Ramapithecus* and *Sivapithecus* and their phylogenetic relationships—Morphology and comparison of the crania. *Acta Anthropol. Sin.* **2:**1–10.

Wu, R., Xu, Q., and Lu, Q. 1986. Relationship between Lufeng *Sivapithecus* and *Ramapithecus* and their phylogenetic position. *Acta Anthropol. Sin.* **5:**1–30.

Xu, Q., and Lu, Q. 1979. The mandibles of *Ramapithecus* and *Sivapithecus* from Lufeng, Yunnan. *Vertebr. Palasiat.* **17:**1–13.

Xu, Q., Lu, Q., Pan, Y., Zhang, X., and Zheng, L. 1978. Fossil mandible of the Lufeng *Ramapithecus. Kexue Tongbao* **9:**554–556.

Events in Hominoid Evolution

18

DAVID R. BEGUN, CAROL V. WARD, and MICHAEL D. ROSE

Introduction

The preceding chapters of this volume have described a number of different approaches and solutions to the interpretation of hominoid evolutionary history. Given the breadth of approaches, it is difficult to compare results among researchers. Despite this diversity, however, there seems to be broad agreement on many issues in the complex evolutionary history of the Hominoidea.

Perhaps most importantly, researchers agree that hominoid evolutionary history is resolvable. This is encouraging, as the diversity of opinion presented here and elsewhere could lead one to conclude that the endeavor is hopeless. It is clear that the analysis of fossils will ultimately permit the reconstruction of hominoid phylogeny, although many issues remain to be resolved.

It is also clear that cladistic methodology has become fully established as the major tool in phylogeny reconstruction in the Hominoidea, as has been the case for some time in other areas of vertebrate paleontology. The exact way in which cladistic methods are applied are almost as numerous as the number of researchers applying them, but the basic principles are agreed on by all. Finally, it appears that most researchers agree that integrating func-

---

DAVID R. BEGUN • Department of Anthropology, University of Toronto, Toronto, Ontario M5S 3G3, Canada. CAROL V. WARD • Anthropology and Pathology & Anatomical Sciences, University of Missouri, Columbia, Missouri 65211. MICHAEL D. ROSE • Department of Anatomy, Cell Biology, and Injury Science, University of Medicine and Dentistry of New Jersey, New Jersey Medical School, Newark, New Jersey 07103.

Function, Phylogeny, and Fossils: Miocene Hominoid Evolution and Adaptations, edited by Begun *et al.* Plenum Press, New York, 1997.

tional and phylogenetic perspectives is necessary in hominoid paleobiology, though, again, there is variation in how these domains are combined.

There is also broad agreement on the position of many hominoid taxa discussed in this volume. Most authors who discuss *Proconsul* consider this taxon to be a basal hominoid (but see Harrison and Rook, this volume, and references therein). Most also agree that *Afropithecus* is more derived, and *Kenyapithecus* still more derived. Among the late Miocene forms, there is a great deal of debate, though most at least agree that *Sivapithecus* and *Pongo* form a clade (but see Pilbeam, this volume).

The authors also agree that African apes and humans form a clade, though relations among these forms are also debated. Many authors in this volume avoid addressing further resolution of the *Pan–Gorilla–Homo* clade; an issue that seems thornier to paleoanthropologists than to molecular systematists (see Pilbeam, this volume). Inasmuch as the issue of relations among African apes and humans is the same issue as human origins (from what nonhuman primate or primate group did humans diverge?), this complex topic will undoubtedly receive increasing attention from paleontologists in the coming years.

Many problems in interpreting hominoid evolution remain. Perhaps the most important issue concerns the definition and standardization of characters. Character states are the raw material of cladistic analysis. Problems arise when researchers differ in their interpretation of the character states actually present in taxa. For example, there is disagreement over the definition of the frontal torus and its presence or absence in Miocene taxa. The result is that some researchers claim that *Dryopithecus* has a torus (Bonis and Koufos, 1993; Andrews, 1992; Begun, 1994) while others claim it does not (Moyà-Solà and Köhler, 1993; Schwartz, this volume). Assigning discrete states to continuous characters can also cause discrepancies among researchers. As Pilbeam has recently suggested (personal communication), standardizing the process by which we assign character states to taxa may be the next major advance in hominoid paleobiology.

Overall, however, the chapters assembled here suggest that Miocene hominoid paleobiology is alive and well. Data generated by these analyses will provide the basis for future work on Miocene and Plio-Pleistocene hominoids. An accurate interpretation of the evolutionary origins of Plio-Pleistocene taxa will prove especially important now that the earliest hominid record has become so diverse, with the recent discoveries of new species and even new genera of hominids (Leakey *et al.*, 1995; White *et al.*, 1994, 1995). The Miocene provides a phylogenetic backdrop for interpreting the evolutionary histories of *Ardipithecus*, *Australopithecus*, and *Homo*.

The connection between the Miocene and later times is not limited to phylogeny, however. The Miocene witnessed an explosion of taxa with diverse adaptations throughout its time span of roughly 15 million years. Many of these adaptations are also found in living or recently extinct primates, including Pliocene hominids (see chapters by Leakey and Walker, McCrossin and

Benefit, Rose, C. V. Ward, Bonis and Koufos, and Begun and Kordos). Documenting the parallel occurrence of similar adaptations in Miocene and living apes is relevant to explanations of the origins of these adaptations.

At the same time, evidence from the Miocene shows that differences among extant and fossil taxa are not always what they appear to be. Kay and Ungar (this volume) argue convincingly that differences in occlusal morphology may not translate into differences in function and behavior between Miocene and recent forms. The time depth and anatomical diversity provided by the Miocene hominoid record should serve as an important comparative data base against which diversity in more recent primate groups can be judged.

A Comprehensive Analysis of Hominoid Phylogeny

As a final contribution, we have assembled a large data base of cranial and postcranial characters known for the eight best-known genera of Miocene hominoids discussed in this volume (Tables I and II). Data are primarily derived from our own observations. Supplementary data are derived from chapters in the book, but interpretation of characters and character states is our own. Our character list is not a summary of all of those discussed in this text, nor is it necessarily approved by all authors. Our goal in assembling this data base is partly to generate another hypothesis of relations among hominoids, but primarily to provide as much data as can currently be assembled on these taxa of Miocene hominoids.

We generated our phylogenetic hypothesis (Fig. 1) using methods described in other chapters (see Begun and Kordos, and C. V. Ward, Chapter 6, this volume). The outgroup consists of *Propliopithecus, Aegyptopithecus, Pliopithecus,* Cercopithecoidea, and Platyrrhini. We use a large number of successive outgroups to facilitate polarity determination for the ingroup by constructing an outgroup with exclusively primitive characters. Where Oligocene catarrhines appear to be autapomorphic, *Pliopithecus* or living monkeys are used.

Results

The cladogram presented in Fig. 1 is based on 240 characters, roughly equally represented by cranial and postcranial traits. Our analysis produced only one most parsimonious cladogram with a length of 446 steps with a consistency index of 63. The same tree topology generated only from postcranial data has a higher consistency index (69) than either the tree based on the entire data set (63) or the tree based only on cranial traits (59). Overall, the level of homoplasy is similar in all three data bases, but is lowest in the postcrania.

Table I. Characters and Character States Used in This Analysis

Character	Character states				
	0	1	2	3	4
Scapula					
1 Genoid shape	elongate	round			
2 Infraglenoid tubercle	broad	narrow			
3 Teres minor attachment	axillary	anterior			
4 Spinous process root	robust	gracile			
5 Glenoid–axillary angle	large	small			
6 Scapular notch	narrow	broad			
7 Acromion root	inferior	superior			
Humerus					
8 Entepicondylar foramen	present	absent			
9 Olecranon fossa	shallow	deep			
10 Lateral epicondyle	distal	proximal			
11 Zona conoidea 1	broad	narrow			
12 Zona conoidea 2	shallow	deep			
13 Capitular distal extent	absent	present			
14 Capitular tail	present	absent			
15 Articular olecranon fossa	minimal	large			
16 Coranoid/radial fossae size	smaller	larger			
17 Coranoid fossa	shallow	deep			
18 Trochlear shape	cylindrical	trochleiform			
19 Medial trochlear keel	weak	strong			
20 Anteroposterior shaft curvature	bowed	straight			
21 Intertuberosity angle	<90°	>90°			
22 Head torsion	<15°	>15°			
23 Head shape	ovoid	spherical			
24 Bicipital groove	broad	narrow			
25 Proximal shaft shape	angular	rounded			
26 Medial epicondylar projection	retroflexed	medial			

27	Trochlear keel symmetry	subequal	>medially
28	Medial epicondyle/keel	distinct	merged
29	Superior trochlear border	straight	notched
Radius			
30	Radial head shape	oval	round
31	Radial head beveling	absent	present
32	Radial neck shape	oval	circular
33	Distal radial articular surface	flat	concave
Carpals			
34	Os centrale	separate	fused
35	Centrale facet on scaphoid	not beaked	beaked
36	Scaphoid/capitate articulation	absent	present
37	Scaphoid tuberosity	small	large
38	Scaphoid tubercle	lateral	palmar
39	Centrale/trapezium articulation	present	absent
40	Centrale trapezium–trapezoid facets	angulated	aligned
41	Trapezium facets on centrale & trapezoid	aligned	angulated
42	Lunate mediolateral width	narrow	broad
43	Lunate dorsopalmar length	long	short
44	Scaphoid facet on lunate	restricted	extensive
45	Lunate scaphoid-radial facet	low angle	high angle
46	Hamate facet on lunate	absent	present
47	Triquetral articulation with ulnar styloid	present	absent
48	Pisiform articulation with ulnar styloid	present	absent
49	Trapezium dorsal tubercle	absent	present
50	MC1 facet on trapezium	flat	trochleiform
51	MC3 facet on capitate	short	long
52	MC3 facet on capitate	flat	irregular
53	Palmar MC4 facets on capitate	present	absent
54	MC2 facet on capitate	continuous	divided
55	Trapezoid facet on capitate	absent	present
56	Capitate head mediolaterally	broad	narrow
57	Centrale facet on capitate	flat	concavoconvex

(*continued*)

Table I. *(Continued)*

Character	Character states				
	0	1	2	3	4
58 Hamulus size	small	large			
59 Triquetral facet on hamate	proximal	lateral			
Metacarpals					
60 MC4 facet on hamate	deep	shallow			
61 MC1 head proximodistally	long	short			
62 MC1 head dorsal part	narrow	browed			
63 MC1 trapezium articular surface	dorsal	proximal			
64 MC1 trapezium articular shape	flat	curved			
65 MC4 palmar capitate facet	present	absent			
66 MC facets on palmar hamate	confluent	angulated			
67 MC2–4 sesamoids	present	absent			
68 MC heads broadest	palmarly	dorsally			
Phalanges					
69 Phalangeal secondary shaft features	small	large			
70 Hand–foot phalangeal length	subequal	hand > feet			
71 Proximal phalangeal proximal articular surface	oval	square			
72 Proximal phalangeal palm tubercles	large	small			
73 Ray 1 term phalanx articular surface	convex	ridged			
Trunk					
74 Costal angle	low	high			
75 Vertebral body height	tall	medium	high		
76 Accessory processes	large	small	absent		
77 Transverse processes	ventral	intermed.	dorsal		
78 Tail	present	absent			
79 Sternebrae	narrow	broad			

Character				
Os coxae				
80 Iliac blade breadth	narrow	intermed.	wide	
81 Iliac blade angle	low	intermed.	high	
82 Lower iliac height	short	intermed.	long	
83 Cranial lunate surface	narrow	intermed.	wide	
84 Pubic length	short	medium	long	
Femur				
85 Trochanteric fossa	open	intermed.	deep	unique
86 Femoral head	cylinder	intermed.	sphere	
87 Femoral neck tubercle	present	absent		
88 Femoral condyle depth	deep	shallow		
89 Femoral condyle shape	symmetrical	intermed.	asymmetrical	
Tibia–fibula				
90 Distal tibial facet	square	short		
91 Medial malleolus projection	distal	flare		
92 Fibular robusticity	thin	robust		
93 Lateral malleolus	small	intermed.	large	
Talus				
94 Talar trochlea depth	deep	intermed.	shallow	
95 Talar trochlea shape	symmetrical	intermed.	asymmetrical	
96 Talar neck angle	absent	present		
97 Talus height	tall	short		
Calcaneus				
98 Distal calcaneus	long	intermed.	short	
99 Flexor hallucis longus groove	small	intermed.	large	
100 Posterior calcaneal contact in gait	heel off	heel down		
101 Posterior talar facet long axis	aligned	angled		
102 Anterior talar facet	curved	intermed.	flat	
103 Plantar calcaneal tubercle	small	large		
104 Calcaneo-navicular facet	large	small		
Cuboid–entocuneiform–misc.				
105 Cuboid peg	small	intermed.	large	

(continued)

Table I. (*Continued*)

Character	Character states				
	0	1	2	3	4
106 Cuboid wedging	slight	stronger			
107 Cuboid length	long	short			
108 Entocuneiform MT 1 joint	distal	medial			
109 Cuneiform length	long	short			
Metatarsals					
110 MT 1 size	gracile	intermed.	robust		
111 MT 1 sesamoid grooves	small	large			
112 MT 1 head shape	symmetrical	asymmetrical			
113 MT 1 head position	aligned	twisted			
114 MT 1 length	long	short			
115 MT 1 prehallux facet	present	absent			
116 MT 2–5 robusticity	gracile	robust			
117 Transverse arch	present	absent			
Phalanges					
118 Foot axis runs through	digit 3	digit 2			
119 Phalangeal robusticity	gracile	robust			
120 Phalangeal curvature	straight	curved			
121 Phalangeal flexor ridges	weak	intermed.	strong		
Incisors–canines					
122 Incisors length	broad	narrow			
123 I^2 cingulum	present	absent			
124 Relative male canine size	large	reduced	v. reduced		
125 Male canines	robust	compressed			
126 Canine cingula	narrow	thick, rounded			
Upper premolars					
127 P^3 cusp heteromorphy	strong	reduced			
128 P^3 paracones	tall, narrow	low, rounded			

129 P_3 shape	narrow	broad		
130 P_3 mesiolingual beak	absent	present		
131 P_3 metaconid	absent	present		
132 P_4 shape	short	longer		
133 P_4 talonids	low	high		
Upper molars				
134 M^{2-3} metacones	large	reduced		
135 Molar cusps	tall	low, rounded cusps		
136 Dentine penetrance	high	low		
137 Anterior/posterior dentition	subequal	>postcanine		
Cranial 1				
138 Palatine process	thin	thick		
139 Alveolar premaxilla	flat	biconvex		
140 Incisive fossa position	opposite *C*	distal to *C*	distal to P^3	
141 Maxillary sinus size	small	larger	largest	
142 Maxillary sinus floor	low	high		
143 Nasal aperture base	narrow	broad		
144 Maxillary depth	shallow	deep		
145 Lateral malar surface	curved	flat		
146 Frontozygomatic breadth	narrow	thicker		
147 Orbital breadth	broad	elongated	v. elongated	
148 Nasal bones at nasion	broad	narrower		
149 Glabella	indistinct	inflated		
150 Supraorbital torus	absent	present	strong	
151 Supraciliary ridges	absent	present		
152 Supratoral sulcus	shallow/absent	broad		
153 Frontal sinus/nasion	above	above & below		
154 Frontal sinuses	large	small	absent	v. large
155 Frontal squama	vertical	horizontal		

(continued)

Table I. (*Continued*)

Character	Character states				
	0	1	2	3	4
156 Facial profile	convex/flat	concave			
157 Temporal fossa	narrow	broader			
158 Inion/glabella	above	lower	lower		
159 External occipital protuberance	strong	reduced			
160 Nuchal plane orientation	superior	posterior	inferior		
161 Brain size	small	larger			
162 Neurocranial length	short	elongated			
163 Glenoid fossa	shallow	deep			
164 Articular tubercle	large	small			
165 Entoglenoid process	low	prominent			
166 Entoglenoid process	narrow	broad			
167 Articular/tympanic temporal	unfused	fused			
Miscellaneous postcranial					
168 Deltopectoral plane	prominent	reduced			
169 Lateral trochlear keel	small	large			
170 Trochlear notch	narrow	broad			
171 Ulnar shaft	deep	shallow			
Cranial 2					
172 Zygomatic root pneumatization	solid	hollow			
173 Inferior orbital foramen/nasal aperture	near apex	below apex			
174 Zygomatic depth	shallow	deep			
175 Zygomatic orientation	lateral	anterior			
176 Maxillary sinus/canine position	posterior	anterior			
177 Maxillary nasal process	robust	hollow			
178 Orbital/nasal distance	large	small	huge		
179 Orbital/nasal surface	concave	flat			
180 Interorbital distance	broad	narrow			
181 Inferior orbital border	crested	rounded			
182 Zygomatic arch/orbit	same level	below			

183	Zygomatic temporal process	shallow	deep	
184	Zygomatic arch angle	inclined	vertical	horizontal
185	Nasal aperture breadth	narrow	broad	
186	Nasal aperture/orbit position	low	v. low	high
187	Nasal aperture/alveolar	low	high	
188	Nasal aperture edges	vertical	inclined	horizontal
189	Lacrimal fossa visible	no	yes	
190	Nasal bone length	short	long	v. long
191	Nasal aperture/malar surface	anterior	intermed.	flat
192	Inferior orbital margin/M^1	above	posterior	
193	Nasal apex/M^2	above	posterior	
194	Maxillary surface	lateral	anterior	
195	Canine fossa	shallow	deeper	
196	*C* root angulation	vertical	medial	
197	*C* root rotation	in line	externally rotated	
198	Maxillary alveolar process	solid	collapsed	inflated
199	Incisor orientation	vertical	horizontal	
200	Greater palatine foramen	round	more elongated	
201	Greater palatine position	anterior	posterior	
202	Lesser palatine foramena	none	small	large
203	Horizontal palatine	broad	narrow	
204	Palatine crest	strong	weak	
205	Posterior palate	shallow	deep	
206	Pyramidal process position	superior	inferior	
207	Pterygoid process	robust	compressed	
208	Alveolar process depth	shallow	deep	
209	Zygomatico-alveolar crest	compressed	broad	
210	Zygomatic root height	low	higher	
211	Incisive fossa	absent	shallow	deep
212	Subnasal floor	fenestrated	stepped	smooth
213	Nasal clivus	short	long	
214	Nasoalveolar clivus length	short	intermed.	long
215	Nasoalveolar clivus orientation	vertical	horizontal	

(*continued*)

Table I. *(Continued)*

Character	Character states				
	0	1	2	3	4
216 Incisive canal caliber	absent	large	intermed.	small	v. small
217 Mandibular canine roots	buccoling.	int. rotated			
218 Lower central incisors	large	small			
219 Lower incisors labiolingually	small	large			
220 Lower canine crown height	tall	short	v. short		
221 Lower molar cingula	strong	weak/absent			
222 Lower M_1/M_2 size ratio	small	large			
223 Lower M_3/M_2 size ratio	$M_3 \leq M_2$	$M_3 > M_2$			
224 I^1/I^2 size heteromorphy	lower	high	v. high		
225 I^2 morphology	peg-shaped	spatulate			
226 Upper canine height	tall	low	v. low		
227 Upper canine cervical flare	weak	strong			
Postcanine 2					
228 P^4 shape	broad	long			
229 P^3 shape	triangular	rectangular			
230 Premolar buccal flare	weak	stronger			
231 P^4 lingual flare	weak	strong			
232 M^1 shape	broad	long			
233 M^2 shape	broad	long			
234 Upper molar cingula	strong	weak/absent			
235 M^1/M^2 size ratio	small	large			
236 Upper molar crowns	low	high			
237 Upper molar sides	bulging	vertical			
238 Molar enamel	smooth	crenulated			
239 M^3 size	large	reduced			
Miscellaneous					
240 Life history	rapid	prolonged			

Table II. Character States for Each Taxon Used in This Analysis[a,b]

Australopithecus

1 2 3 4 5 6 7 8 9
1234567890 1234567890 1234567890 1234567890 1234567890 1234567890 1234567890 1234567890 1234567890

11101?1111111111111111111110111111001111011111111111100011111111111011111122212220223210 20
110220121112111100021000110110201211111111111111111221110011121113001212111111111111110111100
100202111010011012001210111111210203001211001201101111111001

Pan

1 2 3 4 5 6 7 8 9
1234567890 1234567890 1234567890 1234567890 1234567890 1234567890 1234567890 1234567890 1234567890

111010111111111111111111101111110011110111111111111100111111111111111111222122222221121
112221122112111111120111111111201111111111110001122111001112111310121211111111111110011100
010202111010010002001211111111210203001011001001001111111011

Gorilla

1 2 3 4 5 6 7 8 9
1234567890 1234567890 1234567890 1234567890 1234567890 1234567890 1234567890 1234567890 1234567890

1110101111111111111111111110111111001111011111111111110011001111111111111111222122222221121
112221122112111111120111111111201111111111110001112111010112111310111211111111111100011000
010210111110000002001111111111211102001011000001000111111001

Pongo

1 2 3 4 5 6 7 8 9
1234567890 1234567890 1234567890 1234567890 1234567890 1234567890 1234567890 1234567890 1234567890

111010111110111111111111111011111000011101111111111111101011101111111111111122212222222 1121
112221122102111211102011111111120111111111111100101210111210010020101111000000111110111111
110112120121111101111010110101121214101011010001000111111111

(continued)

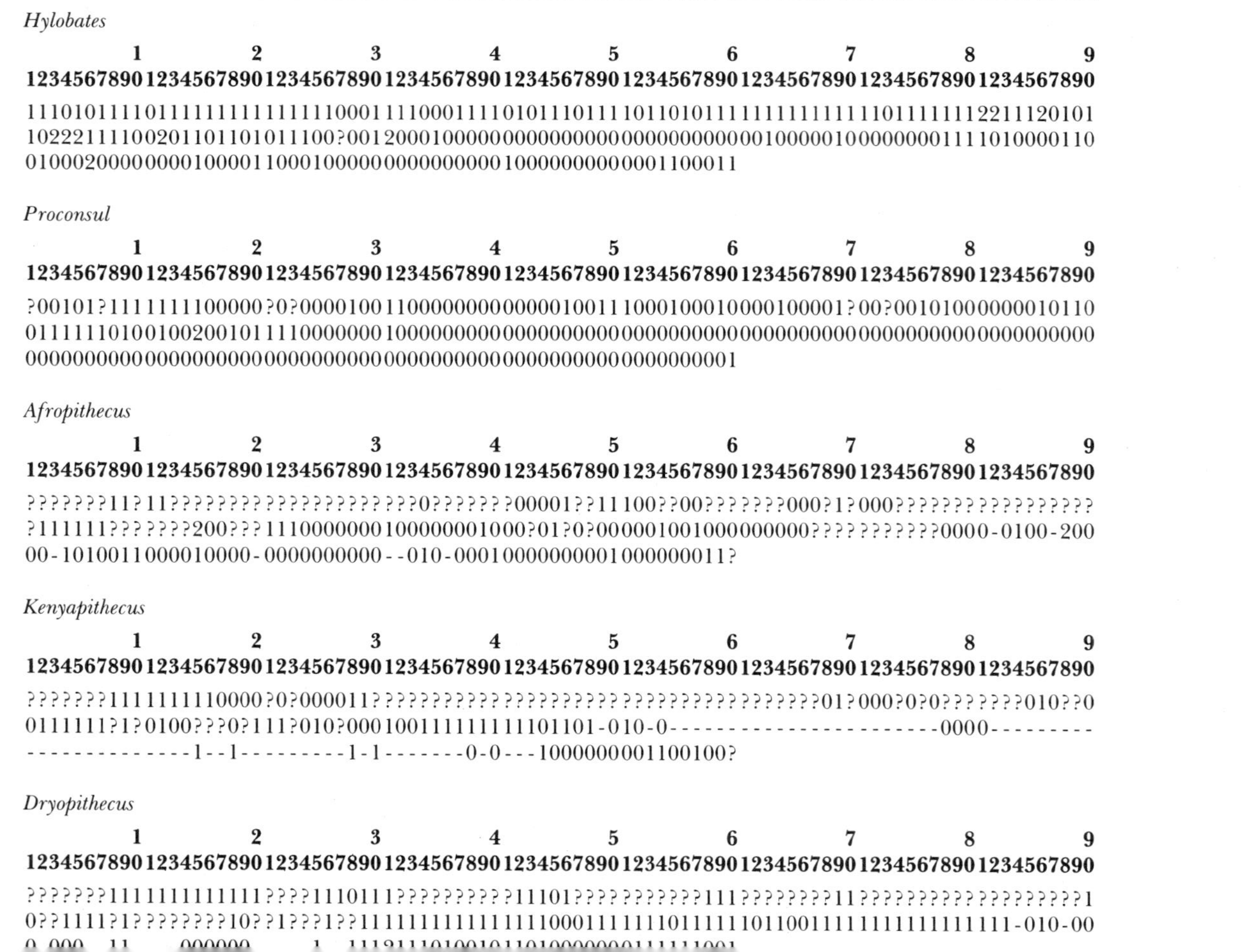

Table II. *(Continued)*

Hylobates

```
         1         2         3         4         5         6         7         8         9
1234567890123456789012345678901234567890123456789012345678901234567890123456789012345678901234567890
111010111101111111111111111000111100011110101110111101101011111111111111011111112211120101
102221111002011011010111 00?00120001000000000000000000000000000001000001000000001111010000110
010002000000001000011000100000000000000010000000000001100011
```

Proconsul

```
         1         2         3         4         5         6         7         8         9
1234567890123456789012345678901234567890123456789012345678901234567890123456789012345678901234567890
?00101?1111111100000?0?0000100110000000000000100111000100010000100001?00?00101000000010110
011111101001002001011110000000100000000000000000000000000000000000000000000000000000000000
000000000000000000000000000000000000000000000000000000000001
```

Afropithecus

```
         1         2         3         4         5         6         7         8         9
1234567890123456789012345678901234567890123456789012345678901234567890123456789012345678901234567890
???????11?11???????????????????????0???????00001??11100??00????????000?1?000???????????????????
?111111???????200???1110000000100000001000?01?0?0000010010000000000???????????0000-0100-200
00-1010011000010000-0000000000--010-00010000000001000000011?
```

Kenyapithecus

```
         1         2         3         4         5         6         7         8         9
1234567890123456789012345678901234567890123456789012345678901234567890123456789012345678901234567890
??????11111111110000?0?000011??????????????????????????????????????01?000?0?0????????010??0
0111111?1?0100???0?111?010?0001001111111111101101-010-0----------------------0000----------
--------------1--1---------1-1-------0-0---1000000001100100?
```

Dryopithecus

```
         1         2         3         4         5         6         7         8         9
1234567890123456789012345678901234567890123456789012345678901234567890123456789012345678901234567890
???????1111111111111????11101 11??????????11101???????????111????????11???????????????????1
0??1111?1????????10??1???1??11111111111111111110001111111011111101100111111111111111-010-00
[illegible]
```

Sivapithecus

1 2 3 4 5 6 7 8 9

123456789012345678901234567890123456789012345678901234567890123456789012345678901234567890

???????1111111111110?0?000001??1???????????????????????????000??11????1?00????????????11????

0??11?11200211201?0201?0110?111011111111111111010111011021001002010-----00000011-001111011

100101120221011111111-101011011212141010100200000001111110 01

Oreopithecus

1 2 3 4 5 6 7 8 9

123456789012345678901234567890123456789012345678901234567890123456789012345678901234567890

111010111?11111??1111111110?11??????????????????????????11?????????????????122212222?212?12?

11?221122002112111?????01111111001 1?11101101000???000?000?1110?100????0?1011011111?00??00

?01102100??00000000???????0102??10100001111000010011011100?

Lufengpithecus

1 2 3 4 5 6 7 8 9

123456789012345678901234567890123456789012345678901234567890123456789012345678901234567890

1???0????????????????????????1???????????????????????????????????????1?00??????????????????

???????????????????????????????1011?1111111?1?0?1?1?0?01000000??00????????????????101?1202

0001020110100011?00?????????1?1???10?00101101000010011111111??

Ouranopithecus

1 2 3 4 5 6 7 8 9

123456789012345678901234567890123456789012345678901234567890123456789012345678901234567890

??

????????????????????????????111012111111111111111-11101111110--00-----------------1-0---00

0-000-111110-001110-----1--11021-101001211110110101111111001

---

[a]*How to use this table*: The first two rows for a taxon each contain 90 characters. To find the character state for a specific character in a specific taxon, obtain the character number from Table I. Go to the taxon in question. If the character is between 1 and 90, it will appear in row 1. If it is between 91 and 180, subtract 90 from the character number and find the result in row 2. If it is between 181 and 240, subtract 180 from the character number and find the result in row 3.

[b]Bold numbers refer to character numbers in Table I.

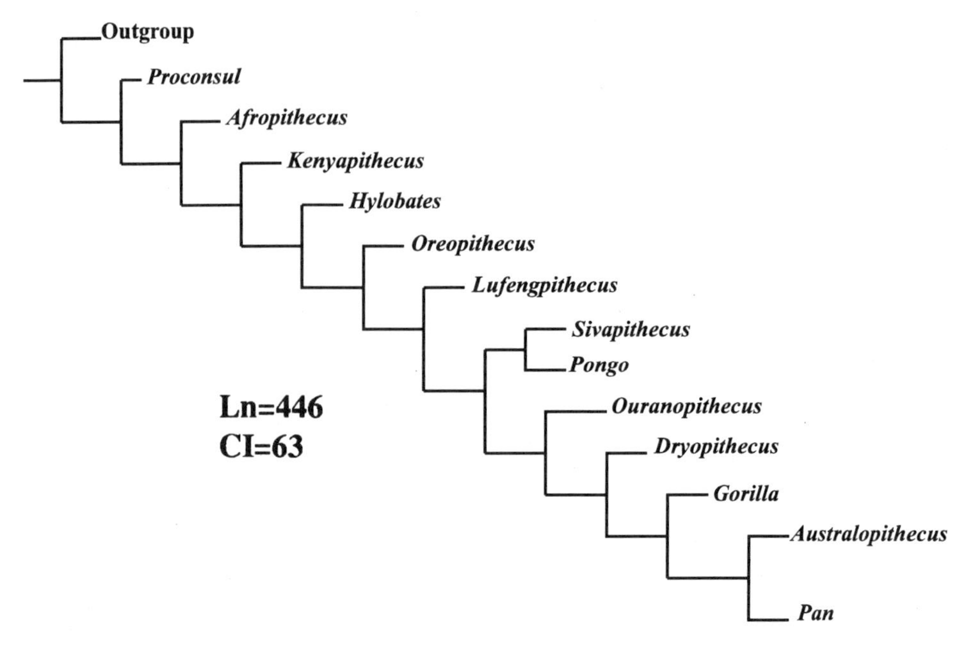

Fig. 1. Cladogram depicting the most parsimonious topology based on the characters listed in Table I. See text for discussion.

Although there is only one most parsimonious cladogram, there are many alternative hypotheses that are minimally less parsimonious than the cladogram in Fig. 1 (fewer than 1% more steps). These alternatives must be evaluated by tracing the history of each character in each cladogram and comparing the proposed transformation sequences with what is known in other groups of primates, and with what could be predicted from functional morphology.

Given the size of this data base, a thorough analysis of this hypothesis is a long-term project, and so only our initial results are presented here. We are not availing ourselves of the common "preliminary results" excuse, however. Our results are not preliminary because the taxa are well known and there is broad agreement on the character states for each (with a few exceptions). We are confident that our phylogenetic tree represents a viable, fully formed hypothesis, but it remains to be tested by the more rigorous methods described in this book.

Discussion

The cladogram in Fig. 1 is interesting in a number of ways. In contrast to many previous phylogenies (e.g., Andrews, 1992; Andrews and Martin, 1987; Groves, 1989; Begun, 1992; Bonis and Koufos, 1993), our hypothesis places *Kenyapithecus* in a more primitive position than *Hylobates*. This aspect of our hypothesis is consistent with that of McCrossin and Benefit (this volume), though other parts differ substantially (see below). The most parsimonious cladogram in which the positions of *Hylobates* and *Kenyapithecus* are reversed is 457 steps in length, about 4% less parsimonious than the hypothesis depicted in Fig. 1. Though this is not an overwhelming increase, it is significant, and at face value suggests that *Kenyapithecus* is more primitive than most researchers have believed. Considering the fact that *Kenyapithecus* was once regarded a direct ancestor of humans (e.g., Leakey, 1962; Simons, 1964), the conclusion that it might be more distantly related to humans than are gibbons is striking. On the other hand, the *Kenyapithecus* cranium is poorly known. The recovery of additional cranial remains, which have had a strong influence on interpretations of late Miocene hominoids (see S. Ward, Bonis and Koufos, Harrison and Rook, and Begun and Kordos, this volume), will undoubtedly serve as an important test of this hypothesis.

Another interesting implication of the position of *Hylobates* is the primitive position of *Proconsul*. Our hypothesis differs from both the views of Walker (this volume), that *Proconsul* is a basal great ape, and Harrison (1987), that *Proconsul* is a stem catarrhine. *Proconsul* is known from more characters than most other taxa, so our results cannot be attributed to the absence of data. The most parsimonious cladogram placing *Hylobates* in a more primitive position than *Proconsul* has a length of 487 steps (Fig. 3G), almost 10% less parsimonious than the one depicted in Fig. 1. The position of *Hylobates* in Fig. 1 is consistent with the view that early and middle Miocene hominoids are not members of the clade of extant hominoids, and are more appropriately considered as stem hominoids. This aspect of our hypothesis is consistent with Rae's view (this volume), except that our hypothesis also includes *Kenyapithecus* among stem hominoids. Extant hominoids and the fossil taxa within that clade would be classified in this phylogeny as euhominoids.

Although the positions of *Proconsul* and *Hylobates* relative to *Kenyapithecus* and other taxa are relatively stable in Fig. 1, only a small number of additional steps are needed to alter the positions of *Afropithecus* and *Kenyapithecus*. Interestingly, placing *Afropithecus* as the sister clade to *Kenyapithecus* requires only one additional step. This alternative is only one of three tree topologies with a length of 447, one step longer than the most parsimonious cladogram (Fig. 2). Both Leakey and Walker (this volume) and McCrossin and Benefit (this volume) suggest a pithecinelike adaptation for both of these fossil taxa, which both pairs of coauthors propose is convergent. Figure 2A suggest that, while it may be convergent with pithecines, a scerlocarp feeding adaptation could be interpreted as a synapomorphy of *Kenyapithecus* and *Afropithecus*. In Fig. 2B, *Proconsul* is the outgroup to an *Afropithecus–Kenyapithecus* clade, and all three taxa form a monophyletic clade that is the sister group to hylobatids and the clade that includes fossil and living great apes and humans.*

*The authors of this chapter disagree on what to call this group. Begun (1994) prefers to call them *hominids*. Ward and Rose prefer to restrict the latter term to humans and fossil forms more closely related to humans than to any other primate. Discussion of the relative merits of each of these points of view is beyond the scope of this chapter. The reader is referred to discussions of this topic in Groves (1986, 1989), Tattersall *et al.* (1988), and Begun (1992).

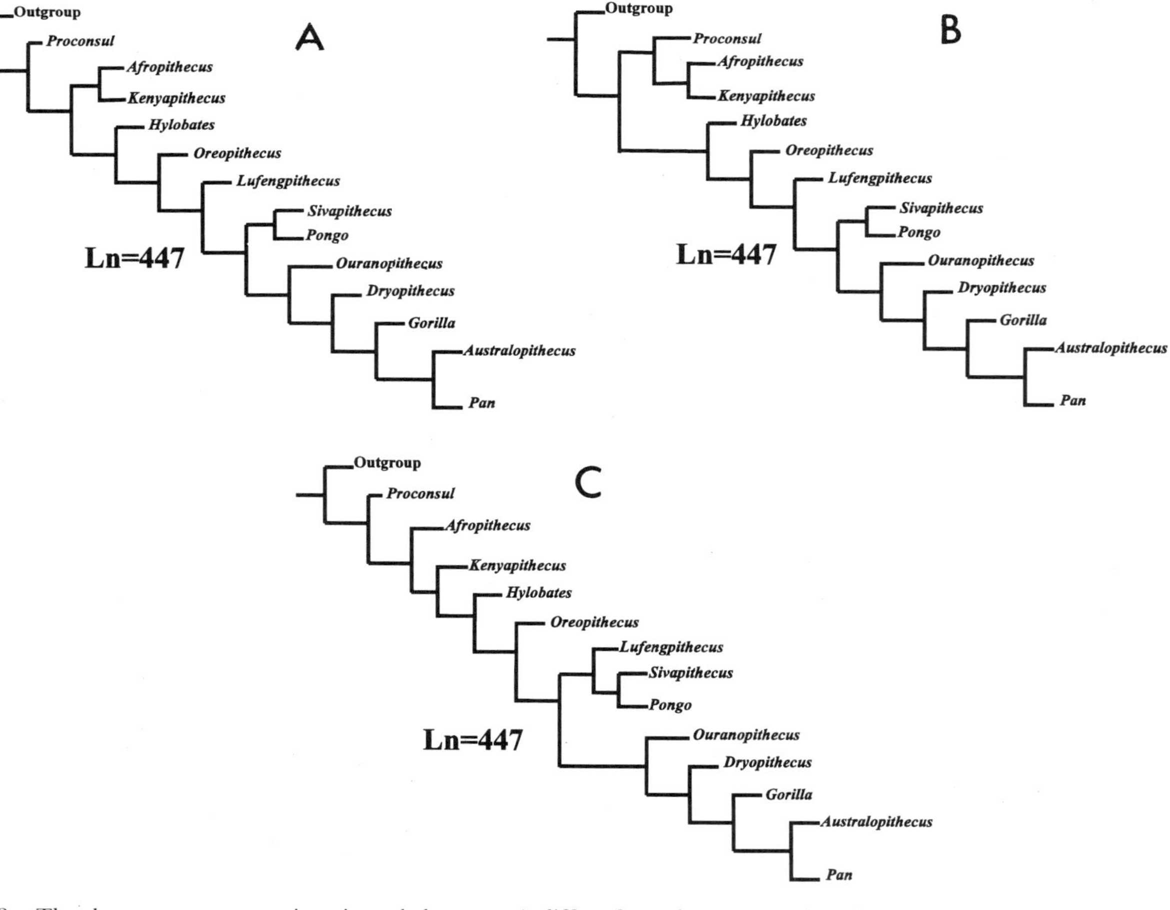

Fig. 2. The three next most parsimonious cladograms. A differs from the most parsimonious cladogram in depicting an *Afropithecus–Kenyapithecus* clade. B differs in having a *Proconsul–Afropithecus–Kenyapithecus* with *Proconsul* as the sister to the latter two. C differs in having an Asian great ape clade, with *Lufengpithecus* as the sister to the *Sivapithecus–Pongo* clade. All of these

Figure 2C also has 447 steps, and involves a change in the position of *Lufengpithecus*. *Lufengpithecus* is often excluded from phylogenetic analyses (e.g., Andrews, 1992; Bonis and Koufos, 1993; Begun, 1994) because it has not been studied directly by many researchers. A number of scholars have suggested that *Lufengpithecus* is closely related to other Asian great apes (e.g., Schwartz, 1990, this volume; Kelley and Pilbeam, 1986) or even to humans (Wu *et al.*, 1984, 1985; Xu and Lu, 1979). In Fig. 1, *Lufengpithecus* is the sister clade to *Ouranopithecus, Dryopithecus,* African apes, and humans. The next sister clade includes *Sivapithecus* and *Pongo*. Only one additional step is required to place *Lufengpithecus* as the sister clade to the *Pongo–Sivapithecus* clade (Fig. 2C), a position suggested by Schwartz (this volume). Unfortunately, many character states are unknown for *Lufengpithecus*. The addition of more postcranial characters, in particular, may help resolve the possible relations of this taxon.

Other hypotheses from the literature (Fig. 3) can be tested with our more comprehensive data set. Figure 1 is consistent with several studies that used a more limited number of fossil taxa, or living forms only. It is consistent with molecular studies that hypothesize a closer relationship between *Pan* and *Homo* than between *Pan* and *Gorilla* (see Ruvolo, 1994, for a recent review). This hypothesis is also consistent with a few studies based on a more limited sample of fossil evidence (e.g., Begun, 1992, 1994), though most paleoanthropologists consider the issue unresolved (e.g., Harrison and Rook, this volume). A *Pan–Gorilla* clade requires a minimum of 4 additional steps (length = 450) (Fig. 3A).

An *Ouranopithecus–Australopithecus* clade with a *Pan–Gorilla* clade as the outgroup (Bonis and Koufos, this volume) requires an additional 3 steps (length = 453) (Fib. 3B). Alternately, a *Gorilla–Ouranopithecus* clade (Dean and Delson, 1992) requires 8 steps more than the most parsimonious hypothesis (length = 454) (Fig. 3C). The most parsimonious cladogram placing *Sivapithecus–Pongo* as the sister clade to African apes and humans (e.g., Andrews, 1992) requires 6 more steps (length = 452), and includes *Lufengpithecus* as the sister to *Sivapithecus–Pongo* (Fig. 3D). Adding *Dryopithecus* to the Asian great ape clade (Moyà-Solà and Köhler, 1993, 1996) is also 6 steps longer than the most parsimonious cladogram (length = 452) (Fig. 3E). An *Oreopithecus Dyropithecus* clade (Harrison and Rook, this volume) is 8 steps longer (length = 454), with this clade as the sister clade to African apes and humans (Fig. 3F). Two final cladograms are considerably longer than the others. A great ape clade that includes *Proconsul* (Walker, this volume) requires 41 more steps (length = 487) (Fig. 3G), and the hypothesis presented by McCrossin and Benefit (this volume) requires 58 additional steps (length = 504) (Fig. 3H).

Six of the eight hypotheses reviewed here are not strongly contradicted by our data. While Fig. 1 represents the most parsimonious, and therefore the most straightforward, explanation for the pattern of character states in Table I, it remains to be seen if it is the most likely hypothesis. Parsimony as a criterion for choosing among competing hypotheses assumes that all other

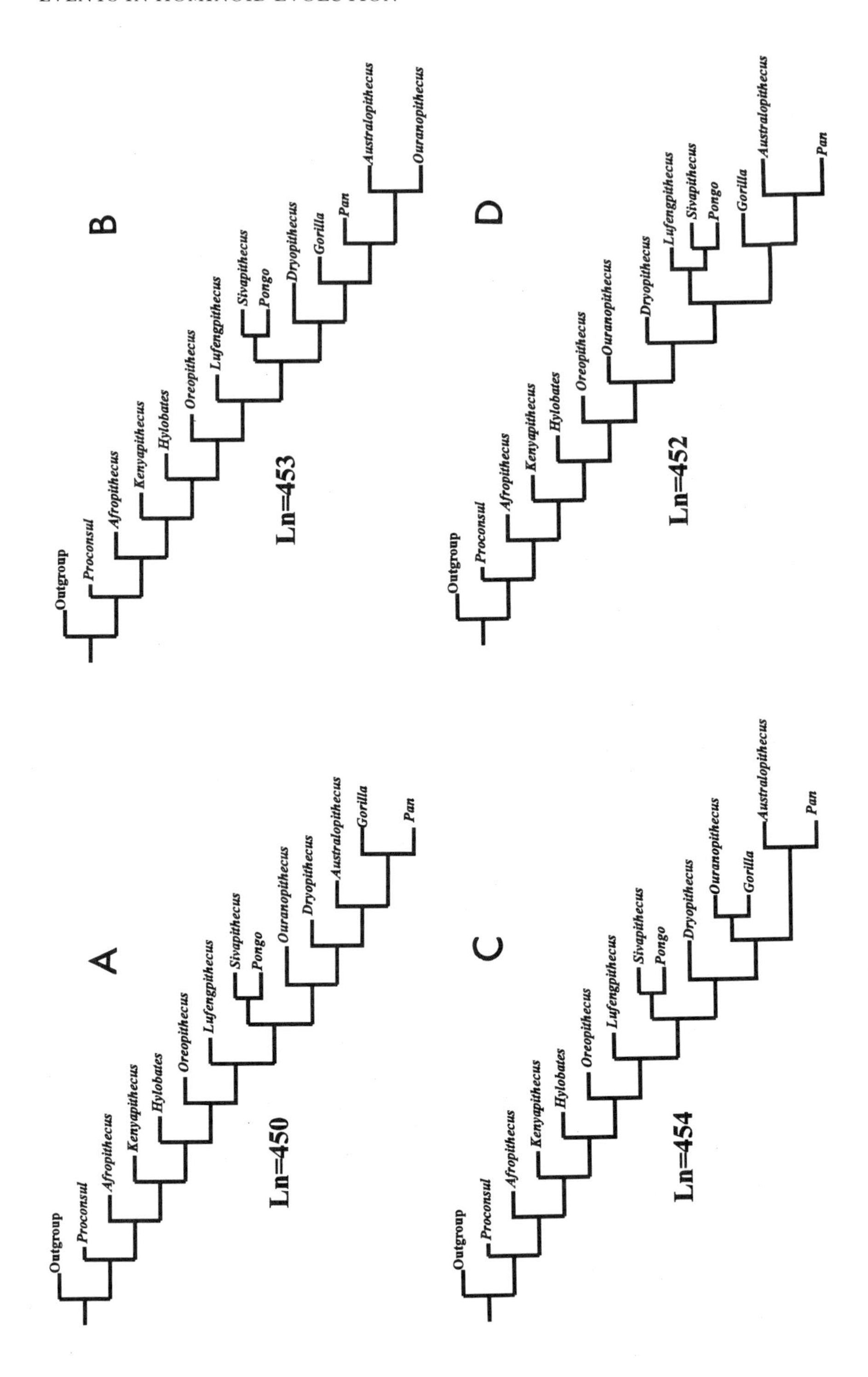
A
Outgroup
Proconsul
Afropithecus
Kenyapithecus
Hylobates
Oreopithecus
Lufengpithecus
Sivapithecus
Pongo
Ouranopithecus
Dryopithecus
Australopithecus
Gorilla
Pan
Ln=450
B
Outgroup
Proconsul
Afropithecus
Kenyapithecus
Hylobates
Oreopithecus
Lufengpithecus
Sivapithecus
Pongo
Dryopithecus
Gorilla
Pan
Australopithecus
Ouranopithecus
Ln=453
C
Outgroup
Proconsul
Afropithecus
Kenyapithecus
Hylobates
Oreopithecus
Lufengpithecus
Sivapithecus
Pongo
Dryopithecus
Ouranopithecus
Gorilla
Australopithecus
Pan
Ln=454
D
Outgroup
Proconsul
Afropithecus
Kenyapithecus
Hylobates
Oreopithecus
Ouranopithecus
Dryopithecus
Lufengpithecus
Sivapithecus
Pongo
Gorilla
Australopithecus
Pan
Ln=452

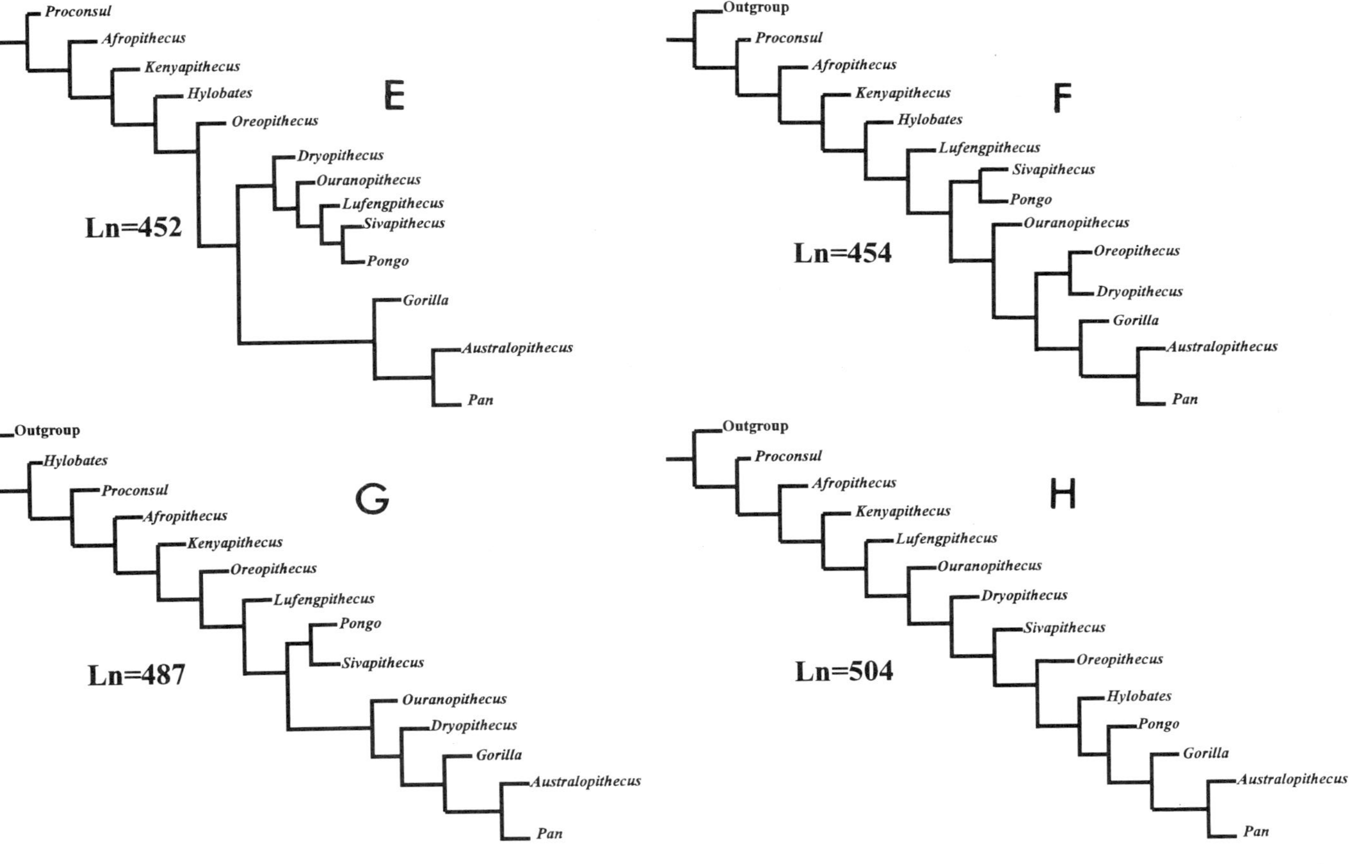

Fig. 3. Eight more cladograms testing various hypotheses presented in the volume and elsewhere in the literature. A is the most parsimonious cladogram with an African ape clade. B is the most parsimonious cladogram with an *Australopithecus–Ouranopithecus* clade. C is the most parsimonious cladogram with an *Ouranopithecus–Gorilla* clade. D is the most parsimonious cladogram with any Asian great ape clade as the sister clade to the African apes and humans. E is the most parsimonious cladogram with *Dryopithecus* as a member of the Asian great ape clade. Excluding *Ouranopithecus* from this clade would be even less parsimonious (Ln = 456). F is the most parsimonious cladogram with *Oreopithecus* as the sister clade to *Dryopithecus*. G is the most parsimonious clade with *Proconsul* in the great ape clade to the exclusion of *Hylobates*. H is the most parsimonious cladogram with a monophyletic living hominoid clade that excludes all fossil taxa. See text for discussion.

considerations are equal. Tracing the character histories in each of the cladograms that differ by fewer than 2% of their length (most of the phylogenies considered above) should provide evidence that allows for a more accurate assessment. It may be that a less parsimonious phylogeny is actually a preferred explanation if character histories are more internally consistent. For example, a cladogram with more steps within character complexes but fewer steps across anatomical regions may be preferable, even if less parsimonious (e.g., Begun and Kordos, this volume).

For the time being, we consider the phylogenetic hypothesis presented in Fig. 1 to be a strong candidate for the correct answer to the question of the phylogenetic relations of these taxa. It represents an advance over previous hypotheses because it is based on a large number of characters from all regions of the skeleton. It also includes many taxa, even those that are often ignored or dismissed such as *Hylobates, Oreopithecus,* and *Lufengpithecus.* Future work analyzing character histories and addition of more fossil data will serve to test and refine this hypothesis.

Evolutionary Implications of Our Phylogenetic Hypothesis

If the phylogeny in Fig. 1 is correct, what are its implications for the chronology of biological events in hominoid evolution? The morphological diversity of the stem hominoids *Proconsul, Afropithecus,* and *Kenyapithecus* suggests an early radiation of hominoids into a broad variety of locomotor and dietary niches. Such diversity may imply a longer evolutionary history for this group than is currently assumed, an idea supported by the recent recognition of *Kamoyapithecus* (Leakey *et al.*, 1995), a new, more primitive, and much older possible hominoid from Kenya that may be related to the early Miocene stem hominoids.

Proconsul and *Afropithecus* were medium- to large-sized pronograde, tailless arboreal quadrupeds that differed from Old World monkeys in having greater limb mobility, and enhanced manual and pedal grasping capabilities. These adaptations foreshadowed the orthograde positional behavior of extant apes, a behavioral adaptation that is interpreted here as a synapomorphy of the Hominoidea. In stem hominoids, we see the initial phases of a specialized arboreal life-style in combination with the retention of a more ancient pattern of overall frugivory, although with diverse dietary specializations in each taxon. As the stem hominoids diversified, they became more varied in their diets and positional behavior. *Kenyapithecus* and *Afropithecus,* it appears, developed or retained an adaptation to process sclerocarp vegetation, and *Kenyapithecus* seems to have become significantly more terrestrial than other Miocene apes. It is not clear whether these stem hominoids were also characterized by the pattern of delayed maturation described by Kelley (this volume) for living hominoids and *Sivapithecus.*

There is no good fossil evidence for the origin of the euhominoids. The most primitive euhominoids, according to Fig. 1, are *Hylobates* and *Oreopithecus*. Both had well-developed, forelimb dominated orthograde arboreality. This suggests that this form of locomotion, usually considered specialized, is actually primitive for the euhominoid clade. If *Oreopithecus* is a member of the great ape and human clade, then the major anatomical differences between greater and lesser apes are body size and sexual dimorphism, both of which are probably autapomorphic for hylobatids. A number of postcranial characters shared by extant large hominoids and missing from hylobatids are arguably related to differences in body size among these forms. If *Oreopithecus* is not included in the extant large hominoid clade, the clade including other fossil great apes and all extant large hominoids differs from more primitive euhominoids in numerous respects. These include relative encephalization, increases in facial length and pneumatization, and a number of changes in the dentition, including molarization of the premolars and increases in incisor robusticity. Harrison and Rook (this volume) suggest that the small brain and short face of *Oreopithecus* are autapomorphies that converge on hylobatids and more primitive catarrhines. However, this is a less parsimonious hypothesis than assuming they are plesiomorphic. Their explanation that these characters are related to the extreme folivory of *Oreopithecus*, however, is an interesting possibility that warrants further exploration.

The origin of the euhominoids provides evidence of the first major anatomical transformation toward an adaptation typical of living hominoids, forelimb-dominated orthograde climbing and/or suspension. Once again, as with early Miocene stem hominoids, the first euhominoids of the middle and late Miocene are diverse in their dietary adaptations. A new form of positional behavior and a new location (Eurasia) seems to have permitted a second adaptive radiation within the Hominoidea. Dietary diversity ranges from highly folivorous *Oreopithecus* to the specialized hard-object feeder *Ouranopithecus*, with less specialized hard-object feeders like *Sivapithecus* and soft-frugivores like *Dryopithecus* in between. In fact, as noted by Kay and Ungar (this volume), this dietary diversity of the middle and late Miocene apes exceeds that observed in living apes.

Late Miocene hominoids are also characterized by positional behavior diversity, as was the case in the early Miocene. Forms for which good data exist were at least partly suspensory (*Oreopithecus, Dryopithecus*). *Lufengpithecus*, although poorly known postcranially, was probably similarly adapted, judging from the small number of postcranial elements known, as well as the paleoenvironment of the locality (see Rose and Andrews *et al.*, this volume). *Sivapithecus*, in contrast, appears to have had a different mode of positional behavior, although it is not clear what this could have been. Most of the anatomy of *Sivapithecus* suggests aboreality, but the humeral shafts suggest the absence of trunk and forelimb characters typical of broad-chested, suspensory hominoids. *Sivapithecus* may be specialized for a novel mode of locomotion; its apparently primitive characters suggest habitually pronograde posture. If these postcranial characters are primitive in *Sivapithecus*, then according to the hypothesis in Fig. 1, *Pongo* must have evolved more typical great ape humeral characters in parallel with the African ape–*Dryopithecus* clade, hylobatids, and *Oreopithecus*, which is less parsimonious but not impossible. If *Sivapithecus* is autapomorphic for these characters, as suggested by Fig. 1, *Pongo* would simply be retaining great ape humeral characters secondarily lost in its sister clade. *Ouranopithecus* is known postcranially from only two phalanges. They are long and powerfully built, and do indicate a certain level of arboreality, but it is impossible to provide more details at this point. The sheer size of *Ouranopithecus* (males were about the size of female *Gorilla*) suggests that they were not active suspensory arborealists.

There is no evidence in the Miocene of the specific types of terrestrial positional behaviors characteristic of African apes or humans. However, the cladogram in Fig. 1 does suggest an evolutionary history for these characters associated with knuckle-walking terrestrial quadrupedalism and bipedality. If *Pan* and *Australopithecus* do form a clade, then it is more parsimonious to suggest that African apes retain knuckle-walking as a primitive character of the African ape and human clade, than it is that gorillas and chimps evolved knuckle-walking independently. A *Pan–Australopithecus* clade thus suggests that humans evolved from knuckle-walkers, a view consistent with some reconstructions (Washburn, 1950; Begun, 1994). The less parsimonious view of a *Pan–Gorilla* clade is more consistent with the view that knuckle-walking is a specialized form of positional behavior that has always been restricted to chimps and gorillas and their immediate ancestors (Tuttle, 1967). The anticipated analysis of the newly discovered postcranial material of *Ardipithecus ramidus* should reveal important information for this debate, but even earlier specimens preceding the divergence of humans from African apes would be most helpful in this regard.

Biogeographic Implications

Figure 1 suggests a complex paleobiogeographic model of the origins and diversification of the great ape and human clade. The great ape and human clade probably originated from a stock of unknown hominoids that may have been *Kenyapithecus*-like in dental anatomy and hylobatid, *Oreopithecus*, or *Dryopithecus*-like in postcranial anatomy. *Kenyapithecus* and other dentally and postcranially similar forms, like *Griphopithecus*, probably precede this radiation, despite the fact that this group had already migrated to Eurasia by MN 6 times, about 15 Ma (Mein, 1986; Steininger, 1986; Bernor and Toblen, 1990). There are two possible explanations. More suspensory arboreal taxa may have entered Eurasia in successive waves along forest corridors between Africa and Eurasia (Bernor, 1983). One of these eventually led to the evolution of hylobatids, but the details of this migration and transformation are mysterious.

If the relations depicted in Fig. 1 are correct, then it is unlikely that the early middle Miocene Asian hominoid *Dionysopithecus* has anything to do with the origin or evolution of gibbons (Barry *et al.*, 1986; Bernor *et al.*, 1988). This is the case given the more recent age of the apparently more primitive genera *Kenyapithecus* and *Griphopithecus* (Pickford, 1986; Bernor and Tobien, 1990), and the morphological resemblances between *Dionysopithecus* and the stem hominoid *Micropithecus* (Fleagle, 1984; Barry *et al.*, 1986).

A second clade leaving Africa may have led to the Asian great ape clade, including *Sivapithecus, Pongo,* and also *Ankarapithecus* (Begun and Güleç, 1996), while a third clade, more closely related to African apes and humans, may have followed. *Oreopithecus* and *Lufengpithecus* may represent other distinct clades, as suggested by Fig. 1, or *Oreopithecus* may be associated with the *Dryopithecus–Ouranopithecus*–African ape and human clade and *Lufengpithecus* with the Asian great ape clade (see above). This is a centrifugal view of hominoid origins and diversification, with more primitive clades being displaced to the peripheries of the range of the Hominoidea, and more derived clades evolving in the center of the range, and is consistent with the model proposed by Groves (1989) for other primate taxa.

Another possibility is that euhominoids evolved in Eurasia, with the African ape and human clade returning recently to Africa. This would explain the poor fossil record of African Miocene euhominoids, and the apparent persistence of *Proconsul* or *Kenyapithecus*-like forms at a few later localities (Hill and Ward, 1988). *Griphopithecus* from Slovakia and Turkey may represent the ancestral stock from which hominoids diverged, again in three major divisions: hylobatids, Asian great apes, and African apes and humans. The last group may have further subdivided into European and African branches, with the more terrestrial African branch returning to Africa sometime during or after MN 10, about 9 Ma, when the area was becoming drier (Steininger and Rögl, 1979; Steininger *et al.*, 1985). There is no compelling paleogeographic evidence to suggest one of these views over the other. Connections between Africa and Eurasia were intermittent throughout the middle and late Miocene, and appropriate ecological conditions were apparently available for either of these two scenarios to have occurred (Steininger *et al.*, 1985).

Summary

Assembling a large data base on hominoid skeletal anatomy has allowed us to generate an internally consistent hypothesis of relations among hominoids and possible details of their diversification. As always, it should be noted that these ideas are testable by refinement and revision of the character list (Tables I and II), and, of course, collection of more data. Of particular interest with regard to stem hominoids would be more data on the face and postcranium of *Kenyapithecus*. Fossil sister clades of the hylobatids, *Pan* and *Gorilla,* as well as a more proximate sister clade to *Sivapithecus* would go far toward resolving a number of questions related to euhominoid diversification. In addition, attention to areas through which hypothesized migrations took place, northeastern Africa, the Middle East, and the Gulf states, may afford exciting new opportunities to test these ideas.

The data presented here represent the first phase in a cooperative effort to understand hominoid evolution. We are confident that continued work testing and expanding this hypothesis, and integrating phylogeny with functional anatomy, will soon lead to resolution of many of the questions regarding the evolution and adaptations of the Hominoidea.

References

Andrews, P. 1992. Evolution and environment in the Hominoidea. *Nature* **360:**641–646.

Andrews, P., and Martin, L. 1987. Cladistic relationships of extant and fossil hominoids. *J. Hum. Evol.* **16:**101–118.

Barry, J. C., Jacobs, L. L., and Kelley, J. 1986. An Early Middle Miocene catarrhine from Pakistan with comments on the dispersal of catarrhines into Eurasia. *J. Hum. Evol.* **15:**501–508.

Begun, D. R. 1992. Miocene fossil hominids and the chimp–human clade. *Science* **247:**1929–1933.

Begun, D. R. 1994. Relations among the great apes and humans: New interpretations based on the fossil great ape *Dryopithecus. Yearb. Phys. Anthropol.* **37:**11–63.

Begun, D. R., and Güleç, E. 1996. Restoration of the type and palate of *Ankarapithecus meteai:* Taxonomic, phylogenetic, and functional implications. (Submitted for publication).

Bernor, R. L. 1983. Geochronology and zoogeographic relationships of Miocene Hominoidea. In: R. L. Ciochon and R. S. Corruccini (eds.), *New Interpretations of Ape and Human Ancestry,* pp. 21–64. Academic Press, New York.

Bernor, R. L., and Tobien, H. 1990. The mammalian geochronology and biogeography of Paşalar (middle Miocene, Turkey). *J. Hum. Evol.* **19:**551–568.

Bernor, R. L., Flynn, L. J., Harrison, T., Hussain, S. T., and Kelley, J. 1988. *Dionysopithecus* from southern Pakistan and the biochronology and biogeography of early Eurasian catarrhines. *J. Hum. Evol.* **17:**339–358.

Bonis, L. de, and Koufos, G. 1993. The face and mandible of *Ouranopithecus macedoniensis:* Description of new specimens and comparisons. *J. Hum. Evol.* **24:**469–491.

Dean, D., and Delson, E. 1992. Second gorilla or third chimp? *Nature* **359:**676–677.

Fleagle, J. G. 1984. Are there any fossil gibbons? In: D. J. Chivers, H. Preuschoft, W. Y. Brockelman, and N. Creel (eds.), *The Lesser Apes,* pp. 431–447. Edinburgh University Press, Edinburgh.

Groves, C. P. 1986. Systematics of the great apes. In: D. R. Swindler and J. Erwin (eds.), *Comparative Primate Biology, Vol. 1: Systematics, Evolution, and Anatomy,* pp. 187–217. Liss, New York.

Groves, C. P. 1989. *A Theory of Primate and Human Evolution.* Clarendon Press, Oxford.

Harrison, T. 1987. The phylogenetic relationships of the early catarrhine primates: A review of the current evidence. *J. Hum Evol.* **16:**41–80.

Hill, A., and Ward, S. 1988. Origin of the Hominidae: The record of African large hominoid evolution between 14 my and 4 my. *Yearb. Phys. Anthropol.* **32:**48–83.

Kelley, J., and Pilbeam, D. R. 1986. The dryopithecines: Taxonomy, comparative anatomy, and phylogeny of Miocene large homionoids. In: D. R. Swindler and J. Erwin (eds.), *Comparative Primate Biology, Vol. 1: Systematics, Evolution, and Anatomy,* pp. 361–411. Liss, New York.

Leakey, L. S. B. 1962. A new Lower Pliocene fossil primate from Kenya. *Ann. Mag. Nat. Hist.* **13:**689–696.

Leakey, M. G., Ungar, P. S., and Walker, A. 1995. A new genus of large primate from the late Oligocene of Lothidok, Turkana District, Kenya. *J. Hum. Evol.* **28:**519–531.

Mein, P. 1986. Chronological succession of hominoids in the European Neogene. In: J. G. Else and P. C. Lee (eds.), *Primate Evolution,* pp. 59–70. Cambridge University Press, London.

Moyà-Solà, S., and Köhler, M. 1993. Recent discoveries of *Dryopithecus* shed new light on evolution of great apes. *Nature* **365:**543–545.

Moyà-Solà, S., and Köhler, M. 1996. A *Dryopithecus* skeleton and the origins of great ape locomotion. *Nature* **379:**156–159.

Pickford, M. 1986. Geochronology of the Hominoidea: A summary. In: J. G. Else and P. C. Lee (eds.), *Primate Evolution,* pp. 123–128. Cambridge University Press, London.

Ruvolo, M. 1994. Molecular evolutionary processes and conflicting gene trees: The hominoid case. *Am. J. Phys. Anthropol.* **94:**89–113.

Schwartz, J. H. 1990. *Lufengpithecus* and its potential relationship to an orang-utan clade. *J. Hum. Evol.* **19:**591–605.

Simons, E. L. 1964. On the mandible of *Ramapithecus. Proc. Natl. Acad. Sci. USA* **51:**528–535.

Steininger, F. 1986. Dating the Paratethys Miocene hominoid record. In: J. G. Else and P. C. Lee (eds.), *Primate Evolution,* pp. 71–84. Cambridge University Press, London.

Steininger, F. F., and Rögl, F. 1979. The Paratethys history—A contribution towards the Neogene geodynamics of the Alpine Orogene. An abstract. *Ann. Geol. Pays. Hel. Tome Hors Ser.* **III:**1153–1165.

Steininger, F. F., Radeber, G., and Rögl, F. 1985. Land mammal distribution in the Mediterranean Neogene: A consequence of geokinematic and climatic events. In: D. J. Stanley and F. C. Wezel (eds.), *Geological Evolution of the Mediterranean Basin,* pp. 559–571. Springer-Verlag, Berlin.

Tattersall, I., Delson, E., and Van Couvering, J. 1988. *Encyclopedia of Human Evolution and Prehistory.* Garland, New York.

Tuttle, R. H. 1967. Knuckle-walking and the evolution of hominoid hands. *Am. J. Phys. Anthropol.* **26:**171–206.

Washburn, S. L. 1950. The analysis of primate evolution with particular reference to the origin of man. *Cold Spring Harbor Symp. Quant. Biol.* **15:**67–78.

White, T., Suwa, G., and Asfaw, B. 1994. *Australopithecus ramidus:* A new species of early hominid from Aramis, Ethiopia. *Nature* **371:**306–312.

White, T., Suwa, G., and Asfaw, B. 1995. *Australopithecus ramidus:* A new species of early hominid from Aramis, Ethiopia. *Nature* **375:**88.

Wu, R., Xu, Q., and Lu, Q. 1984. Morphological features of *Ramapithecus* and *Sivapithecus* and their phylogenetic relationships–Morphology and comparison of the mandibles. *Acta Anthropol. Sin.* **3:**1–10.

Wu, R., Xu, Z., and Lu, Q. 1985. Morphological features of *Ramapithecus* and *Sivapithecus* and their phylogenetic relationships—Morphology and comparison of the teeth. *Acta Anthropol. Sin.* **4:**197–204.

Xu, Q., and Lu, Q. 1979. The mandibles of *Ramapithecus* and *Sivapithecus* from Lufeng, Yunnan. *Verteb. Palasiat.* **17:**1–13.

Geological/Geographic Index

Subject Index

Essays on Saving, Bequests, Altruism, and Life-Cycle Planning